MANAGEMENT OF BIOLOGICAL NITROGEN FIXATION

Developments in Plant and Soil Sciences

VOLUME 65

The titles published in this series are listed at the end of this volume.

Management of Biological Nitrogen Fixation for the Development of More Productive and Sustainable Agricultural Systems

Extended versions of papers presented at the Symposium on Biological Nitrogen Fixation for Sustainable Agriculture at the 15th Congress of Soil Science, Acapulco, Mexico, 1994

Edited by

J.K. LADHA and M.B. PEOPLES

Reprinted from *Plant and Soil*, Volume 174, Nos. 1–2 (1995)

KLUWER ACADEMIC PUBLISHERS
DORDRECHT / BOSTON / LONDON

A C.I.P. Catalogue record for this book is available from
the Library of Congress.

ISBN 0-7923-3413-2

Published by Kluwer Academic Publishers,
P.O. Box 17, 3300 AA Dordrecht, The Netherlands.

Kluwer Academic Publishers incorporates
the publishing programmes of Martinus Nijhoff,
Dr W. Junk, D. Reidel, and MTP Press.

Sold and distributed in the U.S.A. and Canada
by Kluwer Academic Publishers,
101 Philip Drive, Norwell, MA 02061, U.S.A.

In all other countries, sold and distributed
by Kluwer Academic Publishers Group,
P.O. Box 322, 3300 AH Dordrecht, The Netherlands.

Printed on acid-free papers

Printed in the Netherlands

Contents

Preface 1

1. Biological nitrogen fixation: An efficient source of nitrogen for sustainable agricultural production?
M.B. Peoples, D.F. Herridge and J.K. Ladha 3

2. Sustainable agriculture in the semi-arid tropics through biological nitrogen fixation in grain legumes
S.P. Wani, O.P. Rupela and K.K. Lee 29

3. Enhancing crop legume N_2 fixation through selection and breeding
D.F. Herridge and S.K.A. Danso 51

4. Enhancing legume N_2 fixation through plant and soil management
M.B. Peoples, J.K. Ladha and D.F. Herridge 83

5. Role of legumes in providing N for sustainable tropical pasture systems
R.J. Thomas 103

6. Management of biological N_2 fixation in alley cropping systems: Estimation and contribution to N balance
J. Sanginga, B. Vanlauwe and S.K.A. Danso 119

7. Manipulation of rhizobia microflora for improving legume productivity and soil fertility: A critical assessment
J. Brockwell, P.J. Bottomley and J.E. Thies 143

8. Green manure technology: Potential, usage, and limitations. A case study for lowland rice
M. Becker, J.K. Ladha and M. Ali 181

9. Biological nitrogen fixation associated with sugar cane and rice: Contributions and prospects for improvement
R.M. Boddey, O.C. de Oliveira, S. Urquiaga, V.M. Reis, F.L. de Olivares, V.L.D. Baldani and J. Döbereiner 195

10. Biological N_2 fixation by heterotrophic and phototrophic bacteria in association with straw
M.M. Roper and J.K. Ladha 211

11. Potential and pitfalls of trying to extend symbiotic interactions of nitrogen-fixing organisms to presently non-nodulated plants, such as rice
F.J. de Bruijn, Y. Jing and F.B. Dazzo 225

12. New techniques for studying competition by Rhizobia and for assessing nitrogen fixation in the field
K.J. Wilson, M.B. Peoples and R.A. Jefferson 241

13. Future benefits from biological nitrogen fixation: An ecological approach to agriculture
K.E. Giller and G. Cadisch 255

14. Research evaluation and impact analysis of biological nitrogen fixation
M.C.S. Bantilan and C. Johansen 279

Author index 287

Plant and Soil **174:** 1, 1995.

Preface

Management of biological nitrogen fixation for the development of more productive and sustainable agricultural systems

Nitrogen is the most important nutrient input required for crop production. Soil and biological nitrogen fixation (BNF) and chemical sources supply most of that N. Soil organic N in natural or man-made ecosystems is continually lost through plant removal and loss processes such as leaching, denitrification, and ammonia volatilization. Biological nitrogen fixation – the microbial conversion of atmospheric N to a plant-usable form – helps replenish the soil N pool.

The subsistence agriculture of the pre-chemical era efficiently sustained the N status of soils by maintaining a balance between N loss and N gain from BNF. This was possible with less intensive cropping, adoption of rational crop rotations and intercropping schemes, and use of legumes as green manure. The agriculture of the modern chemical era concentrates on maximum output but overlooks input efficiency; it has not been sustainable. Intensive monocropping with no or inadequate crop rotations or green manuring and excessive use of chemical N fertilizers results in an imbalance between N gains and N losses. The losses are often larger than the gains, and soil N status goes to a lower level. Agricultural systems, however, differ in levels of N sustainability. An irrigated double rice cropping system, for example, can be sustained at a higher level than can a rainfed single rice cropping system. The challenge is to sustain soil N fertility in cropping systems operating at high productivity levels. This requires judicious integration of BNF components, which keeps a good balance between N losses and gains.

Sustainability of N cannot be considered in isolation; it must be considered holistically within the framework of system sustainability. A BNF technology must be economically viable, ecologically sound, and socially acceptable to be successful. Although all of these aspects are important, economics often determine the value of a technology. There are frequent conflicts between economy and ecology: a technology that is more sound ecologically may not be viable economically. *Rhizobium* inoculation and green manuring are examples of sound BNF technology but farmers' adoption of them has been disappointing to scientists, extension workers, and policymakers. Inoculating *Rhizobium* into legumes mostly results in increased biological yield, except for soybean where economic yield is often increased. Although higher biological yield improves long-term soil fertility, it does not provide immediate cash benefits to farmers. The result is non-adoption. Likewise, green manuring has tremendous potential as a substitute for chemical fertilizers but is often found to be uneconomical because governments commonly subsidize fertilizer. Many resource management technologies face a similar dilemma. Our research needs to take a holistic rather than a reductionist approach. Admittedly, a holistic approach is more expensive.

A single technology – improved crop genotypes – has contributed the most to productivity increase in modern agriculture. The adoption of a new genotype by farmers is usually very high because no additional cost is involved and existing cropping systems and soil and water management practices are not affected. Although the goal to make cereals fix their own nitrogen is of long-term nature and the funding for research is greatly constrained by concerns of failure, if achieved, would have a significant impact on agriculture world-wide.

This volume is devoted to discussions on the role of BNF in agricultural sustainability. Papers presented on BNF in crop forage and tree legumes are augmented with discussions on integrated farming systems involving BNF, soil and N management, and recycling of legume residues. BNF by non-legumes are discussed and attempts to transform cereals into nodulating plants are critically reviewed. Advances on the development of novel methodologies to understand symbiotic interactions and to assess N_2 fixation in the field are described and means of enhancing BNF through plant and soil management, or breeding and selection are presented. Problems encountered in exploiting BNF under farmers' field conditions and promising approaches to improve BNF exploitation are examined.

J K Ladha
Chairman of the Symposium

Plant and Soil **174:** 3–28, 1995.

Biological nitrogen fixation: An efficient source of nitrogen for sustainable agricultural production?

M.B. Peoples[1], D.F. Herridge[2] and J.K. Ladha[3]
[1]*CSIRO Division of Plant Industry, GPO Box 1600 Canberra, ACT 2601, Australia,* [2]*NSW Agriculture, RMB 944 Tamworth, NSW 2340, Australia and* [3]*International Rice Research Institute, PO Box 933, 1099 Manila, Philippines*

Key words: Azolla, Casuarina, legume, nitrogen fertilizer, rhizobia, symbiotic N_2 fixation

Table of contents

Abstract

Introduction

Fertilizer-N inputs into agriculture

Measured inputs of fixed N

- Food legumes
- Forages legumes, green manures and N_2-fixing trees
- BNF in farmers' fields

Contributions of BNF to the soil N pool and associated rotational effects

Utilization of fertilizer or organic sources of N

- Cropping systems
- Forage systems
- Agroforestry systems

Can the use of BNF as a N source improve farm profitability?

Conclusions

Acknowledgements

References

Abstract

A fundamental shift has taken place in agricultural research and world food production. In the past, the principal driving force was to increase the yield potential of food crops and to maximize productivity. Today, the drive for productivity is increasingly combined with a desire for sustainability. For farming systems to remain productive, and to be sustainable in the long-term, it will be necessary to replenish the reserves of nutrients which are removed or lost from the soil. In the case of nitrogen (N), inputs into agricultural systems may be in the form of N-fertilizer, or be derived from atmospheric N_2 via biological N_2 fixation (BNF).

Although BNF has long been a component of many farming systems throughout the world, its importance as a primary source of N for agriculture has diminished in recent decades as increasing amounts of fertilizer-N are used for the production of food and cash crops. However, international emphasis on environmentally sustainable development with the use of renewable resources is likely to focus attention on the potential role of BNF in supplying N for agriculture. This paper documents inputs of N via symbiotic N_2 fixation measured in experimental plots and in farmers' fields in tropical and temperate regions. It considers contributions of fixed N from legumes (crop, pasture, green manures and trees), *Casuarina*, and *Azolla*, and compares the relative utilization of N derived from these sources with fertilizer N.

Introduction

Globally, cereal cropping dominates cultivated land use (around 50% of total area, Table 1). The remaining arable land is used for production of oilseed, fibre, or food and cash crops. In addition, vast areas are maintained under temporary or permanent pasture for forage production (2–3 fold greater than the total area under cultivation and permanent crop; Table 1, Fig. 1). All cultivated crops, except for legumes (pulses and legume oilseeds) require the soil to provide relatively large amounts of nitrogen (N). It is necessary for the three most important cereals, wheat (*Triticum aestivum*), rice (*Oryza sativa*) and maize (*Zea mays*), to take up 20 to 40 kg soil N ha^{-1} over a period of 3 to 5 months to satisfy the N requirements of the seed and supporting vegetative structure for each tonne of grain produced (e.g. Fig. 2; Myers, 1988). Productive pastures on the other hand may assimilate > 100 kg N ha^{-1} each annum, of which 50 to 90% will be consumed by livestock in intensively grazed systems (Ledgardy, 1991; Thomas, 1995). Even though 75 to 95% of the N ingested is returned as excreta, large amounts of N can be leached or lost as gaseous emissions (Peoples et al., 1994e; Steele and Vallis, 1988), so that the annual demand for N can be continuing and substantial.

The problem facing farmers everywhere is that the capacity of their soils to supply N declines rapidly once agricultural activities commence and N derived from the breakdown of soil organic matter must be supplemented from other sources (Herridge et al., 1994a). During the 1990s, around 1 billion additional people will be added to the 1990 population of 5.3 billion. Over 90% of the increase will occur in the developing countries of Asia, Africa and Latin America, where, already, nearly 1 billion people endure some degree of malnutrition. For productivity to be simply sustained at current levels, let alone improved in the future, the N removed in agricultural produce or lost from the system, must be replaced by N derived either from nitrogenous fertilizers, or biological N_2 fixation (BNF). It is difficult to judge whether farmers are mindful of this concept and manage their N resource accordingly. Evidence suggests that inputs of fertilizer N might well exceed N removal in intensively managed arable soils of the USA and Europe (Sánchez, 1994); however, amounts of N removed from farms in grain or hay and lost through runoff, erosion, leaching and denitrification have been calculated to exceed inputs by fertilizer and BNF in the agricultural systems of the Canadian prairies (on average by 24 kg N ha^{-1} yr^{-1}, Doyle and Cowell, 1993) and Kenya (by 112 kg N ha^{-1} yr^{-1}, Sánchez, 1994). Clearly where agricultural activities are currently "mining", soil reserves, external nutrient inputs in the form of fertilizer and BNF must be increased if farmers are to have any prospect of meeting the food and fibre requirements of a growing world population.

Fertilizer N is a convenient and (currently) cheap source of N for crop growth which provides opportunities for strategic and rapid applications of plant nutrients. However, the use of fertilizer-N in different agricultural systems is ultimately regulated by economic considerations (e.g. per capita income, credit facilities, the current commodity value and expected return for investment at a farmer level, and availability of for-

Table 1. Global allocation of arable land between different commodities

Commodities[a]	Proportion of land area (%)
Cereals	
Wheat	16
Rice	10
Maize	9
All other cereals	13
Total	48
Legumes	
Legume pulses	5
Legume oilseeds	6
Total	11
Other crops	
Other oilseeds	6
Beverages / Tobacco	7
Roots and tubers	4
Sugars	2
Vegetables and fruits	2
Fibres / Rubber / Oil palm	1
Total	22
Temporary pastures / Fodder crops	14
Other	6

[a] Distribution of 1442 million ha estimated to be arable land and under permanent crop in 1991–1992 (FAO, 1992b, 1993).

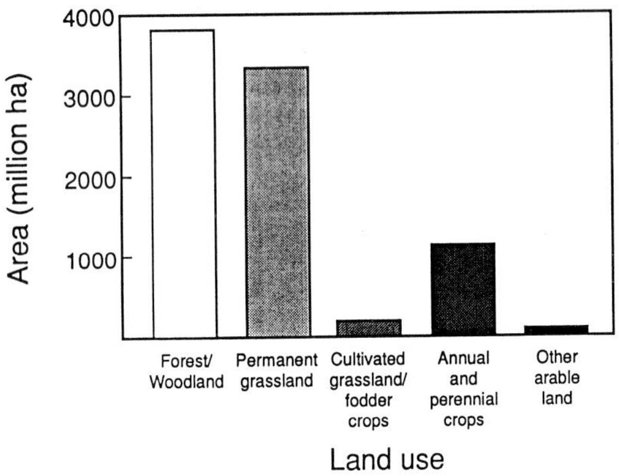

Fig. 1. Global patterns of land use indicating the relative areas under cropping and temporary or permanent pasture (used for forage production for 5 years or more), or containing forests and woodlands (data from FAO, 1993).

eign exchange at a national level), and the presence of effective infrastructures for fertilizer production and distribution. This taken in conjunction with the poor efficiency of utilization of fertilizer N by crops (seldom exceeding 50%) and increasing awareness of the environmental costs of N lost from fertilizers (Bohlool et al., 1992; Craswell and Godwin, 1984; Peoples et al., 1994e), suggests that there is likely to be a limit to the amounts of fertilizer N that farmers' might be willing to apply to improve agricultural production in the long-term. The contribution of BNF to the N-cycle on the other hand can be controlled by manipulating various physical, environmental, nutritional or biological factors (Hansen, 1994) and may therefore be more open to management than fertilizer N.

There is growing international concern about issues of global warming, environmental degradation and loss of natural resources. This concern is summarized in the sentiments expressed by Suzuki (1990) : "We no longer inherit the Earth from our parents, we borrow it from our children". With the current emphasis on the use of renewable resources for environmentally sustainable development, it is timely to assess the potential for BNF to complement or replace fertilizer inputs, and to consider its contribution to the N fertility of agricultural land. Yet we need to achieve more than simply ensuring that inputs balance outputs. Among other things, we should evaluate the capacity of BNF-based farming systems to maintain or enhance agricultural production in the long term in a way that is economically viable and socially acceptable in the short-term. We should also be aware that in some socio-economic frameworks food security can be more important than productivity, and consider the role that BNF can play in reducing the level of production risk and managing agriculture's resource base (Giller and Cadisch, 1995).

If BNF is to represent a reliable renewable resource for sustainable agriculture, its rate of use will be limited by its rate of regeneration and use by non-N_2-fixing crops and grasses. Therefore, it is necessary to know:

(a) The amounts of N fixed by different N_2-fixing systems in a range of environments,
(b) The contribution of BNF to the soil N pool,
(c) The recovery and losses of N derived from BNF sources, and
(d) Options that are available to enhance BNF inputs.

Examples of N_2 fixation inputs observed by researchers at experimental sites and in farmers' fields will be described, and the ultimate fate of that N will be discussed and compared with N from fertilizer in the following sections. Various approaches that are available to increase BNF are described in following papers in this volume.

Most attention will be directed towards N_2 fixation inputs by legumes because of their proven ability to fix N_2 and contribute to integrated agricultural production systems in both tropical and temperate environments. Crop legumes alone occupy 11% of land under cultivation or permanent crop (Table 1). However, legumes are also present in some of the 3541 million ha estimated to be temporary and permanent pastures (Fig. 1), and may provide valuable inputs of N in natural plant communities, forest ecosystems, tree plantations, and intensive cropping systems (Becker et al., 1995; Giller and Cadisch, 1995; Hansen, 1994). Discussion concerning BNF by non-legumes will be restricted to symbiotic systems involving either:

(a) *Anabaena* and the aquatic fern *Azolla*, or
(b) Actinorhizal associations: There are far fewer plant species nodulated by N_2-fixing actinomycetes (*Frankia* spp.) than *Rhizobium*-legume associations, but their numeric inferiority is offset by their capacity to occupy ecological niches where legumes do not thrive. In this particular discussion we specifically consider N_2-fixation by the actinorhizal association with the tree *Casuarina*.

Inputs of N by free-living diazotrophs in soil and plant residues, and non-symbiotic N_2-fixing associations with cereals and C4 grasses will not be discussed here; however, readers are referred to several recent publications (e.g. Chalk, 1991; Peoples and Craswell, 1992; Roger and Ladha, 1992) and papers presented

Table 2. Global changes in consumption of nitrogenous fertilizers between 1960 and 1990[a]

Region	1960	1970	1980	1990	Change 1960–1990
	(million t N)				(%)
Developed economies					
North America	2.6	7.1	11.7	11.3	335
Europe	3.3	9.1	14.3	13.6	312
former USSR	0.7	3.8	8.3	8.7	1,143
Oceania	<0.05	0.1	0.3	0.5	900
Other	0.7	1.0	1.1	1.0	43
Total developed	7.2	21.1	35.7	35.1	388
Developing economies					
Africa	0.1	0.3	0.6	0.8	700
Latin America	0.3	1.2	2.9	3.5	1,067
Asia	1.9	6.2	21.5	37.5	1,874
Other	<0.1	<0.1	0.1	0.2	233
Total developing	2.4	7.7	25.1	42.0	1,650
Total global	9.6	28.8	60.8	77.1	703

[a] Source: FAO (1992a) and Bumb (1994).

elsewhere in the volume (Boddey et al., 1995; Roper and Ladha, 1995) which review these topics.

Fertilizer-N inputs into agriculture

There has been a substantial increase in the use of nitrogenous fertilizer in the past 30 years (Table 2). A near doubling of global cereal production between the early 1960s and 1990 has been attributed in part to a seven-fold increase in fertilizer-N use over the same period (Table 2). Globally the consumption of fertilizer-N increased from 8 to 17 kg N ha^{-1} of agricultural land in the 15 year period from 1973 to 1988 (FAO, 1990). Significant growth in fertilizer-N usage was experienced in both developed and developing countries (Table 2). But while fertilizer usage in Asia and the developing world continued to expand at around 6 to 7% per annum during the, 1980s, fertilizer consumption either remained static or declined in the developed economies of North America and Europe, as a result of grain surpluses and declining, crop prices.

Table 3. Trends in domestic price of fertilizer-N in various geographical regions[a]

Region and country	Currency	Price (local currency t N^{-1}) 1980	1990	Change 1980–1990 (%)
Asia				
Bangladesh	Taka	5,127	10,826	111
Tukey	Lira	21,739	552,174	2,440
Africa				
Ghana	Cedi	762	223,819	29,273
Zambia	Kwacha	506	16,696	3,200
Latin America				
Mexico	Peso	7,679	434,783	5,562
Venezuela	Bolivar	952	3,333	235

[a] Bumb (1994).

Urea is the major form of N-fertilizer used so trends in urea prices reflect trends in N prices. International urea prices decreased steadily in real terms throughout much of the 1980s from around US$ 310 t^{-1} (adjusted to 1990 US$) in 1980 to US$155 t^{-1} in 1990 (Bumb, 1994). The economic and political reforms in Eastern Europe and West Asia have led to a fall in domestic fertilizer use in these parts and requirements for foreign exchange have resulted in the diversion of domestic supplies to international markets leading to further decreases in price from 1990 levels to around US$ 100 t^{-1} by mid 1993. Despite the decrease in urea prices in international markets, farmers in many developing countries were not able to benefit from such low prices because domestic prices increased astronomically (Table 3). Devaluation of domestic currencies, removal of fertilizer subsidies, and domestic inflation have contributed to the rapid increase in prices. In the 1980's, prices in Ghana, for example, increased by over 29,000% and by between 100 and 5,000% in other countries (Table 3).

Mean rates of fertilizer consumption range from < 1 to > 200 kg N per ha of agricultural land for individual countries (Table 4). However, such figures can be misleading since there are wide variations in the extent of use of N-fertilizer for individual crops and agricultural commodities. When considered on a regional basis, a high proportion of the major food and cash crops and intensively grazed pastures and forage crops are heavily fertilized to promote growth (Table 5; Peoples et

Table 4. Average rates of application of fertilizer-N to agricultural land[a]

Consumption rate (kg N ha^{-1})	Examples of countries within category
< 1	Australia, Argentina, Bhutan, Bolivia, Ethiopia, Paraguay, Senegal
10 – 25	Canada, Mexico, Papua New Guinea, Syria, Thailand, former USSR, USA, Venezuela, Zimbabwe
40 – 60	Bangladesh, China, Cuba, El Salvador, Greece, India, Indonesia, Ireland, Italy, Malaysia, Pakistan, Philippines, Sri Lanka, Vietnam
75 – 100	France, Hungary, Poland, United Kingdom
100 – 200	Belgium, Denmark, Germany, Japan
> 200	Egypt, Korea, Netherlands

[a] FAO (1990).

al., 1994e). Applications of fertilizer-N to wheat, rice and maize alone accounted for around 53% of the total global consumption in 1991 (Table 5).

Current assessments of fertilizer usage indicate that two conflicting forces are in operation (Bumb, 1994). Fertilizer use is being depressed in Eastern Europe and West Asia as a result of economic restructuring and political reform, and in Western Europe because of environmental concerns and subsidy removal. Population growth and food security considerations in Asia, Latin America and Africa, on the other hand, are stimulating fertilizer consumption. Fertilizer use in North America is expected to remain stable unless prospects for grain exports improve significantly. Projections of future demand for fertilizer suggest that the net effect of these forces will result in a gradual increase in global fertilizer demand to between 80 to 90 million t N by the year 2,000 (Bumb, 1994). Projections of fertilizer production indicate that fertilizer-N supply should exceed demand throughout the 1990s, therefore stabilizing international fertilizer prices at current low levels. It is anticipated however, that on a regional basis, there will fertilizer surpluses in some regions (Eurasia, North America and North Africa.) and deficits in others (Asia, sub-Saharan Africa and Latin America) where the availability of foreign exchange and domestic fertilizer prices will be critical in ultimately determining the levels of fertilizer use. The likely decrease in fertilizer consumption in Europe and Eurasia could also have an adverse impact on global food production, and therefore, may affect the supply of grains available through aid and trade to developing nations (Bumb, 1994).

In the future, potential to redress this projected shortfall in the supply of fertilizer-N, or replace fertilizer inputs with N derived from BNF sources is likely to be greatest in rain-fed systems where, depending upon the timing and intensity of rainfall, the risk of crop failure is high compared with irrigated crops, and fertilizer responses are more unreliable (Craswell and Godwin, 1984; Myers, 1988). Substantial areas of arable land in all regions of the world support rain-fed agriculture (Table 6), and the challenge to researchers and farmers alike is to improve the contributions of BNF in these systems.

Measured inputs of fixed N

Estimates of the total annual terrestrial inputs of N from BNF range from 139 to 170 million t N (Burns and Hardy, 1975; Paul, 1988), with symbiotic associations growing in arable land accounting for 25 to 30% (35–44 million t N) and another 30% (45 million t N) coming from permanent pasture. While the accuracy of these figures may be open to question, they do help illustrate key points:

(1) The relative importance of BNF in cropping and pasture systems,

(2) The magnitude of the task necessary if BNF is to be improved to replace a proportion of the 80 to 90 million t N of fertilizer-N expected to be applied annually to agricultural land by the end of the decade (Bumb, 1994).

Table 5. Patterns of production of major cereal crops indicating the proportion of each crop receiving fertilizer-N, the average rate of application and estimates of total amounts of N applied in various geographical regions[a]

Region	Wheat				Rice				Maize			
	Global prod^n (%)	Crop fertilized (%)	Rate of applic^n (kg N ha^{-1})	Fertilizer consumed (10^3 N)	Global prod^n (%)	Crop fertilized (%)	Rate of applic^n (kg N ha^{-1})	Fertilizer consumed (10^3 t N)	Global prod^n (%)	Crop fertilized (%)	Rate of applic^n (kg N ha^{-1})	Fertilizer consumer (10^3 t N)
East Asia	18	100	146	4569	41	100	127	4690	19	100	100	2241
Southeast Asia	< 1	-[b]	-	-	22	72	40	1077	3	63	59	316
South Asia	13	96	83	2613	28	80	67	3047	2	72	62	347
Middle East	7	62	83	115	< 1	-	-	-	1	85	97	84
Africa	2	99	52	476	3	23	74	116	5	61	81	1012
Eastern Europe and former USSR	21	88	50	2516	< 1	-	-	-	7	89	136	1169
Western Europe	15	100	134	2250	< 1	-	-	-	3	100	173	4016
North America	17	80	56	2003	2	97	117	124	47	97	157	5062
Latin America	4	96	47	390	3	95	37	222	12	65	88	1523
Oceania	3	66	23	150	< 1	-	-	-	< 1	-	-	-
Total global				16082				9276				15770

[a] Data for 1989–1992 (FAO, 1992b, 1993). Total N applied = 41.1 million t N, or 53% of total global consumption of fertilizer-N in 1991.
[b] No data presented if regional contribution to total global grain production < 1%.

Table 6. Relative areas of irrigated and rainfed agricultural land in the various geographical regions[a]

Region	Proportion of arable land	
	(Irrigated %)	(Rainfed %)
East Asia	52	48
Southeast Asia	19	81
South Asia	34	66
Middle East	56	44
Africa	7	93
Eastern Europe and former USSR	10	90
Western Europe	14	86
North America	8	92
Latin America	12	88
Oceania	4	96
Total global	18 (242)[b]	82 (1105)[b]

[a] Data for 1991 (FAO, 1993).
[b] Values in parenthesis represent millions of ha of arable land either irrigated or rainfed.

Food legumes

Experimental estimates of the proportion of plant N derived from N_2 fixation (P_{fix}) and the amounts of N_2 fixed by important tropical and cool season crop legumes are presented in Table 7. Although experimental treatments and environmental or nutritional variables have generated a large range of P_{fix} values (0–98%) and inputs of fixed N, it appears that potential BNF for most species is in the range of 200 to 300 kg N ha^{-1} crop^{-1} (Table 7). However, since crop N is partitioned either into seed, or vegetative parts at crop maturity, not all of the N_2 fixed will be available for return to the soil. The final contribution of fixed N to the soil following harvest will depend upon the N-balance at harvest, which is determined by the difference between the amounts of N_2 fixed and seed N removed:

$$\text{N} - \text{balance} = (\text{N}_2\text{fixed}) - (\text{seed N}) \quad (1)$$

With protein levels of 20 to 40%, legume seeds have a high demand for N and up to 60 kg N ha^{-1} can be removed with every tonne of seed harvested. Global

Table 7. Range of experimental estimates of the proportion (P_{fix}) and amount of N_2 fixed by important pulses and legume oilseeds[a]

Species	P_{fix} (%)	Amount N_2 fixed (kg N ha^{-1})
Cool-season legumes		
Chickpea (*Cicer arietinum*)	8 - 82	3 - 141
Lentil (*Lens culinaris*)	39 - 87	10 - 192
Pea (*Pisum sativum*)	23 - 73	17 -244
Faba bean (*Vicia faba*)	64 - 92	53 - 330
Lupin (*Lupinus angustifolius*)	29 - 97	32 - 288
Warm-season legumes		
Soybean (*Glycine max*)	0 - 95	0 - 450
Groundnut (*Arachis hypogaea*)	22 - 92	37 - 206
Common bean (*Phaseolus vulgaris*)	0 - 73	0 - 125
Pigeon pea (*Cajanus cajan*)	10 - 81	7 - 235
Green gram (*Vigna radiata*)	15 - 63	9 - 112
Black gram (*V. mungo*)	37 - 98	21 - 140
Cowpea (*V. unguiculata*)	32 - 89	9 - 201

[a] Collated from Peoples and Craswell (1992), Herridge et al. (1993), and Peoples et al. (1994a). Additional data were derived from Jensen (1987) - pea, Evans et al. (1989) - lupin and faba bean, Hardarson et al. (1993) - common bean, and Ladha et al. (1995) - pigeon pea.

average yields of around 2 t ha^{-1} for crops such as soybean (Table 8) mean that at least 120 kg N ha^{-1} must be fixed before the soil can receive any net benefit from BNF (Peoples et al., 1994c).

When the quantities of N involved in plant growth, in N_2 fixation, and in the seed are calculated for crop legumes, it appears that levels of fixation achieved by many crops might not always be sufficient to offset the N removed with seed (Peoples and Craswell, 1992; Peoples et al., 1995). Comparisons of N_2 fixation and amounts of seed N harvested have been undertaken for many crops and examples of final N-balances range from as little as -132 kg N ha^{-1} to as much as +80 kg N ha^{-1} in soybean (Bergersen et al., 1989; Hughes and Herridge, 1989), -34 to +64 in groundnut (Bell et al., 1994; McDonagh et al., 1993), -42 to +34 in chickpea (Doughton et al., 1993), -11 to +25 in lentil (Bardarneh and Ghawi, 1994), -32 to +96 in pea and -41 to +135 kg N ha^{-1} in lupin (Evans et al., 1989), depending upon the amounts of N_2 fixed, the harvest index for N (the proportion of total crop N removed in seed), and whether the vegetative residues are removed from the field (Bell et al., 1994; Ying et al., 1992).

Forage legumes, green manures and N_2-fixing trees

Far fewer measures of N_2 fixation are available for the forage legumes, cover crops, and N_2-fixing trees than for crop and green manure legumes (Giller and Wilson, 1991; Becker et al., 1995). However, the available data indicate that levels of N_2 fixation can be similar to crop legumes (Tables 8 and 9). Levels of P_{fix} have often been reported to be >60%, and inputs of fixed N can be considerable (eg 120–290 kg N ha^{-1} in 45–55 days by shrub legumes; >350 kg N ha^{-1} per year by lucerne). However, in the case of forage legumes, small amounts of fixed N have been measured in some instances (<20 kg N ha^{-1}, Table 8), despite high levels of P_{fix}. This reflects a low legume component in a pasture sward (Peoples et al., 1994d, 1995; Thomas, 1995).

One of the advantages of using BNF of trees, forages, or green manures as a N source is that a larger proportion of the N accumulated is generally available for return to the soil because much less N is removed from the system in agricultural produce compared to the amounts of N harvested in the seed of crop legumes. In these systems, the potential for the return of BNF to soil either through leaf fall, mulch, or via a gazing animal and undetermined turnover of roots and nodules may be considerable.

BNF in farmers' fields

Food legumes

By comparison with the gains made in yields of cereals between 1981 and 1991, global yield increases of major crop legumes were relatively small (Table 10). That a large yield gap (1–3 t ha^{-1}) exists between crop legume yields in experimental plots and those in farmers' fields (particularly in, Asia; e.g. Bhatnagar and Tiwari, 1989) suggests that N_2 fixation of the food legumes could be increased through management practices that remove growth constraints.

The limited surveys of N_2 fixation in farmers' crops indicate that values for P_{fix} (6–91%) and amounts of N_2 fixed (15–267 kg N ha^{-1}, Table 11) are similar to experimental observations (Table 7). The range of N-balance determinations calculated following seed harvest for lupin and soybean (Table 11), are almost identical to those described above for experimental crops.

However, experimental studies are generally designed to generate ranges of N_2 fixation as a means of

Table 8. Range of experimental estimates of the proportion (P_{fix}) and amount of N_2 fixed by important forage legumes[a]

Species	P_{fix} (%)	Amount N_2 fixed (kg N ha^{-1})	Period of measurement
Temperate forages			
Lucerne / alfalfa (*Medicago sativa*)	46 - 92	90 - 386	Annual
Strand medic (*M. littoralis*)	51 - 82	52 - 102	144d
M. truncatula	70	90	na
Birdsfoot trefoil (*Lotus cornicuatus*)	30 - 85	49 - 109	Annual
White clover (*Trifolium repens*)	62 - 93	54 - 291	Annual
Red clover (*T. pratense*)	35 - 87	69 - 373	Annual
Subterranean clover (*T. subterraneum*)	50 - 93	2 - 206	Annual
Crimson clover (*T. incarnatum*)	75 - 81	124 - 185	Annual
Vetch (*Vicia sativa*)	75	106	na
Tropical forages			
Arachis pintoi	72 - 87	1 - 7	84d
Calopogonium spp.	na[b]	64 - 182	Annual
Centrosema spp.	82 - 83	41 - 43	119d
	na	67 - 280	Annual
Clitoria ternatea	77 - 81	197 - 249	190 - 195d
Desmodium spp.	80 - 100	24 - 380	Annual
Desmanthus virgatus	77 - 80	193 - 228	190 - 195d
Siratro (*Macroptilium atropurpureum*)	78 - 92	15 - 167	Annual
Pueraria spp.	75 - 88	9 - 115	72 - 199d
Stylosanthes spp.	60 - 95	2 - 75	63 -77d
	na	20 - 263	Annual
Zornia glabra	88	61	119d

[a]Collated from Giller and Wilson (1991), Ledgard and Steele (1992), Peoples and Craswell (1992) and Thomas (1995). Additional data were derived from Brockwell et al. (1994), and Gault et al. (1995) lucerne / alfalfa, Peoples et al. (1994d) subterranean clover, Heichel et al. (1985) red clover and birdsfoot trefoil, and Ladha et al. (1995) Siratro, *Clitoria*, and *Desmanthus*.
[b]na=information not available.

investigating factors regulating the N_2-fixing process. It is in farmers' fields where issues of sustainability need to be addressed. The presence of low P_{fix} values, poor N yields or negative N-balances in many surveys of commercial crops (Table 11), indicates that farmers do not always fully exploit the potential benefits of BNF.

While there are often large gaps between N_2 fixation measured in farmers' fields and the highest research values, the upper end of the range of N_2 fixation achieved by some farmers approach or exceed those from experimental crops (Tables 7 and 11). The wide range of levels of N_2 fixation found in farmers' fields is encouraging in itself and suggests that management opportunities are available for farmers' to improve BNF inputs (Peoples et al., 1995).

Forage systems

While mean determinations of P_{fix} (69–81%) reported in a survey of BNF of several temperate legume species in annual pastures across southern Western Australia (243 measurements from 81 farms, Sanford et al., 1994) are similar to research results (Table 8); low values (0–65%) were detected at one-third of the sites studied. Low P_{fix} values have also been measured in farmers' fields in the tropics (see data for *Centrosema* in Table 12). This is in stark contrast to the relatively narrow range of values usually reported

Table 9. Range of experimental estimates of the proportion (P_{fix}) and amount of N_2 fixed by important N_2-fixing trees, green manures, and cover crops[a]

Species	P_{fix} (%)	Amount N_2 fixed (kg N ha^{-1})	Period of measurement
Trees			
Acacia holosericea	30	3 - 6	6.5 months
Calliandra (*Calliandra calothyrsus*)			
- hedgerow for forage[b]	14 - 48	11 - 101	3 - 6 months
Casuarina equisetifolia	39 - 65	9 - 440	6 - 12 months
Gliricidia (*Gliricidia sepium*)	52 - 64	86 - 309	Annual
- hedgerow for forage	69 - 75	99 - 185	3 - 6 months
- alley crop hedgerow	43	170	Annual
Leucaena (*Leucaena leucocephala*)	34 - 78	98 - 230	3 - 6 months
Green manures and cover crops			
Aeschynomene afraspera	68 - 76	105 - 145	56d
A. indica	93 - 100	75 - 127	116d
Azolla spp.	52 - 99	22 - 40	30d
Crotalaria (*Crotalaria juncea*)	80 - 96	146 - 221	102 - 190d
Indigo (*Indigofera tinctoria*)	70	79	225d
Calopogonium / Peuraria spp.	50	150	Annual
Sesbania cannabina	70 - 93	126 - 141	Seasonal av
	93	119 - 188	45 - 55d
S. rostrata	68 - 94	70 - 324	45 - 65d
	65 - 78	147 - 281	116d
S. sesban	13 - 18	7 - 18	2 months

[a] Collated from Giller and Wilson (1991), Ladha et al. (1992), Peoples and Craswell (1992), Roger and Ladha (1992), Kumarasinghe and Eskew (1993), and Peoples et al. (1995). Additional data were derived from Ladha et al. (1993) - Gliricidia, Yoneyama et al. (1991) - *Aeschynomene, Crotalaria, S. rostrata* and Ladha et al. (1995 and unpublished) - *Crotalaria*, indigo.

[b] Trees maintained in hedgerows for forage are usually planted at a 0.5 m spacing, with 1.5 m between rows. In alley-cropping systems, trees may be spaced from 0.25 to 0.5 m apart within hedgerows with 4 to 5 m wide alleys.

for experimental plots (50–90%, Table 8). There are a number of possible reasons for this:

(i) Although some investigations have monitored legume growth and N_2 fixation over several years (e.g. Brockwell et al., 1994; Gault et al., 1995; Heichel et al., 1985; Peoples et al., 1994d; Vallis and Gardner, 1985), most measures of N_2 fixation by forages have been determined shortly after establishment of a legume-dominant sward. Unfortunately, the high legume contents common in research studies are not necessarily representative of farmers' pastures. The botanical composition of pastures can fluctuate widely in response to rainfall patterns, grazing pressure, and elapsed time after pasture establishment (Wilson and Simpson, 1993). The data shown in Table 12 reflect, in part, the effect of pasture age (3–30 years after initial establishment) and pasture quality on BNF.

(ii) Experimental determinations have almost invariably been undertaken in the absence of grazing animals. Apart from being an integral part of most- farming systems, livestock can affect pasture growth and composition, and negatively impact upon BNF by introducing localized N-rich sites via urine patches and faeces (Ledgard and Steele., 1992; Peoples et al., 1994d).

Green manures

Azolla. The aquatic fern *Azolla* is probably used as a green manure on <2% of the world's rice crop, but

Table 10. Global change in yield of major cereals and crop legumes during the past decade[a]

Crop	Yield 1981 (t ha^{-1})	Yield 1991 / 1992 (t ha^{-1})	Change in yield over 10 years (%)
Cereals			
Wheat	1.86	2.51	35
Rice	2.75	3.55	29
Maize	3.34	3.98	19
Crop legumes			
Soybean	1.70	1.92	13
Groundnut	0.99	1.16	16
Chickpea	0.62	0.71	13
Common bean	0.55	0.64	16

[a] FAO (1993).

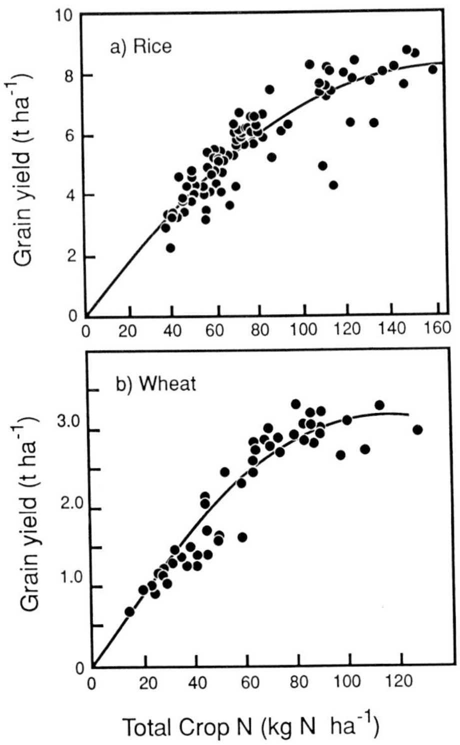

Fig. 2. Relationship between total crop N and grain yield for (**a**) rice (data from 12 rice genotypes; Ladha, unpubl.) and (**b**) wheat (data derived from Strong et al., 1986; Herridge and Doyle, 1988; Doyle et al., 1988; McDonald, 1992).

this still represents around 2 to 3 million ha (Giller and Wilson, 1991). Under optimal conditions, *Azolla* doubles in biomass every 3 to 5 days and one crop can be expected to accumulate between 70 and 110 kg N ha^{-1} (Ventura and Watanabe, 1993). With experimental values of P_{fix} commonly >70% (Kumarasinghe and Eskew, 1993; Roger and Ladha, 1992), *Azolla* represents a potentially important source of N for flooded rice. However, there is little information available concerning inputs of N by *Azolla* in farmers' fields. Since growth and N_2-fixing capacity of *Azolla* can be affected by many environmental variables, mineral nutrition (particularly phosphorus), insect predators and pathogens, it is uncertain whether experimental potentials are ever realized in farmers' paddies (Giller and Wilson, 1991).

Legumes. The principal source of information on legume green manures in farmers' fields comes from measures of P_{fix} for perennial cover-crops in commercial rubber and oilpalm plantations in Malaysia (Faizah and Peoples, unpublished data). The range of values were wide (9–91%, Table 12), but the mean (57%) was close to previous experimental determinations in plantation systems (Giller and Wilson, 1991). The levels of P_{fix} observed tended to be inversely related to the age of the cover-crop, and presumably reflected the change in light penetration through the plantation canopy and the accumulation of organic N in the interrow space between the trees.

Agroforestry systems

Almost all measurements of BNF by N_2-fixing trees, either on farms or as part of natural communities, have been restricted to estimates of P_{fix} from studies or surveys undertaken in Africa, Australia, Brazil, Indonesia, the Philippines, Thailand or USA (Table 12). Although amounts of N_2 fixed cannot be calculated in most cases because tree biomass was not determined, the inputs of N could be substantial because of the large annual yields of dry matter and N of woody species (from 150-850 kg N ha^{-1}) in natural communities (Sandhu et al., 1990; Torrey, 1982), in alley cropping systems (Ladha et al., 1993; Sanginga et al., 1995), or where trees and shrubs are cut for forage (Blair et al., 1990).

Values of P_{fix} for the non-legume tree *Casuarina*, were uniformly high, ranging from 65 to 90% (mean 81%) under different environments and soil types (Table 12). Estimates of P_{fix} for tree legumes on the other hand ranged from 0 to 100%. Although P_{fix} values for a number of legume trees and shrubs indicated that 60% or more of their N requirements

Table 11. Proportion (P_{fix}) and amount of plant N derived from N_2 fixation by crop legumes growing in farmers' fields, and potential contributions of fixed N to soil reserves following seed harvest[a]

Species	Crop N	N_2 fixation		Seed N	N-balance	Reference
		P_{fix}	Amount			
	(kg N ha^{-1})	(%)	(kg N ha^{-1})	(kg N ha^{-1})	(kg N ha^{-1})	
Cool-season						
Pea	220 - 227	60 - 81	133 - 183	135 - 162	-2 to +21	Peoples et al. (1995)
Lupin	46 - 199	65 - 91	36 - 181	nd[b]	nd	Unkovich (1991)
	241 - 372	78 - 88	188 - 327	126 - 170	+62 to +157	Unkovich et al. (1994)
Faba bean	61 - 171	62 - 76	38 - 130	42 - 74	-460 + 56	Rochester, Gault and Peoples, unpubl.
Warm-season						
Soybean	33 - 132	45 - 74	15 - 98	nd	nd	Bhromsiri and Peoples, unpubl.
	330 - 348	13 - 72	44 - 250	178 - 181	-134 to +69	Peoples et al. (1994b)
Green gram	nd	6 - 81	nd	nd	nd	Bushby and Lawn (1992)

[a] Apart from the data of Bhromsiri and Peoples, which relate to soybean grown in farmers' fields during the late rainy season in northwest Thailand, all other measurements come from commercial crops in Australia.
[b] nd = not determined.

were met by N_2 fixation, some fell below 25% (Table 12). Many of these low measurements were reported for *Acacia* spp. In one instance (Hamilton et al., 1993), low P_{fix} values and small inputs of fixed N reflected recovery of the understory legumes after burning and a low density of acacia within the temperate forest. Other examples of low reliance of *Acacia* upon N_2 fixation come from arid, harsh conditions (e.g. Schulze et al., 1991; Shearer et al., 1983), however, species such *Prosopis glandulosa* demonstrated a higher symbiotic capacity in these same environments (Table 12). In the study of Shearer et al. (1983), in a desert ecosystem in California, estimates of P_{fix} for *P. glandulosa* were 40 to 60% at 6 of the 7 sites examined. Growth and N accumulation was monitored at one of the sites where 60% of the plant N was calculated to be from N_2 fixation, and an annual input of 40 kg N ha^{-1} was calculated. This represented a remarkable contribution by BNF in an inhospitable environment.

In general, levels of N_2 fixation tended to be highest in humid, more favorable environments where trees were actively managed to provide mulch or animal forage and where, presumably, N demand was greater than in undisturbed or other low productivity ecosystems.

Contributions of BNF to the soil N pool and associated rotational effects

The increased crop yields that are widely observed following N_2-fixing associations may arise from breaking cycles of pests and diseases, through changes in soil microbial, chemical or physical characteristics or through encouraging the activity of soil macrofauna such as earthworms (Kundu and Ladha, 1994; Peoples and Craswell, 1992; Wani et al., 1995). But most often rotational benefits can be attributed to an improvement in the N economy of soils. The key to the long-term sustainability and productivity of soils is organic matter, in particular organic C and N. Examples are presented of the use of legumes in Australian agriculture to illustrate the role played by BNF in contributing N to soil.

In the western and southern parts of the Australian wheat belt, annual pasture legumes (predominantly subterranean clover and medics) were introduced into pastures grown in rotation with crops in the 1940's and 50's (Reeves, 1991). More recently perennial legumes, in particular lucerne, have also been used. These legume-based pasture systems can be very successful in increasing total (organic) soil N (Dalal et al., 1994). Annual increments of soil N of between 25 and 100 kg N ha^{-1} appear to be common in subterranean clover-based pastures (e.g. Fig. 3; Simpson et al., 1973a), but average rates of soil N accretion much greater than 100 kg N ha^{-1} have been report-

Table 12. Proportion (P_{fix}) and amount of plant N derived from N_2 fixation for legume forages and cover crops, and for N_2-fixing trees growing in farmers' fields or commercial plantations

Species	Legume N-yield (kg N ha^{-1})	N_2 fixation P_{fix} (%)	N_2 fixation Amount (kg N ha^{-1})	Reference
Forage legumes				
Lucerne - grazed	174	42	74[b]	Peoples and Gault, unpubl.
- cut for hay	185	57	105[b]	
Medicago spp.	nd[a]	7 - 100		Sanford et al. (1994)
Subterranean clover	nd	0 - 100		Sanford et al. (1994)
	11 - 141	68 - 90	10 - 95[c]	Peoples and Gault, unpubl.
Trifolium michelianum var. *balansae*	nd	0 - 100		Sanford et al. (1994)
Lotus spp.	nd	1 - 100		Sanford et al. (1994)
Centrosema spp.	nd	37 - 71		Peoples, unpubl.
Legume cover-crops				
Calopogonium/ *Peuraria* spp.	nd	9 - 91		Faizah and Peoples, unpubl.
Trees and shrubs				
Acacia spp.	<1	21 - 48[d]	<1[d]	Hamilton et al. (1993)
	nd	34 - 95		Peoples, unpubl.
	nd	2 - 71		Schulze et al. (1991)
	nd	0		Yoneyama et al. (1993) / Shearer et al. (1983)
Albizia falcataria	nd	43 - 80		Peoples, unpubl.
Aeschynomene spp.	nd	75 - 94		Yoneyama et al. (1990)
Casuarina spp.	nd	65 - 90		Yoneyama et al. (1990)
Calliandra	nd	20 - 47		Peoples, unpubl.
Gliricidia	nd	32 - 88		Peoples, unpubl.
	nd	0		Yoneyama et al. (1993)
Leucaena	nd	59 - 100		Yoneyama et al. (1990, 1993)
Sesbania spp.	nd	61 - 100		Yoneyama et al. (1990, 1993)
Prosopis glandulosa	nd	31		Schulze et al. (1991)
	nd	2 - 61	40[c]	Shearer et al. (1983)

[a] nd = not determined.
[b] Dryland lucerne, one year after establishment under a wheat crop receiving different management in adjacent fields. Measurements of N_2 fixation represent cumulative data from 5 samplings taken during a growth period of 135 during spring and summer (September to February).
[c] Averaged over entire growing season.
[d] The range in P_{fix} values reflect different periods after fire (12–27) months). Amount of N_2 fixed were low due to low plant density and slow growth of understorey legumes in the temperate forest.

ed under lucerne despite the removal of large amounts of shoot N by grazing animals or as hay (Gault et al., 1995; Holford, 1981). However, once the pasture phase is replaced by cereal cropping, N fertility quickly declines (Fig. 3). Thus, the system demands alternating pasture-cropping phases.

Effects of annual crop legumes on soil N are not so clear cut. A number of trials (Dalal et al., 1994; Reeves et al., 1984; Strong et al., 1986) were unable to detect consistent effects of prior legume crop on levels of total soil N. On the other hand, Rowland (1987) reported that total soil N increased in long-term lupin-wheat rotations. There are a number of reasons why

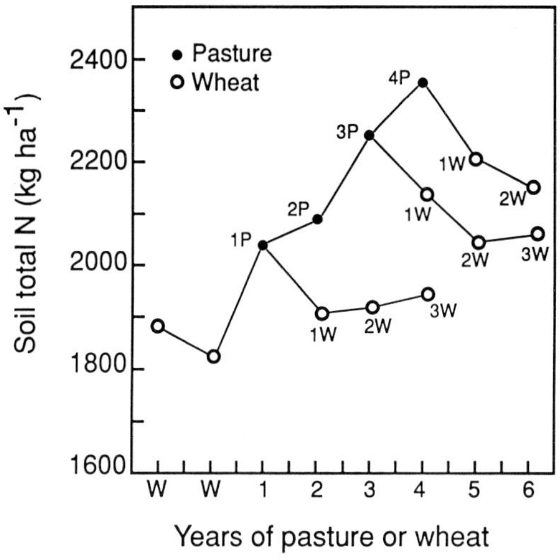

Fig. 3. Changes in levels of total soil N in the southern wheat belt of Australia following different numbers of years of legume-based pasture (P), or wheat (W) (redrawn from Reeves, 1991).

the net effects of crop legumes on total soil N may be difficult to observe:

(i) In virtually all studies of crop legume-cereal rotations, the legumes have been grown for a single season. By comparison, pasture legumes are more likely to be grown for a number of seasons. Holford (1981) found that 2 to 3 years of lucerne was required to increase levels of total soil N significantly. The study of Rowland (1987), in which lupin increased total soil N, was long-term, not a one-year experiment. Thus, it may. be unreasonable to expect to detect significant differences in soil N following a single legume crop.

(ii) Measurements of the small increments in total soil N, likely to accrue a in one-year or a short-term experiment, is technically very difficult.

(iii) Soils are normally cultivated in the fallow period between the legume crop and the following cereal crop, thereby encouraging rapid transformations of the N in legume residues to mineral forms.

(iv) Grain legumes are usually grown in soils that have been cultivated and fallowed. Consequently, nitrate levels may be sufficiently high to inhibit or delay nodulation and so depress N_2 fixation potential.

(v) A large proportion of the N accrued by the legume during growth is removed with the harvested seed (see equation (1) and related discussion above).

There may be more appropriate indices of the benefits of legumes on the soil N pool than measurement of total soil N. A number of rotational trials has demonstrated that legumes can increase the capacity of soils to supply plant-available N regardless of whether there are detectable changes in total soil N or not. The ability of legumes to improve soil reserves of readily mineralizable organic N have been reported for crop legumes (Dalal et al., 1994; Herridge et al., 1993; Wani et al., 1994), temperate and tropical pasture species (Bromfield and Simpson, 1973; Dalal et al., 1994; Thomas and Lascano, 1994), and alley cropping systems (Haggar et al., 1993).

One of the most consistent effects of both crop and pasture legumes is to increase plant-available (nitrate) N in the soil (Table 13). In the studies summarized in Table 13, nitrate-N levels in soil immediately following legumes were between 14 and 77 kg N ha^{-1} greater than the levels after non-legumes (usually measured to a depth of 60 or 120 cm). This extra nitrate, detectable even during growth of the legume, results from a reduced use of soil nitrate ("nitrate-sparing", Evans et al., 1991; Herridge et al., 1994b; Herridge and Bergersen, 1988), the possible release of products of N_2 fixation from nodulated roots (Ofosu et al., 1992, 1993; Pothet al., 1986; Sawatsky and Soper, 1991), or from N mineralized from fallen leaves or roots and nodules lost during growth and development. After a period of time the differences in levels of soil nitrate between legume and non-legume plots usually increase as N contained in the legume residues is released (Table 13).

Few studies have attempted to quantify the relative benefits of fixed or spared N, but one investigation (Chalk et al., 1993) concluded that in the case of a lupin crop, fixed N from fallen leaves and roots, and unutilized soil nitrate contributed in approximately equal proportions to the N benefit to a subsequent cereal crop. However, another study of the residual benefit of vetch (Danso and Papastylianou, 1992), attributed the 61% improvement in N accumulated in barley (*Hordeum vulgare*) following vetch compared to barley after oats (*Avena sativa*), to a reduced uptake of soil N by vetch rather than a release of fixed N.

The combination of conserved soil N, greater mineralization potential and return of fixed N might explain why the benefits of crop legumes to subsequent non-legume crops can be considerable (e.g. Doyle et al., 1988; Peoples and Craswell, 1992; Wani et al., 1995), even when there are apparently only modest returns of fixed N in vegetative residues (Evans et al., 1991; Herridge et al., 1994b). However, rotational benefits might also be greater than expected from calculations of the apparent return of fixed N to soil because estimates of

Table 13. Examples of the increased levels of soil nitrate often detected after growth of a legume

Species	Additional soil nitrate[a] (kg N ha^{-1})		Reference
	Post-harvest	Pre-sowing	
Crop legumes			
Chickpea	+14	+46	Herridge et al. (1994b)
Lupin	+34	nd[b]	Reeves et al. (1984)
Pea	+28 to +38	nd	Jensen and Haahr (1960)
Soybean	+23	+62	Herridge (1987)
Green gram	+26	+57	Doughton and Mackenzie (1984)
Black gram	+38	+68	Doughton and Mackenzie (1984)
Pigeon pea	+15 to +59	nd	Ladha et al. (1995)
Forage legumes			
Subterranean clover	nd	+100 to +150[c]	Angus, Koetz and Peoples, unpubl.
Medicago scutellata	+24[d]	+46[d]	Ladd et al. (1986)
Siratro	+26[e]	nd	Ladha et al. (1995)
Crotalaria	+19 to +77[e]	nd	Ladha et al. (1995)
Clitoria	+25 to +54[e]	nd	Ladha et al. (1995)

[a] Calculated as the difference between the levels of soil nitrate after a legume and after a cereal crop or a period of fallow. Measurements were taken either immediately after growth of a legume, or just prior to sowing a following crop.
[b] nd = not determined.
[c] Difference between autumn buildup of soil nitrate after growth of subterranean clover-dominant pasture or following a period of bare-fallow the previous spring (i.e. no legume growth). The range in soil nitrate levels reflect different clover contents (70–100%) that resulted from use of selective herbicides.
[d] Results were similar irrespective of whether levels of soil nitrate after a pure legume sward were compared with a mixed legume-grass pasture, or wheat.
[e] Levels of soil nitrate at the end of the growing season (following the last cut) were compared with soil under a weedy fallow.

N_2 fixation have almost always been based solely on measurements of shoot N. The contributions of roots and nodules have often been overlooked as a potential source of N. Estimates of amounts of N present in nodulated roots determined from field excavations of root systems of various crop legumes have ranged from < 15 kg N ha^{-1} (Bergersen et al., 1989; Kumar Rao and Dart, 1987) to between 30 and 50 kg N ha^{-1} (Chapman and Myers, 1987; Unkovich et al., 1994). Commonly, such amounts of root N represent < 15% of total plant N. Yet a recent study on lupin using ^{15}N has indicated that the total below-ground N may be almost three-fold higher than might be calculated from N contained in recoverable root material (Russell and Fillery, 1994a). If this finding is applicable to a range of other legumes, then past determinations of the return of fixed N to soil have been greatly underestimated. There could also be considerable rotational benefits from using abundantly nodulated legume genotypes (Song et al., 1994; Wani et al., 1995) simply because they partition relatively more crop N below-ground (Song et al., 1994).

Although this discussion has focussed on the positive aspects of legumes on the N-economy of soils, it should also be acknowledged that there are potential risks associated with a buildup of soil mineral N. The accelerated acidification of many soils under legume-based improved pastures and crop legume-cereal rotations in the winter-dominant rainfall regions of Australia (Coventry and Slattery, 1991; Ridley et al., 1990ab; Willmans, 1980) has largely been attributed to the mismanagement of legume-derived N, inefficient utilization of plant-available soil N and leaching of nitrate (Helyar, 1991). Based on the assumption that nitrate leaching accounts for that proportion of observed soil acidification not attributable to the addition of organic acids or ion removal in agricultural produce, it has been estimated that between 14 and 80 kg nitrate-N ha^{-1} yr^{-1} might be leached from soils in the

500 to 800 mm rainfall belt of southeastern Australia. This proposition is supported by:

(i) The finding that soils acidify more slowly under perennial than annual pastures (Ridley et al., 1990b) - it is presumed that perennials have a greater capacity to utilize soil nitrate and water throughout the year and so the opportunity for nitrate loss is minimized.

(ii) The pattern of nitrate distribution under annual subterranean clover-based pastures - there are regular flushes of mineral N in the soil surface as legume residues decompose following each growing season (Table. 13). The appearance of "nitrate bulges" at depth (Peoples et al., 1995) strongly suggests leaching processes.

That the mismanagement of legume N could have such an impact on soil chemistry in a temperate environment is reason for caution in the humid tropics where there is rapid decomposition of legume residues and a high potential for nitrate leaching (George et al., 1993). Where it is intended that BNF will be used as a primary source of N to improve soil fertility, the legume-derived N will need to be managed well and attention should be given to closing "windows for nitrate loss" (Helyar, 1991).

Utilization of fertilizer or organic sources of N

Maximizing N gains through the use of legumes or *Azolla* not only entails maximizing N_2 fixation, but also requires that recovery of N is optimized. The N transformations occurring during breakdown of leguminous or *Azolla* material are influenced by residue management and soil physical and chemical properties. In rice cropping systems, the dominant N transformations will be depend on whether soils are flooded immediately or remain aerobic (George et al., 1992). The rate of decomposition may be modified by lignin and polyphenol, content (Becker et al., 1994b; Fox et al., 1990; Handayanto et al., 1994) and soil pH, but it is the tissue N concentration (C:N ratio), soil temperature and water status which are the primary factors influencing mineralization and release of legume and *Azolla* N (Peoples and Craswell, 1992; Sandhu et al., 1990; Watanabe et al., 1991). This section compares the recovery of N derived from *Azolla* or legume sources with fertilizer N for a number of important agronomic systems.

Cropping systems

Tropical and warm season crops

Upland rice. In a study in Indonesia, Sisworo et al. (1990) compared the utilization by upland rice of 2 sources of legume N (cowpea and soybean) and urea. Up to 28% of the cowpea N was recovered by a following crop of upland rice and a further 45% was recovered by 5 successive crops over a 2-year period (Table 15). This compared with a total recovery of the soybean or urea N of only 23% over 2 years. Utilization of fertilizer by upland crops elsewhere in the tropics have ranged from 12 to 74%, with poor recoveries reportedly occurring under both dry and very wet conditions (Myers, 1988).

Flooded rice. When a legume is grown as a green manure, either the entire crop is returned, or quantities of leaf mulch are applied to soil. The amount of N and the N concentration in such material is generally high and rates of decomposition can be very rapid following incorporation in tropical flooded soils (Becker et al., 1994a; Ventura and Watanabe, 1993). Studies comparing recoveries of ^{15}N-labelled leguminous green manure with urea fertilizer indicate that between 21 and 62% of the legume N and 24 to 51% of the fertilizer N can be taken up by a following rice crop (Table 14). Not only was the recovery of legume N by rice either comparable or superior to that of inorganic fertilizer N, but it appeared also to be less subject to loss than urea-N (Table 14).

Decomposition of *Azolla* is also rapid and up to 66% of *Azolla*-N has been recovered in subsequent rice crops depending upon the method and time of incorporation, and the amount of *Azolla* applied. In some experiments, rice uptake of N derived from *Azolla* was greater than fertilizer (39–63% cf 27–48% respectively) because of reduced losses of *Azolla*-N (0–11% cf 30–32%; Watanabe et al., 1989), but a series of co-ordinated studies across 6 rice-growing countries showed that the recoveries of N from both sources were similar when averaged across a wide range of environmental and soil conditions (Kumarasinghe and Eskew, 1993; Table 14).

In flooded systems, fertilizer-N is predominantly lost via ammonia volatilization and denitrification. Factors such as fertilizer composition, rate, time and method of application, floodwater depth, and algal growth exert their influences on these two loss processes through the primary variables - ammoniacal N

Table 14. Recovery of N from organic and inorganic sources by tropical-warm season cereal crops

Crop	Organic residue	Plant uptake of N (% of applied)	N Loss (% applied)	Fertilizer[a]	Plant uptake of N (% of applied)	N Loss (% applied)	Reference
Rice							
Upland	Cowpea	28 (1st crop)		Ur	19 (1st crop)		Sisworo et al. (1990)
		73 (over 2 yr)			23 (over 2 yr)		
	Soybean	14 (1st crop)					
		23 (over 2 yr)					
Flooded	*Crotalaria*	21	29	Ur	24	36	Roa and Shinde (1991)
	Aeschynomene	42 - 49	7 - 9	Ur	28 - 31	31 - 50	Diekmann et al. (1993)
	Sesbania spp.	40 - 47	13 - 16				
		12 - 34	0 - 15	Ur	20 - 31	30 - 33	Becker et al. (1994a)
		34	28	Ur	24	36	Rao and Shinde (1991)
		62[b]		Ur	51[b]		Ventura and Watanabe (1993)
	Green manures		14[c]	Ur		35[c]	Becker et al. (1995)
	Azolla	66[b]					Ventura and Watanabe (1993)
		39 - 63	0 - 11	Ur	27 - 48	30 - 32	Watanabe et al. (1989)
		40[d]		Ur	36[d]		Kumarasinghe and Eskew (1993)
				Ur	5 - 57	14 - 85	Peoples et al. (1994e)
Maize							
Broad-acre	Groundnut	6–10					McDonagh et al. (1993)
	Lucerne	17 - 25	25 - 32				Harris and Hesterman (1990)
	Lucerne - shoot	51 - 63[e]		AS	45 - 65		Hesterman et al. (1987)
	Lucerne - root	36 - 44[e]					
	Soybean	18 - 19[e]					
				AN	45 - 53	14 - 41	Peoples et al. (1994e)
Alley crop	*Erythrina*	12					Haggar et al. (1993)
	Gliricidia	27					
	Leucaena	5 - 9	25 - 34	AS	43 - 50	14 - 35	Xu et al. (1993a,b)
		6 - 18					Sanginga et al. (1995)

[a] Ur - urea; AS = ammonium sulphate; AN = ammonium nitrate; KN = potassium nitrate.
[b] Mean of 9 rice crops.
[c] Mean of 18 observations from 10 studies using various legume green manure sources.
[d] Mean under a wide range of environmental and soil conditions across 6 countries.
[e] Uptake of fixed N represented 33–40% and 3–5% of the N applied as lucerne or soybean respectively.

concentrations, the pH and temperature of floodwater, and windspeed (Buresh and DeDatta, 1991; Peoples et al., 1994e). The extent of losses from fertilizer N in rice-based systems can range from 14 to 85% (Table 15). Several field experiments have shown that adding leguminous or *Azolla* green manure with urea to flooded rice can enhance grain yield and reduce, but not eliminate, loss of urea N (Becker et al., 1994a; Buresh and De Datta., 1991; Diekmann et al., 1993; Kumarasinghe and Eskew, 1993). Application of green manure reduced floodwater pH and partial pressure of ammonia, and consequently reduced the potential for ammonia volatilization. The lower floodwater pH was attributed to production of CO_2 during decomposition of organic matter (Diekmann et al., 1993).

Maize. While uptake of fertilizer N by maize can be in the range of 40 to 50% regardless of whether it is grown as an alley crop or not (Table 14), the potential for losses of fertilizer N can be substantial (particularly in tropical soils, e.g. 36–153 kg N ha^{-1} leached below 150 cm, Poss and Saragoni, 1992), and up to 30 to 40% of the N applied might be lost from the plant-soil system (Table 14). Direct comparisons of total recoveries of legume N and fertilizer N (plant uptake of N + N remaining in soil) have been undertaken for tree legume prunings (Xu et al., 1993ab), and lucerne

Table 15. Recovery of N from organic and inorganic sources by temperate cereals

Crop	Organic residue	Plant uptake of N (% of applied)	N Loss (% applied)	Fertilizer[a]	Plant uptake of N (% of applied)	N Loss (% applied)	Reference
Wheat							
	Lupin	9 - 27	0 - 13				Russell and Fillery (1994b)
	Pea	26		AN	42		Rees et al. (1993)
	Lentil residues	6	0	AS	34	30	Bremer and van Kessel (1992)
	Lentil green manure	19	24				
	Medics	20 - 28	3 - 20				Ladd et al. (1983)
		16 - 19	15 - 17	Ur/KN/AS	41 - 50	16 - 22	Ladd and Amato (1986)
	Subterranean clover	6	0				Peoples et al. (1994a)
				KN/AS/AN	38 - 45	7 - 40	Peoples et al. (1994e)
				Ur	20 - 58	23 - 46	Peoples et al. (1994ae)
Barley							
	Pea	15	10				Jensen (1994b)
		6 - 16	5 - 19				Jans-Hammermeister et al. (1994)
	Common bean	17	1				Müller and Sundman (1988)
	White clover	24	11				
	Red clover	20	6				
	Subterranean clover	18	5				
				KN	53 - 62	10 - 14	Peoples et al. (1994e)
				Ur	13 - 54	9 - 36	
Oats							
	Soybean	6 - 8					Bergersen et al. (1992)

[a] Ur = urea; AS = ammonium sulphate; AN = ammonium nitrate; KN = potassium nitrate.

(Harris and Hesterman, 1990): These studies suggested that losses of N from both the tree legume mulch and lucerne were similar to fertilizer N (Table 14). However, in alley-cropping systems, some of the N lost from the rooting zone of the interrow maize might be captured by the roots of neighbouring tree hedgerows and so be retained within the system (Sanginga et al., 1995).

While uptake of fertilizer and legume N can be similar in some maize cropping systems, the utilization of legume N by maize may be relatively poor in other circumstances (commonly <20% of the N applied, Table 14). This appears to contradict the higher productivity and increased N harvest in maize crops commonly observed where legume residues have been returned (Giller and Wilson, 1991; McDonagh et al., 1993; Xu et al., 1993b). It would seem that even though the immediate transfer of legume N to a following crop might be restricted, the accumulation of a readily-mineralizable organic N pool in soil and higher N availability becomes a more important determinant of productivity than timing of release (Haggar et al., 1993).

Temperate crops

Wheat, barley and oats. Although utilization of N from crop and pasture legume residues by temperate cereals (6–28%, Table 15) are less than those reported for warm season and tropical crops (Table 14), this might still represent 20 to 40 kg N ha^{-1} (Ladd et al., 1983; Russell and Fillery, 1994b), and contributions of legume N to the N-economy of a following crop may be high (up to 33% of a cereal's N requirements; Bergersen et al., 1992). A comparison of recoveries of pasture legume residues and fertilizer N under Mediterranean climatic conditions (i.e. cool wet winter, hot dry summer, Ladd and Amato, 1986) indicated that a following wheat crop utilized more fertilizer N (41–50%) than legume N (16–19%, Table 15). Similar conclusions were obtained from comparisons of wheat recoveries of fertilizer or crop legume N (Bremer and van Kessel, 1992; Rees et al., 1993). However, despite a smaller uptake of legume N in a following crop (Table 15), large quantities of the applied legume N appear to be immobilized in microbial biomass and semi stable organic materials in soil during decomposition (Ladd et al., 1986; Jensen, 1994a,b), so that losses of N from plant and soil pools during growth of a subsequent crop (typically < 15%) can be smaller than from fertilizer in similar environments (up to 30–50% of the N applied, Table 15). Nonetheless, some studies indicate that significant losses of legume N can occur in unplanted soils (Jensen, 1994a; Russell and Fillery, 1994b), or when applied as a green manure mulch rather than returned as crop residues (Bremer and van Kessel, 1992; Table 15).

Forage systems

Appreciable amounts of organic N can accumulate in soils under legume-based pastures (Holford, 1981; Fig. 3; Simpson et al., 1974a; Vallis, 1972), but legumes also improve forage quality and intake and increase overall productivity of, and nutrient cycling within, pastures (Peoples and Craswell, 1992; Thomas and Lascano, 1994). Legumes also have the ability to rehabilitate degraded land by improving the physical, chemical and biological characteristics of soil (Thomas, 1995). Many of the benefits have been attributed to the ability of legumes to release N to soil and transfer N to grasses in a pasture (Ledgard and Steele, 1992). Determinations of the importance of transfer of fixed N to associated grasses range from agronomically insignificant amounts (< 10 kg N ha^{-1}) to > 100 kg N ha^{-1}; representing between 1 and 48% of the N_2 fixed (Table 16). Some of the upper estimates of N transfer seem surprisingly large; however, between 40 and 70% of total plant N can be below-ground in pasture legumes (Zebarth et al., 1991), and there may be a considerably larger pool of fixed N available for release than might be determined from measurements of shoot N.

There are some methodological problems in interpreting N-isotope data from experiments such as those presented in Table 16 (Chalk and Smith, 1994), but it appears that transfer of legume N could play an important role in meeting a proportion of the N requirements of grasses in pasture systems (Table 16). However, the rate at which transfer occurs may not always be rapid or the amount large (Vallis, 1983; Ledgard et al., 1985). The extent of N transfer is likely to be influenced by environmental factors, pasture nutrition and growth potential, and grazing management (Simpson et al., 1974a,b).

Fixed N could be returned to soil in a pasture or transferred to neighboring grasses via a number of different pathways.

Below-ground:

(i) Excretion of N into the legume rhizosphere (Ta et al., 1986)

(ii) Decomposition of roots and nodules sloughed off during growth (Russelle et al., 1994)

(iii) Direct interconnection of grass and legume roots via mycorrhizal fungi (Haystead et al., 1988)

(iv) Via the action of soil macro- and micro-fauna grazing legume roots and nodules.

Above-ground:

(i) Decomposition of leaf litter on the soil surface

(ii) Leaching of compounds from herbage as rainwater passes through the canopy

(iii) Volatile losses of ammonia from legume foliage reabsorbed by grass (Denmead et al., 1976)

(iv) Via excreta from grazing animals.

Only one of the studies summarized in Table 16 considered the return of fixed N in dung and urine from grazing animals (Ledgard, 1991). Despite the opportunities for losses ranging from 3 to 50% of the N excreted by the annual which arise from localized return of high rates of N (Bussink, 1992; Peoples et al., 1994e; Steel and Vallis, 1988; Vallis et al., 1985), some 60 kg N ha^{-1} yr^{-1} was estimated to have been transferred (22% of BNF and 48% of total N transferred, Table 16) from white clover to perennial ryegrass (*Lolium perenne*) under grazing by dairy cattle.

N-fertilizers are often applied to pastures in intensively grazed systems to promote growth during the late autumn to early spring. Rates exceeding 200 kg N ha^{-1} are not uncommon (particularly in Europe, Bussink, 1992; FAO, 1992b). Annual losses from 4 to 90 kg fertilizer-N ha^{-1} have been reported from fertilized pastures under a range of different environmental conditions, depending upon the form of fertilizer used and its application rate (Black et al., 1985; Bussink, 1992; Catchpoole et al., 1983; Jordan, 1989; Ruz-Jerez et al., 1994). A number of grazing trials have compared N losses from fertilized grass swards with those from clover-based pastures. Although consistent differences in leaching losses between fertilizer or legume N sources have not been observed (Cuttle et al., 1992), less N was lost in absolute terms as gaseous emissions from the clover-grass swards (Jarvis et al., 1989; Parsons et al., 1990; Ruz-Jerez et al., 1994),

Table 16. Transfer of legume N to associated grasses in mixed pasture swards via shoot residues, belowground organs, or the action of grazing animals

Species	N_2 fixed	N transferred			Reference
		Amount	As % BNF	% grass N	
	($kg\ N\ ha^{-1}$)	($kg\ N\ ha^{-1}$)	(%)	(%)	
Lucerne	54	7	13	68	Brophy et al. (1987)
	58 - 258	1 - 27	1 - 10	5 - 56	Ledgard and Steele (1992)
White clover	227 - 283	11 - 52	5 - 18	14 - 55	Boller and Nosberger (1987)
	269	70	26	27	Ledgard (1991)
(*via animals*)		*(60)*[a]	*(22)*	*(23)*	
Red clover	165 - 373	14 - 42	4 - 25	47 - 48	Boller and Nosberger (1987)
Birdsfoot trefoil	31	5	10	17	Brophy et al. (1987)
Desmodium spp.	na[b]	5 - 7	na	1 - 2	Vallis (1982)
Siratro	na	7 - 9	na	2 - 3	

[a] Italicised data in parentheses represent additional transfer via animal excreta, i.e. total transfer (below-ground + animal contribution) = 130 kg N or 48% of N estimated to be fixed.
[b] na = not available.

which has important implications for minimizing agriculture's contribution to the release of "greenhouse" and ozone depleting gases in the form of oxides of N (Peoples et al., 1994e). However, if denitrification was expressed as a proportion of annual N inputs, losses may be similar irrespective of whether N was supplied to pastures by N_2 fixation or fertilizer (Ruz-Jerez et al., 1994).

Agroforestry systems

Apart from the use of N_2-fixing trees as a N source in the alley cropping systems (above) woody species are an invaluable source of high-quality forage, fuel and timber, and are used to regenerate degraded, or erodable agricultural land (Giller and Wilson, 1991). N_2-fixing trees can also play an important role in providing N in traditional farming practices. *Casuarina* for example is planted in cleared areas of the highlands in Papua New Guinea and grown for 5 to 10 years until cleared for firewood or timber. The land is then planted to yams and other crops which benefit from the N buildup (Torrey, 1982).

While comparisons between the recovery and utilization of fertilizer N or leaf litter N have not be undertaken for agroforestry systems outside alley cropping, some information is available concerning the return and release of N from woody species. Measurements taken in stands of *Leucaena* in the dry tropics and *Casuarina* in temperate environments indicate that leaf and litterfall can contribute 10 to 25 $t\ ha^{-1}$ of organic matter to soil annually, representing 250 to 290 kg N (Sandhu et al., 1990; Torrey, 1982). Although decomposition of this litter under generally harsh conditions is likely to be slower than in the intensive agricultural systems described above, a high proportion of the total litterfall N might be expected to be released to the soil each year (Sandhu et al., 1990), representing a major contribution to the cycling of N in degraded and infertile environments.

Can the use of BNF as a source of N improve farm profitability?

Yield increases by crops following *Azolla* or legumes are often equivalent to applications of between 30 and 80 kg fertilizer N ha^{-1} (Jensen and Haahr, 1990; McDonagh et al., 1993; Peoples and Craswell, 1992; Watanabe, 1982). Equivalence values of 100 kg fertilizer N ha^{-1} or greater have also been reported (Becker et al., 1995; Herridge, 1987; Ladha et al., 1988; Wani et al., 1995), depending upon the amount of N returned to the soil in organic material, or whether the N_2-fixing system is grown for only 1, or for 2 or more consecutive years prior to planting a cereal (Paré et al., 1993). These measures of fertilizer equivalence can provide a site- and season-related estimate of the potential economic value of BNF in a rotation. There are however, other factors that should be considered such as costs of production, opportunity costs of alternative actions, and return from saleable produce. When such an exercise was undertaken for *Azolla* in flooded rice systems, it was concluded that *Azolla* was not a cost-effective substitute for urea fertilizer while fertilizer prices remain low (Rosegrant and Roumzasset, 1988). High labor costs and high opportunity costs of land use were identified as two of the major constraints to the economic feasibility of using *Azolla* as a green manure. This may also be true for other sources of green manures, if grown at the expense of an alternative cash or food crop. Nonetheless, there are situations where the potential contributions of green manures may be very important, and where fertilizer N is not a viable option. One such example may be in single-crop rainfed lowland systems where flood-tolerant legumes can provide valuable sources of green manure N. In these areas, short-term waterlogging occurs during the transition between dry and wet seasons. Alternative crops cannot be grown and farmers are reluctant to apply fertilizer N because of adverse climatic conditions and lack of consistent fertilizer response (Ladha et al., 1992). Another example might be in rubber and oil palm plantations where the beneficial effects of perennial legume cover crops can extend for up to 20 years after planting and be equivalent to the total application of between 840 and 1100 kg fertilizer N ha^{-1} (Giller and Wilson, 1991; Peoples and Herridge, 1990).

It should be much easier to demonstrate the financial benefits of utilizing BNF as a source of N in systems where the N_2-fixing component itself contributes directly to the production of a saleable commodity (seed from crop legumes, livestock production from legume-based pastures). To illustrate this point, a case study is presented below from an important cropping region of Australia.

In the northern wheat belt of New South Wales, an area of around 1 million ha, farmers have traditionally produced large tonnages of high protein wheat with little use of either pasture or grain legumes, or fertilizer N. Many of the soils of the region, particularly the deep, friable black earths, were initially naturally fertile and there seemed no need to add additional N (McGarity, 1975). However, evidence that soils were becoming progressively less able to supply N for cereal production can be found in the results of fertilizer N trials conducted over a 30 year period. In the early 1960's, there was only a low frequency (22%) of economic responses in wheat to applied fertilizer (Colwell and Esdaile, 1966). Ten years later, this figure had almost doubled to 40% (Doyle, 1977), and by the 1980's, responses had risen to 70% (Holford et al., 1992).

A survey of farms in the region between 1983–85 estimated that cereal yield was depressed by 45% because of N deficiency (Martin et al., 1988). Investigations of management practices revealed that 45% of farmers used fertilizer N, but rates of application were generally low (8–11 kg N ha^{-1}). Not more than 8% of farmers surveyed used fertilizer N at realistic rates of 30 to 60 kg N ha^{-1}. Farmers practiced rotations, but for the most part, legumes were not included.

Thus, there was a need to change the agronomic practices that had become inappropriate once soil fertility was no longer high, and to promote the concepts of N management. In response to the crisis of widespread N-deficiency, and to farmer reluctance to use fertilizer N, a series of on-farm trials were commenced to examine the potential of chickpea as a commercial crop in the region and to assess its role as a rotation crop in wheat production systems. In these trials, yields of wheat were increased by 26 to 144% following chickpea (Herridge et al., 1994a). Following chickpea, and also long fallow, benefits were observed in the first wheat crop, but not in the second. These rotational experiments showed clearly that the improved wheat yields following chickpea could be explained in terms of increased availability of soil N (Herridge et al., 1994b).

The average yield figures for experiments conducted in the region between 1987 and 1992 were used to prepare an economic comparison of cropping systems based on wheat monoculture (W-W-W), chickpea (CP-W-W), or long fallow (F-W-W) (Table 17). Although the return from wheat following fallow was almost similar to wheat following chickpea, no income was generated in the first year so that the gross margin over 3-year period was identical to a continuous wheat system (Table 17). The economic analysis of the benefits of chickpea on wheat production indicated that the gross margin more than doubled over 3 years when chickpea was included (Table 17). For an average farmer in the region cropping 200 ha, replacement of every third crop with chickpea would, on current commodity prices, result in the overall farm gross margin increasing from A$26,600 to A$61, 900 (A$1.00 = US$ 0.74).

While similar average yield benefits from chickpea (around 40%) could be achieved with fertilizer applications of 60 to 80 kg N ha^{-1}, the net financial return from the additional wheat (A$ 125–130 ha^{-1} less fertilizer costs of A$50–65 ha^{-1} each annum) would still fall short of the impact of chickpea on farm profitability because of the price differential between the value

Table 17. Returns, costs and gross margins for three cropping systems in the northern cereal belt of New South Wales, Australia. Yields are averaged from on-farm experiments conducted over 5 years[a]

Item	Rotations based on					
	Wheat (W-W-W)		Chickpea (CP-W-W)		Fallow (F-W-W)	
	Yield ($t\ ha^{-1}$)	Value (A$)	Yield ($t\ ha^{-1}$)	Value (A$)	Yield ($t\ ha^{-1}$)	Value (A$)
Year 1	2.33	314	2.02	727	-	-
Year 2	2.24	302	3.18	429	3.37	455
Year 3	2.47	333	2.51	339	2.59	350
Total		949[b]		1495[b]		805[b]
Less fixed costs[c]		105		105		105
Less variable costs[d]		444		461		296
Gross margin over 3 years		400		929		404

[a] Wheat yields ranged from 1.43 to 4.34 $t\ ha^{-1}$ and chickpea yields ranged from 1.29 to 2.90 $t\ ha^{-1}$ (Herridge et al., 1994a).
[b] Based on 1993 on-farm selling prices for Australian standard white (A$135 t^{-1}), and chickpea (A$360 t^{-1}). [A$1 = US$0.74]
[c] Calculated at A$35 ha^{-1} $annum^{-1}$.
[d] Calculated at A$148 ha^{-1} for wheat and A$165 ha^{-1} for chickpea.

of wheat and chickpea (Table 17). However, even if total farm profits resulting from the use of fertilizer N matched or exceeded that achieved with chickpea or other legume rotations (Dalal et al., 1994), there would be eventual problems with buildup of cereal diseases, herbicide resistance of weeds and soil structural decline associated with a N-fertilized monoculture, which would provide sufficient economic justification in the long term to choose the BNF option.

Conclusions

Considerable inputs of biologically-fixed N can be achieved in almost all agricultural ecosystems through the activity of many different symbiotic associations. While the short term recovery of N from these biological sources may not always match fertilizer (particularly in temperate environments), there are consistent rotational benefits to subsequent cereal crops and evidence of significant transfers of fixed N to associated grasses in pastures. Without doubt, BNF improves the N economy of soils. This does not mean that these systems will always make large net contributions of N to soils in which they grow. What it does mean is that the N balance for a legume-cereal sequence for example will be more positive than for a cereal-cereal sequence in the same soil.

Evidence indicates that N derived from legume or *Azolla* sources might be less susceptible to losses than fertilizer N, and that long-term use of these organic materials results in the build-up of a reserve of readily mineralizable organic N. The use of BNF in a farming system can represent a profitable approach to arrest the decline of soil N fertility that inevitably accompanies intensive agriculture.

The nitrate "spared" and N released from crop legume residues and short duration green manures may be capable of meeting only part of the N demand of high-yielding cereal crops and supplementary fertilizer N applications may be required to provide optimal nutrition. However, the amount of N accrued in soil under pasture-ley systems, or where perennial legume cover-crops and tree legume leaf mulch are used, may be sufficient to satisfy subsequent crop requirements. Therefore, it should be possible to manage BNF to provide a renewable source of N to supplement or replace fertilizer N, and redress the deterioration of agriculture's resource base.

Acknowledgements

We thank our colleagues for allowing unpublished data to be presented. Much of this information resulted from experimentation supported by the Australian Centre for International Agricultural Research (ACIAR), Australian International Development Assistance Bureau (AIDAB), and the Grains Research and Development Corporation (GRDC) of Australia. The assistance of these funding bodies is gratefully acknowledged. We also wish to express our appreciation to Australian, Indonesian, Malaysian and Thai farmers for their invaluable co-operation in allowing measurements to be undertaken on their crops, forages and plantations.

References

Badarneh D M D and Ghawi I O 1994 Effectiveness of inoculation on biological nitrogen fixation and water consumption by lentil under rainfed conditions. Soil Biol. Biochem. 26, 1–5.

Becker M, Ladha J K and Ottow J C G 1994a Nitrogen losses and lowland rice yields as affected by residue nitrogen release. Soil Sci. Soc. Am. J. 58, 1660–1665.

Becker M, Ladha J K, Simpson I C and Ottow J C G 1994b. Parameters affecting residue nitrogen mineralization in flooded soils. Soil Sci. Soc. Am. J. 58, 1666–1671.

Becker M, Ladha J K and Ali M 1995 Green manure technology: potential, usage, limitations - A case study for lowland rice. Plant and Soil 174.

Bell M J, Wright G C, Suryantini and Peoples M B 1994 The N_2-fixing capacity of peanut cultivars with differing assimilate partitioning characteristics. Aust. J. Agric. Res. 45, 1455–1468 .

Bergersen F J, Brockwell J, Gault R R, Morthorpe L, Peoples M B and Turner G L 1989 Effects of available soil nitrogen and rates of inoculation on nitrogen fixation by irrigated soybeans and evaluation of δ^{15}N methods for measurement. Aust. J. Agric. Res. 40, 763–780.

Bergersen F J, Turner G L, Gault R R, Peoples M B, Morthorpe L J and Brockwell J 1992 Contributions of nitrogen in soybean crop residues to subsequent crops and to soils. Aust. J. Agric. Res. 43, 155–169.

Bhatnagar P S and Tiwari S P 1989 Yield gap and constraint analysis for harnessing productivity potential of soybean in India. *In* Proc. World Soybean Research Conference IV. Ed. A J Pascale. pp 811–819. Asoc. Argentina De La Soja, Buenos Aires, Argentina.

Black A S, Shelock R R, Smith N P, Cameron K C and Goh K M 1985 Effects of form of nitrogen, season, and urea application rate on ammonia volatilisation from pastures. N. Z. J. Agric. Res. 28, 469–474.

Blair G, Catchpoole D and Horne P 1990 Forage tree legumes: their management and contribution to the nitrogen economy of wet and humid tropical environments. Adv. Agron. 44, 27–54.

Boddey R M, de Oliveira O C, Urquiaga S, Reis V M, de Olivares F L, Baldani V L D and Döbereiner J 1995 Biological nitrogen fixation associated with sugar cane and rice: Contributions and prospects for improvement. Plant and Soil 174..

Bohlool B B, Ladha J K, Garrity D P and George T 1992 Biological nitrogen fixation for sustainable agriculture: a perspective. Plant and Soil 141, 1–11.

Boller B C and Nosberger J 1987 Symbiotically fixed nitrogen from field grown white and red clover mixed with rye-grasses at low levels of ^{15}N-fertilization. Plant and Soil 104, 219–226.

Bremer E and van Kessel C 1992 Plant-avaliable nitrogen from lentil and wheat residues during a subsequent growing season: Soil Sci. Soc. Am. J. 56, 1155–1160.

Brockwell J, Gault R R, Peoples M B, Turner G L, Lilley D M and Bergersen F J 1994 Nitrogen fixation in irrigated lucerne grown for hay. Soil Biol. Biochem. 26 (*In press*).

Bromfield S M and Simpson J R 1973 Effects of management on soil fertility under pasture. 2. Changes in nutrient availability. Aust. J. Exp. Agric. Anim. Husb. 14, 479–486.

Brophy L S, Heichel G H and Russell M P 1987 Nitrogen transfer from forage legumes to grass in a systematic planting design. 1. Crop ecology, production, and management. Crop Sci. 27, 753–758.

Bumb B L 1994 World nitrogen supply and demand: an overview. *In* Nitrogen Fertilization and the Environment. Ed. P E Bacon. Marcel Dekker Inc., New York, USA (*In press*).

Buresh R J and De Datta S K 1991 Nitrogen dynamics and management of rice-legume cropping systems. Adv Agron. 45, 1–59.

Burns R C and Hardy R W F 1975 Nitrogen fixation in bacteria and higher plants. pp 1–189. Springer-Verlag Berlin, Heidelberg, New York. 189p.

Bushby H V A and Lawn R J 1992 Nitrogen fixation in mungbeans - expectation and reality. *In* Proc. 6th Australian Agronomy Conference. pp 161–164. The Aust. Soc. Agron., Melbourne, Australia.

Bussink D W 1992 Ammonia volatilization from grassland receiving nitrogen fertilizer and rotationally grazed by dairy cattle. Fert. Res. 33, 257–265.

Catchpoole V R, Oxenham D J and Harper L A 1983 Transformation and recovery of urea applied to a grass pasture in south-eastern Queensland. Aust. J. Agric. Anim. Husb. 28, 80–86.

Chalk P M 1991 The contribution of associative and symbiotic nitrogen fixation to the nitrogen nutrition of non-legumes. Plant and Soil 132, 29–39.

Chalk P M and Smith C J 1994 ^{15}N isotope dilution methodology for evaluating the dynamics of biologically fixed N in legume-non-legume associations. Biol. Fert. Soils. 17, 80–84.

Chalk P M, Smith C J, Hamilton S D and Hopmans P 1993 Characterization on the N benefit of a grain legume (*Lupinus augustifolius* L.) to a cereal (*Hordeum vulgare* L.) by an in situ ^{15}N isotope dilution technique. Biol. Fert. Soils. 15, 39–44.

Chapman A L and Myers R J K 1987 Nitrogen contributed by grain legumes to rice grown in rotation on the Cununurra soils of the Ord imgation area, Western Australia. Aust. J. Exp. Agric. 27, 155–163.

Colwell J D and Esdaile R J 1966 The application of production function analysis for the estimation of fertilizer requirements of wheat in northern New South Wales. Aust. J. Exp. Agric. Anim. Husb. 6, 418–424.

Coventry D R and Slattery K J 1991 Acidification of soil associated with lupins grown in a crop rotation in north-eastern Victoria., Aust. J. Agric. Res. 42, 391–397.

Craswell E T and Godwin D C 1984 The efficiency of nitrogen fertilizers applied to cereals in different climates. Adv. Plant Nutr. 1, 1–55.

Cuttle S P, Hallard M, Daniel G and Scurlock R V 1992 Nitrate leaching from sheep-grazed grass/clover and fertilized grass pastures. J. Agric. Sci. 119, 335–343.

Dalal R C, Strong W M, Weston E J, Cahill M J, Cooper J E, Lehane K J, King A J and Gaffney J 1994 Evaluation of forage and grain legumes, no-till and fertilisers to restore fertility degraded soils. *In* Transactions of the 15th World Congress of Soil Science, Volume 5a. pp 62–74. The International Soc. Soil Sci. and Mexican Soc. Soil Sci., Acapulco, Mexico.

Danso S K A and Papastlyianou I 1992 Evaluation of the nitrogen contribution of legumes to subsequent cereals. J. Agric. Sci. 119, 13–18.

Denmead O T, Freney J R and Simpson J R 1976 A closed ammonia cycle within a plant canopy. Soil Boil. Biochem. 8, 161–164.

Diekmann K H, De Datta S K and Ottow J C G 1993 Nitrogen uptake and recovery from urea and green manure in lowland rice measured by ^{15}N and non-isotope techniques. Plant and Soil 148, 91–99.

Doughton J A and MacKenzie J 1984 Comparative effects of black and green gram (mung beans) and grain sorghum on soil mineral nitrogen and subsequent grain sorghum yields on the eastern Darling Downs. Aust. J. Exp. Agric. Anim. Husb. 24, 244–249.

Doughton J A, Vallis I and Saffigna P G 1993 Nitrogen fixation in chickpea 1. Influence of prior cropping or fallow, nitrogen fertilizer and tillage. Aust. J. Agric. Res. 44, 1403–1413.

Doyle A D 1977 Nitrogen and phosphorus fertilizer responses to wheat in northern New South Wales in relation to soil testing. New South Wales Dept. Agric. Tech. Bull. 14.

Doyle A D, Moore K J and Herridge D F 1988 The narrow-leafed lupin (*Lupinus angustifolius*) as a nitrogen-fixation rotation crop for cereal production. 3. Residual effects of lupins on subsequent cereal crops. Aust. J. Agric. Res. 39, 1029–1037.

Doyle P J and Cowell L E 1993 Balances of nutrient inputs and exports (grain) in Alberta, Manitoba and Saskatchewan. *In* Impacts of Macronutrients on Crop Responses and Environmental Sustainability on the Canadian Prairies. Eds. D A Rennie, C A Campbell and T L Roberts. pp 1–25. Can. Soc. Soil Sci., Ottawa, Canada.

Evans J, Fettell N A, Coventry D R, O'Connor G E, Walsgott D N, Mahoney J and Armstrong E L 1991 Wheat response after temperate crop legumes in south-eastern Australia. Aust. J. Agric. Res. 42, 31–43.

Evans J, O'Connor G E, Turner G L, Coventry D R, Fettell N, Mahoney J, Armstrong E L and Walsgott D N 1989 N_2 fixation and its value to soil N increase in lupin, field pea and other legumes in south-eastern Australia. Aust. J. Agric. Res 40, 791–805.

FAO 1990 Fertilizer yearbook, Vol. 39. FAO, Rome.

FAO 1992a Fertilizer yearbook, Vol. 41. FAO, Rome.

FAO 1992b Fertilizer use by crop. Report by the International Fertilizer Industry Association, International Fertilizer Development Center, and Food and Agriculture Organization of the United Nations. FAO, Rome.

FAO 1993 Production yearbook, Vol. 46. FAO, Rome.

Fox R H, Myers R J K and Vallis I 1990 The nitrogen mineralisation rate of legume residues in soil as influenced by their polyphenol, lignin, and nitrogen contents. Plant and Soil 129, 251–259.

Gault R R, Peoples M B, Turner G L, Lilley D M, Brockwell J and Bergersen F J 1995 Nitrogen fixation by irrigated lucerne during the first three years after establishment. Aust. J. Agric. Res. (*In press*).

George T, Ladha J K, Buresh R J and Garrity D P 1992 Managing native and legume-fixed nitrogen in lowland rice-based cropping systems. Plant and Soil 141, 69–91.

George T, Ladha J K, Buresh R J and Garrity D P 1993 Nitrate dynamics during the aerobic soil phase in lowland rice-based cropping systems. Soil Sci. Soc. Am. J. 57, 1526–1532.

Giller K E and Wilson K J 1991 Nitrogen Fixation in Tropical Cropping Systems. CAB International, Wallingford, UK. 313 p.

Giller K E and Cadisch G 1995 Future benefits from biological nitrogen fixation: an ecological approach to agriculture. Plant and Soil 174.

Haggar J P, Tanner E V J, Beer J W and Kass D C L 1993 Nitrogen dynamics of tropical agroforestry and annual cropping systems. Soil Biol. Biochem. 25, 1363–1378.

Hamilton S D, Hopmans P, Chalk P M and Smith C J 1993 Field estimation of N_2 fixation by *Acacia* spp. using ^{15}N isotope dilution and labelling with ^{35}S. For. Ecol. Manage. 56, 297–313.

Handayanto E, Cadisch G and K E Giller 1994 Nitrogen release from prunings of legume hedgerow trees in relation to the quality of the prunings and incubation method. Plant and Soil 160, 237–248.

Hansen A P 1994. Symbiotic N_2 Fixation of Crop Legumes. Achievements and Perspectives. Hohenheim Tropical Agriculture Series no. 2. Margraf Verlag, Weikersheim, Germany. 248p.

Hardarson G, Bliss F A, Cigales-Rivero M R, Henson R A, Kipe Nolt J A, Longeri L, Manrique A, Pena-Cabriales J J, Pereira P A A, Sanabria C A and Tsai S M 1993 Genotypic variation in biological nitrogen fixation by common bean. Plant and Soil 152, 59–70.

Harris G H and Hesterman O B 1990 Quantifying the nitrogen contribution from alfalfa to soil and two succeeding crops using nitrogen 15. Agron. J. 82, 129–134.

Haystead A, Malajczuk N and Grove T S 1988 Underground transfer of nitrogen between pasture plants infected with vescicular-arbuscular mycorrhizal fungi. New Phytol. 108, 417–423.

Heichel G H, Vance C P, Barnes D K and Henjum K I 1985 Dinitrogen fixation, and N and dry matter distribution during 4 year stands of birdsfoot trefoil and red clover. Crop Sci. 25, 101–105.

Helyar K R 1991 The management of acid soils. *In* Plant-Soil Interactions at Low pH. Eds. R J Wright, V C Baligar and R P Murrmann. pp 365–382. Kluwer Academic Publishers, Dordrecht.

Herridge D F 1987 Nitrogen fixation dynamics by rain-fed legume crops: potential for improvement. *In* Proc. XIII International Soc. Soil Sci. Congress. Trans Vol VI, pp 794–804. Hamburg, Germany.

Herridge D F and Bergersen F J 1988 Symbiotic nitrogen fixation. *In* Advances in Nitrogen Cycling in Agricultural Ecosystems. Ed. J R Wilson. pp 46–65. CAB International, Wallingford, UK.

Herridge D F and Doyle A D 1988 The narrow-leafed lupin (*Lupinus angustifolius* L.) as a nitrogen-fixing rotation crop for cereal production. 2. Estimates of fixation by field-grown crops. Aust. J. Agric. Res. 39, 1017–1028.

Herridge D F, Rupela O P, Serraj R and Beck D P 1993 Screening techniques and improved biological nitrogen fixation in cool season legumes. Euphytica 1, 1–14.

Herridge D F, Marcellos H, Felton W L and Peoples M B 1994a Legume N_2 fixation-an efficient source of N for cereal production. *In* Nuclear Related Methods in Soil/Plant Aspects of Sustainable Agriculture, FAO/IAEA, Vienna. (*In press*).

Herridge D F, Marcellos H, Felton W L, Turner G L and Peoples M B 1994b Chickpea increases soil N fertility in cereal systems through nitrate sparing and N_2 fixation. Soil Biol. Biochem. 26 (*In press*).

Hesterman O B, Russelle M P, Sheaffer C C and Heichel G H 1987 Nitrogen utilizatlon from fertilizer and legume residues in legume-corn rotations. Agron. J. 79, 726–731.

Holford I C R 1981 Changes in nitrogen and organic carbon of wheat-growing soils after various periods of grazed lucerne, extended fallowing and continuous wheat. Aust. J. Soil Res. 19, 239–249.

Holford I C R, Doyle A D and Leckie C C 1992 Nitrogen response to characteristics of wheat protein in relation to yield responses and their interactions with phosphorus. Aust. J. Agric. Res. 43, 969–986.

Hughes R M and Herridge D F 1989 Effect of tillage on yield, nodulation and nitrogen fixation of soybean in far north coastal New South Wales. Aust. J. Exp. Agric. 29, 671–677.

Jans-Hammermeister D C, McGill W B and Jensen T L 1994 Dynamics of ^{15}N in two soil-plant systems following incorporation of 10% bloom and full bloom field pea. Can. J. Soil Sci. 74, 99–107.

Jarvis S C, Hatch D J and Lockyer D R 1989 Ammonia fluxes from grazed grassland: annual losses from cattle production systems and their relation to nitrogen inputs. J. Agric. Sci., Camb. 113, 99–108.

Jensen E S 1987 Seasonal patterns of growth and nitrogen fixation in field-grown pea. Plant and Soil 101, 29–37.

Jensen E S 1994a Dynamics of mature pea residue nitrogen turnover in unplanted soil under field conditions. Soil Biol. Biochem. 26, 455–464.

Jensen E S 1994b Availability of nitrogen in ^{15}N-labelled mature pea residues to subsequent crops in the field. Soil Biol. Biochem. 26, 465–472.

Jensen E S and Haahr V 1990 The effect of pea cultivation on succeeding winter cereals and winter oilseed rape nitrogen nutrition. Appl. Agric. Res. 5, 102–107.

Jordan C 1989 The effect of fertiliser type and application rate on denitrification losses from cut grassland in Northern Ireland. Fert. Res. 19, 45–55.

Kumar Rao J V D K and Dart P J 1987 Nodulation, nitrogen fixation and nitrogen uptake in pigeonpea (*Cajanus cajan* (L.) Millsp) of different maturity groups. Plant and Soil 99, 255–266.

Kumarasinghe K S and Eskew D L 1993 Isotopic Studies of *Azolla* and Nitrogen Fertilization of Rice. Kluwer Academic Publishers, Dordrecht. 145p.

Kundu D K and Ladha J K 1994 Efficient management of soil and biologically fixed nitrogen in intensively cultivated rice soils. Soil Biol. Biochem. 26 (*In press*).

Ladd J N and Amato M 1986 The fate of nitrogen from legume and fertilizer sources in soils successively cropped with wheat under field conditions. Soil Biol. Biochem. 18, 417–425.

Ladd J N, Amato M, Jackson R B and Butler J H A 1983 Utilization by wheat crops of nitrogen from legume residues decomposing in soils in the field. Soil Biol. Biochem. 15, 231–238.

Ladd J N, Butler J H A and Amato M 1986 Nitrogen fixation by legumes and their role as sources of nitrogen for soil and crop. Biol. Agric. Hortic. 3, 269–286.

Ladha J K, Watanabe I and Saono S 1988 Nitrogen fixation by leguminous green manure and practices for its enhancement in tropical lowland rice. *In* Sustainable Agriculture. Green Manure in Rice Farming. pp 165–183. IRRI, Manila, Philippines.

Ladha J K, Pareek R P and Becker M 1992 Stem-nodulating legume-*Rhizobium* symbiosis and its agronomic use in lowland rice. Adv. Soil Sci. 20, 147–192.

Ladha J K, Peoples M B, Garrity D P, Capuno V T and Dart P J 1993 Estimating dinitrogen fixation of hedgerow vegetation in an alley-crop system using the ^{15}N natural abundance method. Soil Sci. Soc. Am. J. 57, 732–737.

Ladha J K, Kundu D K, Angelo-Van Coppenolle M G, Peoples M B, Carangal V R and Dart P J 1995 Legume productivity and soil nitrogen dynamics in lowland rice-based cropping systems. Soil Sci. Soc. Am. J. (*In press*).

Ledgard S F 199 Transfer of fixed nitrogen from white clover to associated grasses in swards grazed by dairy cows, estimated using ^{15}N method. Plant and Soil 131, 215–223.

Ledgard S F and Steele K W 1992 Biological nitrogen fixation in mixed legume/grass pastures. Plant and Soil 141, 137–153.

Ledgard S F, Freney J R and Simpson J R 1985 Assessing nitrogen transfer from legumes to associated grasses. Soil Biol. Biochem. 17, 575–577.

Martin R J, McMilian M G and Cook J B 1988 Survey of farm management practices of the northern wheat belt of New South Wales. Aust. J. Exp. Agric. 28, 499–509.

McDonagh J F, Toomsan B, Limpinutana V and Giller K E 1993 Estimates of the residual nitrogen benefit of groundnut to maize in Northern Thailand. Plant and Soil 154, 267–277.

McDonald G K 1992 Effects of nitrogenous fertilizer on the growth, grain yield and grain protein concentration of wheat. Aust. J. Agric. Res. 43, 949–967.

McGarity J W 1975 Soils of the Australian wheat-growing areas. *In* Australian Field Crops. Vol 1. Wheat and Other Temperate Cereals. Eds. A Lazenby and E M Matherson. pp 227–255. Angus and Robinson, Sydney.

Müller M M and Sundman V 1988 The fate of nitrogen (^{15}N) released from different plant materials during decomposition under field conditions. Plant and Soil 105, 133–139.

Myers R J K 1988 Nitrogen management of upland crops: from cereals to food legumes to sugar cane: *In* Advances in Nitrogen Cycling in Agricultural Ecosystems. Ed. J R Wilson. pp 57–273. CAB International, Wallingford.

Ofosu-Budu K G, Ogata S and Fujita K 1992. Temperature effects on root nodule activity and nitrogen release in some sub tropical and temperate legumes. Soil Sci. Plant Nutr. 38, 717–726.

Ofuso-Budu K G, Fujita K, Gamo T and Akao S 1993 Dinitrogen fixation and nitrogen release from roots of soybean cultivar Bragg and its mutants Nts 1116 and Nts 1007. Soil Sci. Plant Nutr. 39, 497–506.

Paré T, Chalifour F P, Bourassa J and Antoun 1993 Forage-corn production and N-fertilizer replacement following 1 or 2 years of legumes. Can. J. Plant Sci. 73, 477–493.

Parsons A J, Orr R J, Penning P D, Lockyer D R and Ryden J C 1990 Uptake, cycling and the fate of nitrogen in grass-clover swards continuously grazed by sheep. J. Agric. Sci. 116, 47–61.

Paul E A 1988 Towards the year 2000: directions for future nitrogen research. *In* Advances in Nitrogen Cycling in Agricultural Ecosystems. Ed. J R Wilson. pp 417–425. CAB International, Wallingford.

Peoples M B and Herridge D F 1990 Nitrogen fixation by legumes in tropical and sub-troical agriculture. Adv. Agron. 44, 155–223.

Peoples M B and Craswell E T 1992 Biological nitrogen fixation: investments, expectations and actual contributions to agriculture. Plant and Soil 141, 13–39.

Peoples M B, Bergersen F J, Brockwell J, Fillery I R P and Herridge D F 1994a Management of nitrogen for sustainable agricultural systems. *In* Nuclear Related Methods in Soil/Plant aspects of Sustainable Agriculture, FAO/IAEA, Vienna. (*In press*).

Peoples M B, Gault R R, Brockwell J, Lean B and Sykes J D 1994b Nitrogen fixation by soybean in commercial irrigated crops of central and southern New South Wales. Soil Biol. Biochem. 26 (*In press*).

Peoples M B, Herridge D F, Rerkasern B, Bhromsiri A, Sampet C and Brockwell J 1994c Improving nitrogen fixation by soybean. *In* Proc. World Soybean Research Conference V, Chiang Mai, Thailand. (*In press*).

Peoples M B, Lilley D M, Burnett V F, Ridley A M and Garden D L 1994d Effects of surface application of lime and superphosphate to acid soils on the growth and N_2 fixation by subterranean clover in mixed pasture swards. Soil Biol. Biochem. 26 (*In press*).

Peoples M B, Mosier A R and Freney J R 1994e Minimizing gaseous losses of nitrogen. *In* Nitrogen Fertilization and the Environment. Ed. P E Bacon. Marcel Dekker Inc., New York pp 565–602.

Peoples M B, Ladha J K and Herridge D F 1995 Enhancing legume N_2 fixation through plant and soil management. Plant and Soil 174.

Poss R and Saragoni H 1992 Leaching of nitrate, calcium and magnesium under maize cultivation and oxisol in Togo. Fert. Res. 33, 123–133.

Poth M, La Favre J S and Focht D D 1986 Quantification by direct ^{15}N dilution of fixed N_2 incorporated into soil by *Cajanus cajan* (pigeon pea). Soil Biol. Biochem. 18, 125–127.

Rao K V and Shinde J E 1991 Uptake and balance of ^{15}N labelled green manures and urea in tropical wetlaad rice culture. *In* Stable Isotopes in Plant Nutrition, Soil Fertility and Environmental Studies. pp 317–325. IAEA, Vienna.

Rees R M, Yan L and Ferguson M 1993 The release and plant uptake of nitrogen from some plants and animal manures. Biol. Fert. Soils 15, 285–293.

Reeves T G 1991 The introduction, development, management and impact of legumes in cereal rotations in southern Australia. *In* Soil and Crop Management for Improved Water use Efficiency in Rainfed Areas. Eds. H C Harris, P J M Cooper and M Pala. pp 274–283. ICARDA, Aleppo, Syria.

Reeves T G, Ellington A and Brooke H D 1984 Effects of lupin-wheat rotations on soil fertility, crop disease and crop fields. Aust. J. Exp. Agric. Anim. Husb. 24, 595–600.

Ridley A M, Helyar K R and Slattery W J 1990a Soil acidification under, subterranean clover (*Trifolium subterraneum* L.) pastures in north-eastern Victoria. Aust. J. Exp. Agric. 30, 195–201.

Ridley A M, Slattery W J, Helyar K R and Cowling A 1990b Acidification under grazed annual and perennial grass based pastures. Aust. J. Exp. Agric. 30, 539–544.

Roger and Ladha J K 1992 Biological N_2 fixation in wetland rice fields: estimation and contribution to nitrogen balance. Plant and Soil 141, 41–55.

Roper M M and Ladha J K 1995 Biological nitrogen fixation by heterotrophic and phototrophic bacteria in association with straw. Plant and Soil 174.

Rosegrant M W and Roumasset J A 1988 Economic feasibility of green manure in rice-based cropping systems. *In* Sustainable Agriculture. Green Manure in Rice Farming. pp 11–27. IRRI, Manila, Philippines.

Rowland I C 1987 The Esperance rotation trial. Effect of rotations on crop yields and soil fertility. Division of Plant Research Technical. Report No. 8. 49p.

Russell C A and Fillery I R P 1994a Estimates of lupin below-ground biomass N, drymatter and N turnover to wheat. Aust. J. Agric. Res. 45 (*In press*).

Russell C A and Fillery I R P 1994b Nitrogen mineralisation of crop-lupin stubble in the field in southwest Australia. Aust. J. Agric. Res. 45 (*In press*).

Russelle M P, Allan D L and Gourley C J P 1994 Direct assessment of symbiotically fixed nitrogen in the rhizosphere of alfalfa. Plant and Soil 159, 233–243.

Ruz-Jerez B E, White R E and Ball P R 1993 Long term measurement of denitrification in three contrasting pastures grazed by sheep. Soil Biol. Biochem. 26, 29–39.

Sánchez P A 1994 Tropical soil fertility research: towards the second paradigm. *In* Transactions of the 15th World Congress of Soil Science. Vol. 1, pp 65–88. The International Soc. Soil Sci. and Mexican Soc. Soil Sci., Acapulco, Mexico.

Sandhu J, Sinha M and Ambasht R S 1990 Nitrogen release from decomposing litter of *Leucaena leucocephala* in the dry tropics. Soil Biol. Biochem. 22, 859–863.

Sanford P, Pate J S and Unkovich M J 1994 A survey of proportional dependance of subterranean clover on other pasture legumes on N_2 fixation in south-west Australia utilizing ^{15}N natural abundance. Aust. J. Agric. Res. 45, 165–181.

Sanginga N, Vanlauwe B and Danso S K A 1995 Management of biological N_2 fixation in alley cropping systems: Estimation and contribution to N balance. Plant and Soil 174.

Sawatsky N and Soper R J 1941 A quantitative measurement of the nitrogen loss from the root system of field peas (*Pisum avense* L.) grown in the soil. Soil Biol. Biochem. 23, 255–259.

Schulze E D, Gebaurer G, Ziegler H and Lange O L 1991 Estimates of nitrogen fixation by trees on an aridity gradient in Namibia. Oecologia 88, 451–455.

Shearer G, Kohl D H, Virginia R A, Bryan B A, Skeeters J L, Nilsen E T, Sharifi M R and Rundel P W 1983 Estimates of N_2-fixation from variation in the natural abundance of ^{15}N in Sonoran Desert ecosystems. Oecologia 56, 365–373.

Simpson J R, Bromfield S M and Jones O L 1973a Effects of management on soil fertility under pasture. 3. Changes in total soil nitrogen, carbon, phosphorus and exchangeable cations. Aust. J. Exp. Agric. Anim. Husb. 14, 487–493.

Simpson J R, Bromfield S M and McKinney 1973b Effects of management on soil fertility under pasture 1. Influence of experimental grazing and fertilizer systems on growth, composition and nutrient status of the pasture. Aust. J. Exp. Agric. Anim. Husb. 14, 470–478.

Sisworo W H, Mitrosuhardjo M M, Rasjid H and Myers R J K 1990 The relative roles of N fixation, fertilizer, crop residues and soil in supplying N in multiple cropping systems in a humid, tropical upland cropping system. Plant and Soil 121, 73–82.

Song L, Carroll B and Herridge D F 1994 Field assessments of supernodulating genotypes of soybean for yield, N_2 fixation and benefit to subsequent crops. Soil Biol. Biochem. 26 (*In press*).

Steele K W and Vallis I 1988 The nitrogen cycle in pastures. *In* Advances in Nitrogen Cycling in Agricultural Ecosystems. Ed. J R Wilson. pp 274–291. CAB International, Wallingford, UK.

Strong W M, Harbison J, Nielsen R G H, Hal l B D and Best E K 1986 Nitrogen availability in a Darling Downs soil following cereal, oilseed and grain legume crops. 1. Soil nitrogen accumulation. Aust. J. Exp. Agric. 26, 347–351.

Suzuki D T 1990 Inventing the Future. Reflections on Science, Technology and Nature. Allen and Unwin Pty Ltd, St Leonards, Australia. 247p.

Ta T C, Macdowall F D H and Faris M A 1986 Excretion of nitrogen assimilated from N_2 fixed by nodulated roots of alfalfa (*Medicago sativa*). Can. J. Bot. 64, 2063–2067.

Thomas R J 1995 Role of legumes in providing N for sustainable tropical pasture, systems. Plant and Soil 174.

Thomas R J and Lascano C E 1994 The benefits of forage legumes for livestock production and nutrient cycling in pasture and agropastoral systems of acid-soil savannas of Latin America. *In* Livestock and Sustainable Nutrient Cycling in Mixed Farming Systems of Sub-Saharan Africa. Eds. J M Powell, S Fernandez-Rivera, T O Williams and C Renard. ILCA, Ethiopia (*In press*).

Torrey J G 1982 Casuarina: actinorhizal nitrogen-fixing tree of the tropics. *In* Biological Nitrogen Fixation Technology for Tropical Agriculture. Eds. P H Graham and S C Harris. pp 427–439. CIAT, Cali, Colombia.

Unkovich M J 1991 The potential of the $\delta^{15}N$ natural abundance method for estimating the contribution of fixed nitrogen to the nitrogen economy of broadacre lupins (*Lupinus angustifolius*). MSc thesis, University of Western Australia.

Unkovich M J, Pate J S and Hamblin J 1994 The nitrogen economy of broadacre lupin in southwest Australia. Aust. J. Agric. Res. 45, 149–164.

Vallis I 1972 Soil nitrogen changes under continuously grazed legume-grass pastures in subtropical coastal Queensland. Aust. J. Exp. Agric. Anim. Husb. 12, 495–501.

Vallis I 1983 Uptake by grass and transfer to soil of nitrogen from ^{15}N-labelled legume materials applied to a rhodes grass pasture. Aust. J. Agric. Res. 34, 367–376.

Vallis I and Gardener C J 1985 Effect of pasture age on the efficiency of nitrogen fixation by 10 accessions of *Stylosanthes* spp. Aust. J. Exp. Agric. 25, 70–75.

Vallis I, Peake D C I, Jones R K and McCown R L 1985 Fate of urea-nitrogen from cattle urine in a pasture-crop sequence in a seasonally dry tropical environment. Aust. J. Agric. Res. 36, 809–817.

Ventura W and Wantanbe I 1993 Green manure production of *Azolla microphulla* and *Sesbania rostrata* and their long-term effects on the rice yields and soil fertility. Boil. Fert. Soils 15, 241–248.

Wani S P, Rupela O P and Lee K K 1995 Sustainable agriculture in the semi-arid tropics through biological nitrogen fixation in grain legumes. Plant and Soil 174.

Watanabe I 1982 *Azolla Anabaena* symbiosis-its physiology and use in tropical agriculture. *In* Microbiology of Tropical Soils and Plant Productivity. Eds. Y R Dommergues and H G Diem, pp 169–185. Martinus Nijhoff, The Hague, The Netherlands.

Watanabe I, Padre B Jr and Ramirez C 1991 Mineralization of azolla N and its availability to wetland rice. Soil Sci. Plant Nutr. 37, 679–688.

Watanabe I, Ventura W, Mascariña G and Eskew D L 1989 Fate of *Azolla* spp. and urea nitrogen applied to wetland rice (*Oryza sativa* L.). Biol. Fertil. Soils 8, 102–110.

Williams C H 1980 Soil acidification under clover pasture. Aust. J. Exp. Agric. Anim. Husb. 20, 561–567.

Wilson A D and Simpson R J 1993 The pasture resource base: status and issues, *In* Pasture Management: Technology for the 21st Century. Eds. D R Kemp and D L Michalk. pp 1–25. CSIRO, Melbourne.

Xu Z H, Saffigna P G, Myers R J K and Chapman A L 1993a Nitrogen cycling in leucaena (*Leucaena leucocephala*) alley cropping in semi-arid tropics. I. Mineralization of nitrogen from leucaena residues. Plant and Soil 148, 63–72.

Xu Z H, Saffigna P G, Myers R J K and Chapman A L 1993b Nitrogen cycling in leucaena (*Leucaena leucocephala*) alley cropping in semi-arid tropics. II. Response of maize growth to addition of nitrogen fertilizer and plant residues. Plant and Soil 148, 73–82.

Ying J, Herridge D F, Peoples M B and Rerkasem B 1992 Effects of N fertilization on N_2 fixation and N balances of soybean grown after lowland rice. Plant and Soil 147, 235–242.

Yoneyama T, Murakanzi T, Boonkerd N, Wadisirisuk P, Siripin S and Kouno K 1990 Natural ^{15}N abundance in shrub and tree legumes, Casuarina, and non N_2 fixing plants in Thailand. Plant and Soil 128, 287–292.

Yoneyama T, Uchiyama T, Sasakawa H, Gamo T, Ladha J K and Watanabe I 1991 Nitrogen accumulation and changes in natural ^{15}N abundance in the tissues of legumes with the emphasis on N_2 fixation by stem-nodulating plants in upland and paddy fields. Soil Sci. Plant Nutr. 37, 75–82.

Yoneyama T, Muraoka T, Murakami T and Boonkerd N 1993 Natural abundance of ^{15}N in tropical plants with emphasis on tree legumes. Plant and Soil 153, 295–304.

Zebarth B J, Alder V and Sheard R W 1991 In situ labelling of legume residues with a foliar application of a ^{15}N-enriched urea solution. Commun. Soil Sci. Plant Anal. 22, 437–447.

Plant and Soil **174:** 29–49, 1995.

Sustainable agriculture in the semi-arid tropics through biological nitrogen fixation in grain legumes

S.P. Wani, O.P. Rupela and K.K. Lee
International Crops Research Institute for the Semi-Arid Tropics (ICRISAT), Patancheru 502 324, Andhra Pradesh, India

Key words: inoculation, legumes, nonnodulation, rhizobia, residual effects, symbiotic N_2 fixation

Table of contents

Abstract
Introduction
Contribution of BNF to N balance
Beneficial effects of legumes
  Nitrogen effect
  Non-N rotational effects
Ways to improve BNF in the SAT
  Host-related aspects
    Host-variability for nodulation and nitrogen fixation
    Breeding for increased BNF and nitrate tolerance
  Management practices
    Nitrogen
    Intercropping
    Tillage
    Land form
    Deep sowing
    Other nutrients
    Insects
  Use of inoculants
    Need for inoculation
    Competitive and effective strains
    Factors affecting performance of inoculant strains
    Yield responses to inoculation.
Conclusion and future areas of research
Acknowledgements
References

Abstract

Sustainable agriculture relies greatly on renewable resources like biologically fixed nitrogen. Biological nitrogen fixation plays an important role in maintaining soil fertility. However, as BNF is dependent upon physical, environmental, nutritional and biological factors, mere inclusion of any N_2-fixing plant system does not guarantee increased contributions to the soil N pool. In the SAT where plant stover is also removed to feed animals, most legumes might be expected to deplete soil N. Yet beneficial legume effects in terms of increased yields in succeeding cereal crops have been reported. Such benefits are partly due to N contribution from legumes through BNF and soil N saving effect. In addition, other non-N rotational benefits, for example, improved nutrient availability, improved soil structure, reduced pests and diseases, hormonal effects are also responsible. In this paper we have reviewed the research on the contribution of grain legumes in cropping systems and the factors affecting BNF. Based on the information available, we have suggested ways for exploiting BNF for developing sustainable agriculture in the semi-arid tropics (SAT). A holistic approach involving host-plant, bacteria, environment and proper management practices including need based inoculation for enhancing BNF in the cropping systems in the SAT is suggested.

Introduction

Sustainable agriculture involves the successful management of agriculture resources to satisfy changing human needs while maintaining or enhancing the environment quality and conserving natural resources (TAC, CGIAR, 1988). Sustainable agriculture relies greatly on renewable resources and on-farm nitrogen contributions are achieved largely through biological nitrogen fixation (BNF). Biological nitrogen fixation helps in maintaining and/or improving soil fertility by using N_2 which is in abundance in the atmosphere. Above every hectare of land at sea level, there is 78,000 tones of inert nitrogen gas (N_2). Intensive agricultural systems are characteristically expanded nutrient cycles involving the export of crops from a farm and require continued import of nutrients to the farm.

Nitrogen is the most limiting nutrient for increasing crop productivity. Input efficiency of N fertilizer is low (Prasad et al., 1990) and in turn, contributes substantially to environmental pollution. The continued and unabated use of N fertilizers would further deplete stocks of nonrenewable fossil fuels used in fertilizer production.

Annually, BNF is estimated to be around 175 million tones N of which close to 79% is accounted for by terrestrial fixation (Fig. 1). This illustrates the importance of BNF in the context of the global N cycle. The BNF offers an economically attractive and ecologically sound means of reducing external N inputs and improving the quality and quantity of internal resources. In this paper we deal with the BNF systems involving upland grain legume crops grown in the semi-arid tropics (SAT). The SAT are the areas located in the seasonally dry tropical climates, spread over four continents. The mean annual temperature in the SAT is > 18 °C; rainfall exceeds potential evapotranspiration for only 2 to 4.5 months in the dry SAT and for 4.5 to 7 months in the wet/dry SAT (Troll, 1965).

Contribution of BNF to N balance

Accurate estimation of the amount of N_2 fixed by different crops in a particular agro-ecosystem is a prerequisite for assessing and improving the contribution of BNF to a given cropping system. However, as nitrogen fixation is dependent upon physical, environmental, nutritional and biological factors (Chalk, 1991; Nambiar et al., 1988; Peoples and Crasswell, 1992) it can not be assumed that any N_2-fixing system will automatically contribute to the N cycle. In general while estimating BNF, plant roots and fallen leaf material are not taken into account which results in underestima-

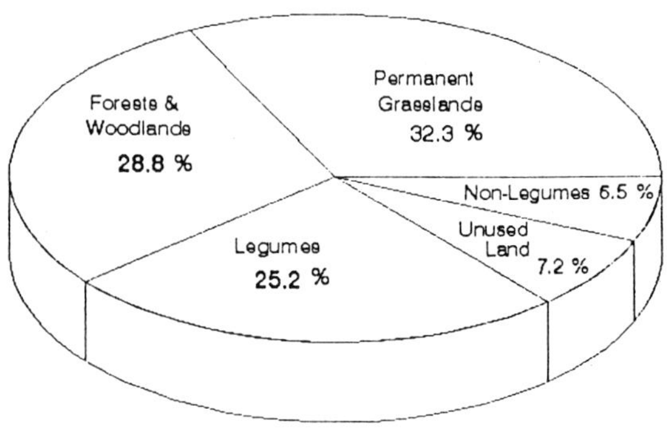

Fig. 1. Distribution of 139 million tonnes of N_2 estimated to be biologically fixed in various terrestrial systems. Source: Burns and Hardy (1975).

Table 1. Examples of estimates of nitrogen fixed by some legumes

Crop	Nitrogen fixed (kg ha^{-1})
Alfalfa	100–300
Black gram	119–140
Clover	100–150
Chickpea	23–97
Cluster bean	37–196
Common bean	3–57
Cowpea	9–125
Groundnut	27–206
Lentil	35–100
Greengram	50–66
Pigeonpea	4–200
Rice bean	32–97
Soybean	49–450
Peas	46
Fenugreek	44

Sources: Derived from Wani and Lee (1992) and Peoples and Crasswell (1992).

Table 2. Net nitrogen balance for pigeonpea and chickpea cultivars grown at Patancheru and Gwalior (India) respectively

	Total plant N uptake (kg ha^{-1})	Plant N derived from fixation (kg ha^{-1})	Net N balance (kg ha^{-1})[a]
Pigeonpea[b]			
Prabhat	69	4	−49
UPAS 120	92	27	−39
T 21	108	43	−39
BDN 1	118	53	−32
Bhedaghat	101	36	−20
JA 275	78	13	−33
Bhandara	108	43	−22
NP (WR) 15	114	50	−27
Chickpea[c]			
Annigeri	110	31	−77
G 130	104	26	−75
ICC 435	102	29	−72
ICCC 42	88	23	−64
ICCV 6	107	30	−76
K 850	104	40	−63

Source: Derived from Kumar Rao and Dart (1987) and Rupela et al. (pers. commun.).

[a] Net N balance calculated as total plant N uptake − (N derived from BNF + N derived form fertilizer + N added to the soil through plant roots and fallen plant parts).

[b] BNF was estimated by N difference method. N derived from fixation calculated for roots also.

[c] BNF was estimated by ^{15}N based A-value method. N derived from fixation calculated for above ground plant parts only.

tion of the quantity of N_2 fixed. It is essential that BNF in roots and fallen plant material e.g. leaves should be considered when estimating the amount of N_2 fixed by legumes.

Legumes have been an important component of agriculture since ancient times. It is widely believed that legumes improve soil fertility because of their N_2-fixing ability. In support of this argument, the reported amounts of N_2 fixed by legumes are cited (Table 1). However, in order to assess the role of BNF in the sustainability of different SAT cropping systems not only the amount of nitrogen fixed by the legume component crop in the system is important, but the overall nitrogen balance of the system needs to be considered. The SAT is characterized by a harsh environment with erratic seasonal rainfall and dense human and animal population and it has unique problems in agriculture also. Due to heavy pressure on land for production to feed a large human and animal population, it is a common practice that along with legume grains, plant material is also often taken away from the field for feeding the animals. In such a case only nodulated roots and fallen leaves go back to the soil.

Net nitrogen balances calculated for different cultivars of pigeonpea grown at Patancheru, India (Kumar Rao and Dart, 1987) and chickpea grown at Gwalior, India (Rupela et al., pers. commun.) indicated that all studied varieties depleted soil nitrogen (Table 2). In all these cases above ground plant materials were removed from the field. In the case of pigeonpea for computing nitrogen fixation, N in plant roots and fallen plant parts also was accounted for. Different maturity groups of pigeonpea cultivars fixed 4–53 kg N ha^{-1} $season^{-1}$ while depleting 20–49 kg N ha^{-1} from the soil. In the case of chickpea, different cultivars fixed 23–40 kg N ha^{-1} $season^{-1}$ and removed 63–77 kg N ha^{-1} $season^{-1}$ from the soil (Table 2). Groundnut fixed 190 kg N ha^{-1} $season^{-1}$ when pod yields were around 3.5 t ha^{-1} at Patancheru (Nambiar et al., 1986), however, groundnut relied for its 20–40% (47–127 kg N ha^{-1} $season^{-1}$) of the N requirement on soil or from fertilizer (Giller et al., 1987), obviously resulting in a negative N balance. Positive net N balances of up to 136 kg ha^{-1} for several legume crops following seed harvest have been shown by Peoples and Craswell (1992). However,

Table 3. Nitrogen balance sheet[a] for different cropping systems for Alfisol, Patancheru, India

Cropping system[b] by year		Import (kg ha^{-1})[c] (A) Fertilizer		Leguminous[d] N_2-fixation		Export (kg ha^{-1})[c] (B) Harvest[e]		Balance (kg ha^{-1})(A)-(B)
1991	1992	1991	1992	1991	1992	1991	1992[f]	
S/P	C	60	60	0+80	0	88+68	66	−22
C	S/P	60	60	0	0+46	64	93+46	−37
G/P	C	18	60	90+50	0	108+56	72	−18
C	G/P	60	18	0	102+82	65	141+75	−19
P	C	18	60	121	0	115	66	+18

[a] N balance calculated based on main import and export sources of N.
[b] S/P = Sorghum intercropped with pigeonpea, C = castor, G/P = groundnut intercropped with pigeonpea, and P = sole pigeonpea.
[c] Each value within a binomial corresponds to the crop in intercrop.
[d] Including atmosphere-derived N (fixed N) in leguminous roots.
[e] Assumed that groundnut roots were exported by harvest.
[f] N contents in mini-plot grown sorghum, pigeonpea, and groundnut were used to calculate total N in the harvest for 1992.
Source: Lee et al. (1993).

if crop residues were removed from the field then net N balances for groundnut are −27 to −95, for soybean −28 to −104, common bean −28, greengram −24 to −65 and cowpea −25 to −69 kg ha^{-1}. Similarly, for soybean grown with different starter N levels after rice which received different fertilization levels, the N balances with seed and stover removed ranged from −12 to −35 kg ha^{-1} in northern Thailand (Jefing et al., 1992). For different cropping systems where pigeonpea and groundnut are grown as intercrops, nitrogen balances were negative (Lee et al., 1993). In the case of sole pigeonpea grown in rotation with sole castor, a positive balance of 18 kg N ha^{-1} during two years crop rotation was observed at Patancheru (Table 3). These results show that legumes also mine the soil N as cereals do. However, total plant N yields from legumes are far higher than the cereal plant N yields. We reach the conclusion that in general, grain legumes, where crop residues are removed, slow the decline of, rather than enhance, the N fertility of the soil in comparison with cereal systems.

Beneficial effects of legumes

Despite the negative N balances for grain legumes grown in rotation or as intercrops, reported benefits of legumes to succeeding non-legume crops have been observed consistently (Table 4). Improvement in cereal yield following monocropped legumes lie mainly in the 0.5 to 3 t ha^{-1} range, representing around 30 to 350% increase over yields in cereal-cereal cropping sequences (Peoples and Crasswell, 1992). In a long-term crop rotation experiment conducted since 1983 at ICRISAT Center, Patancheru, mean residual effects of legume-based crop rotations over the last ten years were observed on sorghum yield as compared to the yield of sorghum from sorghum + safflower (S+F)-S+F plots (Fig. 2). Such increased cereal yields following legume crops were attributed to the N contribution from legumes in crop rotation (De et al., 1983; Kumar Rao et al., 1983; Nambiar, 1990). This opinion is not held by all (Cook, 1988; Danso and Pappastylianou, 1992; Fyson and Oaks, 1990; Russelle et al., 1987; Wani et al., 1991a, 1994a).

Nitrogen effect

Terms like "N residual effect" (De et al., 1983) and "Fertilizer N replacement value" or N equivalent (Hesterman et al., 1987) are used to describe the role of legumes in crop rotations. They refer to the amount of inorganic N required following a non-legume crop to produce another non-legume crop with an equivalent yield to that obtained following a legume. This comparison provides a quantitative estimate of the amount of N that the legume supplies to the non-legume crop. This concept does not distinguish between BNF and the "N-conserving effect" which results from substitution by legumes of biologically fixed N for soil N. Fertilizer N replacement value (FRV) methodology has been widely used but it overestimates the N contribu-

Table 4. Residual effect of preceding legume on cereal yield in terms of fertilizer N equivalents

Preceding legume	Following cereal	Fertilizer N equivalent (kg ha^{-1})
Berseem	Maize	123
Sweet clover	Maize	83
Winged bean	Maize	70
Blackgram	Sorghum	68
Greengram	Sorghum	68
Greengram (monocrop)	Wheat	68
Chickpea	Maize	60–70
Cowpea	Maize	60
Groundnut	Pearl millet	60
Cowpea	Pearl millet	60
Chickpea	Pearl millet	40
Lentil	Pearl millet	40
Peas	Pearl millet	40
Pigeonpea	Wheat	40
Cowpea (monocrop)	Wheat	38
Lathyrus	Maize	36–48
Lablab bean	Maize	33
Pigeonpea	Pearl millet	30
Greengram	Pearl millet	30
Groundnut (monocrop)	Wheat	28
Pigeonpea	Maize	20–67
Peas	Maize	20–32
Lentil	Maize	18–30
Greengram (intercrop)	Wheat	16
Cowpea (intecrop)	Wheat	13
Groundnut (intercrop)	Wheat	12
Groundnut	Maize	9–60
Soybean	Maize	7

Source: Derived from Ahlawat et al. (1981), Bandyopadhyay and De (1986), Chandra and Ali (1986), Dakora et al. (1987), De and Goutam (1987), Doughton and MacKenzie (1984), MacCol (1989), Nambiar et al. (1988), Roy Sharma and Singh (1969), and Weil and Samaranayake (1991).

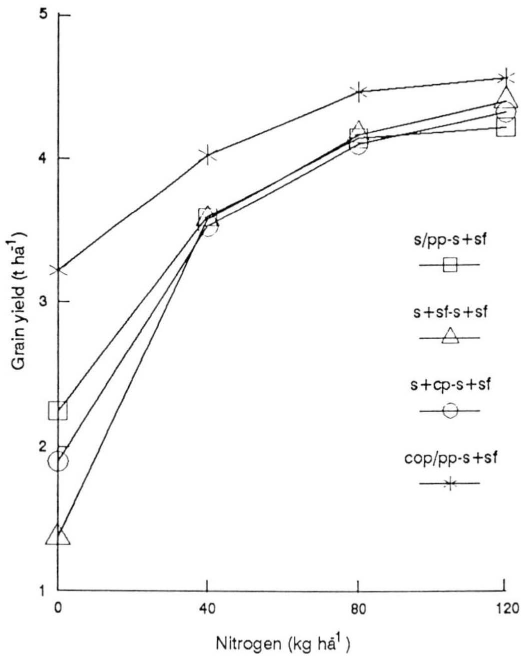

Fig. 2. Mean grain yield of sorghum grown in rainy seasons (1983–92) succeeding different cropping system in previous year, ICRISAT, Patancheru. (2 year crop rotation) S - sorghum, PP - pigeonpea, SF - safflower, CP - chickpea, COP - cowpea, / - intercropped, + - sole crop grown during postrainy season. Source: Rego and Burford (1992).

tion of legumes in a crop rotation. The FRV methodology gives variable estimates depending on the test crop used. The N contribution from hairy vetch and big flower vetch was estimated to be 65 and 75 kg N ha^{-1} respectively with maize as test crop and 125 and 135 kg N ha^{-1} using grain sorghum (Blevins et al., 1990). Recently, ^{15}N methodology has been used to measure the residual effects of legumes to circumvent problems with non-isotopic methods (Danso and Papastylianou, 1992; Senaratne and Hardarson, 1988; Wani et al., 1991a). Based on the estimates obtained via ^{15}N methodology, Hesterman et al. (1987) argued that the amount of N credited to legumes in a crop rotation in the north-central US may be inflated by as much as 123% due to the use of the FRV method. Based on ^{15}N methodology it is reported that only 7.3 to 28% of the ^{15}N in legume crops is taken up by a following grain crop (Ladd et al., 1981, 1983; Vallis, 1983; Wani, unpubl. data). The overestimation is because the FRV method confounds the non-N rotation effect with the N contribution, and this method assumes that use efficiency of fertilizer and legume N is similar.

Growing legumes in rotation does improve mineral N content in soil as compared with the cultivation of non-legume crops. At ICRISAT-Asia Center, Patancheru, India, a long-term rotation experiment is being conducted on a Vertisol since 1983 using two-year crop rotation treatments. The surface soil (0–20 cm) samples collected after harvest of 9th season crop showed in general higher amounts of mineral N contents in the soil from the legume-based cropping system than the non-legume based cropping system (Wani et al., 1994a; Fig. 3). Inclusion of greengram in the crop-

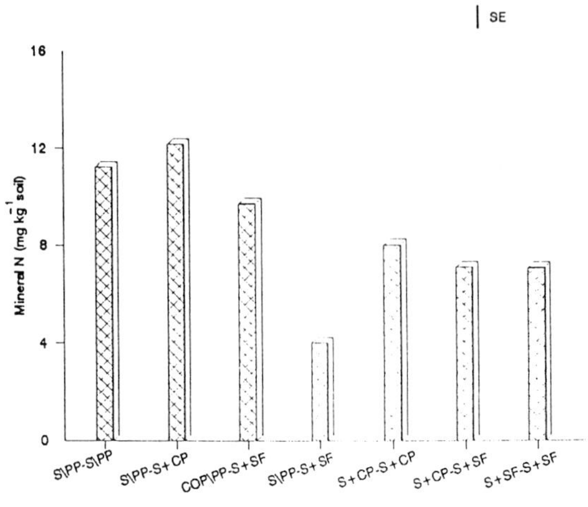

Fig. 3. Mineral N content in surface soil samples (0–20 cm) from plots under different cropping systems since last nine years. S - sorghum, PP - pigeonpea, SF - safflower, CP - chickpea, COP- cowpea, / - intercropped, + - sole crop grown during postrainy season.

Table 5. Nitrogen mineralization potential (N_o), active N fraction, and time (wk) required to mineralize 25 mg N kg^{-1} soil for the soil samples under different cropping systems

Treatment[a]	N_o (mg kg^{-1} soil)	Active N fraction(%)	Time (wk) to mineralize 25 mg N kg^{-1} soil
S/PP-S/PP	94.6 ± 15.98	13	10.3
F+S-F+S	40.5 ± 8.06	9	21.4
COP/PP-S+SF	86.1 ± 19.90	17	1.5
S/PP-S+CP	100 ± 10.04	16	13.8
S/PP-S+SF	67.3 ± 13.46		10.1
S+SF-S+SF	-[b]	-[b]	-[b]
S+CP-S+SF	-[b]	-[b]	-[b]
S+CP-S+CP	56.1 ± 20.98		19.6

[a] S = sorghum, PP = pigeonpea, F = fallow during rainy season, COP = cowpea, SF = safflower, CP = chickpea, / = intercropped, +- sole crop grown during postrainy season.
[b] Not estimated as exponential model was not superior over linear model.
Source: Wani et al. (unpubl. data).

ping sequence increased available nitrogen in the soil at harvest to the extent of 12.6% in the non-fertilized control plot (Rao and Singh, 1991). Similarly, a five times higher mineral N content in the soil under an eight year rotation using fababeans as green manure (agro-ecological rotation) was observed than from the soil under continuous barley treatment which was fertilized with 90 kg N ha^{-1} y^{-1} (Wani et al., 1991a).

In addition to mineral N content in the soil from the long-term rotation experiment, N mineralization potential (N_o) of the soils under pigeonpea-based cropping system was almost two times higher as compared to the fallow-sorghum treatment (Table 5). The "active N fraction", the quotient of N_o and N_{total} and expressed as percentage, varied between 9–17 % with higher values reported for the soil under pigeonpea-based cropping systems (Wani et al., unpubl. data). Using N_o and k (N mineralization rate constant) values the cropping systems were ranked based on the time required to mineralize 25 mg N kg^{-1} soil. Time required to mineralize a fixed quantity of N was less in the case of cropping systems which contained pigeonpea than the time required in the case of cropping systems which involved chickpea or no legume or which was left fallow for one season (Table 5). Such benefits in terms of increased "mineralizable N (N_o)" in the soil were observed even when all the above ground plant parts except fallen leaves were removed. Such increased N_o values at Patancheru were not associated with chickpea which is grown during the post rainy season on residual moisture. Mineralizable soil N (N_o) following one cycle of an eight year rotation using fababeans as green manure was about double that following 60 years of a 5-year rotation involving forage and cereal crops but without returning the crop residues to the soil (Wani et al., 1994b).

The analysis of field soil samples collected prior to the start of the experiment in 1983 and later in 1993 showed that, in the case of Fallow+Sorghum (F+S) system, total soil N content was decreased by 72 μg g^{-1} soil after ten years. S+CP-S+SF and S+SF-S+SF plots also showed decreased total soil N. The continuous greengram + sorghum maintained the soil N while a substantial increase in total N was observed in S/PP-S+SF and cowpea/pigeonpea-sorghum+safflower (COP/PP-S+SF) systems. (Table 6; Wani et al., 1994a). These results demonstrated that pigeonpea-based cropping systems increased the total soil N substantially during ten years.

Sorghum was grown in the greenhouse using surface soil samples collected from the field plots which were under different cropping systems during the last 9 years. Sorghum grown in the soil from the COP/PP-S+SF plots yielded 63% higher as compared to the sorghum grown in the soil from the S+SF-S+SF plots. In other pigeonpea-based cropping systems, sorghum yielded 36–56% higher than that of sorghum yield from

Table 6. Soil total N ($\mu g\ g^{-1}$ soil) in 0–15 cm and 15–30 cm layer under different cropping systems during 1983 and 1993

	Soil depth			
	0–15 cm		15–20 cm	
Cropping system	1983	1993	1983	1993
S/PP-S+SF	559	629	437	480
S+CP-S+SF	540	517	407	443
C/PP-S+SF	543	645	419	501
S+SF-S+SF	537	530	397	438
F+S-F+S	563	491	422	426
F+CP-F+S	567	507	399	446
M+S-M+S	558	559	422	461
	NS	**	NS	**
	±18.4	±13.2	±15.0	14.4

S - sorghum, PP - pigeonpea, SF - safflower, CP - chickpea, C - cowpea, F - fallow M - mungbean, / - intercrop, + - sequential crop, - – rotation, NS - Not significant.
** $p = \leq 0.01$.
Source: Rego et al. (unpubl. data).

the S+SF-S+SF treatment. In the case of chickpea-based cropping systems sorghum yields were lowered by 18–24.5% over the S+SF-S+SF plot yields (Wani et al., unpubl. results). Using ^{15}N methodology it was estimated that 8.4 to 20% of total sorghum plant N in the case of pigeonpea-based cropping systems was derived from the N that was either fixed previously and had accumulated, or the soil N that was made more available due to the presence of pigeonpea in the rotation. This was clear evidence of greater N availability in the case of pigeonpea-based cropping systems over the S+SF-S+SF system. These results were in conformity with the findings of increased N_o potential of these soils reported in Table 5. The A values for the soil from pigeonpea-based cropping system plots were higher by 25.6 to 76.3 mg pot^{-1} (4.5–13.3 kg N ha^{-1} equivalent) than that of the S+SF-S+SF treatment. The fertilizer N replacement values calculated for these treatments using soil from the S+SF-S+SF treatment ranged from 65–161 mg N pot^{-1} (24–28 kg N ha^{-1} equivalent). All these results indicated that increased sorghum yields from the pigeonpea-based cropping systems over the S+SF-S+SF system were partly due to the increased soil N availability and all the benefits can not be explained in terms of the N effects (Wani et al., unpubl. results).

In the agroecological eight year rotation (which included barley, fababean, barley, fababean, barley undersown to red clover and brome grass, forage, forage, forage) barley grown following fababeans (AER 1) yielded 105% higher than that of the barley grown after continuous barley (CG) for eight years with 90 kg N ha^{-1} y^{-1}. Using ^{15}N methodology it was estimated that 48.5% (405 mg N pot^{-1}) of total barley plant N in the case of the AER 1 treatment was derived from the N source that was not present in the soil from the CG treatment. The presence of legumes in the rotation gave an increased N supplying capacity (A value) of the soils over those in the soil from the CG system (Wani et al., 1991a). These authors concluded that the soil N availability to plants contributed significantly to the higher soil fertility in the legume-based systems. However, increased N availability contributed partly to the increased barley yields from legume-based rotations and other mechanisms than the N effect were also responsible for increased barley yields in these plots (Wani et al., 1991a). Similarly, non-N rotational benefits of the legumes towards yield of subsequent crop have been observed by many researchers (Cook, 1988; Danso and Papastylianou, 1992; Peoples and Craswell, 1992; Weil and Samaranayake, 1991).

Non-N rotational effects

If the benefits of crop legumes in rotations cannot be solely explained in terms of the residual fixed N, then what are the sources of the benefits demonstrated in Table 4? Several factors can be involved, the relative importance of each dictated by site, season, and crop sequences.

Crop rotations increased the availability of nutrients other than N through increased soil microbial activity (Kucey et al., 1988; Ladha et al., 1989; Wani et al., 1991 a, b). A two fold increased microbial biomass C, in the AER soil than in the CG soil was observed. The concentration of microbial N g^{-1} soil; the proportions of soil N, or the proportion of soil ^{15}N present as microbial N, and microbial activity as indicated by the respiration rate, were all greater in the agroecosystem than in the CG system (Wani et al. 1991a). These results indicated that higher proportion of soil or fertilizer N was in the labile fraction in the case of AER than in the case of the CG system. Wani et al., (1991b) observed that in an eight year agro-ecological rotation containing fababeans and forage, mycorrhizal colonization of barley roots was increased as compared to a CG system. Further, through positive relationships between levels of mycorrhizal colonization and K, Ca, Mg, Zn, S, and Fe accumulations and barley yields it was inferred that increased mycorrhizae acted as agents to mediate enhanced soil fertility in the rotations over that of a continuous barley system.

Improvements in the soil structure following legumes, mainly improved soil aggregate formation, after three years of alfalfa, clover and hairy vetch mixture (Latif et al., 1992) or with numerous years of a Sod pasture, or hay crop (Olmstad, 1947; Power, 1990; Strickling, 1950) have been observed. Incorporation of legume residues improved soil water-holding (Wani et al., 1994c) and buffering capacity (Buresh and De Datta, 1991).

Ries et al. (1977) suggested that growth promoting substances in legume residues are responsible for the rotation effect. The rotations break the cycles of cereal pests and diseases, and phytotoxic and allelopathic effects of different crop residues (Francis et al., 1986). The effect of crop rotation on pest pressure varies widely, but in general the literature supports Francis and Clegg (1990) who stated that "the greater the differences between crops in a rotation sequence, the better cultural control of pests can be expected". Crop rotation is an effective tool against certain pests, and that efficacy may contribute to the rotation effect, but rotation does not control all pests and diseases. For example, Johanson et al. (1984) reported that black cutworms (*Agrotis ipsilon*) are more of a problem when maize is rotated with either soybean or wheat than when maize is grown continuously. Similarly, Wani et al., (1991b) observed no reduction in the common root rot (*Bipolaris sorokiniana*) of barley grown in rotation plots than the continuous barley plots. On the contrary, marginally higher root rot incidence was recorded from the eight year rotation plots containing fababeans and forages.

Ways to improve BNF in the SAT

Host-related aspects

Host variability for nodulation and nitrogen fixation

Presence of a large genotypic variability for BNF traits like nodule number, nodule mass and acetylene reduction activity (ARA) per plant has been known since early eighties for chickpea, groundnut and pigeonpea (Nambiar et al., 1988), soybean (Wacek and Brill, 1976), cowpea (Zari et al., 1978), common bean (Graham and Rosas, 1977). Using ^{15}N isotope-based methods, differences among cultivars have been detected in soybean (Hardarson et al., 1989; Rennie et al., 1982), common bean (Rennie and Kemp, 1982; Westermann et al., 1981), groundnut (Giller et al., 1987), greengram and blackgram (Sampet and Peoples, unpubl. data cited by Peoples and Crasswell, 1992), pigeonpea (J V D K Kumar Rao, pers. commun.) and chickpea (Rupela et al., unpubl. data). However, efforts to use this variability in breeding for improved BNF has been limited or non existent in many of these legumes. Arunachalam et al. (1984) found that ARA and nodule mass have good predictive value for plant growth and yield related traits in groundnut. After analysis of a six parent diallel cross, Nigam et al. (1985) observed that non-additive genetic variance for ARA was predominant in groundnut. The groundnut line NC Ac 2821 had the highest general combining ability for ARA, total nitrogen, leaf area and was proposed as a good parent for breeding programs. The crosses made between the high- and low-nodulating chickpea lines to investigate the inheritance of nodulation indicated segregation for nodulation in F_2 populations from nonnodulating to nodulating (O P Rupela, unpubl.). These studies thus indicate the complexity of the BNF related traits. Most of the studies reported above for chickpea and groundnut were made in the field. Legumes like pigeonpea offer another difficulty for BNF studies because their nodules are loosely attached to roots and generally fall off during excavation of the field grown plants. It is perhaps due to this reason that there are no reports in pigeonpea so far on studies of the type reported above for groundnut.

Indication of plant to plant variability for nodulation within chickpea cultivars was further investigated. It was observed that not only consistent low- and high-nodulating plants were present within chickpea cultivars (Rupela, 1994), even nonnodulating plants occurred in normal cultivars or land races (Rupela, 1992). Consistent variability for nodulation extent was also subsequently detected within the pigeonpea cultivars. Unlike in chickpea, however, nonnodulating plants in pigeonpea were found in segregating populations at F_2 (Rupela and Johansen, 1995). It is perhaps due to the absence of any natural selection pressure for nodulation or BNF during development of a cultivar that the different nodulation types continue to exist within a material up to release stage. This gained strength from the fact that during a screening for high-nodulating plants at high mineral N in soil, we observed

Table 7. Different nodulation types of chickpea and pigeonpea plants identified at ICRISAT Center, India

Chickpea
- Nonnodulating with native root nodulating bacteria (RNB) (rn6)
- Nonnodulating with IC 59, low nodulating with native RNB
- Low nodulating at low N
- High nodulating at low N
- High nodulating at low N but low nodulating at high N
- High nodulating at high N

Pigeonpea
- Nonnodulating with native RNB
- Low nodulating at low N
- High nodulating at low N

Parenthesis has the name of the identified gene.
Source: Rupela (1994).

the desired plants in 85 out of 90 advanced breeding lines of chickpea that were studied (Rupela, 1994).

Using appropriate screening procedures several different nodulation types have been identified within several chickpea and pigeonpea cultivars (Table 7) since 1985. Preliminary studies of Venkateswarlu and Katyal (1994) also indicated plant to plant variability within cultivars of groundnut. Intracultivaral variability for nodulation may be present in other legumes also. Obviously the Nod$^-$ (NN) and the low-nodulating (LN) selections are of academic interest and serve as an important reference base in BNF quantification studies. High-nodulating (HN) selections are expected to improve yield in low soil N conditions. In our screening studies the HN selection generally grew better than the NN and LN selections of a given cultivar, but large plot yield trials have been conducted only with the LN and HN chickpea selections of ICC 4948 and ICC 5003. The HN-selection of cultivar ICC 4948 produced 31% more grains than its LN-selection at low soil N(N1) level (Fig. 4). The HN-selection of ICC 4948 yielded better even at high soil N(N2) level. But the LN and HN selections of another cultivar ICC 5003 yielded the same under N1 and N2 levels. In a previous pot trial the root length density of LN-ICC 5003 was 32 m plant^{-1} which was 2-times greater than that of the LN-ICC 4948. Perhaps the cultivar ICC 5003 could scavenge the soil N more efficiently than that of ICC 4948 due to its high root length density and as a result both the HN and LN lines of ICC 5003 yielded similarly.

These studies thus suggest a great scope for enhancing BNF in legumes through host plant selection. Most

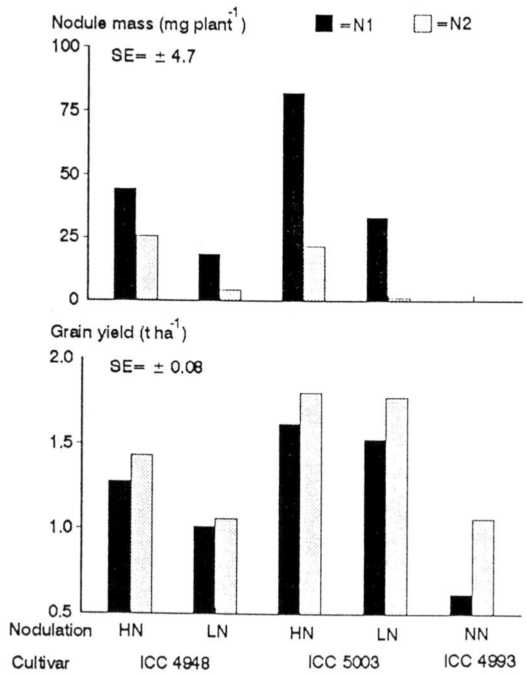

Fig. 4. Nodule mass at 45 days after sowing and grain yield of chickpea cultivars of different nodulation ratings (HN = high nodulating, LN = low nodulating, NN = nonnodulating; grown at two mineral N levels in soil low N (N1, about 10 mg kg^{-1} soil) and high N (N2, about 20 mg kg^{-1} soil); postrainy season, 1991/92, Vertisol, ICRISAT. Both N-levels and nodulation were significantly different (p = 0.05) for the above parameters. Their interactions were also significantly different for nodule mass. Source: Rupela, (unpubl.).

HN selections yielded higher than the LN selections (Fig. 4 and unpubl. studies). However, it needs to be established in further studies.

Breeding for increased BNF and nitrate tolerance

Soybean cv. Dunadja from Romania showed no reduction in N_2-fixation with application of 100 kg N ha^{-1} while in all the other seven cultivars N_2 fixation was substantially reduced (Hardarson et al., 1989). Similarly soybean cultivars of Korean origin with higher N_2-fixation than the commercial cultivars grown in Australia have been identified and used as donor parents in a breeding program in Australia (Betts and Herridge, 1987; Peoples and Herridge, 1990). Plant mutagenesis has been used to generate NO_3 tolerant N_2-fixing phenotypes e.g. nitrate-tolerant symbiont in soybean (Carroll et al., 1985). Extreme super nodulating mutants of soybean and *Phaseolus vulgaris* produced significantly lower biomass and/or grain yield than their parent lines (Buttery et al., 1990; Hansen et

Table 8. Response of chickpea, pigeonpea and groundnut to fertilizer nitrogen in experiments on farmers' fields in India

State	Chickpea		Pigeonpea		Groundnut	
	No. of trials	kg grain kg^{-1} N	No. of trials	kg grain kg^{-1} N	No. of trials	kg grain kg^{-1} N
Andhra Pradesh	47	16.5	56	17.5	258	18.0
Bihar	77	17.0	–[a]	–	25	25.0
Gujarat	–	–	159	15.5	–	–
Haryana	88	12.0	–	–	–	–
Himachal Pradesh	50	11.0	–	–	–	–
Karnataka	275	11.0	104	8.0	310	14.5
Madhya Pradesh	624	19.0	15	10.5	–	–
Maharashtra	351	8.5	–	–	495	12.0
Orissa	71	8.5	39	19.0	–	–
Punjab	113	10.5	–	–	62	17.0
Rajasthan	267	18.0	159	13.0	38	12.5
Tamil Nadu	–	–	–	–	384	14.0
Uttar Pradesh	408	21.5	–	–	14	12.0
Average		15.4		14.2		14.4

[a] Not conducted.
Source: Tandon (1992).

al., 1989; Wu and Harper, 1991). Species differ considerably in their symbiotic tolerance to mineral N and when sufficient natural variation already exists (Betts and Herridge, 1987; Hardarson et al., 1984) it may not be necessary to resort to mutagenesis procedures for breeding purposes (Gibson and Harper, 1985).

Management practices

Nitrogen

Most of the legumes cannot derive 100% of their N requirement through BNF. In the tropics where legume residues are not returned to the soil, most legumes deplete the soil N (Table 2 and 3). In the long run, such systems cannot be sustainable. Further, large numbers of on-farm experiments in India showed that legumes responded markedly to fertilizer N; such responses are expected as legumes have a high N requirement. The SAT soils are poor in N, N_2-fixation mechanisms do not become functional from day one and all the legume requirement cannot be met from BNF. Significant responses to 20–30 kg N ha^{-1} as starter have been observed under good growth conditions (Table 8). At application rates of 20 kg N ha^{-1}, overall response rate (grain kg^{-1} N) was 14.2 in pigeonpea, 14.4 in groundnut and 15.4 in chickpea all under non-irrigated conditions (Table 8). Responses of such high magnitude point that to achieve increased legumes productivity along with increased BNF and maintaining the soil fertility, we need to adopt need-based mineral N application to legumes. Soil mineral N status at the time of sowing of the legume crop must be taken into account before deciding on the need and rate of N fertilizer application.

In general, high soil nitrogen levels, applied or residual, reduces nodulation and N_2 fixation (Tables 9 and 10). To improve BNF contribution from the legumes under such circumstances soil N must be managed through inclusion of appropriate nitrate tolerant high N_2-fixing legume crop or genotype of a given crop as mentioned earlier and/or appropriate cropping and management practices. It has been observed that application of 200 kg N ha^{-1} decreased N_2 fixation by groundnut only by 18% (from 120–102 kg ha^{-1}) whereas in cowpea by 54% (from 125 to 57 kg ha^{-1}) (Yoneyama et al., 1990). These results suggest that there exists a potential to select appropriate legume crops or cultivars of a given legume for specific areas with high soil N contents without decreasing their BNF contribution to the system.

Intercropping

Legumes are generally grown as intercrops with cereals or other non-legumes in the SAT (Willey, 1979) and application of N to the cereal crop reduced N_2 fixation by the component legume crop (Nambiar et

Table 9. Nitrogen concentrations in root environment where approximately 50% reduction in N_2- fixation was recorded

Suppressive concentration (in reference)[a]	ppm equiv.	BNF as	Crop	Plant culture	Reference
1.43 m*M*	20	Nodule no.	Chickpea	Pot	Rawsthorne et al. (1985)
6 m*M*	84	Nod mass, ARA	Soybean	Pot	Buttery and Dirks (1987)
2 mol m^{-3}	28	^{15}N	Chickpea, Fababean	Pot	Peoples et al. (1987)
5 m*M*	70	ARA	Chickpea	Pot	Sawhney et al. (1989)
200 kg ha^{-1}	89	ARA	Soybean	Field	Wu and Harper (1991)
112 kg ha^{-1}	50	Nod mass	Pigeonpea, Soybean	Field	Buttery et al. (1988)
3 m*M*	42	Nod mass, ARA	Common bean	Pot	Buttery et al. (1990)
112 kg ha^{-1}	50	Nod mass	Common bean	Field	Buttery et al. (1990)
10 m*M*	140	Nod mass, ARA	Fababean	Field	Buttery and Gibson (1990)
5 m*M*	70	Nod mass	Soybean	Pot	Cho and Harper (1991)

[a] In all cases, except for Rawsthorne et al. (1985), the listed nitrate concentration was the lowest level used in different trials.

al., 1983; Ofori and Stern, 1987). Similarly, shading by associated cereals reduced BNF in the component legumes (Nambiar et al., 1983; Wahua and Miller, 1978). Strip cropping of the cereals and legumes can overcome both these problems and improve the systems productivity without reducing BNF contributions in the system from the associated legumes. Indeterminate legumes fix more N than determinate types in intercropping (Fugita et al., 1992). Nitrogen fixation in climbing bean (Francis, 1986; Graham and Rosas, 1978), cowpea (Ofori et al., 1987) and Siratro (Ogata et al., 1986) was unaffected by intercropping with cereals. In cases where strip cropping is not possible, climbing type legumes can be used.

Tillage

Nodulation and N_2 fixation in soybean grown in subtropical Australia were substantially improved under no tillage with N balance of 80 kg N ha^{-1}, compared with the cultivated system with 30 kg N ha^{-1} N balance. Increased N_2 fixation resulted mainly from the higher proportion of plant N derived from fixation since yields were unaffected by tillage practice (Peoples and Crasswell, 1992). Clean cultivation accelerates the oxidation of organic matter in soils and generally results in higher NO_3 in the profile (George et al., 1992; Thomas et al., 1973) which would affect BNF in legumes.

Land form

Greengram, pigeonpea and soybean grown on broad bed and furrows (BBF) on Vertisol improved nodulation than when grown on a flat surface. However, improved nitrogenase activity on BBF was recorded with greengram and pigeonpea only (Wani and Potdar, unpubl. data). However, in Vertisols, chickpeas sown on flat beds nodulated better than those sown on ridges with the same sowing density (Rupela and Saxena, 1987). As the ridged fields had greater evaporation losses due to increased surface area, this may be important when moisture is limiting.

Deep sowing

Deep sowing of groundnut results in the development of an elongated hypocotyl, poor rooting, poor nodulation and nitrogen fixation, notably in spanish types. Virginia types have considerable nitrogenase activity even when sown deep because of their ability to nodulate on the hypocotyl (Nambiar et al., 1988). Farmers tend to sow chickpea at a sufficient depth to ensure good crop stand as it is generally grown on residual moisture. Deep sown chickpea crops in heavy black soils suffer a substantial reduction in nodulation and N_2 fixation. In lighter soils chickpea have been found to nodulate at depth (Rupela et al., 1985).

Other nutrients

It should be realized however, that poor N_2 fixation can be due to poor plant growth resulting from pests, diseases, and nutrient deficiencies. Addition of P stimulated pigeonpea nodulation in both an Alfisol and a Vertisol (Kumar Rao and Dart, 1981). In Karnataka, India, trials on farmer's fields with pigeonpea showed increased nodulation due to application of diammonium phosphate (DAP) alone than to the inoculation with

Table 10. Effect of soil mineral N and N fertilizers on crop N productivity and the proportion (P) and amount of crop N derived from N_2 fixation

Species	Location	Level		Total	N_2 fixed		Reference
		Soil mineral N (kg N ha^{-1})	Fertilizer N (kg N ha^{-1})	crop N (kg N ha^{-1})	P	Amount (kg N ha^{-1} $crop^{-1}$)	
Groundnut	India	–	0	196	0.61	120	Yoneyama et al. (1990)
			100	210	0.47	99	
			200	243	0.42	102	
Chickpea	Australia	10(to 120 cm)		114	0.85	97	Doughton et al. (1993)
		326		194	0.17	33	
			0	97	0.81	79	
			50	114	0.59	59	Herridge et al. (1994
			100	115	0.29	25	
Soybean	Australia	70(to 120 cm)		230	0.34	78	Herridge et al. (1990)
		260		265	0.06	16	
	India	–	0[a]	63	0.29	18	Yoneyama et al. (1990)
			100	108	0.26	28	
		-	0[b]	89	0.48	43	
			100	115	0.24	28	
	Malaysia	-	40 at sowing	169	0.68	115	Norhayati et al. (1988)
		day 45+ further or	20 as nitrate				
			20 as urea	200	0.15	30	
Common bean	Kenya	-	10	149	0.39	58	Ssali and Keya (1986)
			100	158	0.10	16	
Cowpea	Kenya	-	20	116	0.53	62	Ssali and Keya (1984)
			100	137	0.08	11	
	India	-	0	163	0.77	125	Yoneyama et al. (1990)
			100	138	0.67	92	
			200	172	0.33	57	

[a] Uninoculated.
[b] Inoculated .

Rhizobium alone (Chinmulgund and Hegde, 1987). Cassman et al. (1981) found that field-grown soybean had a higher P requirement when it was dependent on BNF for its N supply as compared to the mineral N dependency. Based on the results from 140 on-farm demonstration plots with soybean in Uganda it was observed that on an average 300 kg ha^{-1} yield increase was obtained with 40 kg P_2O_5 ha^{-1} application and further increase of 300 kg ha^{-1} was obtained through inoculation with *Rhizobium* (Keyser and Li, 1992).

In groundnut, fertilization with B, Co, Mo and Zn in a medium calcareous soil, with and without *Rhizobium* inoculation significantly increased nodulation, percentage of effective nodules and plant dry matter (Joshi et al., 1987). It has been reported that Fe deficiency specifically limits nodule development in groundnut grown in the calcareous soils of Thailand (O'Hara et al., 1988). Soil acidity along with Mn and Al toxicities can also restrict N_2 fixation in groundnut. Excess Mn was detrimental to plant growth per se rather than to nodulation, but nitrogenase activity was more affected by Al than plant growth (Nambiar and Anjaiah, 1989a). Application of Co at a rate of 500 mg cobalt nitrate kg^{-1} seed significantly increased grain yield of pigeonpea (Raj, 1987), soil application of 0.45 kg Mo ha^{-1} as sodium molybdate significantly increased nodulation and grain yield of pigeonpea

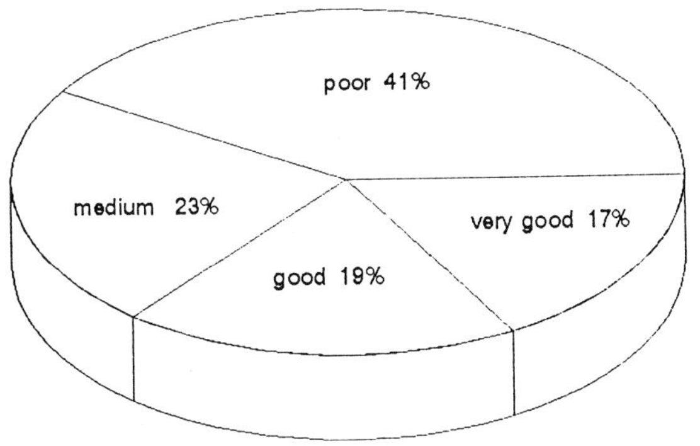

Fig. 5. Nodulation status of chickpea based on 314 fields. (AICPIP data cited by Tauro and Khurana, 1986).

(Khurana and Dudeja, 1981). Soil application of 1 kg cobalt chloride, 1 kg sodium molybdate ha^{-1} and 25 kg $ZnSO_4$ ha^{-1} increased chickpea grain yield by 10, 7 and 4% respectively over the control. Inoculation with *Rhizobium* increased chickpea yield by 26% over the non-inoculated control however, inoculation along with Co, Mo and Zn application increased yield by 41, 39 and 28% respectively over the control (Namdeo and Gupta, 1992).

Insects

Extensive nodule damage to pigeonpea by a Dipteran larva, *Rivellia angulata* was reported in farmers' fields (Sithanantham et al., 1981). The extent of nodule damage was greater in pigeonpea grown in Vertisols (up to 86%) as compared to 20% in Alfisols (Nambiar et al., 1988). Nambiar et al. (1990) reported reduced nodule damage by 50% due to inoculation of pigeonpea with engineered *Bradyrhizobium* carrying an insecticide gene (*Bacillus thuringiensis* subsp. israelensis) in the presence of *Rivellia angulata* larvae under greenhouse conditions. These results suggest the potential benefits from planned introduction of engineered *Bradyrhizobia* carrying insecticide genes into natural environments. Another possible solution is to select pigeonpea genotypes that can resist or tolerate attack by nodule damaging insects. Soil application of a single dose of insecticide (aldrin) prevented nodule damage up to 45 DAS however, during later stages insect damage could not be controlled (Kumar Rao and Sithanantham, 1989).

Use of inoculants

Much of the applied research efforts in studying BNF have gone into identifying efficient strains of bacteria as inoculants. Before inoculation with appropriate strains to be used, it needs to be determined whether inoculation is needed?

Need for inoculation

The most important point is do we need inoculation of the legumes in a region where these crops have been grown over long periods? Development of an inoculation industry in many countries has been largely motivated by the desire to introduce legume species to new areas (Burton, 1982). Most cultivated tropical soils are assumed to have relatively large populations (> 100 g^{-1} dry soil) of rhizobia capable of nodulating the legumes grown in such soils (Nambiar et al., 1988). However, surveys of farmers' grain and fodder legume crops have shown poor nodulation in large areas and good nodulation only in a few pockets (Fig. 5) (IARI, 1980; Kabi and Poi, 1988; Kulkarni and Joshi, 1988; Tauro and Khurana, 1986). In a survey of farmers' chickpea fields around Gwalior, Madhya Pradesh (M.P.), 39% fields had < 100 rhizobia g^{-1} soil, 17% had 10^2–10^3 and 44% fields had a population $> 10^3$ (Rupela et al., 1987). In a similar survey conducted for 43–47 villages from each of the three districts of Madhya Pradesh, India for nodulation of pigeonpea, black gram, green gram and lentil showed poor nodulation (0–10 nodules $plant^{-1}$) in 64 to 100% of the surveyed area (Namdeo and Gupta, 1992). The need to inoculate the legumes grown on cultivated soils must be assessed by considering the interacting factors between the soil, the host plant and *Rhizobium*.

Presence of nodules on plant roots does not necessarily mean that sufficient N_2 is being fixed for maximum benefit to the host plant. In groundnut or pigeonpea nodulation occurs naturally at most locations due to the cross-species promiscuity of the cowpea rhizobia. However, the ability to fix high amounts of N (efficiency) is governed by the symbiotic capability between *Rhizobium* and the host plant. Hence, it may be necessary to introduce superior (more competitive and efficient) strains of *Rhizobium* to ensure adequate N_2 fixation for maximum growth and yield of the host plant. In a survey of groundnut crops grown in farmers' fields in southern India, 52 out of 95 fields showed inadequate nodulation with less than 10 per cent ARA of that which can be obtained under reasonable field conditions (Nambiar et al., 1982). Although, adequate nodulation was observed in some parts, ineffective nodules exceeded the number of effective nodules. Field surveys have shown that proportion of ineffective strains was as high as 40% in chickpea, 53% in green gram and 63% in groundnut (Tauro and Khurana, 1986). In another study 94% strains of rhizobia were observed ineffective in groundnut (Kulkarni and Joshi, 1988). Poor nodulation in farmers' fields could be due to several factors e.g. inadequate soil moisture, lack of appropriate rhizobia in soil, deficiency or toxicity of a particular nutrient, unfavorable conditions like prolonged water logging, unfavorable pH, abundance of bacterial predators, pests and disease attack, etc.

Using the network approach NifTAL initiated Worldwide Rhizobial Ecology Network (WREN) and conducted standardized inoculation trials with extensive environmental data. Thies et al. (1991) developed a mathematical model using native rhizobia numbers (estimated by most probable number method) and soil mineral N data as inputs to predict the inoculation responses at different sites. This approach accounted for 83% of the variation observed due to inoculation. These models have been incorporated into an interactive computer program called "RESPONSE" which reduces the need for costly, site-specific field inoculation trials to determine the need for inoculation with *Rhizobium*. This remains a valid approach to determine the need for inoculation in most of the cases. However, Nambiar (1985) reported significant yield increases from Cameroon, India, and China in the case of groundnut due to inoculation with NC 92 strain from the soils having large populations of native rhizobia. These results indicate that a simulation model using most probable number (MPN) data and mineral N data can not provide reliable answers in all the cases and there is a need to fine-tune the model.

Competitive and effective strains

In soils lacking rhizobia nodulating a particular legume, inoculation with efficient strains increased yields (Nambiar et al., 1988). In soils which contain established native *Rhizobium* populations, the introduced strains should be competitive and efficient. The degree of establishment and persistence of an inoculant strain generally decreased with increase in population density of the native rhizobia (ICRISAT, 1981). However, some inoculant strains have succeeded in forming more nodules even in the presence of active indigenous competing rhizobia eg. NC 92 on groundnut (Nambiar et al., 1984). Little is known of the factors controlling competitiveness but host cultivar, soil properties, soil microflora, environmental factors and the nature of the competing strains influence the success of inoculant strains in nodule formation (Alexander, 1982). The success of the strain NC 92 in terms of nodule formation increased with repeated inoculation (Table 11). Higher inoculum rate of 10^6–10^8 cells per seed at the initial inoculation helped in early establishment (Nambiar et al., 1984). Strains of vesicular arbuscular mycorrhizae (VAM) significantly influenced nodule formation by bradyrhizobia strains. In the absence of any VAM, when mixtures of NC 92 and NC 43.3 were inoculated, strain NC 92 occupied more nodules (89%) than strain NC 43.3 (34%). In the presence of *Acaulospora laevis*, 86% nodules in the NC 92 + NC 43.3 mixture were formed by NC 92, but the presence of *Glomus fasiculatus* reduced the competitive ability of strain NC 92 (49% NC 92 nodules) (Nambiar and Anjaiah, 1989b). Field trials with soybean have demonstrated that to achieve nodule occupancy of greater than 50%, inoculant rhizobia/bradyrhizobia must be applied at a rate at least 1,000 times greater than the estimated number of indigenous bradyrhizobia in soil (Weaver and Frederick, 1974). Competition between inoculated and native *Rhizobium* strains and response to inoculation was less pronounced in the presence of soil mineral N than under conditions where such N was immobilized and made unavailable (Somasegaran and Bohlool, 1990). Use of massive inoculation rates can overcome competition from indigenous strains (Kapusta and Rouwenhorst, 1973), but such a delivery system is not yet economical and practical.

In many rice-growing areas, legumes are grown after paddy, using residual moisture. In such fields, less than 100 cowpea group rhizobia g^{-1} soil were observed and continuous cultivation of paddy had an adverse effect on *Rhizobium* survival. Under such conditions inoculation with effective strains showed significant responses in chickpea and pigeonpea (Nambiar et al., 1988).

Factors affecting performance of inoculant strains

Crop responses to inoculation with biofertilisers are not as visible as those with fertilizer N. Being biological agents, these are subjected to a range of hostile environments and their survival and efficiency is governed by several factors. Generally, there is a decline in the rhizobial population on seeds but conventional wisdom is that multiplication should occur as the rhizosphere forms, so that accelerated germination can also assist in ensuring an adequate population. The seed coat of a dicot is often carried on the top of the cotyledons into

Table 11. Persistence of inoculum strain NC92 over two seasons on groundnut

Season		% nodules formed on groundnut plants	
1st	2nd	72 days after sowing	116 days after sowing
Uninoculated	Uninoculated	9 (5)[a]	11 (8)
Uninoculated	Inoculated	31 (27)	27 (25)
Inoculated	Uninoculated	28 (25)	42 (32)
Inoculated	Inoculated	39 (41)	75 (54)
SE		± 2.5	± 5.4

[a] Data analysed after arsine transformation: original means in parenthesis.
Source: Nambiar (1985).

the open air, so that only a part of the inoculum may be left to multiply within the rhizosphere. In the case of crops grown on residual moisture, such as chickpea, the inoculated rhizobia cannot move downwards with the growing root from the top soil where inoculated, resulting in poor nodulation. Secondly, deep sowing results in a good crop stand but affects nodulation adversely (Nambiar et al., 1988).

Carrier-based inoculants are usually coated on seeds for the introduction of bacterial strains into the soil. However, alternative inoculation methods are necessary where seed treatment with fungicides and insecticides is needed or where seed of crops such as groundnut and soybean can be damaged when inoculated with an adhesive. In addition, use of superphosphate as the P source can be harmful for *Rhizobium* because of contact with the acidic fertilizer. Often the soils themselves are acidic and lime coating of seed has been a popular measure for additional protection. The normal carrier-based inocula can be successfully applied separately from the seed (Bonnier, 1960; Burton, 1982). While all methods of inoculation were successful under favorable conditions, "liquid" and "solid" methods were superior to seed inoculation under adverse conditions (Brockwell et al., 1980). Increased groundnut yields were obtained when inoculation was done by applying a slurry of peat-based inoculum in the seed furrow (Table 12). At ICRISAT, a bullock-drawn seed drill commonly used by farmers has been modified for simultaneous *Rhizobium* application in the seed furrow (Nambiar, 1985).

Soil properties can also affect the survival of inoculated rhizobia. For example, out of 11 locations tested for response of groundnut cv Robut 33–1, inoculation with strain NC 92 failed to increase yields at two locations, namely Tirupathi and Kadiri, India (AICORPO, 1983). Subsequent analysis of soil samples from Tirupathi revealed a high (150 mg kg^{-1}) available manganese content (Nambiar, 1985). Manganese and aluminum can be toxic to symbiotic N_2 fixation even if they are not at a level high enough to affect plant growth (Franco, 1977). Soil acidity and alkalinity can also pose problems for symbiotic N_2 fixation. For such problem areas, specific strains with the ability to overcome such adverse conditions need to be selected as inoculants. Significant differences were observed among pigeonpea rhizobial strains for their ability to nodule and fix N_2 under saline conditions (Subba Rao et al., 1990).

Yield response to inoculation

The field performance of inoculation is variable. Not many on-farm data are available on the impact of inoculation on grain yields. In 12 trials with chickpea, inoculated plots gave on an average 116 kg ha^{-1} more grain as compared to non-inoculated plots. In another set of field demonstrations, inoculation resulted in grain yield increase in the range of 112–227 kg ha^{-1} (Chandra and Ali, 1986). The results of 1500 demonstrations on farmers' fields with pigeonpea conducted in Gulbarga district of Karnataka State in India showed 100% increase in yield (1035 vs. 516 kg ha^{-1}) due to balanced use of DAP and *Rhizobium* inoculation (Chinmulgund and Hegde, 1987). On research stations in 16 trials inoculation of chickpea with *Rhizobium* increased grain yield by 342 kg ha^{-1} (range 30–610). Significant improvement in chickpea grain yield was reported from 7 out of 16 locations (Subba Rao, 1976) and 6 out of 12 locations (Subba Rao and Tilak, 1977), predominantly in central and northern India with yields varying from −14 to 30% compared to the control plots yield. Increase in grain yield of the pigeonpea inoculated with effective *Rhizobium* ranged from 19 to 68% over non-inoculated controls (Nambiar et al., 1988). In groundnut, inoculation responses varied from

Table 12. Effect of fungicide and method of inoculation on nodulation by strain NC92[a] on groundnut

Treatment	Method of inoculation and % nodules formed by strain NC 92[a]		
	Liquid	Seed	Uninoculated
Untreated	30 (27)	22 (20)	4 (2)
Captan	28 (23)	7 (4)	3 (1)
Thiram	25 (18)	6 (4)	7 (2)
Dithane	19 (10)	14 (9)	7 (3)
Bavistin	24 (16)	14 98)	10 (3)
Mean	25 (19)	13 99)	6 (2)

SE mean for comparing inoculation means within a fungicide treatment is ± 5.5.

[a] Nodules typed by ELISA 60 days after sowing. Data analysed after arcsine transformation: original means in parenthesis.
Source: Nambiar (1985).

decreased yields to significantly increased yields over non-inoculated controls (Kulkarni and Joshi, 1988; Nambiar et al., 1988; Subba Rao, 1976). Over 228 inoculation trials were conducted under the International Network of Legumes Inoculation Trials (INLIT) by cooperating scientists in 28 countries over the years. In approximately 52% of the cases, inoculation resulted in significant yield increases (Davis et al., 1985). In summary, yield responses to inoculation were site specific, depending on location, species, fertility, and other factors.

Sometimes, legumes yields are not increased by inoculation but N concentration in grains or plant parts is increased over N concentration in non-inoculated control plants. In cases where both types of responses are not observed, it might simply result in a saving of soil N which might be useful for the succeeding crop.

Conclusion and future areas of research

Biological nitrogen fixation plays an important role in sustaining productivity of the soils in the SAT. Legumes fix substantial amounts of nitrogen (Table 1) through the BNF process and play an important role in the N cycle. However, mere inclusion of legumes in the cropping systems in the SAT will not ensure N contributions to the system through BNF. The important issue is how best we can exploit BNF technology for developing sustainable cropping systems in the SAT?

Until now considerable effort in BNF research has gone in the area of selection of efficient bacterial strains for using as inoculants. For realizing the maximum benefits from BNF we must take a holistic approach (Bantilan et al., 1994; Wani et al., 1994a). There is need to understand the BNF system which includes host, bacterium and environment and ensure that all the partners involved work in harmony to deliver maximum benefit. There is a need to accurately quantify N_2 fixation by legumes in a system after taking into account the N_2 fixed in the roots and fallen plant parts. Such information will help us to identify the systems which really maintain or improve the soil N status. Host controlled factors play an important role in regulating BNF but have not received its due share by researchers. We need to identify type of legume and also genotype of a given legume which yields more and also derive larger part of its N requirement from fixation in a particular cropping system. For example, we need to identify crops and genotypes of legumes which can fix more N_2 under sole cropping and intercropping situations without being affected by high mineral N contents in soil. There is a need to identify host genotypes which can fix well under adverse soil conditions like soil acidity, Al and Mn toxicity, alkalinity, water logging, etc.

At ICRISAT nonnodulating lines of chickpea, pigeonpea and groundnut have been developed from the existing cultivars and/or segregating populations. Natural occurrence of nonnodulating plants ranged from 120 to 490 per million plants and efforts are required to see that occurrence of such plants do not increase. Most of the breeding and testing work is done at the research stations where mineral N contents are far higher than observed on the farmers fields. There is every likelyhood that low- or nonnodulating plants may not be identified as they will grow normally using soil N. To avoid this, appropriate checks during breeding and testing for discarding low-nodulating plants must be built in the breeding programs.

Along with the selection of appropriate host plant and genotypes there is need to provide optimum management practices to ensure maximum contribution from the BNF. Through appropriate management practices soil N should be manipulated in intercropped situations for example appropriate form of fertilizer like slow releasing formulations, organic N and suitable method of application for example placement between cereal rows rather than broadcasting and mixing in the soil must be worked out. Appropriate amendments with nutrients other than N which might limit the plant growth and BNF should be done.

Suitable land management practices which can improve water storage capacity of soils or which can drain excess water away from the plant depending on the situation need to be used to harness maximum benefits from BNF. Efforts for selection of efficient strains of bacteria to use as inoculants and identification of specific host-bacteria combinations must go on. Situations which need inoculation should be identified and efforts for success to inoculation in such areas must be concentrated.

For increasing crop yields through biofertilizers, the following strategy is suggested. Most important constraints to effective exploitation of BNF technology in the SAT are:

- the quality of the inoculants
- lack of knowledge about inoculation technology for the extension personnel and the farmers.
- effective inoculant delivery system
- formulation of the policy to exploit BNF successfully.

The history of inoculant manufacture and of many strain collections is full of examples of organisms which look like rhizobia but are not! Contaminated cultures contribute to the problems which placed the inoculant industry of Australia in peril in the early 1950s. Many inocula of poor quality were sold and the losses at sowings of new legumes into poor soils were enormous (Thompson, 1982). This was repeated in India during the late 1970s and early 1980s. Several rhizobial inocula from the Indian manufacturers were examined at ICRISAT (Thompson, 1982) for their infectivity tests. Irrespective of private or public institution origin, the majority failed to pass the published standards (ISI, 1977). There must be strict quality control mandatory on all biofertilizer producers irrespective of their status as private/public or government organization.

For success of biofertilizers in the SAT concerted efforts right from production, demonstration to distribution will be required. The next step is convincing and educating the farmers regarding the benefits of these inoculants. The pricing of the biofertilizers must be controlled if private agencies are involved, otherwise if farmers don't see the significant effects in term of economic yields, they may not be interested in using the biofertilizers. There is a need to demonstrate the benefits from BNF technology in terms of maintenance or improvement of soil fertility through long-term experiments. At this stage the policy issue arises that biofertilizers should be used or considered as an insurance for harnessing BNF to its maximum potential taking systems approach. As discussed earlier the nonnodulating or LN plants look similar in appearance to well nodulated plants in chickpea but this is at the cost of soil or fertilizer N. We must take the view that in the end we may derive benefit in terms of maintaining or improving the productivity of our soils. We should not be disappointed by not seeing the direct benefits in terms of increased legume yields in some cases. A holistic approach to improve production of legumes is needed and we must ensure that all the constraints for good plant growth other than N nutrition are alleviated and suitable management practices are provided for better performance of BNF technology.

Acknowledgements

We thank Mr P N Murthy for wordprocessing this document, Mrs S Rajeswari for freelance graphics and two anonymous reviewers for their critical comments on the manuscript.

References

AICORPO 1983 All India Coordinated Research Project on Oilseeds.Annual Progress Report, Kharif 1982, 2, 241–248. Rajenderanagar, Hyderabad, India.

Ahlawat I P S, Singh A and Saraf C S 1981 Effects of winter legumes on the nitrogen economy and productivity of succeeding cereals. Exp. Agric. 17, 57–62.

Alexander M 1982 Research to enhance nitrogen fixation misplaced emphasis? *In* Priorities in Biotechnology Research for International Development. pp 209–229. National Academy Press, Washington DC.

Arunachalam V, Pungle G D, Dutta M, Nambiar P T C and Dart P J 1984 Efficiency of nitrogenase activity and nodule mass in predicting the relative performance of genotypes assessed by a number of characters in groundnut (*Arachis hypogaea*). Exp. Agric. 20, 303–309.

Bandyopadhyay S K and De R 1986 Nitrogen relationships and residual effects of intercropping sorghum with legumes. J.Agric. Sci., (Camb) 107, 629–632.

Bantilan M C S, Johansen C and McDonald D 1994 Expectations fromBNF research – Research administrators' point of view. *In* Linking Biological Nitrogen Fixation Research in Asia: Report of a Meeting of the Asia Working Group on Biological Nitrogen Fixation in Legumes. Eds. O P Rupela, J V D K Kumar Rao, C Johansen and S P Wani. pp 61–90. International Crops Research Institute for the Semi-Arid Tropics, Patancheru, Andhra Pradesh, India.

Betts J H and Herridge D F 1987 Isolation of soybean lines capable of nodulation and nitrogen fixation under high levels of nitrate supply. Crop Sci. 27, 1156–1161.

Blevins R L, Herbek J H and Fyre W W 1990 Legume cover crops as a nitrogen source for no-till corn and grain sorghum. Agron. J. 82, 769.

Bonnier C 1960 Symbiose Rhizobium-leguminouses: Aspects particuliers aux regions tropicales. Ann. Inst. Pasteur, Paris 98, 537–556.

Brockewell J, Gault R R, Chase D L, Hely F W, Zorin M and Corbin E J 1980 An appraisal of practical alternatives to legumes seed inoculation: Field experiments on seed bed inoculationwith solid and liquid inoculants. Aust. J. Agric. Res. 31, 47–60.

Buresh R J and De Datta S K 1991 Nitrogen dynamics and management in rice-legume cropping systems. Adv. Agron. 45, 1–59.

Burns R C and Hardy R W F 1975 Nitrogen fixation in bacteria and higher plants. Springer-Verlag, Berlin. 189p.

Burton J C 1982 Modern concepts in legume inoculation. *In* Biological Nitrogen Fixation Technology for Tropical Agriculture. Eds. P H Grahm and S C Harris. pp 105–114. CIAT, Cali Colombia.

Buttery B R and Dirks V A 1987 The effects of soybean cultivar, *Rhizobium* strain and nitrate on plant growth, nodule mass, and acetylene reduction rate. Plant and Soil 98, 285–293.

Buttery B R and Gibson A H 1990 The effect of nitrate on the time course of nitrogen fixation and growth in *Pisum sativum* and *Vicia faba*. Plant and Soil 127,143–146.

Buttery B R, Park S J, Finday W I and Dhanvantari B N 1988 Effects of fumigation and fertilizer on growth, yield, chemical composition and mycorrhizae in white bean and soybean. Can. J. Plant Sci. 68, 677–686.

Buttery B R, Park S J and Dhanvantari B N 1990 Effect of combined nitrogen, *Rhizobium* strain and substrate on a supernodulating mutant of *Phaseolus vulgaris* L. Can.J. Plant Sci. 70, 955–963.

Carroll B J, McNeil D L and Gresshoff P M 1985 A supernodulation and nitrate-tolerant symbiotic (nts) soybean mutant. Plant Physiol. 78, 34–40.

Cassman K G, Whitney A S and Fox R L 1981 Phosphorus requirements of soybean and cowpea as affected by mode of N nutrition. Agron. J. 73, 17–22.

Chalk P M 1991 The contribution of associative and symbiotic nitrogen fixation to the nitrogen nutrition of non-legumes. Plant and Soil 132, 29–39.

Chandra S and Ali M 1986 Recent achievements in pulses production. Tech. Bull. No. 1, Directorate of Pulse Research, ICAR, Kanpur. 44p.

Chinmulgund V K and Hegde S V 1987 Increasing redgram yields through diammonium phosphate (application) and *Rhizobium* (inoculation) treatment. Fert. Mkt. News, 18, 9–13.

Cho M-J and Harper J E 1991 Effect of localized nitrate application on isoflavonoid concentration and nodulation in split-root systems of wild-type and nodulation-mutant soybean plants. Plant Physiol. 95, 1106–1112.

Cook R J 1988 Biological control and holistic plant-health care in agriculture. Am. J. Altern. Agric. 3, 51–62.

Dakora F D, Aboyinga F A, Mahama Y and Apaseku J 1987 Assessment of N_2 fixation in groundnut (*Arachis hypogaea* L.) and cowpea (*Vigna unguicalata* L. Walp) and their relative N contribution to a succeeding maize crop in northern Ghana. MIRCEN J. 4, 389–399.

Danso S K A and Papastylianou I 1992 Evaluation of the nitrogen contribution of legumes to subsequent cereals. J. Agric. Sci. 119, 13–18.

Davis R J, Cady F B, Wood C L and Chan C P Y 1985 Design and analysis of an International Experimental Network: Legume inoculation trials in the NifTAL project. INLIT experiences. Research Series 042, College of Tropical Agriculture and Human Resources, University of Hawaii, Honolulu.

De R and Gautam R C 1987 Management practices to increase and stabilize pearl millet production in India. *In* Proceedings of the International Pearl Millet Workshop. Ed. J R Witcombe. pp 247–253. ICRISAT, Patancheru, India.

De R, Yogeswara Y and Ali W 1983 Grain and fodder legumes as preceding crops affecting the yield and N economy of rice. J. Agric. Sci., Camb. 101, 463–466.

Doughton J A and MacKenzie J 1984 Comparative effects of black and green gram (mung beans) and grain sorghum on soil mineral nitrogen and subsequent grain sorghum yields on the eastern Darling Downs. Aust. J. Exp. Agric. Anim. Husb. 24, 244–249.

Doughton J A, Vallis I and Saffina P G 1993 Nitrogen fixation in chickpea. I. Influence of prior cropping or fallow nitrogenfertilizer and tillage. Aust. J. Agric. Res. 44,1403–1413.

Francis C A 1986 *In* Multiple Cropping Systems. Macmillan, New York. 383 p

Francis C A and Clegg M D 1990 Crop rotations in sustainable production systems. *In* Sustainable Agricultural Systems. Ed. C A Edwards. p 107. Soil and Water Conservation Society, Ankeny, 10.

Francis C A, Harwood R R and Parr J F 1986 The potential for regenerative agriculture in the developing world. Am. J.Altern. Agric. 1, 65.

Franco C A 1977 Contribution of the legume-rhizobium symbiosis to the ecosystem and food production. *In* Exploiting the Legume-*Rhizobium* Symposium in Tropical Agriculture. Eds. J M Vincent et al. pp 237–252. University of Hawaii, USA.

Fujita K, Ofosu-Budu K G and Ogata S 1992 Biological nitrogen fixation in mixed legume-cereal cropping systems. Plant and Soil, 141, 155–175.

Fyson A and Oaks A 1990 Growth promiton of maize by legume soils. Plant and Soil 122, 259–266.

George T, Ladha J K, Buresh R J and Garrily D P 1992 Managing native and legume-fixed nitrogen in lowland rice-based cropping systems. *In* Biological Nitrogen Fixation for Sustainable Agriculture. Eds. J K Ladha, T George and B B Bohlool. pp 69–92. Kluwer Academic Publishers, Dordrecht.

Gibson A H and Harper J E 1985 Nitrate effect on nodulation of soybean by *Bradyrhizobium japonicum*. Crop Sci. 25,497–501.

Giller K E, Nambiar P T C, Srinivasa Rao B, Dart P J and Day J M 1987 A comparison of nitrogen fixation in genotypes of groundnut (*Arachis hypogaea* L.) using ^{15}N-isotope dilution. Biol. Fertil. Soils 5, 23–25.

Graham P H and Rosas J C 1977 Growth and development of intermediate bush and climbing cultivars of *Phaseolus vulgaris* L. inoculated with *Rhizobium* J. Agric. Sci., Camb. 88, 503–508.

Graham P H and Rosas J C 1978 Plant and nodule development and nitrogen fixation in climbing cultivars of *Phaseolus vulgaris* L. grown in monoculture or associated with *Zea mays* L. J. Agric. Sci., Camb. 90, 311–317.

Hansen A P, Peoples M B, Gresshoff P M, Atkins C A, Pate J S and Carroll B J 1989 Symbiotic performance of super-nodulating soybean (*Glycine max* (L.) Merrill) mutants during development on different nitrogen regimes. J. Exp. Bot. 40,715–724.

Hardarson G, Zapata F and Danso S K A 1984 Effect of plant genotype and nitrogen fertilizer on symbiotic fixation by soybean cultivars. Plant and Soil 2, 397–405.

Hardarson G, Golbs M and Danso S K A 1989 Nitrogen fixation in soybean (*Glycine max* L. Merrill) as affected by nodulation patterns. Soil Biol. Biochem. 21, 783–787.

Herridge D F, Bergersen F J and Peoples M B 1990 Measurement of nitrogen fixation by soybean in the field using the ureide and natural abundance methods. Plant Physiol. 93, 708–716.

Herridge D F, Marcellos H, Helton W L, Turner G L and Peoples M B 1994 Chickpea increases soil-N fertility in cereal systems through nitrate sparing and N_2fixation. Soil Biol. Biochem. *(In press).*

Hesterman O B, Russelle M P, Sheaffer C C and Heichel G H 1987 Nitrogen utilization from fertilizer and legume residues in legume-corn rotations. Agron. J. 79, 726-731.

IARI 1980 The energy problems of agriculture and development of renewable resources. *In* Scientific Basis of Indian Agriculture: Seventyfive Years of Service. pp 189–209. Indian Agricultural Research, New Delhi, India.

ICRISAT 1981 ICRISAT Annual Report for 1979–80. pp 91–94. International Crops Research Institute for the Semi-Arid Tropics, Patancheru, AP., India.

ISI 1977 Specifications for *Rhizobium* Inoculants. Indian Standards. 8268p.

Jefing Y, Herridge D F, Peoples M B, and Rerkasem B 1992 Effects of N fertilization on N_2 fixation and N balances of soybean grown after low land rice. Plant and Soil 147,235–242.

Johanson T B, Turpin F T, Schreiber M M and Griffith D R 1984 Effects of crop rotation, tillage, and weed managementsystems on black cutworm (Lepidoptera: Noctuidae) infestations in corn. J. Econ. Entomol. 77, 919.

Joshi P K, Bhat D M and Kulkarni J H 1987 Groundnut root nodulation as affected by micronutriens application and *Rhizobium* inoculation. Int. J. Trop. Agric. 5,199–202.

Kabi M C and Poi S C 1988 Improvement of pulse cultivation in West Bengal through effective use of *Rhizobium*. *In* Biofertilizers Potentialities and Problems. Eds. S P Sen and P Palit. pp 35–44. Naya Prakash, Calcutta.

Kapusta G and Rouwenhorst D L 1973 Influence of inoculum size on *Rhizobium japonium* serogroup distribution frequency in soybean model. Agron. J. 65, 916–919.

Keyser H H and Li F 1992 Potenial for increasing biological nitrogen fixation in soybean. Plant and Soil 141,119–135.

Khurana A L and Dudeja S S 1981 Field populations of rhizobia and response to inoculation, molybdenum and nitrogen fertilizer in pigeonpea. *In* Proceedings of the International Workshop on Pigeonpea. Vol. 2, pp 381–386. 15–19 December 1980, ICRISAT Center, India. Patancheru, AP, ICRISAT, India.

Kucey R M N, Chaiwanakupt P, Arayangkool T, Snitwongse P, Siripaibool C, Wadisirisuk P and Boonkerd N 1988 Nitrogenfixation (^{15}N dilution) with soybeans under Thai field conditions. II. Effect of herbicides and water application schedule. Plant and Soil 108, 87–92.

Kulkarni J H, and Joshi P K 1988 *Rhizobium* inoculation studies in groundnut: Problems and prospects. *In* Biofertilizers Potentialities and Problems. Eds. S P Sen and P Palit. pp 51–56. Naya Prakash, Calcutta.

Kumar Rao J V D K and Dart P J 1981 Effect of different plant growth media on nodulation growth and nutrient uptake of pigeonpea. *In* Proceedings of the International Workshop on Pigeonpeas. Vol 2, pp 403–408. 15–19 December 1980, ICRISAT Center, India. Patancheru, AP, ICRISAT, India.

Kumar Rao J V D K and Dart P J 1987 Nodulation, nitrogen fixation and nitrogen uptake in pigeonpea (*Cajanus cajan* (L.)Millsp.) of different maturity groups. Plant and Soil 99, 255–266.

Kumar Rao J V D K and Sithanantham S 1989 Impact of nodule damage by *Rivellia angulata* on N_2-fixation, growth and yield of pigeonpea (*Cajanus cajan* (L) Millsp.) grown in a Vertisol. Biol. Fertil. Soils 7, 95–100.

Kumar Rao J V D K, Dart P J and Sastry P V S S 1983 Residual effect of pigeonpea (*Cajanus cajan*) on yield and nitrogen response of maize. Exp. Agric. 19, 131–141.

Ladd J N, Oades J M and Amato M 1981 Distribution and recovery of nitrogen from legume residues decomposing in soils sown to wheat in the field. Soil Biol. Biochem. 13, 251–256.

Ladd J N, Amato M, Jackson R B and Butler J H A 1983 Utilization by wheat crops of nitrogen from legume residues decomposing in soils in the field. Soil Biol. Biochem. 15(3), 231–238.

Ladha J K, Tirol-Padre A, Punzalan G C, Garcia M and Watanabe I 1989 Effect of inorganic N and organic fertilizers on nitrogen-fixing (acetylene-reducing) activity associated with wetland rice-plants. *In* Nitrogen Fixation with Non-Legumes. Eds. F A Skinner, R M Boddey and I Fendrik. pp 263–272. Kluwer Academic Publishers, Dordrecht.

Latif M A, Mehuys G R, Mackenzie A F, Ali I and Faris M A 1992 Effects of legumes on soil physical quality in a maize crop. Plant and Soil 140, 15–23.

Lee K K, Wani, S P, and Anders M M 1993 Nutrient and waterbalance study on an Alfisol. *In* Resource Management Program Quarterly Report, July-September, 1993. pp 41–42. International Crops Research Institute for the Semi-Arid Tropics, Patanheru, Andhra Pradesh, India.

MacCol D 1989 Studies on maize (*Zea mays*) at Bunda, Malawi. II. Yield in short rotations with legumes. Exp. Agric. 25, 367–374.

Nambiar P T C 1985 Responses of groundnut (*Arachis hypogaea* L.) to *Rhizobium* inoculation in the field: problems and prospects. MIRCEN J. 1, 293–309.

Nambiar P T C 1990 Nitrogen nutrition of groundnut in Alfisols.Information Bulletin No. 30. pp 1–30. International Crops Research Institute for the Semi-Arid Tropics, Patancheru, AP, India.

Nambiar P T C, Dart P J, Srinivasa Rao B and Ravishankar H N 1982 Response of groundnut (*Arachis hypogaea*) toinoculation. *In* Biological Nitrogen Fixation Technology for Tropical Agriculture. Presented at a Workshop, 9–13 March 1981, Cali, Columbia. Eds. P H Graham and S C Harris. pp 241–248. Centro Internacional de Agricultura Tropical, Cali, Colombia.

Nambiar P T C, Rao M R, Reddy M S Floyd C, Dart P J and Willey R W 1983 Effect of intercropping on nodulation and N_2fixation by groundnut. Exp. Agric. 19, 79–86.

Nambiar P T C, Srinivasa Rao B, and Anjaiah V 1984 Studies on competition, persistence and methods of application of a groundnut *Rhizobium* strain NC92. Peanut Sci. 11,83–87.

Nambiar P T C and Anjaiah V 1989a Effect of manganese toxicity on growth and N_2 fixation in groundnut, *Arachis hypogaea*. Ann. Appl. Biol. 115, 361–366.

Nambiar P T C and Anjaiah V 1989b Competition among strains of *Bradyrhizobium* and vesicular-arbuscular mycorrhizae for groundnut (*Arachis hypogaea* L.) root infection and their effect on plant growth yield. Biol. Fertil. Soils 8, 311–318.

Nambiar P T C, Rego T J and Srinivasa Rao B 1986 Comparison of the requirements and utilization of nitrogen by genotypes of sorghum (*Sorghum bicolor*) and nodulating and non-nodulating groundnut (*Arachis hypogaea* L.). Field Crops Res. 15, 165–179.

Nambiar P T C, Rupela O P and Kumar Rao J V D K 1988 Nodulation and nitrogen fixation in groundnut (*Arachis hypogaea* L.), chickpea (*Cicer arietinum*), and pigeonpea (*Cajanus cajan* L Millsp). *In* Biological Nitrogen Fixation: Recent Developments. Ed. N S Subba Rao. pp 53–70. Oxford and IBH Publishers, New Delhi.

Nambiar P T C, Ma S W and Aiyer V N 1990 Limiting an insect infestation of nitrogen-fixing root nodules of the pigeonpea (*Cajanus cajan*) by engineering the expression of an entomocidal gene in its root nodules. Appl. Environ. Microbiol. 56, 2866–2869.

Namdeo S L and Gupta S C 1992 Response of pulses to microbial inoculants – A review of the work done at Sehore (MP). *In* National Seminar on Organic Farming. Eds. M M Rai and L N Verma. pp 150–161. Jawaharlal Nehru Krishi Vishwa Vidyalaya, Jabalpur, MP, India.

Nigam S N, Dwivedi S L, Nambiar P T C, Gibbons R W, and Dart P J 1985 Combining ability analysis of N_2-fixation and related traits in peanut. Peanut Sci. 12, 55–57.

Norhayati M, Mohd Noor S, Chong K, Faizah A W, Herridge D F, Peoples M B and Bergersen F J 1988 Adaptation of methods for evaluating N, fixation in food legumes and legume cover crops. Plant and Soil 108, 143–150.

O'Hara G W, Dilworth M J, Boonkerd N and Parkpian P 1988 Iron-deficiency specifically limits nodule development in peanut inoculated with *Bradyrhizobium* sp. New Phytol.108, 51–57.

Ofori F and Stern W R 1987 Cereal-legume intercropping systems. Adv. Agron. 41, 41–90.

Ofori F, Pate J S and Stern W R 1987 Evaluation of N_2 fixation and nitrogen economy of a maize/cowpea intercrop system using ^{15}N dilution methods. Plant and Soil 102, 149–160.

Ogata S, Fujita K, Matsumoto K and Saneoka H 1986 Studies on mixed croping of grasses and legumes. 1. Dry matter production and behavior of fixed N in the system of mixed cropping sorghum, soybean and siratro. J. Jpn. Grassl. Sci. 32, 36–43 (*In Japanese with English summary*).

Olmstad L B 1947 The effect of long-time cropping systems and tillage practices upon soil aggregation at Hays, Kansas. Soil Sci. Soc. Am. Proc. 11, 89.

Prasad R, Sharma S N, Singh S and Prasad M 1990 Nitrogen management. *In* Soil Fertility and Fertilizer Use, Vol. IV. Nutrient Management and Supply System for Sustaining Agriculture in 1990s. Eds. Virendra Kumar, G C Shrotriya and S V Kaore. pp 41–51. PR Department, Marketing Division, IFFCO, New Delhi.

Peoples M B and Herridge D F 1990 Nitrogen fixation by legumes in tropical and sub-tropical agriculture. Adv. Agron. 44, 155–223.

Peoples M and Crasswell E T 1992 Biological nitrogen fixation: Investments, expectations and actual contributions to agriculture. Plant and Soil 141, 13–39.

Peoples M B, Sudin M N and Herridge D F 1987 Translocation of nitrogenous compounds in symbiotic and nitrate-fed amide exporting legumes. J. Exp. Bot. 38, 567–579.

Power J F 1990 Legumes and crop rotations. *In* Sustainable Agriculture in Temperate Zones. Eds. C A Francis, C B Flora and L D King. p 178. Wiley, New York.

Raj A S 1987 Cobalt nutrition of pigeonpea and peanut in relation to growth and yield. J. Plant Nutr. 10, 2137–2145.

Rao C C S and Singh K D 1991 Effect of residual greengram(*Vigna radiata* L. Wilczek) after maize-wheat sequence on soil available nitrogen, phosphorus and potassium. Legume Res. 14,125–130.

Rawsthorne S, Hadley P, Summerfield R J and Roberts E H 1985 Effect of supplemental nitrate and thermal regime on the nitrogen nutrition of chickpea (*Cicer arietinum* L.) II. Symbiotic development and nitrogen assimilation. Plant and Soil 83, 279–283.

Rego T J and Burford J R 1992 Sustaining crop productivity on rainfed Vertisols through grain legumes. Agron. Abstr. Annual Meetings, American Society of Agronomy, Crop Science Society of America, Soil Science Society of America, Clay Minerals Society, Minneapolis, Minnesota, USA. Nov 1–6, 1992. 289p.

Rennie R J, Dubetz S, Bole J B and Muendel H H 1982 Dinitrogen fixation measured by ^{15}N isotope dilution in two Canadian soybean cultivars. Agron. J. 74, 725–730.

Rennie R J and Kemp G A 1982 N_2-fixation in field beans quantified by ^{15}N isotope dilution. II. Effect of cultivars and beans. Agron. J. 75, 645–649.

Ries S K, Sweeley V W C C and Leavitt R R 1977 Tricontanol: A new naturally occurring plant growth regulator. Science 195,1339–1341.

Roy Sharma R P and Singh Ambika 1969 Residual effect of legumes in winter legume-maize rotation. Indian J. Agric. Sci. 39, 530–539.

Rupela O P 1992 Natural occurrence and salient characters of nonnodulating chickpea plants. Crop Sci. 32, 349–352.

Rupela O P 1994 Screening for intracultivaral variability for nodulation of chickpea and pigeonpea. *In* Linking Biological Nitrogen Fixation Research in Asia: Report of a meeting of the Asia Working Group on Biological Nitrogen Fixation in Legumes. Eds. O P Rupela, J V D K Kumar Rao, S P Wani and C Johansen. pp 75–83. International Crops Research Institute for the Semi-Arid Tropics, Patancheru, Andhra Pradesh, India.

Rupela O P and Johansen C 1995 Identificaton of nonnodulating, and low and high nodulating plants in pigeonpea. Soil Biol. Biochem. (Special Volume) *(In press)*.

Rupela O P and Saxena M C 1987 Nodulation and nitrogen fixation in chickpea. *In* The Chickpea. Eds. M C Saxena and K B Singh. pp 191–206. CAB International, Wallingford, Oxfordshire, UK.

Rupela O P, Sudarshana M R and Thompson J A 1985 Sowing depth can determine nodulation pattern in chickpea. *In* Nitrogen Fixation Research Progress. Eds. H J Evans, P J Bottomley and W E Newton. p 320. Martinus Nijhoff Publishers, Dordrecht.

Rupela O P, Toomsan B, Mittal S, Dart P J and Thompson J A 1987 Chickpea *Rhizobium* populations: Survey of influence of season, soil depth and cropping pattern. Soil Biol. Biochem. 19, 247–252.

Russelle M P, Hesterman O B, Sheaffer C C and Heichel G H 1987 Estimating nitrogen and rotation effects in legume-cornrotations. *In* The Role of Legumes in Conservation Tillage Systems. Ed. T F Power. p 41. Soil Conservation Society, Washington, DC.

Sawhney V, Amarjit Sheoran I S and Singh R 1989 Effect of short term nitrate application on nodule metabolism in chickpea *Cicer arietinum* L. Indian J. Exp. Biol. 27, 220–223.

Senaratne R and Hardarson G 1988 Estimation of residual N effect of fababean and pea on two succeeding cereals using ^{15}N methodology. Plant and Soil 110, 81–91.

Sithanantham S, Kumar Rao J V D K, Reed W and Dart P J 1981 Studies on nodule damage in pigeonpea. *In* Proceedingsof the International Workshop on Pigeonpeas. Vol. 2, pp 409–415. ICRISAT Center, India. ICRISAT, Patancheru, AP, India.

Somasegaran P and Bohlool B B 1990 Single-strain versus multi strain inoculation. Effect of soil mineral N availability on rhizobial strain effectiveness and competition for nodulation of chickpea, soybean and dry bean. Appl. Environ. Microbiol. 56, 3298–3303.

Ssali H and Keya S O 1984 Nitrogen level and cultivar effects on nodulation, dinitrogen fixation and yield of grain legumes. I. Cowpea cultivars. East. Afr. Agric. J. 45,247–254.

Ssali H and Keya S O 1986 The effects of phosphorus and nitrogen fertilizer level on nodulation, growth and dinitrogen fixation of three bean cultivars. Trop. Agric. (Trinidad) 63, 105–109.

Strickling E 1950 The effect of soybeans on volume weight, and water stability of aggregates, soil organic matter content and crop yield. Soil Sci. Soc. Am. Proc. 15, 30.

Subba Rao N S 1976 Field response of legumes in India to inoculation and fertilizer appliations. *In* SymbioticNitrogen Fixation in Plants. Ed. P S Nutman. pp 265–288. Cambridge University Press, London.

Subba Rao N S and Tilak K V B R 1977 Rhizobial cultures – their role in pulse production. *In* Souvenier Bulletin, Directorate of Pulse Development. Ed. N S Subba Rao. pp 1–19. Oxford and IBH, New Delhi.

Subba Rao G V, Johansen C, Kumar Rao J V D K and Jana M 1990 Response of the pigeonpea-*Rhizobium* symbiosis to salinity stress: Variation among *Rhizobium* strains in symbiotic activity. Biol. Fertil. Soils 9,49–53.

TAC CGIAR 1988 Sustainable agricultural production: Implications for international agricultural research. CGIAR (Consultative-Group on International Agricultural Research) meeting, Berlin.

Tandon H L S 1992 Fertilisers and their integration with organic and biofertilisers. *In* Fertilisers, Organic Manures, Recyclable Wastes and Biofertilisers. Ed. H L S Tandon. pp 12–36. Fertiliser Development and ConsultationOrganisation, 204–204A Bhanot Corner, New Delhi.

Tauro P and Khurana A L 1986 Problems and prospects of growth, extension and promotion of biofertilisers. *In* Proceedings of the FAI Seminar. pp 1–4. Growth and Modernisation of Fertiliser Industry, PS111, Fertiliser Association of India, India.

Thies J E, Singleton P W and Bohlool B B 1991 Modeling symbiotic performance of introduced rhizobia in the field by use of indices of indigenous population size and nitrogen status of the soil. Appl. Environ. Microbiol. 57, 29–37.

Thomas G W, Blevins R L, Philips R E and McMahon M A 1973 Effect of a killed sod mulch on nitrate movement and corn yield. Agron. J. 65, 736–739.

Thompson J A 1982 Inoculants and ecology of *Rhizobium*. *In* Biological Nitrogen Fixation. pp 49–64. Bhabha Atomic Research Center, Trombay, Bombay.

Troll C 1965 Seasonal climates of the earth. *In* World Maps of Climatology. Eds. E Rodenwalt and H Juskatz. p 28. Springer Verlag, Berlin.

Vallis I 1983 Uptake by grass and transfer of soil of nitrogen from ^{15}N-labelled legume materials applied to a Rhodes grass pasture. Aust. J. Agric. Res. 34, 367–376.

Venkateswarlu B and Katyal J C 1994 Improvement of nitrogenfixation in groundnut by host-plant selection. *In* Linking Biological Nitrogen Fixation Research in Asia: Report of a meeting of the Asia Working Group on Biological Nitrogen Fixation in Legumes. Eds. O P Rupela, J V D K Kumar Rao, S P Wani and C Johanson. 6–8 December 1993. International Crops Research Institute for the Semi-Arid Tropics, Patancheru, Andhra Pradesh, India.

Wacek T J and Brill W J 1976 Simple, rapid assay for screening nitrogen-fixing ability in soybean. Crop Sci. 16, 519–523.

Wahua T A T and Miller D A 1978 Effects of shading on the N_2 fixation, yield and plant composition of field-grown soybeans. Agron. J. 70, 387–392.

Wani S P and Lee K K 1992 Role of biofertilisers in upland crop production. *In* Fertilisers Organic Manures Recycle Wastes and Biofertilisers. Ed. H L S Tandon. pp 91–112. Fertiliser Development and Consultation Organization, 204–204A, Bhanot Corner, New Delhi.

Wani S P, McGill W B and Robertson J A 1991a Soil N dynamics and N yield of barley grown on Breton loam using N from biological fixation or fertilizer. Biol. Fertil. Soils 12, 10–18.

Wani S P, McGill W B and Tewari J P 1991b Mycorrhizal and common root infection, and nutrient accumulation in barley grown on Breton loam using N from biological fixation or fertilizer. Biol. Fertil. Soils 12, 46–54.

Wani S P, Rego T J and Kumar Rao J V D K 1994a Contribution of legumes in cropping systems: A long-term perspective. *In* Linking Biological Nitrogen Fixation Research in Asia: Report of a Meeting of the Asia Working Group on Biological Nitrogen Fixation in Legumes. Eds. O P Rupela, J V D K Kumar Rao, S P Wani and C Johanson. pp 84–90 International Crops Research Institute for the Semi-Arid Tropics, Patancheru, Andhra Pradesh.

Wani S P, McGill W P, Haugen-Kozyra K L and Juma N G 1994b Increased proportion of active soil N in Breton loan undercropping systems with forages and green manures. Can. J. Soil Sci.

Wani S P, McGill W B, Haugen-Kozyra K L, Robertson J A and Thurston J J 1994c Improved soil quality and barley yields with fababeans, manure, forages and crop rotation on a grayluvisol. Can. J. Soil Sci. *(In press)*.

Weaver R W and Frederick L R 1974 Effect of inoculum rate on competitive nodulation of *Glycine max* L. Merrill. II Field studies. Agron. J. 66, 233–236.

Weil R R and Samaranayake A 1991 Effects of winged bean on a following maize crop. Exp. Agric. 27, 329–338.

Westermann D T, Kleinkopf G E, Porter L K and Legett G E 1981 Nitrogen sources for bean seed production. Agron. J. 73,660–664.

Willey R W 1979 Intercropping – its importance and research needs. Part 2: Agronomy and research approaches. Field Crops Abstr. 32, 73–85.

Wu S and Harper J E 1991 Dinitrogen fixation potential and yield of hyper nodulating soybean mutants: A field evaluation. Crop Sci. 31, 1233–1240.

Yoneyama T, Nambiar P T C, Lee K K, Srinivasa Rao B and Williams J H 1990 Nitrogen accumulation in three legumes and two cereals with emphasis on estimation of N_2 fixation in the legumes by the natural ^{15}N abundance technique. Biol. Fertil. Soils 9, 25–30.

Zari K W, Miller J C Jr, Weaver R W and Barnes L W 1978 Intraspecific variability for nitrogen fixation in southern pea (*Vigna uniculata*). J. Am. Soc. Hortic. Sci. 103,806–808.

Plant and Soil **174:** 51–82, 1995.

Enhancing crop legume N_2 fixation through selection and breeding

D.F. Herridge and S.K.A. Danso
NSW Agriculture, Agricultural Research Centre, R.M.B. 944, Tamworth, N.S.W., Australia 2340 and Joint FAO/IAEA Division, P.O. Box 100, A-1400 Vienna, Austria

Key words: breeding, crop legume, heritability, nitrate tolerance, nodulation, N_2 fixation, rhizobia, yield

Table of contents

Abstract
Introduction
Operational framework for selection and breeding of legumes for enhanced N_2 fixation
Strategies for selection and breeding of legumes for enhanced N_2 fixation
Host × strain specificity
Measuring N_2 fixation
    N yield
    N difference
    ^{15}N methods
    Xylem N solutes
    Acetylene reduction assay (ARA)
Assessing heritability and repeatability of N_2 fixation
Research to select and breed for enhanced N_2 fixation in legumes
    Common bean
        Soybean
        Studies to define genetic variation in N_2 fixation
        Improvement / breeding programs
    Soybean
        Studies to define genetic variation in N_2 fixation
        Improvement / breeding programs
        Controlled nodulation — selection and breeding
Conclusions
References

Abstract

Legume N_2 fixation is variable, but nonetheless is a valuable process in world agriculture. There is great potential to increase the contribution by the crop legumes to the world's supply of soil N. This will be achieved by (i) increasing the area of legumes sown by farmers; (ii) improved management of the crops in order that the major determinants of productivity, e.g. land area, water availability, are converted to harvested product with maximum efficiency; and (iii) genetic modification of the commonly-grown species to ensure high dependence of the legume crop on N_2 fixation at all levels of productivity. Currently-used methods for measuring N_2 fixation and for assessing heritability and repeatability of N_2 fixation in breeding and selection programs are reviewed. Results from research programs to define genetic variation in N_2 fixation and to enhance N_2 fixation through selection and breeding are presented with particular emphasis on common bean (*Phaseolus vulgaris*) and soybean (*Glycine max*).

Introduction

The demand for nitrogen (N) in world agriculture is increasing at a rate approximately equivalent to the rate of increase in world population, i.e. about 2% p.a. (FAO, 1992a). However, the amount actually available and consumed does not increase at this rate because of the continuing rundown of the N-supplying capacity of agricultural soils, the lack of purchasing power of impoverished communities for production commodities such as nitrogenous fertilizers and the knowledge gaps between researchers and farmers in N management of crops and soils for maximum production efficiency.

In 1992, 905 mill. ha were used globally to produce 2,228 mill. t grains and oilseeds (Table 1). We calculate that about 80 mill. t N was required (consumed) to achieve this level of production. The sources of supply were fertilizer (53 mill. t reduced through losses to 40 mill t. [48% of total consumed]), mineral N from soil sources (31 mill. t or 37%) and plant N fixed via biological nitrogen fixation (BNF) by the pulse and oilseed crop legumes (12 mill. t, or 15%). Of the 31 mill. t N supplied to grain and oilseed crops from the soil, a substantial proportion would be recycled legume residue N. Thus, legumes play a key role in the maintenance of world food and fibre production. The value of N consumed annually for global grain and oilseed production is about US$33 billion (costing N at US$400 t^{-1}). Of this amount, N_2 fixation by the crop legumes supplies US$5 billion worth of N. A 15% improvement in N_2 fixation equates to almost US$1 billion additional N.

The options for improving N_2 fixation are two-fold: management of the legume to maximize growth and minimize stresses (Peoples et al., 1994), and breeding legumes with enhanced capacity for N_2 fixation. The potential for the latter was recognized at least two decades ago (Lie and Mulder, 1971; Phillips et al., 1971). However, progress has been slow. Mytton (1983) noted that little attention had been given to an operational breeding approach to improve N_2 fixation and that the basic genetic information necessary for understanding the expression of desired characters was largely absent (see also Graham and Temple, 1984). To some extent, those observations remain relevant today.

Our foci in this review are the practical 'operational' aspects of selecting elite N_2 fixing genotypes and incorporating genes for enhanced N_2 fixation into other (agronomically desirable) backgrounds through breeding. We pay particular attention to just two species — common bean (*Phaseolus vulgaris*) and soybean (*Glycine max*). They have been subjected to disproportionate research in the area of N_2 fixation improvement, probably for different reasons. In the case of the common bean, the economic importance of the crop as a pulse legume, particularly in Latin America, together with low N_2 fixation activity prompted interest and subsequent research. Soybean, on the other hand, is the most widely-grown legume and enhancement of N_2 fixation of this crop would have very large economic benefits.

Operational framework for selection and breeding of legumes for enhanced N_2 fixation

Bliss (1993) suggested that selection and breeding for enhanced N_2 fixation should be done under conditions of low soil N to allow expression of N_2 fixation potential and discrimination between high- and low-fixing lines. Then the following would form the basis of an operational framework for the breeding program:

- choice of traits (characters) as selection criteria that can be measured precisely and economically, while

Table 1. Global statistics of area and production of crops in 1992 and estimates of amounts of N consumed and supplied from various sources

Crop	Area[a] (mill. ha)	Production[a] (mill. t)	N (mill. t)			
			Consumed[a]	Supplied		
				Fertilizer[b]	Soil[c]	BNF[d]
Cereals	700	1,950	60	50	22	-
Oilseeds	83	107	4	3	2	-
Soybean	55	114	14	0	6	8
Pulses	67	57	5	0	1	4
Total	905	2,228	83	53	31	12

[a] Amounts N required for each tonne grain produced were as follows: cereal - 30 kg N t^{-1}; oilseed - 40 kg N t^{-1}; soybean - 120 kg N t^{-1}; pulses - 80 kg N t^{-1}.
[b] FAO (1992a, b).
[c] Calculated by difference, assuming 75% fertilizer-N ends up as plant-available soil nitrate-N.
[d] Assume average Pfix for soybean = 60%; average Pfix for pulses = 80%.

allowing discrimination between superior and inferior lines;
- variability in legume germplasm and heritability of differences for either Pfix (proportion of plant N derived from N_2 fixation) or other N_2 fixation traits;
- identification of genetically diverse parents, incorporating both agronomic and N_2 fixation traits;
- choice of selection units (i.e. individual plants or families) that facilitate precise quantification of traits of interest and allow production of progeny from selected plants;
- use of a breeding procedure (e.g. mass selection, family selection) that provides maximum genetic gain for N_2 fixation and recombination with other (agronomic) traits.

Strategies for selection and breeding of legumes for enhanced N_2 fixation

Nutman (1984) concluded that high N_2-fixing lines of the temperate forage species, red clover (*Trifolium pratense*), were superior because of an enlarged N_2 fixing system, rather than because of increased efficiency of N_2 fixation. Superior plants had the following characteristics: earlier nodulation, more nodules and larger nodules. Significantly, there were no differences between the high N_2-fixing lines and control lines in indices of nodule efficiency, i.e. plant dry matter, plant N and ARA (acetylene reduction assay) per unit nodule mass. Increased nodulation may be the key to enhanced N_2 fixation of a number of the crop legumes, although the efficiency of nodule function, rather than nodulation per se, may be critical for certain species, e.g. common bean, and particular situations, e.g. soybean in the mid-west of the US. Thus, the general strategies for increasing legume N_2 fixation are aimed at:

- maximizing legume yield within the constraints imposed by agronomic and environmental considerations. This approach has particular application to low-yielding species such as common bean, lentil (*Lens culinaris*), mung bean (*Vigna radiata*) and chickpea (*Cicer arietinum*). It has less relevance for the larger, vigorously-growing species like fababean (*Vicia faba*), pea (*Pisum sativum*) and soybean (Attewell and Bliss, 1985; Herridge and Bergersen, 1988; Jensen, 1986). Increasing legume yield can be achieved by plant breeders in traditional breeding programs. Biological yield largely determines N_2 fixation, particularly in low N soils (Duc et al., 1988; Hardarson et al., 1984; Kumar Rao and Dart, 1987). Breeders, however, normally select for grain yield, rather than for biological yield. Grain yield, the product of biological yield and harvest index, is to a degree dependent upon biological yield. Therefore, breeders who operate in low N soils and select for high grain yield will also select for high N_2 fixation.
- active selection and breeding for symbiotic characteristics in legumes. Examples of these are nitrate tolerance, i.e. the ability of the plant to nodulate and fix N_2 in the presence of soil nitrate, and general nodulation capacity. Natural variation for nitrate

tolerance and nodulation capacity exists and has also been created using plant mutagenesis (Betts and Herridge, 1987; Carroll et al., 1985a, b; Herridge and Betts, 1985, 1988; Jacobsen and Feenstra, 1984; Park and Buttery, 1988). It may be impossible, and even undesirable, to produce a legume that is solely dependent upon N_2 for growth and cannot use nitrate; there is scope, however, to improve the levels of tolerance for the majority of the commonly-grown crop legumes.

– optimizing the numbers and effectiveness of rhizobia in the rooting zone, through strain selection and inoculation techniques, and through plant breeding for promiscuous or selective nodulation (Cregan and Keyser, 1986; Devine, 1984; Kueneman et al., 1984). Continued improvements in the effectiveness of strains of rhizobia used as legume inoculants and in the process of inoculation should also be sought (Brockwell and Bottomley, 1994). There appears to be scope to use strains of rhizobia for specific environmental niches, e.g. acid tolerant strains for acidic soils (Howieson and Ewing, 1986).

Host × strain specificity

Accounting for host × strain specificity when selecting and breeding for enhanced N_2 fixation presents practical problems (Mytton, 1984). There are essentially two approaches. The first approach is to consider host × strain specificity important and to identify highly effective combinations of host cultivar and rhizobial strain (Alwi et al., 1989; Mytton, 1975, 1983; Mytton et al., 1977; Nambiar et al., 1984). An extension of this approach can be found in the program at USDA Beltsville which aims to develop cultivars of soybean that bypass the native soil rhizobia and only nodulate with highly effective inoculant strains (Cregan and Keyser, 1986).

The second approach is to ignore specificity and to screen plant genotypes in a field soil containing high numbers of effective rhizobia or screen under glasshouse or field conditions using strain(s) known to be highly effective with a wide range of genotypes (Betts and Herridge, 1987; Bliss and Miller, 1988; Herridge and Rose, 1994; Kueneman et al., 1984; Nangju, 1980; Phillips and Teuber, 1985). This approach has merit. In the majority of field situations and with current inoculation technology, it is not possible to control the mix of rhizobial strain(s) that nodulate a legume crop, making it difficult to establish the highly effective, specific host-strain combination. There is also substantial evidence that a superior host genotype, selected on the basis of N_2 fixation with one or more highly effective strains, will express that superiority with other strains (Buttery and Dirks, 1987; Pacovsky et al., 1984; Phillips and Teuber, 1985; Rennie and Kemp, 1983b; Wiersma and Orf, 1992; Wolff et al., 1991).

Therefore, the protocol for a breeding program that does not aim to produce specific host-strain combinations would involve screening plant genotypes against the most effective strain(s) available. The strains would constitute a native soil population or would be used as inoculants. The latter strain(s) should also have proven effectiveness with a wide range of genotypes of the particular legume. This strategy should allow maximum expression of N_2 fixation by the host and does not involve the major problem of competition for nodulation between inoculant strain(s) and native soil rhizobia.

Measuring N_2 fixation

There is no single correct method for measuring N_2 fixation by legumes. None of the current methods, N yield, N difference, ^{15}N, acetylene reduction and xylem solute (ureide), can be relied upon to provide an accurate measure of N_2 fixation for every legume species grown under all possible variations of soil type and environment. Each method has unique advantages and limitations. Indeed, each method has been used at various times for assessing variation in N_2 fixation amongst legume genotypes and for identifying elite lines in breeding programs (Table 2). There is a trend with time, however, for the acetylene reduction assay (ARA), commonly-used in the 1970's, to be replaced by more accurate and reliable methods. Following is a short description of current methods.

N yield

The simplest estimates of N_2 fixation are obtained by measuring the amount of N in the legume biomass and are based on the assumption that the legume derives all of its N from N_2 fixation. In virtually all cases involving field-grown plants, the values obtained will overestimate N_2 fixation because the method ignores the contribution of soil N to plant growth. In soils that are extremely low in plant-available N, valid compar-

Table 2. Summary of methods used for plant culture and assessing symbiotic activity in programs to select and breed for enhanced N_2 fixation

Species	Program	Plant culture	Assessment		Reference
			Nodulation	N_2 fixation [a]	
Alfalfa	Screening /breeding	Glasshouse pots, 0 nitrate	Score	ARA, shoot, root DM	Seetin and Barnes, 1977
White clover	Screening	Controlled environ. pots, 0 nitrate	-	shoot DM	Mytton, 1975
Pea	Screening	Controlled environ. pots, 0 nitrate	Leghaem-oglobin	ARA, shoot DM	Hobbs and Mahon, 1982
Red clover	Breeding	Controlled environ. test tubes, 0 nitrate	Time, number, size, mass	ARA, shoot DM, %N, N	Nutman, 1984 (synthesis of 25-year program)
Alfalfa	Breeding	Glasshouse pots, 0 nitrate	Score, nodule enzymes	ARA, shoot, root DM	Barnes et al., 1984 (synthesis of 9-year program)
Alfalfa	Breeding	Glasshouse pots 0 and 8 m*M* nitrate; field, low nitrate	Number, flavenoids	shoot DM, %N, N, ^{15}N (I.D.)	Phillips and Teuber, 1985 Teuber and Phillips, 1988
Common bean	Selection /breeding	Glasshouse pots, 0 nitrate; field low/mod nitrate	Mass, number, carbon	ARA, shoot DM, N	Graham and Temple, 1984 (synthesis of 10-year program)
Common bean	Breeding	Field, low nitrate	Mass, number	ARA, shoot DM, N, grain yield, ^{15}N (I.D.)	McFerson et al., 1982 Attewell and Bliss, 1985
	Breeding	Field, low nitrate	Mass, number	shoot DM, N, grain yield, ^{15}N (I.D)	Bliss, 1993 (synthesis of 15-year program)
Common bean	Selection	Controlled environ. pots, 0 nitrate; field, low/mod nitrate ± fert. N	-	^{15}N (I.D.), ^{15}N (nat. abund.) shoot DM, N	Rennie and Kemp, 1983a, b
Soybean	Screening /breeding	Glasshouse pots, 0 and 5 m*M* nitrate; field, + fert. N	Mass, number, enzymes	ARA, ureides - stem, leaves, xylem. shoot DM, N	Wu and Harper, 1990, 1991
Soybean	Screening	Field, mod. soil nitrate	Occupancy grain yield, N N uptake, NHI	shoot DM, N	Leffel et al., 1992
Soybean	Screening	Glasshouse pots, 0 and 2.5 m*M* nitrate field, low and high nitrate soils	Mass, number	ureides - xylem, stem and root. ^{15} (nat. abund.)	Betts and Herridge, 1987. Herridge and Betts, 1988. Herridge et al., 1990.
	Breeding	Field, low and high nitrate soils	-	ureides - xylem F2 - single plant, non-destructive	Herridge and Rose, 1994
Soybean	Breeding	Field, ± fert. N, low/mod. nitrate soil	Mass, number	ureides - xylem ^{15}N (nat. abund.)	Song et al., 1995

[a] ARA - acetylene reduction assay, I.D. - isotope dilution, nat. abund. - natural abundance.

isons of treatment effects may be possible. Although absolute estimates of total N_2 fixed will still be high in these cases, the error may be negligible particularly if total biomass N is large.

N difference

A true measure of N_2 fixation based on legume N yield can only be obtained when the contribution of soil N to total biomass N is determined. This is usually achieved by growing a non N_2-fixing crop concurrently in the same soil. The difference in total N accumulated by the legume (N_{leg}) and non-fixing control (N_{nonfix}) is regarded as the amount of N_2 fixed. Thus:

$$N_2fixed = N_{leg} - N_{nonfix} \quad (1)$$

The major assumption of the method is that the legume and non-fixing control take up identical amounts of N from the soil. Because of this, the choice of the control is of utmost importance. Ideally, the legume and control should explore the same rooting volume, have the same ability to extract and utilize soil mineral-N, and have similiar patterns of N uptake. The non-fixing control may be a non-legume, an unnodulated legume or a non-nodulating legume, preferably an isoline of the test legume. Unfortunately, there are often substantial differences between N_2-fixing and non-fixing plants in their capacities to use soil N. Even when a supposed ideal non-fixing control is used, e.g. a non-nodulating isoline, erroneous estimates of N_2 fixation may still result from differences in root morphologies (Boddey et al., 1984).

The observation that levels of soil mineral-N were invariably higher following a legume crop than after a non-legume (Doughton and McKenzie, 1984; Evans et al., 1985) led Evans and Taylor (1987) to propose a modification of the N-difference equation to account for differences in the utilization of soil mineral-N by the legume and non-legume (non-fixing control). Additional measurements are made of the amounts of soil mineral-N in the root zones of the two crops at maturity. Thus:

$$\begin{aligned} N_2fixed &= (N_{leg} - N_{nonfix}) \\ &+ (SoilN_{leg} - SoilN_{nonfix}) \quad (2) \end{aligned}$$

With both the N difference and modified N difference methods, greater accuracy will always be achieved when plant-available soil N is low and legume biomass N is high.

^{15}N methods

The stable isotope ^{15}N occurs in atmospheric N_2 at a constant 0.3663 atom% ^{15}N. If ^{15}N enrichment (abundance) in plant-available soil N is different from that in atmospheric N_2, then the proportion of legume N derived from each source can be measured by the isotopic abundances in the legume and in a non-fixing control totally dependent on the same soil N. In many cases, the very small differences in the natural abundance of ^{15}N between plant-available soil N and atmospheric N_2 can be used, provided the samples can be analysed with a very precise mass spectrometer. More usually, the difference between the ^{15}N enrichment of the soil and atmosphere is expanded by incorporating ^{15}N-labelled materials in the soil.

^{15}N enrichment

The ^{15}N enrichment method is generally regarded as the standard method for estimating legume N_2 fixation. Its use has greatly increased over the past decade, a fact that is reflected in the extensive list of recent reviews on the method (e.g. Chalk, 1985; Danso, 1988; Danso et al., 1993; Ledgard and Peoples, 1988; Witty et al., 1988). However, the high cost of instrumentation to measure ^{15}N plus the expense of the ^{15}N-labelled materials are real constraints to even greater use of the method. Its main advantage is that it provides a time-averaged estimate of Pfix, integrated for the period of plant growth to the time of harvest. Thus:

$$Pfix = 1 - \frac{(atom\%^{15}Nexcess_{legume})}{(atom\%^{15}Nexcess_{nonfix})} \quad (3)$$

where, atom% ^{15}N excess = (atom% $^{15}N_{sample}$) — (atom% $^{15}N_{airN2}$), and atom% ^{15}N of air N_2 = 0.3663. The estimate of Pfix is independent of legume yield, although it is necessary to measure dry matter and N yield to estimate the amount of N_2 fixed.

The major assumption of both the ^{15}N enriched and natural ^{15}N abundance methods is that the legume and non-fixing reference plants utilize soil N with the same isotopic composition. With the enriched system, this translates into the legume and non-fixing reference plants utilizing the same relative amounts of N from added ^{15}N and endogenous soil N. This may not always occur and is the major weakness of the method (Witty et al., 1988). Thus, the choice of non-fixing reference plant is of utmost importance. Ledgard et al. (1985) and Witty (1983), amongst others, have shown the effect of the non-fixing reference plant on estimated Pfix. For this effect to be minimized, legume and reference plants should have similar patterns of soil-N use, so that the inevitable shifts in isotopic composition with time and space become inconsequential.

Natural ^{15}N abundance

Almost all transformations in soil result in isotopic fractionation. The net effect is often a small increase in the ^{15}N abundance of soil N compared with atmospheric N_2 (Shearer and Kohl, 1986). Because the differences are so small, data are commonly expressed as parts per thousand (‰ or $\delta^{15}N$). Thus:

$$\delta^{15}N = 1000 \times \frac{(atom\%^{15}N_{sample}) - (atom\%^{15}N_{standard})}{(atom\%^{15}N_{standard})} \quad (4)$$

where the standard is usually atmospheric N_2 (0.3663 atom%). By definition, the $\delta^{15}N$ of air N_2 is zero. The natural abundance method gives an integrated estimate of Pfix over time and has the advantage of being able to be used in already established experiments, provided non-fixing reference plants are also growing in the experimental plots. Plant N derived from N_2 fixation is calculated thus:

$$Pfix = \frac{(\delta^{15}N_{nonfix}) - (\delta^{15}N_{leg})}{(\delta^{15}N_{nonfix}) - B} \quad (5)$$

The $\delta^{15}N$ value of B is a measure of isotopic fractionation during N_2 fixation and is determined by analysis of the $\delta^{15}N$ of total plant N of the nodulated legume grown in N-free media. Isotopic fractionation during N_2 fixation is minimal but not zero and should be taken into account when calculating Pfix (Peoples et al., 1989a).

Although the principles of the natural abundance method are similar to those of ^{15}N enrichment, the major limitations are quite different. An isotope ratio mass spectrometer capable of measuring accurately differences of 0.1 ‰ (about 0.00004 atom% ^{15}N) is needed. Great care is necessary in sample preparation to avoid isotopic fractionation (see review by Bergersen et al., 1990). As well, contamination by ^{15}N enriched material must be rigorously avoided. The accuracy of the method will depend ultimately on the levels and uniformity of the ^{15}N in the soil. Levels of $\delta^{15}N > 6.0$ are preferable, although values as low as 2 ‰ might still be useful, depending on the level of Pfix (Unkovich et al., 1994). Values below 6.0 are often found in pasture and plantation soils and in natural forest systems (Peoples et al., 1991). Fortunately, for soils that are regularly cultivated, $\delta^{15}N$ values of plant-available N tend to range between 6.0 and 16.0 (Peoples and Herridge, 1990), and can be relatively constant with time and depth (Bergersen et al., 1990). Therefore, the major limitation of ^{15}N enrichment, i.e. choice of appropriate non-fixing reference plant, is less critical.

Xylem N solutes

Collection of xylem sap and analysis of its contents has been used widely for assessing the nutritional status of field-grown plants (e.g. Bollard, 1960). Xylem sap was collected either as sap bleeding spontaneously under pressure from the stump of the intact root following decapitation of the shoot (root-bleeding sap), or under mild vacuum applied to freshly-harvested shoot segments (vacuum-extracted sap). The latter technique (Bennet et al., 1927; Bollard, 1953), facilitated the development of the ureide assay of N_2 fixation from one with restricted application (glasshouse-grown plants, analysis of root-bleeding sap; McClure et al., 1980) to an assay that could be applied to a wide range of species and field environments (Herridge, 1984; Herridge et al., 1987, 1988a, 1990; Herridge and Betts, 1985; Norhayati et al., 1988; Peoples et al., 1989a, b; Rerkasem et al., 1988).

The principal underlying the ureide assay is that the composition of N solutes in xylem sap changes from one dominated by the ureide compounds, allantoin and allantoic acid, in N_2-dependent plants to one dominated by nitrate and amino-N in plants utilizing soil N. In calibration experiments, correlations between the relative abundance of ureide-N (ureide-N as a proportion of total sap-N) in xylem sap and Pfix were extremely strong with regression coefficients of almost unity (e.g. Herridge and Peoples, 1990; Peoples et al., 1989b; Rerkasem et al., 1988). In the case of soybean, the following equations are used (Herridge and Peoples, 1990):

$$Pfix(\%) = 1.56(RU - 7.7)$$
for plants in vegetative and flowering stages (6)

$$Pfix(\%) = 1.56(RU - 15.9)$$
for plants during pod – fill (7)

where the % relative abundance of ureide-N in sap (RU) is calculated as:

$$RU = 400a/(4a + b + c) \quad (8)$$

and a, b and c are, respectively, the molar concentrations of ureides, nitrate and α amino-N (Herridge, 1984).

Table 3. Principal N solutes in xylem sap of N_2-dependent food and oilseed legumes

Amides (asparagine, glutamine)	Ureides (allantoin, allantoic acid)
Chickpea (*Cicer arietinum*)	Soybean (*Glycine max*)
Lentil (*Lens culinaris*)	Pigeon pea (*Cajanus cajan*)
Pea (*Pisum sativum*)	Mung bean (*Vigna radiata*)
Fababean (*Vicia faba*)	Black gram (*Vigna mungo*)
Narrow-leafed lupin (*Lupinus angustifolius*)	Cowpea (*Vigna unguiculata*)
White lupin (*Lupinus albus*)	Common bean (*Phaseolus vulgaris*)
Groundnut (*Arachis hypogaea*)	Winged bean (*Psophocarpus tetragonolobus*)

Not all legumes export fixed N_2 as ureides (Table 3). With the cool season food legumes, around 80% of fixed N_2 is exported from the nodules as the amides, asparagine and glutamine, the remainder as amino acids (Herridge et al., 1988b). It would be extremely useful though to extend the principal of the ureide assay of N_2 fixation activity to the amide exporters. Unfortunately, with these species, none of the readily-measured N solutes appears to be specifically associated with N_2 fixation. There may be scope to measure shifts in asparagine: glutamine ratios, or the relative proportions of nitrate in xylem sap. Calibration experiments involving chickpea, fababean, lentil and pea have been reported (Peoples et al., 1987), but as yet, validation of the relationships under field conditions have not been attempted.

Acetylene reduction assay (ARA)

The acetylene reduction assay arose from observations in the 1960's that the N_2-fixing enzyme, nitrogenase, catalyzed the reduction of acetylene (C_2H_2) to ethylene (C_2H_4). Since that time, the ARA has played a major role in N_2 fixation research because of its rapidity, simplicity, high sensitivity and low equipment and resource costs. The standard ARA method involves enclosing detached nodules or nodulated roots in air-tight containers and exposing them to an atmosphere of about 10% C_2H_2. After an incubation period, gas samples are collected and analysed for C_2H_4 using gas chromatography (Turner and Gibson, 1980). The ARA was adapted for field studies of legume N_2 fixation (Hardy et al., 1968) to mainly compare treatment effects on N_2 fixation, rather than to estimate the amount of N_2 fixed for a particular time interval.

Eventually the ARA method lost favour because of a number of major problems. These included: the difficulty in recovering nodules; the need to interpolate between single, short-term measurements to obtain time-integrated measurements; the need to determine correct conversion ratios between C_2H_4 produced and N_2 fixed, which can vary according to environmental, diurnal and plant effects acting independently on N_2 and C_2H_2 assimilation and reduction; non-linearity in the rate of C_2H_2 reduction over the period of the assay; effects of nodule removal and decapitation of plants and difficulties in sampling, particularly in hard-setting soils. The sheer magnitude of the problems and the resultant errors in estimating C_2H_2 reduction (N_2 fixation) activity suggest that even simple comparisons of material in a breeding program may be invalid. Readers are referred to Turner and Gibson (1980), Witty and Minchin (1988), Vessey (1994) and Minchin et al. (1994) for expanded discussion on the use of the ARA.

Assessing heritability and repeatability of N_2 fixation

In a breeding program, the heritability of desired traits or characters must be established. Low heritability of a character can result from error in measurement or from a complex genetic basis. Unfortunately, most of the published studies which quantified the heritability of N_2 fixation used the ARA method for measurement and there are few examples in which more reliable assessments were made. Notwithstanding this lack of information, it appears that the heritability of N_2 fixation can be readily assessed using standard statistical methods. Furthermore, it appears that N_2 fixation is moderately heritable.

Ronis et al. (1985), using ^{15}N, investigated broad-sense heritability of total and percent fixed N in har-

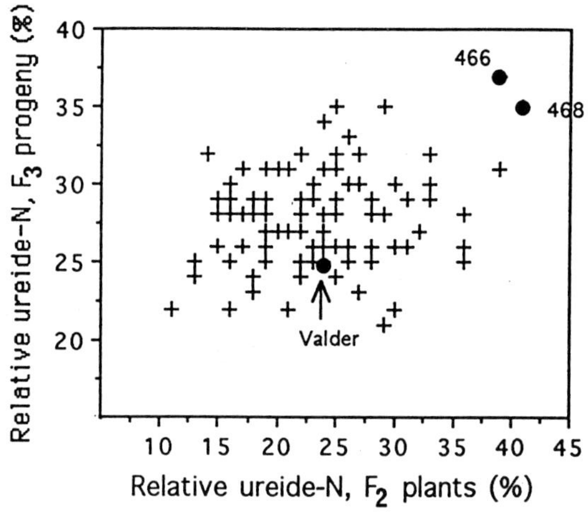

Fig. 1. Relative abundance of ureide-N in xylem sap of 99 individual F_2 plants (+) and their F_3 progeny and low (Valder) and high N_2-fixing (Korean 466 and 468) parents (•) grown in high nitrate soils in successive seasons.

vested seed of three F_2 soybean populations, each of 110 plants. Broad-sense heritability (h^2) was calculated, after utilizing the parental lines to estimate environmental variance, as:

$$h^2 = \frac{\sigma^2_{F2} - 1/2(\sigma^2_{P1} + \sigma^2_{P2})}{\sigma^2_{F2}} \qquad (9)$$

where h^2 = broad-sense heritability,
σ^2_{F2} = variance of the F_2 populations
σ^2_{P1} = variance of the parental (1) population
σ^2_{P2} = variance of the parental (2) population.

Broad-sense heritabilities for fixed N contents of seed ranged from 0.53 to 0.60; estimates for percent fixed N in seed (similar to Pfix) were lower and less consistent, ranging from 0.12 to 0.43. They concluded that improvement of the latter character would be more difficult. However, they didn't concede that the low heritability estimate (0.12) was for progeny of the 'Williams' × 'Calland' cross, both of which had similar, intermediate, N_2 fixation capacity. In other experiments, the parents were more diverse in N_2 fixation, creating greater variation in the progeny. Heritability estimates for those crosses were moderate at 0.37 and 0.43, suggesting that breeding for improved N_2 fixation of soybean using appropriate parents should be possible.

Herridge and Rose (1994) used a similar approach to calculate broad-sense heritabilities of N_2 fixation for 11 F_2 populations of soybean, ranging in size from 14 to 136 plants. Nitrogen fixation was assessed in this study using the xylem ureide technique. Relative ureide-N data for the individual F_2 plants were subjected to analysis of variance to calculate total phenotypic variance (σ^2_P) for the F_2 populations. Single-plant data for the parental and control genotypes was used to estimate environmental variance (σ^2_E). Genetic variance (σ^2_G) was then calculated as ($\sigma^2_P - \sigma^2_E$). Finally, broad-sense heritability (h^2) was calculated as:

$$h^2 = \sigma^2_G/(\sigma^2_G + \sigma^2_E) \qquad (10)$$

Average environmental variance (σ^2_E), calculated from the single-plant values of relative ureide-N for the parent and control genotypes was estimated at 49.4 (Table 4). The resultant estimates of genetic variability ranged from 22.5 to less than zero. Heritability estimates ranged from 0 to 0.31.

Broad-sense heritabilities of single-plant xylem-ureide analysis were assessed also by parent-offspring regression between individual F_2 plants and F_3 progeny means. Heritability of relative ureide-N in xylem saps on a plot basis was calculated from estimations of σ^2_E and σ^2_G from the analysis of variance of the F_3 experiment. The correlation coefficient of the F_2 parent-offspring regression was low ($r = 0.16$) and non-significant ($p > 0.05$) (Fig. 1). These results were disappointing but did not necessarily imply that N_2 fixation was not heritable or that N_2 fixation was not quantified accurately. The consistency of assessments of N_2 fixation on a plot basis for parent lines Valder, Korean 466 and Korean 468 over the two years lent support to both the concept of heritability and the ureide method for measurement of N_2 fixation (Fig. 1).

Nitrate levels are variable in field soils. Nitrogen fixation is particularly prone to variations in soil nitrate (Harper, 1987), a response that is more critical when assessing single plants, i.e. F_2 , than when multi-plant samples are used, i.e. F_3 and parent lines in both years. Thus, Herridge and Rose (1994) concluded that, for uniformity in the assessment of single plants for N_2 fixation, it may be necessary to grow plants in sand or water culture and to supply nitrate in measured, nutrient form, rather than grow plants in the field. They concluded also that the F_2 parent-F_3 offspring correlation of 0.16 indicated that selection of individual F_2 plants for enhanced N_2 fixation was better than random. This was supported by the heritability estimate ($h^2 = 0.36$) obtained from partition of variance in the F_3 experiment where relative ureide-N was estimated on a whole-plot basis. Alternatively, assessments for

Table 4. Relative ureide-N in xylem sap at R3-R5 of individual plants of 11 F_2 populations of soybean and 9 parent and control genotypes, grown in a high-nitrate soil

F_2 population/ genotype	No. plants tested	Mean (%)	Range (%)	Phenotypic variance (%)	Broad-sense heritability
F_2 population					
A	136	27	10–50	63.0	0.22
B	54	27	13–47	67.2	0.27
C	70	28	14–46	71.9	0.31
D	36	28	11–55	65.6	0.25
E	59	28	13–47	58.4	0.15
F	14	26	2–37	36.0	0
G	34	27	15–39	34.6	0
H	102	30	15–55	59.3	0.17
J	110	24	10–48	48.0	0
K	86	29	12–43	37.6	0
L	33	27	14–41	60.8	0.19
Parent and control genotype					
Bragg	20	17	12–32	28.6	-
Bossier	20	18	13–27	12.5	-
Davis	20	22	15–34	21.3	-
Valder	20	24	12–45	70.0	-
Reynolds	20	28	19–34	26.1	-
Korean 464	20	27	16–49	48.0	-
Korean 466	18	39	31–52	113.6	-
Korean 468	20	41	24–60	74.1	-
Korean 469	20	42	30–54	50.3	-

Table 5. Correlation (r) over sites and seasons for relative ureide-N in xylem sap of soybean, pooled for mid and late maturity group experiments

Seasons - site		1989/90 (F_6)		1990/91 (F_7)
		Breeza 1 (low N)	Breeza 2 (high N)	Breeza 1 (low N)
1989/90	Breeza 2 (high N)	0.47**		
1990/91	Breeza 1 (low N)	0.18	0.07	
	Narrabri (high N)	0.19	0.49**	0.14

** Significant at $p = 0.01$, df = 38.

N_2 fixation could be delayed until replication was possible, i.e. in the F_2-derived F_3 generation, or later (see also Bliss, 1993).

In the same program, Herridge and Rose assessed broad-sense heritability of N_2 fixation (relative ureide-N in xylem sap) for F_3-derived F_6 and F_7 lines. Estimates for the F_3-derived F_6 lines ranged from 0.32 to 0.52 and were similar to the value of 0.36, calculated from the F_2-derived F_3 lines. Selection of field sites for evaluation of N_2 fixation was critical for discriminating the high N_2-fixing 'Korean' type of symbiosis. Superiority of the 'Korean-type' genotypes was expressed more consistently in the high-nitrate soils. When correlations of the ureide data for the F_3-derived F_6 and F_7

lines across sites and years were calculated, the results were variable (Table 5).

The 1989/90 data from the low-nitrate (Breeza 1) site was not correlated with either of the 1990/91 data sets (Breeza 1 and Narrabri). However, the 1989/90 data from the high-nitrate (Breeza 2) site was significantly correlated with data from the high-nitrate site in the following season, but not with data from the-low nitrate site. Thus, the F_6-F_7 regression analysis gave a similar estimate of heritability or repeatability as the partitioning of variances, provided the high-nitrate sites were being compared. At this level, Herridge and Rose (1994) concluded that progress to selection should be reasonable provided variation of the environmental factors affecting the symbiosis were minimized.

A number of other studies have focussed on the heritability of nodulation, rather than N_2 fixation (see Sinclair et al., 1991). Greder et al. (1986) evaluated three populations of F_3-derived F_5 and F_6 lines of soybean in the field for ability to nodulate with the native soil rhizobia when the seed was not inoculated, and to nodulate in the same soils with a highly effective inoculant strain, *Bradyrhizobium japonicum* USDA 110. Estimates of broad-sense heritabilities for nodulation (nodule mass) by native soil rhizobia ranged from 0 to 0.66 and were equal to or exceeded 0.55 for each population when averaged over sites. For recovery of USDA 110, i.e. nodulation by the inoculant strain, heritability estimates were smaller, ranging from 0 to 0.57. When averaged over sites, they were less than 0.55 for the three populations. These data, together with correlation analysis of nodulation, agronomic and yield traits, indicated that selection for increased nodule mass was warranted.

Similar conclusions were drawn by Arrendell et al. (1985) in a study of F_2-derived F_5 and F_6 generation progenies from a cross of Virginia and Spanish cultivars of groundnut. Broad-sense heritability estimates for nodule number ranged from 0.25 to 0.57 and averaged 0.47 over the two years and six samplings; estimates for nodule weight ranged from 0.48 to 0.78 and averaged 0.66. Estimates for ARA, shoot weight and fruit weight ranged from 0.53 to 0.85. The moderate to high estimates for these traits indicated that superior nodulating and N_2-fixing genotypes within the populations studied could be readily identified and that selection for enhanced N_2 fixation should result in indirect selection for yield.

Research to select and breed for enhanced N_2 fixation in legumes

Common bean

Of all the commonly-grown agricultural legumes, the common bean is regarded as the most inefficient at fixing N_2. This has been established as a result of numerous studies in the US, Canada and countries of South America in which different genotypes of common bean were compared, or where the comparisons were made between common bean and other grain legumes such as cowpea and soybean.

Studies to define genetic variation in N_2 fixation
Graham and coworkers examined more than 600 cultivars of common bean under short-day subtropical conditions. They reported greatest N_2 fixation in the indeterminate, climbing cultivars (Graham and Rosas, 1977; Graham, 1981; Graham and Temple, 1984).

Piha and Munns (1987a) compared N_2 fixation of 9 genotypes of common bean with soybean and cowpea in the field. They also contrasted growth of the three species without fertilizer N (i.e. dependent on N_2 fixation and soil-derived N) with plants that received 220 kg ha^{-1} fertilizer N. Their results showed that the beans did not accumulate dry matter or N or fix N_2 as well as either cowpea or soybean (Table 6). When the bean genotypes were split into maturity groups, it became apparent that there was a maturity effect on the growth and N_2 fixation traits. The early maturing genotypes had the lowest values for both, the late maturing genotypes the highest. Relative N accumulation, an index of the genotype's capacity to satisfy demand for N through N_2 fixation, indicated that the late maturing genotypes of common bean were similar to cowpea and soybean (values of 0.98, 0.97 and 0.94, respectively) and were slightly better than the mid maturity genotypes (value of 0.91). The early genotypes of common bean had by far the lowest value of 0.73. In other words, N_2 fixation could only meet 73% of N demand. Piha and Munns concluded that N_2 fixation was inadequate for the early maturing common bean but adequate for the later maturing types. However, with all maturity types, the capacity for growth and N_2 fixation did not match up to the capacities of either cowpea or soybean.

In a companion glasshouse study, involving 8 of the 13 genotypes grown in the field, Piha and Munns (1987b) confirmed that N_2 fixation by common bean

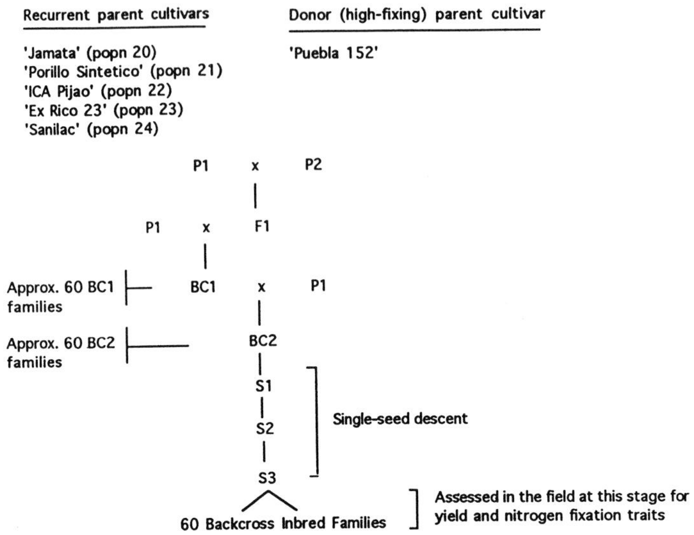

Fig. 2. The backcross inbred method used by Bliss and coworkers to generate populations of common bean with enhanced capacity for N_2 fixation (source: McFerson et al., 1982).

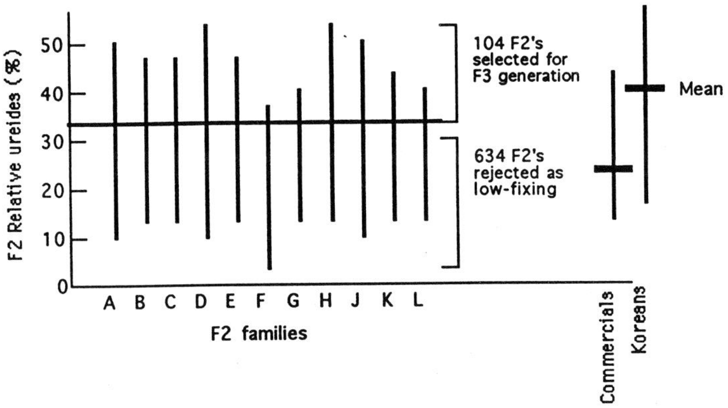

Fig. 3. Ranges of N_2 fixation (relative ureide-N of xylem sap) for the eleven F_2 families and for the commercial and Korean parents. The horizontal line through the families indicates the cut-off point for selection for F_3 generation.

was inadequate but found also that low N_2 fixation activity was not related to low nodulation. They did show, however, that the bean nodules evolved greater amounts of H_2 than those of either soybean or cowpea resulting in reduced efficiency of function (relative efficiency [RE] values of 0.5–0.7 for bean versus 0.95 for soybean and cowpea).

Pacovsky et al. (1984) had previously published RE's of 0.5–0.6 for symbioses of 3 cultivars of common bean and 9 rhizobial strains. They concluded that the copious production of H_2 and the associated energy loss limited the productivity of the bean symbiosis. The RE values indicated that more than 40% of the available electron flow to nitrogenase was used to reduce H^+ to H_2, rather than N_2 to NH_4^+.

Hungria and Neves (1987), in a factorial study of 5 bean cultivars and 6 strains of rhizobia, endorsed these findings. Their data showed a strong inverse relationship between nodule H_2 production and both nodule specific activity of C_2H_2 reduction, i.e. N_2 fixation, and plant N yield, and confirmed that low N_2 fixation was not a result of low nodulation (Table 7). For example, strain SEMIA 487 produced, on average, 33% more nodules than strain C-05, yet fixed about half the N (shoot N values of 96 versus 183 mg plant^{-1}). Hydrogen evolution and therefore nodule RE were affected

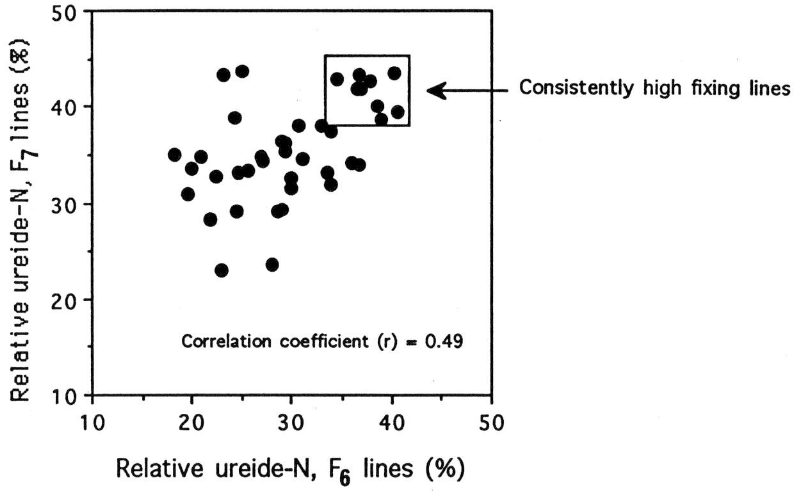

Fig. 4. Relative abundance of ureide-N for F_3-derived F_6 and F_7 lines of soybean, grown in successive years on high nitrate soils at Breeza and Narrabri, Australia.

Table 6. Growth and N_2 fixation of genotypes of cowpea, soybean and early, mid and late maturing common bean in the field in California (data from Piha and Munns, 1987a). Values shown are the means for the following number genotypes: cowpea (2), soybean (2), late bean (3), mid bean (2), early bean (4)

Species	Days to maturity	Total shoot N -N (kg ha^{-1})	Total shoot N +N (kg ha^{-1})	Relative N accumulation[a]	N_2 fixed (-N) (kg ha^{-1})
Cowpea	84	254	261	0.97	196
Soybean	100	252	268	0.94	189
Bean - late maturity	96	171	175	0.98	109
mid maturity	90	113	124	0.91	52
early maturity	84	97	132	0.73	35

[a] N in shoots of +N plants/N in shoots of inoculated (-N) plants.

by both plant genotype and rhizobial strain. Cultivar Negro Argel and strain C-05 had RE values approaching the high values reported above for soybean and cowpea.

These were significant findings because a number of programs aimed at improving bean N_2 fixation were based on the premise that nodulation per se was inadequate (e.g. the CIAT program) and that selection for improved nodulation traits would lead to increased levels of N_2 fixation (CIAT, 1987; Kipe-Nolte et al., 1993). The data referred to above suggest, however, that the efficiency with which the nodules function, rather than the gross number or weight of the nodules, is the major problem. Thus, in selection and breeding programs where nodulation is used as the principal criterion for selection, opportunities for worthwhile progress may be limited. A more direct and useful selection trait would be N_2 fixation (or plant N or seed N yield under low N conditions of growth), with number or weight of nodules serving only to confirm that nodulation had occurred.

Hardarson et al. (1993) reported results of a coordinated program, sponsored by the International Atomic Energy Agency (IAEA), to investigate the N_2-fixing potential of common bean and to identify high-fixing lines which could be used as parents in breeding programs. Experiments were conducted in Austria, Brazil, Chile, Colombia, Guatemala, Mexico and Peru during 1985 to 1991.

Table 7. Nodulation (weight, specific activity for C_2H_2 reduction and H_2 evolution, RE) and N yield (N_2 fixation) of bean cultivar and rhizobial strains, glasshouse cultured in Brazil (data from Hungria and Neves, 1987). Values shown are 3 of the 5 bean cultivars, averaged over the 6 strains, and 3 of the 6 rhizobial strains, averaged over the 5 cultivars

Cultivar/strain	Shoot N	Nodulation			
		Weight	C_2H_2 redn	H_2 evolutn	RE[a]
	(mg plant^{-1})	(g plant^{-1})	(specific act. umol g nod^{-1} h^{-1})		
Bean cultivars					
Negro Argel	117	0.54	16.3	2.5	0.85
Venezuela 350	80	0.64	15.3	4.0	0.74
Rio Tibagi	61	0.42	9.8	4.7	0.52
Rhizobial strains					
C-05	183	0.48	25.5	3.1	0.88
SEMIA 487	96	0.64	19.3	4.0	0.79
CIAT 727	22	0.29	9.0	5.1	0.44

[a] RE = 1 - H_2 evolved (air)/C_2H_2 reduced.

Table 8. Summary of N_2 fixation data for common bean, grown in the field in Central and South America and in Austria

Year/location	No.	Range		Commercial cultivar		
	cultivars	Pfix	N_2 fixed	Name	Pfix	N_2 fixed
		(%)	(kg ha^{-1})		(%)	(kg ha^{-1})
1987 - Austria	29	27–67	25–165	-	-	-
1987 - Brazil (i)	17	12–25	4–12	Rio Tibagi	22	8
- Brazil (ii)	7	19–53	11–53			
1987 - Chile	21	38–60	27–62	Tortola	44	50
1988 - Chile	12	27–60	25–115	Tortola	52	90
1985 - Colombia	9	32–47	18–36			
1989 - Guatemala (i)	10	69–73	92–125	ICTA-San Martin	70	105
- Guatemala (ii)	10	22–57	12–50	ICTA-Tamazulapa	31	24
1987 - Mexico (i)	20	5–58	7–108	Flor de Mayo	51	89
1988 - Mexico (ii)	17	0–50	0–70	Flor de Mayo	20	25
1986 - Peru	20	24–56	15–59	Canario Divex	54	42
1988 – Peru	22	3–56	7–51	Canario	47	31

Large effects on Pfix and N_2 fixed were recorded for genotype, location and year of experiment (Table 8). The high values were for adapted cultivars and breeding lines grown under favourable conditions. In four of the experiments, individual genotypes fixed in excess of 100 kg ha^{-1}, through the combined effects of high yield and moderate to high Pfix values (50 to 73%). Conversely, low N_2 fixation, e.g. Brazil (i), tended to be associated with unfavourable conditions of growth, i.e. high temperatures and dry soil, rather than with inferior genotypes. Average Pfix and N_2 fixed for the commercial cultivars was 43% and 52 kg ha^{-1}, respectively. The participants in the program concluded that the best of the cultivars in each country's trials could be considered for commercial release if all other traits were acceptable. If not, they certainly could be used as high-fixing parents in breeding programs.

Improvement/breeding programs

One program that has made progress in breeding for increased N_2 fixation in common bean is that of Bliss and co-workers at the University of Wisconsin. The genotypes chosen as the recurrent parents were the well adapted, commercially-acceptable cultivars 'Jamapa', 'Porrillo Sintético', 'ICA Pijao', 'Ex Rico 23' and 'Sanilac'. The late maturing, indeterminate, climbing type 'Puebla 152', was used as the high N_2 fixing donor parent (McFerson et al., 1982). Using an inbred backcross method to develop populations of bean which resembled the recurrent parent in desirable agronomic traits but differed in N_2 fixation and yield in low N soils, they produced hybrid lines with enhanced N_2 fixation, acceptable seed yield and other agronomic traits of the recurrent parents (Fig. 2).

The various populations of inbred lines associated with the five recurrent parents (Fig. 2) have been extensively studied during the past decade (see Bliss, 1993). Data for two of the populations are presented in Table 9. Sanilac is a white seeded navy bean, grown extensively in the northern U S until recently. It produced only modest yields and had low N_2 fixation. The inbred lines from the cross of Sanilac and Puebla 152 showed substantial improvements in yield and N_2 fixation, which were combined in some cases with other desirable (agronomic) traits. St Clair et al. (1988) reported lines 24–17, 24–21 and 24–55 fixed 6 to 10 times as much N as Sanilac, showed substantially higher rates of growth and produced up to 40% more grain N. None of the inbred lines could match Puebla 152, the high-fixing donor for plant and seed yield and N_2 fixation. One line, 24–21, also retained the desirable characteristics of short maturity and determinacy. The enhanced yield and N_2 fixation of line 24–17 were linked to the undesirable traits of late maturity and indeterminacy. Line 24–55, although agronomically-acceptable, did not perform as well as 24–21. These results were significant because they indicated that N_2 fixation was heritable and could be combined with other agronomically-desirable traits.

The cross of the black seeded, Porrillo Sintético and Puebla 152, on the other hand, did not produce lines with enhanced yield and N_2 fixation, i.e. superior to Porrillo Sintético (Table 9) (Attewell and Bliss, 1985; St. Clair et al., 1988). Reasons for this could be that the original selection for high N_2 fixation, based on the ARA, was not effective or that the two parents did not differ genetically in N_2 fixation.

Bliss and coworkers used a variety of methods for evaluating and measuring N_2 fixation by the hybrid lines and parents in the breeding program, including plant N yield, seed yield, seed N yield, ^{15}N methods (enriched and depleted formulations), acetylene reduction assay and nodulation indices (Attewell and Bliss, 1985; Pereira et al., 1989; St Clair et al., 1988). They concluded that selection of breeding lines based on plant and/or seed N was most effective and most cost-effective. In their studies, these parameters were always highly correlated with N_2 fixation, determined using ^{15}N (St Clair et al., 1988). The only proviso was that the plants were grown in a low N soil and that the plants were well nodulated by highly effective rhizobia, either introduced by inoculation or already present in the soil.

They did not consider that the xylem ureide method was appropriate for assessing N_2 fixation in the program, mainly because of lack of familiarity with the method rather than for technical reasons (Bliss and Miller, 1988). The xylem ureide method was essentially developed for soybean (Herridge, 1982, 1984; McClure et al., 1980) but was subsequently extended to other crop legumes. In a preliminary experiment, the relative abundance of ureide-N in root-bleeding (xylem) sap of common bean was shown to vary as the plants' proportional dependence on N_2 fixation and mineral N varied (Peoples and Herridge, 1990). These results were confirmed in calibration experiments (Hansen et al., 1993; Peoples and Herridge, unpubl. data). Thus, the ureide method could be used in breeding programs involving the common bean in much the same way as the method was used to assess N_2 fixation by single plants and plots of breeding lines of soybean (Herridge and Rose, 1994).

Bliss and co-workers have now released five high N_2-fixing lines of common bean, designated WBR22–3, WBR22–8, WBR22–34, WBR22–50 and WBR22–55 (Bliss et al., 1989; Bliss, 1993). They are all derived from the cross of ICA Pijao, a black seeded, indeterminate, bush type, bred for sub-tropical conditions,

Table 9. Nitrogen fixation, yield and agronomic traits of recurrent parents Sanilac and Porrillo Sintetico, high-fixing parent Puebla 152, and inbred backcross progeny (mean of 2 experiments) (source: St. Clair et al., 1988)

Parent or line	N yield (g plant^{-1})	Pfix[a] (%)	N_2 fixed (mg plant^{-1})	Rate of N_2 fixation (g plant^{-1} d^{-1})	Maturity (days)	Seed N yield (mg plant^{-1})
Sanilac	739	7	46	0.6	96	636
24–17	1073	44	475	6.5	107	900
24–21	987	31	306	4.1	93	842
24–55	836	35	292	4.3	100	702
Puebla 152	1222	56	674	6.0	114	1083
Porrillo Sintetico	1197	50	597	8.5	107	991
21–16	1072	49	517	6.2	107	844
21–38	997	49	493	6.9	107	757
21–43	1030	48	491	5.2	107	832

[a] Pfix — proportion of plant N derived from N_2 fixation.

and Puebla 152 (Fig. 2). In field tests in Brazil during 1987, one of the lines (WBR22–34) fixed twice as much N as the standard Brazilian cultivar, Rio Tibagi, and about 20% more than Negro Argel, identified in a previous study as a high fixer (Table 7; Hungria and Neves, 1987). Negro Argel had the highest Pfix value (54%), followed by WBR22–34 (44%) then Rio Tibagi (35%). Line WBR22–3 performed favourably in a field experiment in the same year in Mexico. It had the fourth highest value for Pfix (50%) and the second highest value for N_2 fixed (90 kg ha^{-1}). Interestingly, the best performing line was 21–58, a line from Porrillo Sintético x Puebla 152 cross. The five released lines outyielded Rio Tibagi consistently in experimental and national yield trials. In 1984–6 trials in Brazil, yields of the lines exceeded those of Rio Tibagi by 10 to 25%.

The five released lines as well as other breeding lines selected for high N_2 fixation and important agronomic traits have undergone further evaluation during the past five years. Bliss cautions that although the lines usually yield well under low N conditions (because of good N_2 fixation), their utility as commercial cultivars will depend just as much on other traits such as disease resistance and seed type (Bliss, 1993). Thus, breeding programs are underway in countries like Brazil to combine traits like disease resistance and high N_2 fixation into adapted breeding lines, using the general protocols outlined over a decade ago (McFerson et al., 1982; Fig. 2).

When considering future directions for breeding of common bean, Bliss (1993) suggested that breeding plants with capacity to nodulate and fix N_2 in the presence of soil nitrate, i.e. nitrate tolerance, should be a priority. There has been very little emphasis on selection and breeding for this character in common bean, even though it has been a major priority for soybean (see below). There is evidence also that variation exists for nitrate tolerance in both natural populations and in mutant lines (Park and Buttery, 1988, 1989; St. Clair et al., 1988) and that common bean is particularly intolerant of the suppressive effects of soil nitrate (George and Singleton, 1992).

Other selection traits that may have merit in a breeding program for common bean are early and late nodulation (Chaverra and Graham, 1992; Kipe-Nolt et al., 1993; Kipe-Nolt and Giller, 1993), although the practicalities of repeated assessments of large numbers of plants for nodulation may prove impossible. The more direct protocol of Bliss (1993) of growing plants in low N soils and measuring dry matter and N yields of plants and grain is far simpler and has the added advantage that the effects of the nodulation traits on accumulation of fixed N by the plant have been effectively integrated by the characters measured.

Table 10. Published estimates of N_2 fixation, crop N and grain yields of soybean

Country	Grain yield ($t\ ha^{-1}$)	Total crop N ($kg\ ha^{-1}$)	N_2 fixation: Pfix (%)	N_2 fixation: N_2 fixed ($kg\ ha^{-1}$)	Method of assessment	Reference
Australia	0.18–3.31	33–302	0–83	0–233	ureide	1
"	2.16–3.96	346–406	34–67	139–204	^{15}N	2
"	n.a.	329	95	312	^{15}N	3
US	1.38–2.76	100–187	13–40	14–75	N diff.	4
"	1.84–2.89	246–273	42–78	116–192	^{15}N	5
"	3.41–4.49	368–387	71–80	263–31 1	Ndiff.	6
Thailand	n.a.	121–643	14–70	17–450	^{15}N	7
"	n.a.	50–188	54–78	27–147	^{15}N	8
"	n.a.	157–251	0–45	0–113	^{15}N	9
Canada	n.a.	150–334	14–62	33–151	^{15}N	10
France	2.10–2.70	147–182	26–38	38–70	^{15}N	11

n.a. — not available; 1. Herridge and Holland, 1992; 2. Bergersen et al., 1985; 3. Chapman and Myers, 1987; 4. Weber, 1966; 5. Vasilas and Ham, 1984; 6. Bezdicek et al., 1978; 7. Rennie et al., 1988: 8. Kucey et al., 1988a; 9. Kucey et al., 1988b; 10. Rennie, 1984; 11. Armager et al., 1979.

Soybean

Soybean is the most widely-grown of the crop legumes with 55 million ha sown in 1992, representing 47% of the total area sown to the oilseed and pulse legumes (Table 1). We calculate soybean to fix 8 million t N annually, worth US$3 billion. A 15% improvement in N_2 fixation equates to increased inputs into world agriculture of 1.2 million t N each year, valued at US$500 million (see also economic analysis of improving soybean and alfalfa N_2 fixation in the US (Tauer, 1989)).

Numerous studies have been reported which show the variation in N_2 fixation of soybean. In those studies summarized in Table 10, total N_2 fixed ranged from 0 to 450 kg ha^{-1}; the range for Pfix was 0 to 95%. Variations in N_2 fixation were partly linked to variations in yield, in turn linked to maturity and other genetic traits and to environmental constraints. Soybean clearly has the capacity to produce large yields (up to 4.55 t ha^{-1} grain and >600 kg ha^{-1} total crop N) and fix large amounts of N in low N soils under optimum environmental conditions. Thus, in contrast to the common bean, improving N_2 fixation in soybean will not result primarily from genetically-based improvements in yield.

Variations in the amounts of N_2 fixed can also be linked to variations in Pfix. When the soils in which the plants are grown have high nitrate fertility, plants use moderate to large amounts of soil N, Pfix is reduced to a low level resulting in low total N_2 fixed. In two of the studies in Table 10, Pfix values of zero were recorded (Herridge and Holland, 1992; Kucey et al., 1988a). There is now substantial evidence that improved N_2 fixation of soybean will result from incorporation of nitrate tolerance into commercial cultivars. Soil nitrate inhibition of nodulation and N_2 fixation can affect other soil and crop factors in a number of ways. During early crop growth, elevated levels of nitrate suppress nodulation and N_2 fixation. With continued presence of nitrate, the crop's reliance upon N_2 fixation can remain low, leading to reduced plant N, seed yield and seed N if the two sources of N (nitrate and N_2 fixation) together can't satisfy the crop's demand (Herridge and Brockwell, 1988; Wiersma and Orf, 1992). Low N_2 fixation can also mean a net loss of N from the soil when the high protein seed is harvested (Herridge and Holland, 1992), and increased need for fertilizer N inputs.

Other approaches for increasing soybean N_2 fixation involve optimizing the numbers and effectiveness of rhizobia in the rooting zone, through strain selection and inoculation techniques, and through plant breeding for promiscuous or selective nodulation (Cregan and Keyser, 1986; Devine, 1984; Greder et al., 1986; Kueneman et al., 1984; Kvien et al., 1981).

Studies to define genetic variation in N_2 fixation

Variation in nitrate tolerance within a large and diverse germplasm collection of soybean (489 genotypes) was reported by Betts and Herridge (1987). In the program, initiated in 1980, plants were assessed for growth, nodulation and N_2 fixation (relative ureide-N in xylem sap and plant parts). The first two cycles of screening involved culturing the plants in sand-filled pots in a glasshouse, supplied with either nitrate-free nutrients or nutrients containing 2.5 m*M* nitrate. A further two cycles were conducted in high-nitrate field soils.

There were large variations in responses to nitrate (Table 11; see also Betts and Herridge, 1987). From the original 489 genotypes, 66 'nitrate-tolerant' lines were identified on the basis of an overall index which combined the three ureide indices and the nodulation value. The second screen was similar to the first in identifying variation and confirmed the consistency of 32 of the original 66 'tolerant' lines. Genotypes of Korean origin displayed higher than average levels of nodulation and N_2 fixation in the presence of nitrate. Of the original 19 Korean lines, 15 (80%) were included in the second screening, and nine (47%) were selected for the third round of (field) screening. Only 5% of the remaining 470 genotypes were selected as high-fixing after the two glasshouse screenings. It became apparent also that substantial differences in tolerance to nitrate occurred within the group of commercial cultivars e.g. Davis and Lee had greater tolerance than Bragg.

In the third year, 40 genotypes were sown into a high nitrate soil in the field (Herridge and Betts, 1988). The genotypes showing the highest levels of nodulation and N_2 fixation under these conditions were all Korean lines (Table 12). They had similiar shoot yields to commercial cultivars, Bragg and Davis, suggesting that increased N_2 fixation reduced their use of soil N. Post-harvest measurements of soil nitrate confirmed this; up to 34 kg ha^{-1} additional N was recovered from the Korean plots immediately after grain harvest compared with the Bragg plots. Seed yield of the Korean lines was, on average, 30% less than that of Bragg, due to a combination of shattering, early maturity and poor agronomic type. Correlation matrices among the indices of nodulation and N_2 fixation and plant growth and grain yield revealed independence between the symbiotic- and yield-related characters. Therefore, the Korean lines appeared to be suitable for use as high-fixing donor parents in a breeding programme with selection for both grain yield and N_2 fixation.

Subsequent comparisons of Korean genotypes, 466 and 468, with commercial cultivars, Bragg and Davis, and mutants of Bragg, nts1007 and nts1116 (Carroll et al., 1985b), at five field sites showed that the Korean genotypes nodulated better than Bragg, Davis and nts1116 and were about equal to nts1007 (Herridge et al., 1990). Values for Pfix, estimated using both xylem ureide and natural ^{15}N abundance methods, were similar for the two Korean genotypes, nts1007 and Davis. Bragg had the lowest values for Pfix, with nts 1116 intermediate between Bragg and the other four. These levels of symbiotic activity of the Korean genotypes were in spite of low plant and seed yields and early maturity.

Results from the four years of screening indicated that differences in nodulation between the Korean genotypes and commercial cultivars were observed only when the symbioses were stressed, i.e. moderate to high nitrate supply in glasshouse sand culture and in the field or low numbers of soybean rhizobia in the field. In the absence of stress, nodulation of the two groups was similar. Thus, enhanced nodulation of the Korean genotypes was not mediated through a loss of the autoregulatory processes that limit nodulation, e.g. nts mutants (Delves et al., 1986, 1987), or through an altered ability to assimilate and metabolize nitrate (Betts and Herridge, 1987), but more likely resulted from more efficient rhizobial infection and/or nodule initiation.

Serraj et al. (1992) also examined a large population of soybean genotypes for tolerance to nitrate. From the initial screening of 158 lines, they selected five 'tolerant' lines and compared them with sensitive cultivar Kingsoy for nodulation, ARA, nodular nitrate reductase activities and ARA response to oxygen. The most tolerant line, Tielingbaime, maintained enhanced nodulation and N_2 fixation (ARA) in the presence of 3 m*M* nitrate, compared with the other 'tolerant' lines and the sensitive Kingsoy (Table 13). The nodulation index was about twice that of Jiling 13 and about 15 times that of Kingsoy. The variation in ARA was similar. Tielingbaime also had the lowest nodule NRA, suggesting that its capacity to maintain N_2 fixation in

Table 11. Mean values for the ureide indices of N_2 fixation and for nodulation of genotypes of soybean screened for nitrate tolerance (2.5 m*M* nitrate-N, supplied with nutrients). A total of 489 genotypes were included in the first screen and 87 in the second. Data shown are for selected groups of high and low-fixing lines (source: Betts and Herridge, 1987)

	Relative ureide-N values (%)			Nodulation index[a]
	Xylem sap	Shoots	Roots	(%)
First glasshouse screen				
High N_2-fixing lines (66)	43	39	35	3.3
Low N_2-fixing lines (9)	10	5	3	1.2
Second glasshouse screen				
High N_2-fixing lines (32)	38	20	29	3.7
Low N_2-fixing lines (2)	15	4	10	1.8
Bragg	16	6	8	1.8

[a] (nodule mass/shoot mass) × 100.

Table 12. Assessments of nodulation, N_2 fixation and yield of selected 'nitrate tolerant' Korean genotypes of soybean and commercial cultivars in a high nitrate soil at Breeza, New South Wales, 1985 (source: Herridge and Betts, 1985, 1988)

Genotype	Nodulation		Pfix	Shoot DM	Grain yield
	Wt (mg plant^{-1})	No plant^{-1}	(%)[a]	(g plant^{-1})	(t ha^{-1})
Nitrate tolerant					
Korean 466	376	34.5	31	45.9	1.6
Korean 468	254	16.8	18	43.3	1.7
Korean 469	176	19.5	22	41.6	1.4
Korean 464	319	16.5	11	48.1	1.5
Commercial					
Bragg	24	2.0	0	39.7	2.2
Davis	40	1.3	0	48.5	2.2

[a] Assessed during mid podfill using the xylem ureide technique (Herridge and Peoples, 1990).

Table 13. Nodulation, N_2 fixation and nodule nitrate reductase activity (NRA) of nitrate tolerant and sensitive lines of soybean, grown in pots supplied with 3 m*M* nitrate (source: Serraj et al., 1992)

Line	Nodulation index[a] (%)	ARA		Shoot dry wt	NRA	
		per plant	sp.act.		cytosol	bacteroid
		(μmol C_2H_2 h^{-1})		(g plant^{-1})	(μmol NO_2 h^{-1} g^{-1} f. wt)	
Tielingbaime	3.0	72.6	53.0	46.2	1.42	0.03
Adepta	2.3	42.0	41.5	43.1	1.40	0.43
Jiling 13	1.6	18.8	29.4	40.8	1.58	1.60
Kingsoy	0.2	4.8	-	32.0	n.m.	n.m.

[a] (nodule mass/shoot mass) × 100; n.m. not measured

the presence of nitrate could be due to both enhanced nodulation and reduced uptake of nitrate.

Plant mutagenesis was first used to generate pea with greatly enhanced nodulation and with a degree of nitrate tolerance (Jacobsen and Feenstra, 1984). Initial studies with soybean at the University of Illinois, USA, focussed on the generation and assessment of induced mutants of Williams soybean for defective nitrate reductase (NR) activity (Nelson et al., 1983; Ryan et al., 1983). Mutants were produced with defective constitutive NR activity but with unaltered N_2 fixation and nitrate tolerance.

Carroll et al. (1985a), working with Bragg soybean, produced 15 ethyl methanesulfonate (EMS)-induced mutants from 2,500 families of M_2 seedlings that formed up to 40 times the number of nodules as the parent and displayed increased acetylene reduction (N_2 fixation) activity in the presence of nitrate. These mutants [termed nts (nitrate-tolerant symbiotic)] were described also as supernodulators because they produced greater numbers of nodules in the absence of nitrate and appeared to be defective in the autoregulatory control of nodulation (Carroll et al., 1985a; Delves et al., 1986, 1987). Two categories of mutants were described, namely intermediate and extreme supernodulators. Genetic analysis indicated that the increased nodulation was controlled by a single Mendelian recessive gene operating, in the case of soybean, through the shoot (Delves et al., 1986; Lee et al., 1991). There was no evidence of host × rhizobial strain specificity affecting expression of the supernodulation trait (Carroll et al., 1985b; Gremaud and Harper, 1989). Similar mutant phenotypes have now been selected from other cultivars of soybean, e.g. Williams (Gremaud and Harper, 1989), Elgin 87 (Buzzell et al., 1990), and Enrei (Akao and Kouchi, 1992).

The initial assessments of nodulation and N_2 fixation of the supernodulating mutants under glasshouse conditions indicated substantial nitrate tolerance compared with the parent cultivars (Table 14). In the presence of either 5 or 5.5 m*M* nitrate, nodulation and ARA were increased in young plants 10 to 20 fold. Yield data, however, indicated that the increased symbiotic activity was associated with a 30–40% reduction in total growth and with restricted root growth (Day et al., 1986; Gremaud and Harper, 1989). Ohyama et al. (1993) subsequently showed that the restricted root growth resulted in a reduced capacity to absorb nitrate.

The apparent contradiction of greatly increased nodulation and N_2 fixation of the mutants coupled with depressed growth suggested inefficiencies of nodule function and raised doubts about the validity of comparing ARA values of mutants and parent cultivars. Indeed Day et al. (1987) reported that the nodules of nts283 had an altered morphology in having less haemoglobin and a smaller infected area and lower specific nitrogenase activity, i.e. on a nodule dry wt basis. They did find, however, that nitrogenase activity per unit bacteroid protein was identical with the parent Bragg. A number of studies indicated that the concentration of N in shoots of the mutants was 30–50% greater than that of the parent, which could partially account for the discrepancy of similar or enhanced AR (nodule) activity and depressed growth (Day et al., 1986; Hansen et al., 1992a; Ohyama et al., 1993).

Subsequent assessments of the soybean mutants over whole growth cycles using more reliable methods for measuring N_2 fixation, e.g. ^{15}N and ureide analysis, indicated partial nitrate tolerance of N_2 fixation and reduced uptake of mineral N (Eskew et al., 1989; Hansen et al., 1989; Wu and Harper, 1991). It is likely that both traits are associated with supernodulation (Hansen et al., 1992b). Hansen et al. (1989) reported a pot study showing that enhanced N_2 fixation of the supernodulating mutants during early vegetative growth, considered to be associated with the characteristically rapid and profuse nodulation, was generally not maintained during reproductive growth. The exception was at the highest level of nitrate (10 m*M*) in which mutants nts 1116 and nts1007 fixed 31% and 250% more N, respectively, than Bragg.

Evaluations of the supernodulators in the field in the US and in Australia have been restricted to just four studies, involving mutants of Bragg and Williams soybean (Herridge et al., 1990; Pracht et al., 1994; Song et al., 1995; Wu and Harper, 1991). Grain yields of the supernodulators in the US studies were reduced by 20 – 33% (Wu and Harper, 1991) and 28 – 41% (Pracht et al., 1994) compared to Williams, the parent cultivar, when averaged over years and treatments. Data on N_2 fixation confirmed results of the glasshouse studies by showing that the enhanced early N_2 fixation of the mutants was not maintained during later reproductive growth. Results indicated also greater tolerance to fertilizer N, but infrequent increases in N_2 fixation on an area basis because the higher or similar Pfix values of the mutants could not compensate fully for whole plant yields that were commonly 20–30% below those of normally-nodulating cultivars. Results of the Herridge et al. (1990) were similar.

Table 14. Nodulation, C_2H_2 (N_2) reduction and growth of supernodulating mutants of Bragg and Williams soybean and their normally-nodulating parents in the presence and absence of nutrient-supplied nitrate

Nitrate/genotype	Nodulation		ARA (umol h^{-1})		Growth	
	No (plant^{-1})	mass (mg plant^{-1})	Plant (plant^{-1})	Sp. Act. (g nod^{-1})	Shoot (g plant^{-1})	Root
Bragg - Australia (source: Carroll et al., 1985a)[a]						
0 nitrate						
Bragg	69	886	1.2	n.d.	21.1	n.d.
nts 382	431	1583	1.0	n.d.	11.8	n.d.
5.5 m*M* nitrate						
Bragg	29	174	0.2	n.d.	40.5	n.d.
nts 382	414	1886	2.1	n.d.	28.5	n.d.
Williams - USA (source: Gremaud and Harper, 1989)[b]						
0 nitrate						
Williams	187	95	5.3	56	0.52	0.32
NOD1–3	473	137	9.8	72	0.32	0.12
5.0 m*M* nitrate						
Williams	7	2	0	0	5.19	1.89
NOD1–3	122	88	2.0	16	3.42	1.13

[a] Harvested 64 days after sowing; growth (whole plant) and nodule mass fresh wt basis; ARA data from second experiment, +N plants suplied with 2.75 m*M* nitrate, harvested at 28 days.
[b] Harvested 21 days after sowing.
n.d not determined.

The experiments reported by Song et al. (1995) are noteworthy (Table 15). They showed higher N_2 fixation activity for the supernodulators, relative to commercial cultivar, Centaur, and either reduced grain yields (original mutants) or equivalent yields (supernodulators derived from crossing the mutants with commercial cultivars). However, cultivar Manark was superior to both supernodulators and other commercial cultivars with a combination of high grain yield and high N_2 fixation activity. The major advantage of the supernodulators was to increase the yield of a following cereal crop, sown immediately after soybean harvests.

Wu and Harper (1990) had previously observed that naturally-occurring soybean lines exhibiting the supernodulation phenotype had not been identified, even though the gene control for the trait was relatively unstable and pressure of moderate to high levels of soil nitrate in plant improvement programs could favour selection of such a phenotype. They concluded that the soybean's N metabolism may not limit production of biomass and grain and therefore nitrate tolerance through supernodulation was of no direct advantage to the plant (i.e. for yield).

Results of Song et al. (1995) indicate that a major advantage of supernodulation may be the rotational or carry-over effect for a subsequent cereal crop. The exact cause of the carry-over effect warrants further study but it is likely associated with higher levels of plant-available N in the soil after grain harvest because of reduced uptake of soil nitrate by the supernodulators together with release through mineralization of greater quantities of N from the legume residues. Assessments of N_2 fixation, using xylem ureide and natural ^{15}N abundance methods indicated some potential for reduced uptake of soil N by the supernodulators, compared with commercial cultivars. Low nitrogen harvest indices (not determined in this study), increased concentrations of N in plant tissues (e.g. Hansen et al., 1992a) and the relatively large nodule weights would contribute to increased N in the residues following growth of the supernodulators. Nodule N alone could amount to 15–20 kg N ha^{-1} which represents a significant proportion of the 20–30 kg N ha^{-1} required for the increased cereal yields. Thus, although the large

Table 15. Nodulation, N_2 fixation (relative ureide-N in xylem sap), grain yield and yields of a cereal crop sown immediately after soybean harvest of normally-nodulating and supernodulating genotypes, relative to commercial cultivar, Centaur (source: Song et al., 1995). Data are summarized from results of 6 years of field experiments in Queensland, Australia

Genotype	Nodulation No.	Nodulation Mass	Relative ureide-N	Grain yield	Yield of subsequent cereal
	(% of centaur)				
Centaur	100	100	100	100	100
Manark	n.d.	n.d.	120	120	102
Bragg	107	113	n.d.	96	91
nts1116	245	236	121	96	131
777–36[a]	n.d.	n.d.	119	102	n.d.
nts1007	338	245	113	84	121
T89238[b]	n.d.	n.d.	115	101	n.d.

n.d. not determined

Absolute values for Centaur were: nodule no. = 28; nodule mass = 380 mg plant^{-1}; relative ureide-N = 43%; grain yld = 2.3 t ha^{-1}; subsequent cereal yld = 3.9 t ha^{-1}.

[a] Line derived from nts1116 × Nessen cross.

[b] Line derived from nts1007 × Bossier cross.

Table 16. Pedigrees of the 11 populations of soybean and numbers of single plants or lines assessed at each generation for either plant and seed traits (F_2-F_7 generations), yield (F_4-F_7 generations) or N_2 fixation (F_2, F_6 and F_7 generations)

Population - pedigree	F_2 single plants 1986/87	F_2-derived in F_3 gen. 1987	F_3-derived in F_4 gen. 1987/88	F_3-derived in F_5 gen. 1988/89	F_3-derived in F_6, F_7 gen. 1989–91
	(No.)				
A. 464 × Valder	161	22	144	36	3
B. Valder × 464	61	6	45	5	0
C. Valder × 466	87	16	89	26	1
D. Valder × 468	49	8	58	14	1
E. Reynolds × 466	72	9	73	25	8
F. Reynolds × 464	14	1	11	1	0
G. 464 × Forrest	38	5	49	10	3
H. Forrest × 469	119	14	112	37	6
J. Bossier × 464	121	6	46	10	0
K. 468 × Bossier	93	12	109	29	10
L. Bossier × 469	34	5	47	7	1
Total	849	104	783	200	33

number of nodules on the roots of the supernodulators are to some degree parasitic on the host (e.g. Hansen et al., 1992c), they may represent a source of N for succeeding crops.

All programs that aim to breed legume cultivars with enhanced N_2 fixation must meet certain criteria. Such criteria for commercial application of the supernodulators include:

- heritability and stability of the nodulation phenotype,
- improved N_2 fixation related to increased nodulation,
- increased N benefits to the soil-plant system,
- grain yields equal to current high-yielding cultivars.

The Song et al. (1995) study indicates that progress has been made towards these goals and the breeding program, conducted by Dr Song at Pacific Seeds in southern Queensland, Australia, is continuing. Similarly, the Williams mutants are currently being backcrossed with high yielding lines at the University of Illinois in an attempt to improve performance.

Improvement/breeding programs

We report details of the breeding program of Herridge and Rose, Australia, in which the four Korean genotypes, 464, 466, 468 and 469 (Table 12) were used as high-fixing parents in crosses with commercial cultivars, Valder (maturity group [MG] IV), Reynolds (MG VI), Forrest (MG VI) and Bossier (MG VIII) (Herridge and Rose, 1994). The breeding protocol differed from that used by Bliss and co-workers with common bean (Fig. 2) in a number of ways: material was screened for the most part in high nitrate, rather than low nitrate soils; the xylem ureide method was used to assess N_2 fixation; initial assessments of N_2 fixation were with individual F_2 plants, although later assessments (F_6 and F_7) involved populations of plants. A summary of activities are presented in Table 16.

Nitrogen fixation was assessed on individual F_2 plants using the xylem ureide method. Sap was extracted from the top half of each plant leaving the lower half to continue growth and to produce seed for harvest (Herridge et al., 1988a; Herridge and Rose, 1994). The relative abundance of ureide-N of the F_2 plants varied between 2 and 55%, indicating segregation for N_2 fixation activity (Fig. 3). There was no evidence of heterosis, in contrast to results reported by Seetin and Barnes (1977) for alfalfa, by Hobbs and Mahon (1982) for pea and by Ronis et al. (1985) for soybean. Average relative ureide-N values for the 11 F_2 populations were surprisingly constant (24 – 29%, equivalent to Pfix values of 13 – 20%) and were between the lower values of the commercial parents (17 – 28%; Pfix values 2 –19%) and the higher values of 3 of the 4 Korean parents (39 – 42%; Pfix values 36 – 40%). The relative ureide-N value for fourth Korean parent, 464, was 27%, about the same as for Reynolds, the best commercial parent. Although average N_2 fixation activities of the F_2 populations were below those of the best Korean lines, 35 individual F_2 plants displayed equally high levels of N_2 fixation i.e. relative ureide-N > 40%.

The F_2 populations were culled on the basis of N_2 fixation (xylem relative ureide-N value >31%) (Fig. 3), plant type (agronomic rating >2 on a scale of 1 to 6) and seed colour (yellow, green or yellow-green). Evidence of linkages between N_2 fixation and other more easily determined plant characters was also sought. Correlation matrices of these characters showed no such linkages. Problems could have occurred if N_2 fixation was found to be linked to certain traits of the Korean genotypes e.g. black, brown seeds, poor agronomic type. On the other hand, linkage to other more benign traits could have led to simpler procedures for selecting material.

At the commencement of this study, the genetic control of enhanced N_2 fixation in the Korean genotypes was unknown. Major genes had been identified which influence *Bradyrhizobium* compatibility (Caldwell, 1966; Vest, 1970; Vest and Caldwell, 1972) and hypernodulation (Carroll et al., 1985) of soybean. Frequency distributions of relative ureide-N values in each of the 11 F_2 populations were normal, with no evidence of discontinuities in these distributions that would suggest a major gene segregation.

Culling of F_3-derived lines in the F_4 and F_5 generations was based on yield and agronomic traits. It was only in the F_6 and F_7 generations that N_2 fixation was again assessed. A number of the F_3-derived lines were clearly superior and, importantly, stable in N_2 fixation (Fig. 4). In both the F_6 and F_7 generations, these lines had relative ureide-N of around 40%, compared with consistently lower values or more variable values for other lines. A number of the consistently high-fixing lines (A82–3, D22–8, K78–1 and E68–5) were subsequently used as parents in a backcrossing program (I A Rose and D F Herridge, unpubl. data).

The values for yield and N_2 fixation presented in Table 17 summarize the progress made in the first cycle of selection. At the high nitrate sites in the F_6 and F_7 trials, Forrest was obtaining 27 and 33% of its N

Table 17. Days to flowering, seed yield and Pfix values at F_6 and F_7 for lines selected for backcrossing and for high (Korean 468) and low N_2-fixing (Forrest) parents

Line	Original cross	Flowering	Seed yield mean 3 sites		Pfix[a] high nitrate sites	
			F_6	F_7	F_6	F_7
		(days)	(t ha^{-1})		(%)	
Forrest		50	2.67	2.66	27	33
D22–8	Valder × Korean 468	46	2.08	2.29	47	55
A82–3	Korean 464 × Valder	52	2.00	1.70	49	49
K78–1	Korean 468 × Bossier	57	1.91	2.09	46	54
A46–4	Korean 464 × Valder	58	2.24	2.20	51	56
Korean 468		43	0.95	0.58	47	52

[a] Proportion of plant N derived from N_2 fixation, calulated using formula in Herridge and Peoples (1990) from relative ureide-N values of xylem sap, measured during late vegetative growth/flowering.

from N_2 fixation at the time of sampling. By contrast, N_2 fixation by Korean 468 accounted for 47 and 52% of current inputs of N. Lines D22–8, A82–3, A46–4, and E72–3 were outstanding in respect of N_2 fixation, with Pfix values of around 50%. The Pfix value for commercial cultivar, Reynolds, was around 41%, suggesting that this cultivar may already have some of the symbiotic characteristics being sought for commercial cultivars, i.e. tolerance of the suppressive effects of soil nitrate on nodulation and N_2 fixation.

Grain yields of the F_3-derived lines in the F_6 and F_7 generations, although as much as three times larger than yields of Korean 468, could not match those of the highest-yielding commercial cultivar, Forrest. Some lines gave yields comparable to older commercial cultivars such as Bossier. In particular, lines A46–4, K78–1 and D22–8 had average yields across the 6 trials of >2.0 t ha^{-1} (Table 17).

Herridge and Rose concluded that selection of the field site for assessing N_2 fixation was vital for discriminating the high-fixing, nitrate-tolerant lines. The enhanced capacity for N_2 fixation of these lines and of the Korean parents was expressed only when N_2 fixation of the commercial cultivars, e.g. Forrest, was suppressed by the high nitrate soils. Data from the F_6 and F_7 trials support this by showing that correlations across sites and/or seasons were improved when the soils were high in nitrate (see Table 5).

Yield of the high-fixing, nitrate-tolerant material needs to be improved by about 20% before commercial release. A second cycle of selection was commenced in 1991 with crossing of lines D22–8, A82–3, A46–4 and K78–1 with high yielding genotypes. Single seed descent lines were formed as F_4 single plant progeny and more than 1400 lines from six populations were field-tested in F_5 and F_6 trials for phenology, growth habit, lodging, disease resistance, yield, shattering and seed oil and protein. The best lines from those assessments will be evaluated for N_2 fixation in the F_7 generation.

Controlled nodulation — selection and breeding

Legume inoculation with rhizobia and bradyrhizobia is a long-established and successful practice, especially with particular crops in the more technically-advanced countries. Vincent (1965) and others have argued that it is a desirable practice in most agricultural soils throughout the world although Date (1977) cautioned that the need to inoculate was not universal and should be carefully determined for each individual situation. Because of this and because of the exacting technology required for production, distribution and use of inoculants, the practice of inoculation remains the exception rather than the rule (Brockwell and Bottomley, 1994).

In the less developed countries in particular, there would be substantial advantages in growing legumes that were nodulated by highly effective rhizobia already in the soil. In Africa in the 1970's and 1980's, soybean was considered a new crop that required inoculation for effective nodulation (Nanju, 1980). In response, scientists at the International Institute of Tropical Agriculture (IITA) sought to exploit the fact that small farmers in Nigeria had grown well-nodulated soybean such as local cultivar 'Malayan' for more than 30 years with low inputs and without inoculation (Nangju, 1980). Rapid progress could then be made in establishing soybean as an industry in Africa if 'promiscuous' nodulation could be combined with the yield, quality and disease resistance traits of improved 'American' cultivars.

Nanju conducted field experiments in the late 1970's at two sites in Nigeria in which soybean of Asian and American origin were evaluated for nodulation and yield in the absence of inoculation and when inoculated. The two Asian cultivars, Malayan and Orba, nodulated successfully with the indigenous rhizobia, resulting in effective symbioses (Table 18). They did not respond to inoculation. The US bred cultivars, Bossier and Jupiter, on the other hand, nodulated poorly without inoculation and showed large increases in nodule number and mass and in grain yield with inoculation. The American cultivars had much greater yield potential, particularly when compared with Malayan. The ability of Malayan to nodulate with native rhizobia was confirmed in other experiments in Nigeria and in Tanzania (Nanju 1980).

In other trials conducted by IITA scientists, 400 genotypes were screened for promiscuous nodulation (i.e. effective nodulation without inoculation) on low N soils at five sites in Nigeria. Only 10 of the genotypes were promiscuous nodulators — Malayan, M-351, TGm 120, TGm 119, Obo, Indo 216, Indo 180, Indo 226, Indo 243 and Orba (Pulver et al., 1985). The promiscuous nodulation of a number of these genotypes was confirmed in a pot experiment in which rhizobial isolates from the field sites were used as inoculum. Other promiscuous nodulators were identified in trials in Tanzania, Zambia and the Ivory Coast (Chowdhury, 1977; Kueneman et al., 1984) and, in some cases, had already been used as parents in breeding programs because of their vigour under the local conditions.

The potential of combining promiscuous nodulation and yield traits through breeding was first shown by Chowdhury and Doto (1982). F_3 lines, derived from a Bossier × IH 192 cross segregated for both yield and N_2 fixation (Table 19). Lines 3 and 16 retained the earliness and yield of Bossier together with the promiscuous nodulation and enhanced N_2 fixation of parent IH 192. Lines 1 and 10 were high yielding but segregated for low nodulation and N_2 fixation.

Grafting experiments reported by Pulver et al. (1985) provided further evidence that promiscuous nodulation could be used to genetically enhance N_2 fixation in soybean. The experiments showed that the symbioses between the roots of promiscuous nodulators, Malayan and Orba, and native soil rhizobia could supply the N required for the shoots of high yielding, improved lines, Jupiter and Bossier, to realize full yield potential. Subsequent hybridization of the Asian and US cultivars by IITA scientists successfully combined high yield and desirable agronomic traits with promiscuous nodulation and enhanced N_2 fixation for a large range of progeny (e.g. Table 20).

Methods or assessing nodulation and N_2 fixation of uninoculated breeding material include visual scoring of nodulation, ureide concentrations in leaf tissue and shoot colour and grain and shoot yield (Kueneman et al., 1984).

In other environments where inoculation is an option, populations of infective rhizobia often exist in high numbers in the soil and represent a formidable barrier to the introduction of more effective strains (e.g. Devine, 1984). In the USA, large populations of the soybean bradyrhizobia have developed with cropping so typically less than 10% of nodules on soybean are formed by the inoculant and yield responses to inoculation are rare. In these situations, N_2 fixation and yield may be limited by the low effectiveness of the native soil rhizobia (Greder et al., 1986; Kvien et al., 1981). The more recent study of Vasilas and Fuhrmann (1993) showed that Forrest soybean nodulated with highly-effective strain USDA 122 fixed 29% more N_2 and produced 31% more grain than plants nodulated with the native soil rhizobia.

One approach to this problem was to select genotypes of soybean with affinity for highly effective rhizobial strains. Kvien et al. (1981) screened 1600 genotypes for ability to nodulate with native soil rhizobia in Minnesota, USA, and for ability to preferentially nodulate with a highly effective inoculant strain, USDA 110, in the presence of large numbers of native soil rhizobia. They identified several genotypes which nodulated poorly with the native strains and gave increased recoveries of USDA 110 and increased yields when inoculated with USDA 110. In a subsequent report, however, one of the genotypes which had nodulated most successfully with USDA 110 exhibited nodule occupancy that was no different from the check cultivar (Moawad et al., 1984). Greder et al. (1986) concluded that selection of genotypes with high recoveries of USDA 110 would be difficult and suggested instead

Table 18. Nodulation and yield of Asian and American cultivars, either inoculated or grown without inoculation in the field in Nigeria

Cultivar	Nodulation				Grain yield	
	No.		Mass (mg plant^{-1})		(t ha^{-1})	
	-inoc	+inoc	-inoc	+inoc	-inoc	+inoc
Asian						
Malayan	42	50	175	235	0.59	0.57
Orba	42	43	255	310	1.80	1.88
American						
Bossier	7	47	47	311	1.43	2.33
Jupiter	10	35	86	274	2.12	2.97

Table 19. Yield and symbiotic traits of improved (Bossier) and promiscuously nodulating (IH 192) parent lines and derived lines, grown without inoculation in the field in Tanzania (source: Chowdhury and Doto, 1982)

Line	Nodule mass (mg plant^{-1})	ARA (μmol h^{-1})	Flowering (days)	Seed yield (g plant^{-1})
Bossier	2	1.9	40	5.2
Line 3	38	10.3	38	6.8
Line 16	42	10.3	38	11.2
Line 1	14	2.6	38	8.4
Line 10	15	1.2	38	9.0
IH 192	67	6.8	68	4.9

Table 20. Yields (t ha^{-1}) of soybean lines derived from crosses of improved and promiscuously nodulating genotypes, grown without inoculation in the field in Nigeria (source: Kueneman et al. 1984). Bossier is included as an improved check

Line	- fertilizer N	+ fertilizer N
Promiscuous bred lines		
TG×326–034D	2.55	2.55
TG×330–054D	1.98	2.37
TG×457–060C	2.30	2.47
Improved check		
Bossier	0.90	1.60

that selection of genotypes on the basis of nodule mass may be warranted.

A second approach, initiated during the 1970's and 1980's at the USDA laboratories, Beltsville (USA), was to produce soybean cultivars that were more selective in nodulation. The plant would bypass the resident rhizobia in the soil to be nodulated by selected, highly effective inoculant strains (Cregan and Keyser, 1986; Devine, 1984).

This strategy of exploiting the host's capacity to exclude certain rhizobia from forming nodules had its derivation in the identification of host genes affecting nodulation. Five genes have been identified that restrict nodulation in soybean (Table 21). These host genes control nodulation at the species, serogroup and strain level within the soybean rhizobia (Keyser and Li, 1992). The rj_1 gene, responsible for non-nodulation, has been transferred through conventional breeding to several soybean cultivars for use in inoculation and N_2 fixation research. Dr T E Devine's program at USDA, Beltsville, sought to exploit the rj_1 gene by combining soybean cultivars containing the gene with rhizobia that could overcome its restriction. He had established previously that such rhizobia existed in natural populations in the soil in very low numbers. From examination of 100,000 – 200,000 plants in the field, he isolated around 300 strains of rhizobia and tested them for effectiveness. None of them was particularly effective and the program has been put on hold.

The second program at USDA, Beltsville, involved Drs P Cregan and H Keyser and co-workers. They aimed to improve N_2 fixation of soybean by restricting the plant's capacity to nodulate with the less effective but ubiquitous and highly competitive USDA 123 group of strains (serocluster), thereby leaving the plant free to nodulate with highly effective inoculant or indigenous strains, e.g. USDA 110. Progress achieved has been to:

- identify genotypes (PI 371607 and PI 377578) that restrict USDA 123 under controlled (pot study) conditions and in the field ((Cregan and Keyser, 1986) (Table 22),
- identify genotypes PI 417566 and PI 283326 that, together with PI 377578, restrict nodulation by 20 of 23 strains from serocluster 123,
- identify a specificity between plant genotype and serogroup of the restricted strains,
- determine by classical genetic analysis that the restriction of USDA 123 by PI 377578 and the reduced competitiveness of strain MN1–1c on PI 417566 are dominant traits, similiar to restriction by the Rj2, Rj3 and Rj4 alleles (Cregan et al., 1989a, 1989b).

Later work aimed to determine if the genes responsible for nodulation restriction were independently assorted or closely linked. If the latter were the case, then combining the various restrictions through crossing would prove difficult. There may be other problems which may make the task of breeding a single plant genotype that combines all of the desired levels of restriction (nodulation control) very difficult. For instance: the manufactured genotype may also restrict desirable, effective strains; undesirable, infective strains may build in the soil over time, thereby creating the problem all over again; the expression of the restriction characters influenced by environment (temperature, growth medium), similiar to environmental influences affecting nodulation of pea.

Conclusions

Breeding crop legumes with enhanced capacity for N_2 fixation has been promoted as a worthwhile and economic goal since the 1970's. At a meeting in the US in 1978, Dr Peter Graham presented results of research at CIAT, Colombia, where they had already field tested 600 cultivars of common bean for N_2 fixation and had intensively studied 60 of those. By the early 1980's, a breeding program was in place at CIAT (Graham, 1981). A second program involving common bean was initiated in 1980 by Dr Fred Bliss and collaborators at the University of Wisconsin, USA, using elite material from CIAT. Other programs were also initiated at that time involving a range of species and research organizations, e.g. soybean in Africa and Australia, groundnut in India and the US (see Table 2). From that considerable effort, only a very small number of cultivars have been released with improved capacity for N_2 fixation (Bliss et al., 1989).

There may be two major reasons for this apparent lack of success. Firstly, it is a difficult task to combine a single, desirable trait like N_2 fixation with other agronomic and yield traits. Secondly, techniques for accurate assessment of N_2 fixation by field-grown legumes were not available. In recent years, more use has been made of ^{15}N methods (e.g. St Clair et al., 1988), but this technology has real limitations because of cost and the slow turnaround in sample analysis. In a NSW Agriculture (Australia) program to breed cultivars of soybean with enhanced N_2 fixation, the xylem ureide technique proved invaluable for field assessments.

Table 21. Host genetic control of nodulation in soybean (sources Devine, 1984; Keyser and Li, 1992)

Allele	Phenotype	Reference
rj_1	Non-nodulating with virtually all rhizobial strains	1
Rj_2	Cortical proliferations or ineffective nodules formed by rhizobia in serogroups 6(c1) and 122 with cv. Hardee	2
Rj_3	Small nodule-like structures formed by USDA 33 with cv. Hardee	3
Rj_4	Ineffective cortical proliferations by USDA 61 with cvs Hill and Dunfield	4
Dominant	Rudimentry nodules by USDA 205 with cv. Kent	5

1. Williams and Lynch 1954; 2. Caldwell 1966; 3. Vest 1970; 4. Vest and Caldwell 1972; 5. Devine 1984.

Table 22. Nodule occupancy in two Pl genotypes of soybean and Williams, grown in the field (source Keyser and Li, 1992)

Genotype	Percent of nodules occupied by		
	USDA 123	USDA 122, USDA 138	Other
Commercial check			
Williams	76	20	4
Restrictive genotypes			
P1371607	3	89	8
P1377578	5	92	3

The technique was used as a single plant, non-lethal assay at the F_2 stage and for assessing replicated plots in later generations, i.e. in the F_3-derived F_5, F_6 and F_7 generations. For the F_2 individual plants, it was necessary to sample xylem sap from around 800 units (plants) and to complete the chemical analysis of the saps within a month. This was done successfully. This assessment could not have been achieved using any other method. With the application of new molecular techniques and the development of bioindicator systems using reporter genes to provide information on the relative levels of ureides, the ureide assay has the potential to become an even more powerful tool for the selection and breeding of (ureide producing) legumes for enhanced N_2 fixation (Wilson et al., 1994).

References

Akao S and Kouchi H 1992 A supernodulating mutant isolated from soybean cultivar Enrei. Soil Sci. Plant Nutr. 38, 183–187.

Alwi N, Wynne J C, Rawlings J O, Schneeweis T J and Elkan G H 1989 Symbiotic relationship between *Bradyrhizobium* strains and peanut. Crop Sci. 29, 50–54.

Armager N, Mariotti A, Mariotti F, Durr J C, Bourguignon C and Lagacherie B 1979 Estimate of symbiotically fixed nitrogen in field grown soybeans using variations in ^{15}N natural abundance. Plant and Soil 52, 269–280.

Arrendell S, Wynne J C, Elkan G H and Isleib T G 1985 Variation for nitrogen fixation among progenies of a Virginia × Spanish peanut cross. Crop Sci. 25, 865–869.

Attewell J and Bliss F A 1985 Host plant characteristics of common bean line selected using indirect measures of N_2 fixation. *In* Nitrogen Fixation Research Progress. Eds. H J Evans, P J Bottomley and W E Newton. pp 3–9. Martinus Nijhoff Publ., Dordrecht.

Barnes D K, Heichel G H, Vance C P and Ellis W R 1984 A multiple-trait breeding program for improving the symbiosis for N_2 fixation between *Medicago sativa* L. and *Rhizobium meliloti*. Plant and Soil 82, 303–314.

Bennett J P, Anderssen F G and Milad Y 1927 Methods of obtaining tracheal sap from woody plants. New Phytol. 26, 316–323.

Bergersen F J, Peoples M B, Herridge D F and Turner G L 1990 Measurement of N_2 fixation by ^{15}N natural abundance in the management of legume crops: roles and precautions. *In* Nitrogen Fixation: Achievements and Objectives. Eds. P M Gresshoff, E

Roth, G Stacey and W E Newton. pp 315–322. Chapman and Hall, New York.

Bergersen F J, Turner G L, Gault R R, Chase D L and Brockwell J 1985 The natural abundance of ^{15}N in an irrigated soybean crop and its use for calculation of nitrogen fixation. Aust. J. Agric. Res. 36, 411–423.

Betts J H and Herridge D F 1987 Isolation of soybean lines capable of nodulation and nitrogen fixation under high levels of nitrate supply. Crop Sci. 27, 1156–1161.

Bezdicek D F, Evans D W, Abede B and Witters R E 1978 Evaluation of peat and granular inoculum (*Rhizobium japonicum*) for soybean yield and nitrogen fixation under irrigation. Agron. J. 70, 865–868.

Bliss F A 1993 Breeding common bean for improved biological nitrogen fixation. Plant and Soil 152, 71–79.

Bliss F A and Miller J C Jr 1988 Selecting and breeding grain legumes for enhanced nitrogen fixation. *In* World Crops: Cool Season Food Legumes. Ed. R J Summerfield. pp 1001–1012. Kluwer Academic Publ., London.

Bliss F A, Pereira P A A, Araujo R S, Henson R A, Kmiecik K A, McFerson J R, Teixeira M G and da Silva C C 1989 Registration of five high nitrogen fixing common bean germplasm lines. Crop Sci. 29, 240–241.

Boddey R M, Chalk P M, Victoria R L and Matsui E 1984 Nitrogen fixation by nodulated soybean under tropical field conditions estimated by the ^{15}N isotope dilution technique. Soil Biol. Biochem. 16, 583–588.

Bollard E G 1953 The use of tracheal sap in the study of appletree nutrition. J Exp. Bot. 4, 363–368.

Bollard E G 1960 Transport in the xylem. Ann. Rev. Plant Physiol. 11, 141–166.

Brockwell J and Bottomley P J 1995 Manipulation of rhizobial microflora for improving crop productivity and soil fertility: a critical assessment. Plant and Soil 174.

Buttery B R and Dirks V A 1987 The effects of soybean cultivar, rhizobium strain and nitrate on plant growth, nodule mass and acetylene reduction rate. Plant and Soil 98, 285–293.

Buzzell R I, Buttery B R and Ablett G 1990 Supernodulation mutants in Elgin 87 soybean. *In* Nitrogen Fixation: Achievements and Objectives. Eds. P M Gresshoff, L E Roth, G Stacey and W E Newton. p 726. Chapman Hill, New York.

Caldwell B E 1966 Inheritance of strain-specific ineffective nodulation in soybeans. Crop Sci. 6, 427–428.

Carroll B J, McNeil D L and Gresshoff P M 1985a Isolation and properties of soybean [*Glycine max* (L.) Merr.] mutants that nodulate in the presence of high nitrate concentrations. Proc. Natl. Acad. Sci. USA 82, 4162–4166.

Carroll B J, McNeil D L and Gresshoff P M 1985b A supernodulation and nitrate-tolerant symbiotic (*nts*) soybean mutant. Plant Physiol. 78, 34–40.

Chalk P M 1985 Estimation of N_2 fixation by isotope dilution: an appraisal of techniques involving ^{15}N enrichment and their application. Soil Biol. Biochem. 17, 389–410.

Chapman A L and Myers R J K 1987 Nitrogen contribution by grain legumes to rice grown in rotation on the Cununurra soils of the Ord irrigation area, Western Australia. Aust. J. Exp. Agric. 27, 155–163.

Chaverra M H and Graham P H 1992 Cultivar variation in traits affecting early nodulation of common bean. Crop Sci. 32, 1432–1436.

Chowdhury M S 1977 Response of soybean to *Rhizobium* inoculation at Morogoro, Tanzania. *In* Biological Nitrogen Fixation in Farming Systems of the Tropics. Eds. A Ayanaba and P J Dart. pp 245–253. John Wiley and Sons, Chichester.

Chowdhury M S and Doto A L 1982 Biological nitrogen fixation as a criterion for soybean breeding: preliminary results. *In* Biological Nitrogen Fixation Technology for Tropical Agriculture. Eds. P H Graham and S C Harris. pp 45–48. CIAT, Cali, Colombia.

CIAT 1987 Annual Report of the Bean Production Program, Cali, Colombia. 352p.

Cregan P B and Keyser H H 1986 Host restriction of nodulation by *Bradyrhizobium japonicum* strain USDA 123 in soybean. Crop Sci. 26, 911–916.

Cregan P B, Keyser H H and Sadowsky M J 1989a Soybean genotype restricting nodulation of a previously unrestricted serocluster 123 bradyrhizobia. Crop Sci. 29, 307–312.

Cregan P B, Keyser H H and Sadowsky M J 1989b Host plant effects on nodulation and competitiveness of the *Bradyrhizobium japonicum* serotype strains constituting serocluster 123. Appl. Environ. Microbiol. 55, 2532–2536.

Danso S K A 1988 The use of ^{15}N enriched fertilizers for estimating nitrogen fixation in grain and pasture legumes. *In* Nitrogen Fixation by Legumes in Mediterranean Agriculture. Eds. D P Beck and L A Materon. pp 345–358. Martinus Nijhoff, Dordrecht.

Danso S K A, Hardarson G and Zapata F 1993 Misconceptions and practical problems in the use of ^{15}N soil enrichment techniques for estimating N_2 fixation. Plant and Soil 152, 25–52.

Date R A 1977 The development and use of legume inoculants. *In* Biological Nitrogen Fixation in Farming Systems of the Tropics. Eds. A N Ayanaba and P J Dart. pp 169–180. John Wiley, Chichester.

Day D A, Lambers H, Bateman J, Carroll B J and Gresshoff P M 1986 Growth comparisons of a supernodulating soybean (*Glycine max*) mutant and its wildtype parent. Physiol. Plant. 68, 375–382.

Day D A, Price D G, Schuller K A and Gresshoff P M 1987 Nodule physiology of a supernodulating soybean (*Glycine max*) mutant. Aust. J. Plant Physiol. 14, 527–538.

Delves A, Higgins A and Gresshoff P M 1987 Supernodulation in interspecific grafts between *Glycine max* (soybean) and *Glycine soja*. J. Plant Physiol. 128, 473–478.

Delves A C, Mathews A, Day D A, Carter A S, Carroll B J and Gresshoff P M 1986 Regulation of the soybean-*Rhizobium* nodule symbiosis by root and shoot factors. Plant Physiol. 82, 588–590.

Devine T E 1984 Genetics and breeding of nitrogen fixation. *In* Biological Nitrogen Fixation. Ed. M Alexander. pp 127–154. Plenum Publishing, New York.

Diatloff A, Redden R J and Herridge D F 1991 Correlation between xylem ureide levels and nodulation in field-grown *Phaseolus vulgaris*. Aust. J. Exp. Agric. 31, 679–682.

Doughton J A and McKenzie J 1984 Comparative effects of black and green gram (mung beans) and grain sorghum on soil mineral nitrogen and subsequent grain sorghum yields on the eastern Darling Downs. Aust. J. Exp. Agric. Anim. Husb. 24, 244–249.

Duc G, Mariotti A and Armager N 1988 Measurement of genetic variability for symbiotic dinitrogen fixation in field-grown fababean (*Vicia faba* L.) using a low level ^{15}N-tracer technique. Plant and Soil 106, 269–276.

Eskew D L, Kapuya J and Danso S K A 1989 Nitrate inhibition of nodulation and nitrogen fixation by supernodulating nitrate-tolerant symbiosis mutants of soybean. Crop Sci. 29, 1491–1496.

Evans J, Turner G L, O'Connor G and Bergersen F J 1985 N_2 fixation and conservation of soil mineral nitrogen by *Lupinus angustifolius*. *In* Nitrogen Fixation Research Progress. Eds. H J Evans, P J Bottomley and W E Newton. p 690. Martinus Nijhoff, Boston.

Evans J and Taylor A 1987 Estimating dinitrogen (N_2) fixation and soil accretion of nitrogen by grain legumes. J. Aust. Inst. Agric. Sci. 53, 78–82.

FAO 1992a FAO Production Yearbook, FAO Statistics Series No 112, FAO Rome. 46p.

FAO 1992b FAO Fertilizer Yearbook, FAO Statistics Series No.100, FAO Rome.

George T and Singleton P W 1992 Nitrogen assimilation traits and dinitrogen fixation in soybean and common bean. Agron. J. 84, 1020–1028.

Graham P H 1981 Some problems of nodulation and symbiotic nitrogen fixation in *Phaseolus vulgaris* L.: a review. Field Crops Res. 4, 93–112.

Graham P H and Rosas J C 1977 Growth and development of indeterminate bush and climbing cultivars of *Phaseolus vulgaris* L. inoculated with *Rhizobium*. J. Agric. Sci. Camb. 88, 503–508.

Graham P H and Temple S R 1984 Selection for improved nitrogen fixation in *Glycine max* (L.) Merr. and *Phaseolus vulgaris* L. Plant and Soil 82, 315–327.

Greder R R, Orf J H and Lambert J W 1986 Heritabilities and associations of nodule mass and recovery of *Bradyrhizobium japonicum* serogroup USDA 110 in soybean. Crop Sci. 26, 33–37.

Gremaud M F and Harper J E 1989 Selection and initial characterization of partially nitrate tolerant nodulation mutants of soybean. Plant Physiol. 89, 169–173.

Hansen A P, Martin P, Buttery B R and Park S J 1992c Nitrate inhibition of N_2 fixation in *Phaseolus vulgaris* L. cv. OAC Rico and a supernodulating mutant. New Phytol. 122, 611–615.

Hansen A P, Peoples M B, Gresshoff P M, Atkins C A, Pate J S and Carroll B J 1989 Symbiotic performance of supernodulating soybean (*Glycine max* (L.) Merrill) mutants during development on different nitrogen regimes. J. Exp. Bot. 40, 715–724.

Hansen A P, Rerkasem B, Lordkaew S and Martin P 1993 Xylem-solute technique to measure N_2 fixation by *Phaseolus vulgaris* L.: calibration and sources of error. Plant and Soil 150, 223–231.

Hansen A P, Yoneyama T and Kouchi H 1992a Short-term nitrate effects on hydroponically-grown soybean cv. Bragg and its supernodulating mutant. I. Carbon, nitrogen and mineral element distribution, respiration and effect of nitrate on nitrogenase activity. J. Exp. Bot. 43, 1–7.

Hansen A P, Yoneyama T and Kouchi H 1992b Short-term nitrate effects on hydroponically-grown soybean cv. Bragg and its supernodulating mutant. II. Distribution and respiration of recently-fixed ^{13}C-labelled photosynthate. J. Exp. Bot. 43, 9–14.

Hardarson G, Bliss F A, Cigales-Rivero M R, Henson R A, Kipe-Nolt J A, Longeri L, Manrique A, Pena-Cabriales J J, Pereira P A A, Sanabria C A and Tsai S M 1993 Genotypic variation in biological nitrogen fixation by common bean. Plant and Soil 152, 59–70.

Hardarson G, Zapata F and Danso S K A 1984 Effects of plant genotypes and nitrogen fertilizer on symbiotic nitrogen fixation by soybean cultivars. Plant and Soil 82, 397–405.

Hardy R W F, Holsten R D, Jackson E K and Burns R C 1968 The acetylene-ethylene assay for N_2 fixation: laboratory and field evaluation. Plant Physiol. 43, 1185–1207.

Harper J E 1987 Nitrogen metabolism. *In* Soybeans: Improvement, Production, and Uses. 2nd ed. Agron. Monogr. 16. Ed. J R Wilcox. pp 495–531.ASA, CSSA and SSSA, Madison, WI.

Herridge D F 1982 Relative abundance of ureides and nitrate in plant tissues of soybean as a quantitative assay of nitrogen fixation. Plant Physiol. 70, 1–6.

Herridge D F 1984 Effects of nitrate and plant development on the abundance of nitrogenous solutes in root-bleeding and vacuum-extracted exudates of soybean. Crop Sci. 24, 173–179.

Herridge D F and Betts J H 1985 Nitrate tolerance in soybean: variation between genotypes. *In* Nitrogen Fixation Research Progress. Eds. H J Evans, P J Bottomley and W E Newton. p 32. Martinus Nijhoff, Boston.

Herridge D F and Betts J H 1988 Field evaluation of soybean genotypes selected for enhanced capacity to nodulate and fix nitrogen in the presence of nitrate. Plant and Soil 110, 129–135.

Herridge D F and Bergersen F J 1988 Symbiotic nitrogen fixation. *In* Advances in Nitrogen Cycling in Agricultural Ecosystems. Ed. J R Wilson. pp 46–65. CAB International, Wallingford, UK.

Herridge D F, Bergersen F J and Peoples M B 1990 Measurement of nitrogen fixation by soybean in the field using the ureide and natural ^{15}N abundance methods. Plant Physiol. 93, 708–716.

Herridge D F and Brockwell J 1988 Contributions of fixed nitrogen and soil nitrate to the nitrogen economy of irrigated soybean. Soil Biol. Biochem. 20, 711–717.

Herridge D F and Holland J F 1992 Production of summer crops in northern New South Wales. 1. Effects of tillage and double cropping on growth, grain and N yields of six crops. Aust. J. Agric. Res. 43,105–122.

Herridge D F, O Connell P and Donnelly K 1988a The xylem ureide assay of nitrogen fixation: sampling procedures and sources of error. J. Exp. Bot. 39, 12–22.

Herridge D F and Peoples M B 1990 The ureide assay for measuring nitrogen fixation by nodulated soybean calibrated by ^{15}N methods. Plant Physiol. 93, 495–503.

Herridge D F and Rose I A 1994 Heritability and repeatability of enhanced N_2 fixation in early and late inbreeding generations of soybean. Crop Sci. 34, 360–367.

Herridge D F, Roughley R J and Brockwell J 1987 Low survival of *Rhizobium japonicum* inoculant leads to reduced nodulation, nitrogen fixation and yield of soybean in the current crop but not in the subsequent crop. Aust. J. Agric. Res. 38, 75–82.

Herridge D F, Sudin M N, Pate J S and Peoples M B 1988b Translocation and utilization of nitrogen by pulses. *In* World Crops: Cool Season Food Legurnes. Ed. R J Surnmerfield. pp 793–800. Kluwer Academic Publ., Dordrecht.

Hobbs S L A and Mahon J D 1982 Effects of pea (*Pisum sativum*) genotypes and *Rhizobium leguminosarum* strains on N_2 (C_2H_2) fixation and growth. Can. J. Bot. 60, 2594–2600.

Howieson J G and Ewing M A 1986 Acid tolerance in the *Rhizobium meliloti-Medicago* symbiosis. Aust. J. Agric. Res. 37, 55–64.

Hungria M and Neves M C P 1987 Cultivar and *Rhizobium* strain effects on nitrogen fixation and transport in *Phaseolus vulgaris* L. Plant and Soil 103, 111–121.

Jacobsen E and Feenstra W J 1984 A new pea mutant with efficient nodulation in the presence of nitrate. Plant Sci. Lett. 33, 337–344.

Jensen E S 1986 Syrnbiotic N_2 fixation in pea and field bean estimated by ^{15}N fertilizer dilution in field experiments with barley as a reference crop. Plant and Soil 92, 3–13.

Keyser H H and Li F 1992 Potential for increasing biological nitrogen fixation in soybean. Plant and Soil 141, 119–135.

Kipe-Nolt J A and Giller K E 1993 A field evaluation using the ^{15}N isotope dilution method of lines of *Phaseolus vulgaris* L. bred for increased nitrogen fixation. Plant and Soil 152, 107–114.

Kipe-Nolt J A, Vargas H and Giller K E 1993 Nitrogen fixation in breeding lines of *Phaseolus vulgaris* L. Plant and Soil 152, 103–106.

Kucey R M N, Snitwongse P, Chaiwanakupt P, Wadisirisuk P, Siripaibool C, Arayangkool T, Boonkerd N and Rennie R J 1988a Nitrogen fixation (^{15}N dilution) with soybeans under Thai field conditions. Plant and Soil 108, 33–41.

Kucey R M N, Chaiwanakupt P, Arayangkool T, Snitwongse P, Siripaibool C, Wadisirisuk P and Boonkerd N 1988b Nitrogen fixation (^{15}N dilution) with soybeans under Thai field conditions. Plant and Soil 108, 87–92.

Kueneman E A, Root W R, Dashiell K E and Hohenberg J 1984 Breeding soybeans for the tropics capable of nodulating effectively with indigenous *Rhizobium* spp. Plant and Soil 82, 387–396.

Kumar Rao J V D K and Dart P J 1987 Nodulation, nitrogen fixation and nitrogen uptake in pigeonpea (*Cajanus cajan* (L.) Millsp) of different maturity groups. Plant and Soil 99, 255–266.

Kvien C S, Ham G E and Lambert J W 1981 Recovery of introduced *Rhizobium japonicum* strains by soybean genotypes. Agron. J. 73, 900–905.

Ledgard S F, Morton R, Freney J R, Bergersen F J and Simpson J R 1985 Assessment of the relative uptake of added and indigenous soil nitrogen by nodulated legumes and reference plants in the ^{15}N dilution measurement of N_2 fixation: derivation of method. Soil Biol. Biochem. 17, 317–322.

Ledgard S F and Peoples M B 1988 Measurement of nitrogen fixation in the field. *In* Advances in Nitrogen Cycling in Agricultural Ecosystems. Ed. J R Wilson. pp 351–367. CAB International, Wallingford, UK.

Lee S H, Ashley D A and Boerma H R 1991 Regulation of nodule development in supernodulating mutants and wild-type soybean. Crop Sci. 31, 688–693.

Leffel R C, Cregan P B, Bolgiano A P and Thibeau D J 1992 Nitrogen metabolism of normal and high-seed-protein soybean. Crop Sci. 32, 747–750.

Lie T A and Mulder E G 1971 Biological Nitrogen Fixation in Natural and Agricultural Habitats. Plant and Soil, Special Volume, Martinus Nijhoff, The Hague. 590p.

McClure P R, Israel D W and Volk R J 1980 Evaluation of the relative ureide content of xylem sap as an indicator of N_2 fixation in soybeans. Plant Physiol. 66, 720–725.

McFerson J R, Bliss F A and Rosas J C 1982 Selection for enhanced nitrogen fixation in common bean *Phaseolus vulgaris* L. *In* Biological Nitrogen Fixation Technology for Tropical Agriculture. Eds. P H Graham and S C Harris. pp 29–44. CIAT, Cali, Colombia.

Minchin F R, Witty J F and Mytton L R 1994 Reply to Measurement of nitrogenase activity in legume root nodules: in defense of the acetylene reduction assay' by J K Vessey. Plant and Soil 158, 163–167.

Moawad H A, Ellis W R and Schmidt E L 1984 Rhizosphere response as a factor in competition among three serogroups of indigenous *Rhizobium japonicum* for nodulation of field-grown soybeans. Appl. Environ. Microbiol. 47, 607–612.

Mytton L R 1975 Plant genotype × rhizobium strain interactions in white clover. Ann. Appl. Biol. 80, 103–107.

Mytton L R 1983 Host plant selection and breeding for improved symbiotic efficiency. *In* The Physiology, Genetics and Nodulation of Temperate Legumes. Eds. D G Jones and D R Davies. pp 373–393. Pitman, London.

Mytton L R 1984 Developing a breeding strategy to exploit quantitative variation in symbiotic nitrogen fixation. Plant and Soil 82, 329–335.

Mytton L R, El-Sherbeeny M H and Lawes D A 1977 Symbiotic variation in *Vicia faba*. 3. Genetic effects of host plant, rhizobium strain and of host × strain interaction. Euphytica 26, 785–791.

Nangju D 1980 Soybean response to indigenous *Rhizobia* as influenced by cultivar origin. Agron. J. 72, 403–406.

Nambiar P T C, Dart P J, Rao B S and Ravishankar H N 1984 Response of groundnut (*Arachis hypogaea* L.) to *Rhizobium* inoculation. Oléagineux 39, 150–154.

Nelson R S, Ryan S A and Harper J E 1983 Soybean mutants lacking constitutive nitrate reductase activity. I. Selection and initial plant characterization. Plant Physiol. 72, 503–509.

Norhayati M, Mohd Noor S, Chong K, Faizah A W, Herridge D F, Peoples M B and Bergersen F J 1988 Adaption of methods for evaluating N_2 fixation in food legumes and legume cover crops. Plant and Soil 108, 143–150.

Nutman P S 1984 Improving nitrogen fixation in legumes by plant breeding: the relevance of host selection experiments in red clover (*Trifolium pratense* L.) and subterranean clover (*T. subterraneum* L.). Plant and Soil 82, 285–301.

Ohyama T, Nicholas J C and Harper J E 1993 Assimilation of $^{15}N_2$ and $^{15}NO_3^-$ by partially nitrate-tolerant mutants of soybean. J. Exp. Bot. 44, 1739–1747.

Pacovsky R S, Bayne H G and Bethlenfalvay G J 1984 Symbiotic interactions between strains of *Rhizobum phaseoli* and cultivars of *Phaseolus vulgaris* L. Crop Sci. 24, 101–105.

Park S J and Buttery B R 1988 Nodulation mutants of white bean (*Phaseolus vulgaris* L.) induced by ethyl-methane sulphonate. Can. J. Plant Sci. 68,199–202.

Park S J and Buttery B R 1989 Identification and characterization of common bean (*Phaseolus vulgaris* L.) lines well nodulated in the presence of high nitrate. Plant and Soil 119, 237–244.

Peoples M B, Bergersen F J, Turner G L, Sampet C, Rerkasem B, Bhromsiri A, Nurhayati D P, Faizah A W, Sudin M N, Norhayati M and Herridge D F 1991 Use of the natural enrichment of ^{15}N in plant available soil N for the measurement of symbiotic N_2 fixation. *In* Stable Isotopes in Plant Nutrition, Soil Fertility and Environmental Studies. pp 117–129. IAEA, Vienna.

Peoples M B, Faizah A W, Rerkasem B and Herridge D F 1989a Methods for Evaluating Nitrogen Fixation by Nodulated Legumes in the Field. ACIAR Monograph No. 11. ACIAR, Canberra. 76p.

Peoples M B, Hebb D M, Gibson A H and Herridge D F 1989b Development of the xylem ureide assay for the measurement of nitrogen fixation by pigeon pea (*Cajanus cajan* (L.) Millsp.). J. Exp. Bot. 40, 535–542.

Peoples M B and Herridge D F 1990 Nitrogen fixation by legumes in tropical and subtropical agriculture. Adv. Agron. 44, 155–223.

Peoples M B, Ladha J K and Herridge DF 1995 Enhancing legume N_2 fixation through plant and soil management. Plant and Soil 174.

Peoples M B, Sudin M N and Herridge D F 1987 Translocation of nitrogenous compounds in symbiotic and nitrate-fed pulse legumes. J. Exp. Bot. 38, 567–579.

Pereira P A A, Burris R H and Bliss F A 1989 ^{15}N-determined dinitrogen fixation potential of genetically diverse bean lines (Phaseolus vulgaris L.). Plant and Soil 120, 171–179.

Phillips D A and Teuber L R 1985 Genetic improvement of symbiotic nitrogen fixation in legumes. *In* Nitrogen Fixation Research Progress. Eds. H J Evans, P J Bottomley and W E Newton. pp 11–17. Martinus Nijhoff Publ., Dordrecht.

Phillips D A, Torrey J G and Burris R H 1971 Extending symbiotic nitrogen fixation to increase man's food supply. Science 174, 169–171.

Piha M I and Munns D N 1987a Nitrogen fixation capacity of field-grown bean compared to other grain legumes. Agron. J. 79, 690–696.

Piha M I and Munns D N 1987b Nitrogen fixation potential of beans (*Phaseolus vulgaris* L.) compared with other grain legumes under controlled conditions. Plant and Soil 98, 169–182.

Pracht J E, Nickell C D, Harper J E and Bullock D G 1994 Agronomic evaluation of non-nodulating and hypernodulating mutants of soybean. Crop Sci. 34, 738–740.

Pulver E L, Kueneman E A and Ranga-Rao V 1985 Identification of promiscuous nodulating soybean efficient in N_2 fixation. Crop Sci. 25, 660–663.

Rennie R J 1984 Comparisons of N balance and ^{15}N isotope dilution to quantify N_2 fixation in field-grown legumes. Agron. J. 76, 785–790.

Rennie R J and Dubetz S 1986 Nitrogen-15-determined nitrogen fixation in field-grown chickpea, lentil, faba bean and field pea. Agron. J. 78, 654–660.

Rennie R J and Kemp G A 1983a N_2-fixation in field beans quantified by ^{15}N dilution. I. Effect of strain of *Rhizobium phaseoli*. Agron. J. 75, 640–644.

Rennie R J and Kemp G A 1983b N_2-fixation in field beans quantified by ^{15} dilution. II. Effect of cultivars of beans. Agron. J. 75, 645–649.

Rennie R J, Rennie D A, Siripaibool C, Chaiwanakupt P, Boonkerd N and Snitwongse P 1988 N_2 fixation in Thai soybeans: effects of tillage and inoculation on ^{15}N-determined N_2 fixation in recommended cultivars and advanced breeding lines. Plant and Soil 112, 183–193.

Rerkasem B, Rerkasem K, Peoples M B, Herridge D F and Bergersen F J 1988 Measurement of N_2 fixation in maize (*Zea mays* L.)-ricebean (*Vigna umbellata* (Thunb.) Ohwi and Ohashi) intercrops. Plant and Soil 108, 125–135.

Ronis D H, Sammons D J, Kenworthy W J and Meisinger JJ 1985 Heritability of total and fixed N content of the seed in two soybean populations. Crop Sci. 25, 1–4.

Ryan S A, Nelson R S and Harper J E 1983 Soybean mutants lacking constitutive nitrate reductase activity. II. Nitrogen assimilation, chlorate resistance, and inheritance. Plant Physiol. 72, 510–514.

Seetin M W and Barnes D K 1977 Variation among alfalfa genotypes for rate of acetylene reduction. Crop Sci. 17, 783–787.

Serraj R, Drevon J J, Obaton M and Vidal A 1992 Variation in nitrate tolerance of nitrogen fixation in soybean (*Glycine max*) – *Bradyrhizobium* symbiosis. J. Plant Physiol. 140, 366–371.

Shearer G and Kohl D H 1986 N_2 fixation in field settings: estimations based on natural ^{15}N abundance. Aust. J. Plant Physiol. 13, 699–756.

Sinclair T R, Soffes A R, Hinson K, Albrecht S L and Pfahler P L 1991 Genotype variation in soybean nodule number and weight. Crop Sci. 31, 301–304.

Song L, Carroll B J, Gresshoff P M and Herridge D F 1995 Field assessment of supernodulating genotypes of soybean for yield, N_2 fixation and benefit to subsequent crops. Soil Biol. Biochem. 26 (*In press*).

St Clair D A, Wolyn D J, DuBois J, Burris R H and Bliss F A 1988 Field comparison of dinitrogen fixation determined with nitrogen-15-depleted and nitrogen-15-enriched ammonium sulphate in selected inbred backcross lines of common bean. Crop Sci. 28, 773–778.

Tauer L W 1989 Economic impact of future biological nitrogen fixation technologies on United States agriculture. Plant and Soil 119, 261–270.

Teuber L R and Phillips D A 1988 Influence of selection method and nitrogen environment on breeding alfalfa for increased forage and yield quality. Crop Sci. 28, 599–604.

Turner G L and Gibson A H 1980 Measurement of nitrogen fixation by indirect means. *In* Methods for Evaluating Biological Nitrogen Fixation. Ed. F J Bergersen. pp 111–138. John Wiley and Sons, Chichester.

Unkovitch M J, Pate J S, Sanford P and Armstrong E L 1994 Potential precision of the ^{15}N natural abundance method in field estimates of nitrogen fixation by crop and pasture legumes in south-west Australia. Aust. J. Agric. Res. 45, 119–132.

Vasilas B L and Ham G E 1984 Nitrogen fixation in soybeans: an evaluation of measurement techniques. Agron. J. 76, 759–764.

Vasilas B L and Fuhrmann J J 1993 Field response of soybean to increased dinitrogen fixation. Crop Sci. 33, 785–787.

Vessey J K 1994 Measurement of nitrogenase activity in legume root nodules: in defense of the acetylene reduction assay. Plant and Soil 158, 151–162.

Vest G 1970 Rj_3 — a gene conditioning ineffective nodulation in soybean. Crop Sci. 10, 34–35.

Vest G and Caldwell B E 1972 Rj_4 — a gene conditioning ineffective nodulation in soybean. Crop Sci. 12, 692–693.

Vincent J M 1965 Environmental factors in the fixation of nitrogen by the legume. *In* Soil Nitrogen. Eds. W V Bartholomew and F E Clark. pp 384–435. American Society Agronomy, Madison.

Weber C R 1966 Nodulating and nonnodulating soybean isolines: II. Response to applied nitrogen and modified soil conditions. Agron. J. 58, 46–49.

Wiersma J V and Orf J H 1992 Early maturing soybean nodulation and performance with selected *Bradyrhizobium japonicum* strains. Agron. J. 84, 449–458.

Williams L F and Lynch D L 1954 Inheritance of a non-nodulating character in the soybean. Agron. J. 46, 28–29.

Wilson K J, Peoples M B and Jefferson R A 199 New techniques for studying competition by rhizobia and for assessing nitrogen fixation in the field. Plant and Soil 174.

Witty J F 1983 Estimating N_2-fixation in the field using N^{15}-labelled fertilizer: some problems and solutions. Soil Biol. Biochem. 15, 631–639.

Witty J F and Minchin F R 1988 Measurement of nitrogen fixation by the acetylene reduction assay; myths and mysteries. *In* Nitrogen Fixation by Legumes in Mediterranean Agriculture. Eds. D P Beck and L A Materon. pp 331–344. Martinus Nijhoff, Dordrecht.

Witty J F, Rennie R J and Atkins C A 1988 ^{15}N addition methods for assessing N_2 fixation under field conditions. *In* World Crops: Cool Season Food Legumes. Ed. R J Summerfield. pp 715–730. Kluwer Academic Publ., London.

Wolff A B, Streit W, Kipe-Nolt J A, Vargas H and Werner D 1991 Competitiveness of *Rhizobium leguminosarum* bv. *phaseoli* strains in relation to environmental stress and plant defense mechanisms. Biol. Fertil. Soils 12, 170–176.

Wu S and Harper J E 1990 Nitrogen fixation of nodulation mutants of soybean as affected by nitrate. Plant Physiol. 92, 1142–1147.

Wu S and Harper J E 1991 Dinitrogen fixation potential and yield of hypernodulating soybean mutants: a field evaluation. Crop Sci. 31, 1233–1240.

Plant and Soil **174:** 83–101, 1995.

Enhancing legume N_2 fixation through plant and soil management

M.B. Peoples[1], J.K. Ladha[2] and D.F. Herridge[3]
[1]*CSIRO Division of Plant Industry, GPO Box 1600 Canberra, ACT 2601, Australia,* [2]*International Rice Research Institute, PO Box 933, 1099 Manila, Philippines and* [3]*NSW Agriculture, RMB 944 Tamworth, NSW 2340, Australia*

Key words: management, legumes, rhizobia, symbiotic N_2 fixation

Table of contents

Abstract
Introduction
Strategies to enhance P_{fix} by optimizing the numbers and effectiveness of rhizobia
  Inoculation
Strategies to enhance P_{fix} by either reducing levels of, or sensitivity to, soil nitrate
  Cropping sequence
  Tillage
  Inoculation
  Choice of cultivar or species
Strategies to maximize legume growth within environmental constraints
  Plant nutrition and soil amelioration
  Time of sowing
  Supply of water and tillage
  Choice of cultivar or species
  Pest and disease control
  Maintaining legumes under grazing
Conclusions
Acknowledgements
References

Abstract

Atmospheric N_2 fixed symbiotically by associations between *Rhizobium* spp. and legumes represents a renewable source of N for agriculture. Contribution of legume N_2 fixation to the N-economy of any ecosystem is mediated by: (i) legume reliance upon N_2 fixation for growth, and (ii) the total amount of legume-N accumulated. Strategies that change the numbers of effective rhizobia present in soil, reduce the inhibitory effects of soil nitrate, or influence legume biomass all have potential to alter net inputs of fixed N. A range of management options can be applied to legumes growing in farming systems to manipulate N_2 fixation and improve the N benefits to agriculture and agroforestry.

Introduction

The range of experimentally determined values of N_2 fixation by temperate and tropical legumes (Peoples et al., 1995) reflects the inherent capacities of legumes to accumulate and fix N, the environmental constraints on those capacities, and the effects of cultural practices or experimental treatments on both. Where soil fertility is high, legumes in the field thrive without fixing atmospheric N_2; under such conditions they may derive all their N requirements from soil N. But in the majority of soils, levels of plant-available N are usually insufficient to satisfy fully a legume's requirement for N and, provided effective rhizobia are present, that unfulfilled N demand will be met by biological N_2 fixation (BNF).

There has to be an upper limit on BNF. Herridge and Bergersen (1988), postulated a theoretical upper limit of 635 kg N ha^{-1} for soybean (*Glycine max*) and more than 300 kg N ha^{-1} for pigeon pea (*Cajanus cajan*) and groundnut (*Arachis hypogaea*). Although values approaching the theoretical limits may be achieved under optimal conditions (i.e. high legume yield and low soil nitrate), in practice the levels of N_2 fixation in farmers' fields may often be only a fraction of the potential (Peoples et al., 1995).

The amounts of N_2 fixed by legumes are controlled by two factors : (i) the proportion of plant nitrogen (N) derived from symbiotic N_2 fixation (P_{fix}) and (ii) the amount of N accumulated during growth, and can be expressed as:

$$\text{Amount } N_2 \text{ fixed} = P_{fix} \times \text{legume N yield} \quad (1)$$

Therefore, strategies which influence either P_{fix} or legume growth will affect gross input of fixed N. In crop legumes, the net return of fixed N to soil will depend upon the amount of N removed in seed. Therefore, the potential benefit of fixed N in vegetative residues of crop legumes will be determined by the N-balance at final harvest:

$$\text{N} - \text{balance} = (N_2 \text{ fixed}) - (\text{seed N}) \quad (2)$$

However, the contribution of fixed N will be reduced even further if crop legume residues are subsequently removed from the field to be used for animal fodder, organic mulch, or fuel, or if the trash and stubble remaining after seed harvest are burnt. All these practises are common in different parts of the world and represent important losses of legume N from farming systems (Peoples et al., 1994a; Wani et al., 1995; Ying et al., 1992).

Approaches which might be used to improve BNF can be classified as:

- Protocols which increase P_{fix} by:
  - (a) maximizing the numbers and effectiveness of rhizobia in soil,
  - (b) reducing levels of soil nitrate,
  - (c) reducing sensitivity of the symbiosis (through the plant, the bacteria and / or their interaction) to soil nitrate.
- Protocols which maximize legume growth.
- Protocols which exploit genotypic variability in:
  - (a) the rhizobia,
  - (b) the host,
  - (c) the host-strain interactions.

The following sections describe a range of management procedures which have the potential to enhance BNF by legumes in diverse farming environments. Details of strategies that might be applied to specific agricultural systems may be found in other contributions to this volume and are the subject of several reviews (e.g. Gibson et al., 1982; Ladha et al., 1988; Ledgard and Steele, 1992; Peoples et al., 1994c). Rhizobial strain selection, inoculant technology and genetic manipulation of the legume host will not be addressed since these topics have either been reviewed recently (e.g. Brockwell and Bottomley, 1994; Hardarson, 1993; Herridge et al., 1993), or are covered elsewhere in this volume (Bantilan and Johansen, 1995; Brockwell and Bottomley, 1995; Herridge and Danso, 1995).

Strategies to enhance P_{fix} by optimizing the numbers and effectiveness of rhizobia

Inoculation

Wherever a particular legume has long been established in an agricultural system or is considered a traditional crop, it is likely that there will be adequate numbers of indigenous rhizobia for nodulation (e.g. Thompson et al., 1991). Beyond this generalization, it is difficult to predict situations where inoculation might not be necessary. Even the notion that it is unnecessary to inoculate "promiscuous" legumes in tropical soils may be flawed (Singleton et al., 1992).

The results of a series of inoculation trials coordinated by the University of Hawaii's Nitrogen Fixation by Tropical Agricultural Legumes NifTAL) Project indicate that even legumes belonging to the "cowpea" cross-inoculation group (cowpea, green gram, black gram, groundnut, pigeon pea, etc.), can often benefit from inoculation (Table 1). In almost 30% of cases, improving soil fertility further increased yield of inoculated plants (Singleton et al., 1992). Detailed evaluation of the results suggested that legume species that responded to inoculation in one country did not necessarily do so in another. Subsequent analysis by NifTAL of 305 soil samples collected from 17 countries in the tropics indicated that while rhizobia in the cowpea group of promiscuous legumes were present at a large number of sites, the populations were extremely variable in size (Table 2). Almost one-half of the soils sampled had fewer than 100 rhizobia g $soil^{-1}$. Rhizobia for several of the important legumes grown in the tropics (soybean, common bean, leucaena) were either not present at many sites, or occurred in such low numbers that it would be prudent to inoculate (Table 2). Clearly, it simply cannot be taken for granted that tropical soils have sufficient numbers of resident root-nodule bacteria to meet a legume's demand for N.

There are a number conditions under which soils may be devoid of effective rhizobia (Brockwell and Bottomley, 1995; Peoples and Craswell, 1992) but perhaps the most common situations are those where soil factors are unfavorable for survival of sufficient numbers of rhizobia. A study of the persistence of *Bradyrhizobium japonicum* in farmers' fields in Australia showed that > 1,000 rhizobia per gram of soil could be detected up to 6 years after the previous soybean crop in red, acidic (pH 5.7, H_2O) soils, but the rhizobia failed to survive 5–6 months following well-nodulated soybean crops in grey alkaline (pH>8.0) clays so that inoculation was essential each time soybean was grown (Peoples et al., 1994b). In other rhizobial species such as *Rhizobium meliloti* (which nodulates annual medics; Brockwell et al., 1991), *R. leguminosarum* bv. *viciae* (which nodulates pea and faba bean; Carter et al., 1994; Marshall et al., 1993), and *R. trifolii* (which nodulates subterranean clover *(Trifolium subterraneum*; Richardson and Simpson, 1988), low soil pH results in poor survival and persistence of rhizobia, and can affect the numbers and distribution of indigenous rhizobia.

The success of inoculation in the field depends upon inoculant quality (Brockwell and Bottomley, 1994), the procedure used (Brockwell et al., 1988), operator competence, compatibility with fertilizers (e.g. direct contact of rhizobia with acidic fertilizers such as superphosphate may effect survival), and the presence of toxic agrichemicals (Corbin et al., 1977). Even with high quality inoculant and good inoculation practice, failures can occur because environmental factors influence survival of rhizobia (Brockwell and Bottomley, 1995). For example, 4–5% of a soybean inoculum was recovered from soil 24 hours after sowing at 28°C, but < 0.2% survived sowing at 38°C (Brockwell et al., 1987).

Typically, responses to inoculation are most likely to occur when specific rhizobia are absent or numbers are low, when soil nitrate is low, and when the legume has a high potential for growth (i.e. adequate moisture and nutrients). If populations of infective rhizobia already exist in the soil, they may represent a formidable barrier to the establishment of inoculant strains (Thies et al., 1991; Wani et al., 1995). Inoculation responses may be affected with as few as 10 to 100 rhizobia per g of soil (Singleton et al., 1992). A population of 100 indigenous rhizobia per g of soil represents 1.5×10^{11} rhizobia per ha (0–10cm), equivalent to inoculating each seed of a soybean crop with 300,000 organisms (Brockwell and Bottomley, 1995).

Strategies to enhance P_{fix} by either reducing levels of, or sensitivity to, soil nitrate

Cropping sequence

High levels of soil nitrate can be a potent inhibitor of N_2 fixation (e.g. soybean data of Table 3, Fig. 1; Streeter, 1988). The amount of plant-available soil N can be influenced by recent cropping and cropping sequence. The examples presented in Table 4 for pea (*Pisum sativum*) and soybean illustrate that legumes can fix greater amounts of N and contribute more fixed N to soil if sown immediately after a cereal crop than if grown in previously fallowed soil (see also Bergersen et al., 1989) or immediately after a legume-based pasture. In addition to effects on soil nitrate, different crop rotations can also influence BNF through effects on legume growth (Peoples et al., 1992).

The importance of cropping sequence on N_2 fixation was demonstrated in a survey of commercial soybean crops in Australia (Peoples et al., 1994b). Only 11 of 33 irrigated crops fixed more N than was removed in seed (as determined using Equation 2 above). All those crops with the highest N_2 fixation and greatest potential for a net return of fixed N (range +16 to +69 kg N ha^{-1}) followed several crops of cereal, or were double-cropped with wheat, so that levels of soil nitrate at sowing were low. Similar observations have also been reported for dryland legume crops (Herridge

Table 1. Summary of inoculation responses of a range of legume species[a]

Species	Number of trials	Significant response to inoculation (% of total)
Cowpea - (*Vigna unguiculata*)	9	56
Green gram (*V. radiata*)	40	70
Black gram - (*V. mungo*)	15	53
Groundnut - (*Arachis hypogaea*))	26	50
Pigeon pea - (*Cajanus cajan*)	8	13
Soybean - (*Glycine max*)	40	65
Common bean - (*Phaseolus vulgaris*)	10	10
Chickpea - (*Cicer arietinum*)	31	48
Lentil - (*Lens culinaris*)	27	48
Leucaena - (*Leucaena leucocephala*)	8	38

[a]Data collated from over 20 different countries for local farmers' practice. Adapted from Singleton et al. (1992).

Table 2. Frequency distribution of rhizobial populations in tropical soils[a]

Rhizobial population (number g soil^{-1})	Test host (% of samples)				
	"Cowpea" group[b]	Lima bean	Soybean	Common bean	Leucaena
0	16	26	47	35	51
1–10	9	26	13	11	9
10–100	21	21	15	19	15
100–1000	12	5	10	15	9
> 1000	40	21	14	19	15

[a]After Singleton et al. (1992).
[b]Hosts included cowpea, green gram, groundnut, and siratro (*Macroptilium atropurpureum*).

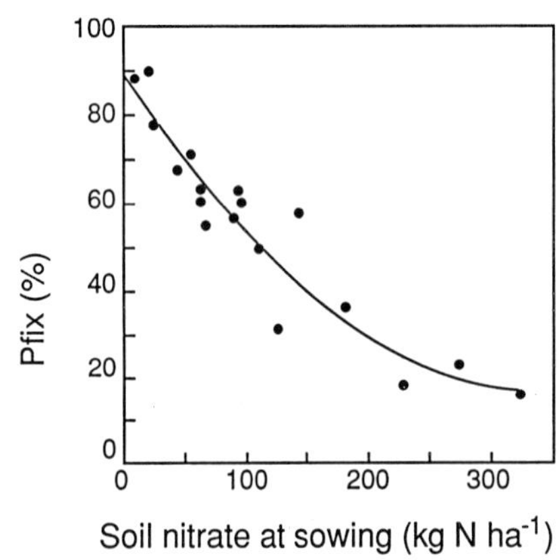

Fig. 1. Effects of soil nitrate of sowing (to 1.2 m depth) on the proportion (P_{fix}) of N_2 fixed by chickpea. Data derived from Doughton et al. (1993) and Herridge et al. (1994).

and Holland, 1992). On the other hand, levels of BNF were depressed and N-balances were poor (range -134 to -17 kg N ha^{-1}) when soybean followed 2 or 3 consecutive years of grain legume, at sites with a recent history of legume pasture, or where N-fertilizer had been applied (Peoples et al., 1994b). Thus, with the proper choice of cropping sequence, farming systems can be managed for improved N_2 fixation, although other factors such as storage of water in fallow soil and nutritional aspects must also be considered.

In the farming systems described above, individual crops are grown in sequence one after the other, but in many parts of the world two or more crop species are often grown simultaneously in the same field during a growing season. Although many combinations of crops are practised, cereal-legume intercropping systems are most common (Ofori and Stern, 1987). These

Table 3. Interaction between soil nitrate, choice of cultivar, and rate of inoculation on the proportion (P_{fix}) of plant N derived from N_2 fixation, and the amount of N fixed by food legumes

Species and cultivar	Soil nitrate at sowing[b] (kg N ha^{-1})	Rate of inoculation	Crop N (kg N ha^{-1})	N_2 fixation	
				P_{fix} (%)	Amount (kg N ha^{-1})
Soybean[a]					
Bragg	260	Normal[c]	228	5	11
Bragg		10 × normal	204	36	73
Davis		Normal	198	20	40
Bragg	70	Normal	230	42	97
Bragg		10 × normal	195	59	115
Davis		Normal	281	64	180
Common bean[d]					
Flor de Mayo	na[e]	na	125	20	25
Azufrado	na	na	140	50	70

[a]Data derived from Herridge et al. (1990).
[b]Measured to a depth of 1.2 m.
[c]Normal rate of inoculation (*B. japonicum* strain CB1809) = 5×10^5 bacteria per seed.
[d]Data derived from Hardarson et al. (1993).
[e]na = not available.

mixed crops tend to have higher yields and greater land use efficiency per unit land area than comparable monocrops. In intercropping systems, competition for soil mineral N between legumes and cereals can result in stimulation of BNF and increased harvest of total N of interplanted crops (e.g. Chandel et al., 1989; Rekasem et al., 1988). Legumes of indeterminate growth habit appear to be more efficient in terms of N_2 fixation than determinate types when utilized for mixed cropping (Ofori and Stern, 1987).

Tillage

Cultivation accelerates the oxidation of organic matter in soils which results in high levels of nitrate-N in the profile. Therefore, reduced tillage usually results in lower levels of soil nitrate. Experimentation has indicated that nodulation and N_2 fixation by legumes can often be improved under no-tillage, compared with conventionally cultivated systems (Herridge and Holland, 1992; Hughes and Herridge, 1989; Rennie et al., 1988). Increased BNF usually results from increased P_{fix}, although yields can also be influenced by tillage practice in drier environments. As a net result, the residual return of fixed N can be increased by up to 50 kg N ha^{-1} with direct-drill techniques (Herridge and Holland, 1992; Hughes and Herridge, 1989), although the final result may depend upon the effectiveness of weed control under no-tillage and the resultant impact of weeds on legume growth (Shafiq et al., 1994).

Inoculation

Inoculation can lead to the establishment of large populations of rhizobia in the plant rhizosphere and to improved nodulation and N_2 fixation, even in high nitrate soils (e.g. soybean data of Table 3, Fig. 1). However, the establishment of rhizobial populations can be improved, and nodulation and N_2 fixation increased in the presence of high nitrate, by high rates of inoculation (compare N_2 fixation by soybean cultivar Bragg at normal and 10 times normal inoculation rate in Table 3 ; see also Bergersen et al., 1989; Brockwell et al, 1989). Multiple regression analysis of relationships between P_{fix} of soybean and *B. japonicum* numbers and soil nitrate (Herridge and Brockwell, 1988) indicate that they can be highly correlated ($R^2 = 0.80$; Peoples and Herridge, 1990). This serves to illustrate the importance of the interaction between soil nitrate and rhizobial numbers on the regulation of N_2 fixation.

Table 4. Effect of cropping sequence or growing season on N_2 fixation by crop legumes and potential residual value of fixed N after seed harvest

Cropping sequence	Crop N	N_2 fixation		Seed N	Residual
		P_{fix}	Amount		benefit[a]
	(kg N ha^{-1})	(%)	(kg N ha^{-1})	(kg N ha^{-1})	(kg N ha^{-1})
Soybean[b] *(Glycine max)*					
Previously fallowed	417	34	143	187	-44
Previously cropped	363	67	244	205	+39
Pea[c] *(Pisum sativum)*					
Following pasture	220	60	133	135	-2
Previously cropped	227	81	183	162	+21

[a]N-balance = (N_2 fixed) - (Seed N).
[b]Data derived from Bergersen et al. (1985).
[c]Sampled directly from 2 crops growing in adjacent fields on a single farm (Peoples and Gault, unpubl.).

Choice of cultivar or species

There is now substantial evidence that the choice of cultivar or species influences potential contributions of fixed N to farming systems (Wani et al., 1995). The data shown in Table 3 illustrate that different lines of a legume species growing in the same soil can vary considerably in their ability to fix N. This is the basis of the selection criteria used in several breeding programs aimed at enhancing the genotypic capacity of legumes to fix N in the presence of soil nitrate (Herridge and Danso, 1995). It has been suggested that there are distinct differences between species in the relative tolerance of BNF to nitrate and in their innate ability to fix N (Hardarson, 1993); however, the evidence for this is conflicting (Herridge et al., 1993). It appears that legume species can differ in their reliance upon N_2 fixation for growth, even when they accumulate similar amounts of total plant N (Table 5), although the ranking of the symbiotic performance of particular species found in one environment or soil type may not apply to other situations (e.g. see lupin and pea data of Table 5).

Strategies to maximize legume growth within environmental constraints

Plant nutrition and soil amelioration

Nutritional constraints may control a legume's capacity for growth and N_2 fixation (Craswell et al., 1987; Giller and Wilson, 1991). Constraints may be in the form of phytotoxic concentrations of nutrients which inhibit root or shoot growth, such as excessive levels of aluminium and manganese which develop in acid soils, and sodium chloride in the case of saline soil or irrigation water (Peoples et al., 1994d; Smith et al., 1993). Alternatively, constraints can result from an inadequate supply of major or minor-nutrients, which can affect growth of the legume host, nodule function, or multiplication and survival of the rhizobial component of the symbiosis (O'Hara et al., 1988). Increased levels of N_2 fixation which result when such nutritional limitations are alleviated often reflect increased legume growth rather than gross changes in P_{fix} (Table 6). Clearly N_2 fixation can be influenced in problem soils by applications of appropriate fertilizers (Table 6; Becker et al., 1991), but other strategies, such as the choice of tolerant symbioses where soils can not be amended or the use of mycorhizas to improve nutrient uptake, can also be viable options to increase growth potential and BNF (Gianinazzi-Pearson and Diem, 1982; Giller and Cadisch, 1995).

While discussing the general area of plant nutrition it is also pertinent to consider the potential suppressive effects of fertilizer N. Between 70 and 100% of pulse and oilseed legume crops are routinely fertilized with "starter-N" in Latin America, Western Europe and East Asia, and 20 to 40% of legume crops receive fertilizer N in North America, South Asia, the Middle East, Eastern Europe and the former USSR (< 10% elsewhere in the world, FAO, 1992). The rates of application of fertilizer N to legumes (5 to 50 kg N ha^{-1}) are generally

Table 5. Comparisons of plant reliance upon N_2 fixation (P_{fix}) and amounts of N_2 fixed by different food legume species at equivalent crop N yields

Location and species	N yield	N_2 fixation	
		P_{fix}	Amount fixed
	(kg N ha^{-1})	(%)	(kg N ha^{-1})
France[a]			
Pea (*Pisum sativum*)	199	65	126
Faba bean (*Vicia faba*)	196	92	181
Australia[b]			
Site 1: soil pH 8.0 / 348mm rainfall			
Lupin (*Lupinus angustifolius*)	57	60	34
Pea	61	87	52
Site 2: soil pH 4.8 / 511mm rainfall			
Lupin	120	78	97
Pea	124	78	95
Australia[c]			
Green gram (*Vigna radiata*)	58	15	9
Cowpea (*Vigna unguiculata*)	62	38	24

[a]Data derived from Beck et al. (1991).
[b]Data derived from Evans et al. (1989).
[c]Data derived from Peoples et al. (1994a).

low compared with levels of N applied to cereals and cash-crops (FAO, 1992; Peoples et al., 1995); however, even relatively low levels of soil nitrate are capable of depressing BNF (Fig. 1). Although small amounts of fertilizer N have been reported to stimulate growth and N_2 fixation in some instances (Becker et al., 1991; Gibson et al., 1982), the use of starter N can jeopardize N_2 fixation inputs in other situations (Jensen, 1986; Peoples et al., 1994b). Unutilized fertilizer N may also be carried-over from previous cereal crops to influence BNF in a following legume. It has been suggested that this may be one reason why levels of N_2 fixation are generally low in soybean crops grown in rotation with fertilized maize (*Zea mays*) in the mid-west of the USA (Harper, 1987).

The potential impact of fertilizer N, either applied directly to legumes as "starter-N" or remaining unused from previous crops, on inputs of fixed N was examined in a study of a rice-soybean rotation in northern Thailand (Ying et al., 1992). The results indicated that urea supplied to rice, and/or broadcast just prior to soybean sowing had only marginal effects on nodulation, BNF and yield of the soybean. While little deleterious effect of N fertilization was observed on BNF, it cannot be assumed that this will always be the case (Jensen, 1987), since it was likely that much of the urea N applied in this particular study was lost following flooding of the paddy to grow rice, or during flood irrigation of the soybean (George et al., 1992).

Strategic applications of fertilizer N to promote forage production in legume-based pastures in intensive grazing systems such as those associated with the dairy industry have been shown to decrease N_2 fixation inputs (Table 7). There are two main reasons for this. Firstly, the additional fertilizer preferentially stimulates grass growth and the legume component declines as a proportion of the botanical composition of the sward and in some instances the legume biomass is also decreased (Table 7). Secondly, the resultant increased levels of soil mineral N can depress P_{fix} (Table 7). The combined effect of these responses is a marked reduction in the contribution of fixed N by forage legumes to the N economy of the pasture (Table 7). This has implications for forage quality and animal live weight gain (Thomas, 1995).

Time of sowing

In many places in the tropics, legumes can be grown throughout much of the year. In Thailand, for instance,

Table 6. Examples of the effect of alleviating nutritional constraints on legume growth, the proportion of plant N derived from N_2 fixation (P_{fix}) and amount of N_2 fixed

Contraint	Species	N yield	N_2 fixation		
			P_{fix}	Amount	Increase
		(kg N ha^{-1})	(%)	(kg N ha^{-1})	(%)
Excess					
Soil acidity	Subterranean clover[a]				
-Lime		20	96	19	–
+ Lime		40	87	35	84
Saline irrigation water	White clover[b]				
High salt		94	89	57	–
Low salt		150	90	135	137
Deficiency					
Iron	Pigeon pea[c]				
-Fe		42	16	7	–
+Fe		81	21	17	143
Phosphorus	Subterranean clover[a]				
+superphosphate		7	81	6	–
-superphosphate		81	89	72	1100
	Soybean[d]				
-rock phosphate		53	75	40	–
+rock phosphate		86	76	65	63

[a]Data for subterranean clover (*Trifolium subterraneum)* are derived from Peoples et al. (1994d).
[b]Data for white clover (*Trifolium repens*) are derived from Smith et al. (1993).
[c]Data derived from Herridge and Holland (1993).
[d]Unpublished data of A Bhromsiri cited by Peoples et al. (1994c). Data represent the mean of 6 responsive sites in northern Thailand.

it is possible to grow soybean during the early wet, the late wet, and the dry seasons. Although soybean's reliance upon N_2 fixation may not alter, the capacity for growth and the amounts of N_2 fixed can differ considerably between these three growing seasons (Table 8). To maximize BNF and contributions of fixed N to soil in northern Thailand, soybean ideally should be grown during the early wet season when water supply and temperatures are both optimal (Table 8). However, in Thailand, as in many other parts of Asia, the most favorable growing period is usually reserved for rice. Soybean is more often grown either during the late wet when the crop may encounter low temperatures, or during the dry season following one or two rice crops when low temperatures are common at the beginning of the season, and water is often in short supply. By growing soybean only during less favorable seasons, farmers can not exploit fully the potential benefits of soybean cropping.

While there is usually limited choice of cropping season outside of the tropics, the time of sowing can affect N_2 fixation inputs. For instance, values for N_2 fixation were high and there was a small net contribution of fixed N to soil following seed harvest of winter sown chickpea in Syria (Table 8), whereas both crop N and P_{fix} were low in spring sowings where drought limited growth as early as anthesis. In contrast, in France, differences between N_2 fixed by winter-sown and spring-sown chickpea were still observed, but were not so marked because of extended moisture availability throughout the later stages of growth (Beck et al., 1991; Table 8).

Supply of water and tillage

Seasonal water supply affects growth potential and P_{fix} (Giller and Wilson, 1991; Peoples et al., 1992). Many studies have shown that BNF can be adversely affected by both waterlogging and soil dehydration at critical times during development and growth (Gibson et al., 1982; Peña-Cabriales and Castellanos, 1993; Sall and Sinclair, 1991). In irrigated systems novel management

Table 7. Examples of the effect of fertilizer-N on N_2 fixation by pasture legumes

Species	Rate of N applied (kg N ha^{-1})	Total pasture dry matter production (t ha^{-1})	Legume N yield (kg N ha^{-1})	N_2 fixation	
				P_{fix} (%)	Amount (kg N ha^{-1})
Subterranean clover[a]	0	4.46 (32%)[d]	40	79	31
	50	5.97 (24%)	34	48	16
	100	8.20 (18%)	44	32	14
Strand medic[b]	0	9.19 (38%)	125	82	102
	50	9.90 (29%)	101	51	52
White clover[c]	5	2.76 (43%)	69	82	57
	50	3.42 (22%)	38	84	32

[a]Data of Quigley and Peoples (unpublished), measured over 155d.
[b]Data for strand medic (*Medicago littoralis*) derived from Butler (1988), measured over 144d.
[c]Data derived from Eltilib and Ledgard (1988), measured over 33d.
[d]Values in parentheses represent the legume component as a proportion of total pasture dry matter.

Table 8. Effect of growing season on N_2 fixation and the potential residual value of fixed N after seed harvest of food legumes

Species/location/ growing season	Crop N (kg N ha^{-1})	N_2 fixation		Seed N (kg N ha^{-1})	Residual benefit[c] (kg N ha^{-1})
		P_{fix} (%)	Amount (kg N ha^{-1})		
Soybean[a]					
Thailand:					
Early wet season	328	72	236	178	+58
Late wet season	107	72	77	83	- 6
Dry season	184	73	135	160	- 25
Chickpea[b]					
Syria:					
Winter sown	142	81	115	110	+5
Spring sown	21	27	6	18	-12
France:					
Winter sown	134	55	74	115	- 41
Spring sown	113	44	49	103	- 54

[a]Means of 5 soybean varieties grown at Chiang Mai in north-west Thailand (Peoples et al., 1994c).
[b]Data derived from Beck et al. (1991).
[c]N-balance = (N_2 fixed) - (Seed N).

strategies (saturated soil culture) have been developed for soybean (Troedson et al., 1989), and common bean (White and Molano, 1994) which have the potential to improve growth and N_2 fixation. In the saturated soil culture system, the water table is maintained between 3 and 15 cm below the surface of the soil. Imposition of this high water table a few weeks after plant emergence causes an immediate yellowing of leaves which then regreen 10 to 15 days later. During this period of acclimation, relatively more assimilates are directed to below-ground parts, resulting in a proliferation of roots and nodules at or above the water table. In an evaluation of the saturated soil culture system for soybean in Thailand, it was found that P_{fix}, the amount of N_2 fixed, and the residual fixed N available for return to the soil were all greater under saturated soil culture than under conventionally irrigation. Although similar results were obtained for both soybean lines used in the study, the level of N_2 fixation attained, and the final N-balance achieved was cultivar dependent (Wang et al., 1993).

The saturated soil culture system appears to have most relevance for soybean and specific lines of common bean which can adapt to a high, permanent water-table (White and Molano, 1994). With other legume species under irrigation, optimizing watering regimes for maximum growth is the best option for enhancing BNF inputs (Table 9). However, the majority of legumes are grown under natural rainfall conditions without irrigation. One strategy available for dryland cropping systems is to minimize tillage. With no-tillage, weeds are controlled with herbicides rather than by cultivation, and the soil surface is protected by the residues of previous crops. Infiltration and accumulation of water under such systems are usually greater than with conventional cultivation, and no-till soils are generally wetter and cooler, and can result in increased growth, yield and N_2 fixation by crop legumes when seasonal rainfall is below average (Herridge and Holland, 1992). Another option for semi-arid areas, or for periods during the dry season where growth can be limited by soil-water availability, might be to choose a legume to match the environment. This can involve using cultivars which have the capacity continue to grow and fix N at low moisture levels (Herridge et al., 1993; Rose et al., 1992; Sall and Sinclair, 1991), or sowing drought tolerant species such as pigeon pea, or groundnut to maximize yield potential. The use of surface mulches to conserve soil moisture might also increase nodulation and BNF (Gibson et al., 1982).

Choice of cultivar or species

Research has established that legume N-yield is the major determinant of N_2 fixation, particularly when the levels of soil nitrate are low and adequate numbers of effective rhizobia are present in soil. Yields in these instances are often related to species or cultivar. Examples of the strong relationship between crop N and N_2 fixed, and thus between N yield and P_{fix}, are presented in Figure 2. Within a species, variations in growth, final N yield and amounts of N_2 fixed may be due to cultivar effects (Table 10) and duration of growth (i.e. maturity group). However, whether a crop legume is ultimately a net contributor to or net exploiter of soil N will be determined by a cultivar's harvest index for N (Table 10; Bell et al., 1994).

Examples of variation in the genetic capacity of different species to grow and fix N under the same environmental conditions are presented for crop and pasture legumes, tree legumes and green manures in Tables 11 and 12. In a comparison of upland food legume crops in Thailand, for instance black gram, green gram and soybean all had similar growth patterns (64–73d to maturity), but black gram appeared to be better adapted to the environment and fixed more N than the other species (Table 11). However, in a related investigation in Australia, soybean had a much longer duration of growth than black gram (140d vs. 95d, respectively), and consequently had greater growth potential and opportunity for N_2 fixation (Table 11, see also green manure comparison in Table 12). By contrast, another long duration crop, pigeon pea, grew and fixed N very poorly (Table 11). This reflected specific nutritional problems with pigeon pea nodulation, N_2 fixation and yield in the soils at the study site (Brockwell et al., 1991; Herridge and Holland, 1993). In addition to BNF being mediated by duration of growth, the potential to fix N may also be related to age in perennial species such as cover-crops in plantations and tree legumes (Table 12; Peoples et al., 1995; Sanginga et al., 1995).

Pest and disease control

It is clear that foliar or root diseases will affect plant vigor and depress growth potential, and hence influence N_2 fixation. Productivity losses from 10 to >90% resulting from disease have been reported for crop and pasture legumes (Johnstone and Barbetti, 1987; Sinclair, 1994). Disease losses can be reduced by implementing a management program that addresses

Table 9. Effect of irrigation frequency on the growth and N_2 fixation by white clover[a]

Pasture type[b]	Irrigation frequency[c]	Clover N yield	N_2 fixation	
			P_{fix}	Amount
		(kg N ha^{-1})	(%)	(kg N ha^{-1})
Clover dominant	Low	108	61	66
	High	145	62	90
Grass dominant	Low	66	67	44
	High	93	71	66

[a]Data of Lattimore, Gault and Peoples (unpubl.).
[b]Clover represented 85% of the forage dry matter in the clover dominant pasture and 40% in the grass dominant pasture.
[c]A low irrigation frequency represented an allowable depletion limit of 120 mm evaporation from a class A pan evaporimeter (8 irrigations over the 109d growing season), while the high irrigation frequency was based on an allowable depletion limit of 60 mm (15 irrigations over 109d).

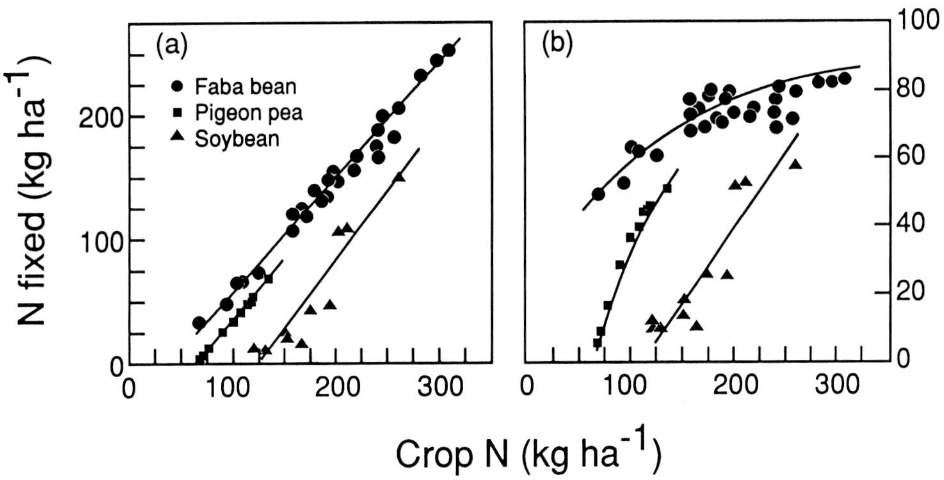

Fig. 2. Relationship between crop N and (**a**) the amount of N_2 fixed and (**b**) the proportion (P_{fix}) of plant N derived from N_2 fixation for faba bean, pigeon pea and soybean. Data derived from Hardarson et al. (1984), Kumar Rao and Dart (1987) and Duc et al. (1988).

the basic principles of disease control - eradication, exclusion and protection. The use of tolerant or resistant legume cultivars is one approach to controlling disease outbreaks (Dear et al., 1993); however, utilization of high-quality seed and appropriate rotations or cultural practices are other practical means of limiting disease incidence (Barbetti and MacNish, 1984; Sinclair, 1994). Suitable fungicides are also available to minimize the incidence of a number of legume diseases (e.g. root rot by *Phytophthora clandestina*; Hochman et al., 1990), but care must be taken with the use of a number of preparations as seed treatments since they can also be toxic to rhizobia if inoculant is seed-coated (Ramos and Ribeiro, 1993; Revellin et al., 1993). The use of alternative inoculation techniques such as liquid inoculation can overcome some incompatability problems between rhizobia and chemicals (Brockwell et al., 1988).

High infestations of sap-sucking, leaf-eating, and stem or pod-boring insects can also drastically reduce legume biomass and yield. Leucaena for example has been devastated by a sucking psyllid insect *Heteropsylla cubana* in many countries (Giller and Wilson, 1991), and redlegged earthmites, *Halotydeus destructor*, can be responsible for dry matter losses of 20–50% in subterranean clover (Allen, 1987). Chemical insec-

Table 10. Examples of the effect of cultivar or provenance on N_2 fixation by legumes

Species/cultivar, line, or provenance	N yield (kg N ha^{-1})	N_2 fixation P_{fix} (%)	N_2 fixation Amount (kg N ha^{-1})	Seed N (kg N ha^{-1})	Residual benefit (kg N ha^{-1})
Food legumes					
Pea[a]					
L82	136	66	90	105	-15
Dundale	126	79	99	82	+17
Groundnut[b]					
TmV2	269	61	164	193	- 29
Early Bunch	326	62	203	237	- 34
Virginia Bunch	319	65	206	198	+ 8
Cowpea[c]					
AFB 1757	130	51	66	82	-16
Ife BPC	200	58	117	85	+32
Common bean[d]					
Pinto Gala	92	27	25	na[g]	na
Araucano INTA	192	60	115	na	na
Pasture legumes					
White clover[e]					
Grasslands Huia	329	68	224	na	na
Grasslands Pitau	383	77	291	na	na
Trees					
Gliricidia[f]					
OFI 12/86	165	52	86	na	na
OFI 14/84	483	64	309	na	na

[a]Mean across three sites (Armstrong et al., 1994).
[b]Data derived from Bell et al. (1994).
[c]Data derived from Awonaike et al. (1990).
[d]Data derived from Hardarson et al. (1993).
[e]Data derived from Ledgard et al. (1990).
[f]Data for Gliricidia (*Gliricidia sepium*) are derived from Liyanage et al. (1994).
[g]na - data either not avialable or not appropriate.

ticides are generally available to most farmers to treat invasions of common insect pests, but management decisions about spraying to control insects must balance the degree of infestation (and potential yield loss) against the cost of the chemical and the expected economic return. It is likely that the economic threshold for spraying will also change as the growing season progresses.

It is difficult to protect below-ground organs from the larvae and nematodes that feed upon roots and nodules. Ultimately such pests may have a bigger impact on a legume's N_2-fixing potential than insects which attack the shoot. Root-knot and cyst nematodes interfere with the nodulation of many legume species (Gibson et al., 1982). In pigeon pea for example, extensive nodule damage can occur (20–86% of nodules) in farmers' field due to the action of a Dipteran larva, *Rivellia angulata* (Wani et al., 1995). Soil application of insecticide can prevent nodule damage for up to 45d after sowing, but can not control insect damage at later stages of growth (Kumar Rao and Sithanantham, 1989). More success has been reported in the control of the obligate nodule-feeding larvae of the *Sitona* spp. weevil. These larvae are known to damage 50 to 100% of lentil nodules during anthesis in parts of West Asia and North Africa. Application of carbofuran (2,3-Dihydro-2, 2-dimethyl-7-benzofuranol methylcarbamate) at sowing controls the pest and both crop N yield

Table 11. Comparisons of the N_2-fixing capacities of different species of food legumes

Location and species	N yield	N_2 fixation	
		P_{fix}	Amount
	(kg N ha^{-1})	(%)	(kg N ha^{-1})
Cyprus[a]			
Chickpea	61	41	25
Pea	127	63	91
Faba bean	220	80	176
France[b]			
Chickpea	134	55	74
Lentil	213	69	147
Thailand[c]			
Green gram	74	89	66
Black gram	125	95	119
Soybean	65	87	57
Australia[c]			
Black gram	56	37	21
Soybean	282	90	254
Pigeon pea	36	44	16
Hawaii[d]			
Soybean	241	69	166
Common bean	142	16	23

[a]Papastylianou (1988).
[b]Beck et al. (1991).
[d]Peoples et al. (1994a).
[d]George and Singleton (1992).

and N_2 fixation may be increased by 20 to 30% (Saxena, 1988). However, it has been suggested that treatment be restricted to situations of high *Sitona* infestations since the chemical may be toxic to both rhizobia and the host plant (Beck et al., 1991).

Maintaining legumes under grazing

The amount of legume required in a pasture to maximize herbage production, optimize animal liveweight gain, and to provide sufficient fixed N to balance the N removed or lost during cycling through the animal-plant-soil system will depend primarily upon P_{fix} and legume biomass. However, this will also be mediated by the extent of forage utilization by grazing animals (Thomas, 1995). The levels of P_{fix} are generally considered to be high for most forage species (Hardarson, 1993; Thomas, 1995), although the presence of animals and return of N-rich excreta can result in localized reductions in N_2-fixing activity (Ledgard and Steele, 1992). On commercial farms, BNF can be influenced by whether the pasture is grazed, or cut for hay (Peoples et al., 1995), but the most important factor determining net inputs of fixed N in a pasture system will be the maintenance of a high legume content within a pasture, and the persistence of that legume component (Tables 9 and 13).

Pastures tend to follow a pattern of high legume content initially with an increase in broad-leaved weeds and grasses with time. A number of farm surveys of the botanical composition of pastures have concluded that the legume component is often below that required for optimum animal production or to sustain reliable inputs of fixed N (reviewed by Wilson and Simpson, 1993). The perception of problems of legume persistence in pastures is mixed, possibly because year-

Table 12. Comparisons of the N_2-fixing capacities of different species of forage legumes, green manures and N_2-fixing trees

Location and species	N yield	N_2 fixation		Period of
		P_{fix}	Amount	measurement
	(kg N ha^{-1})	(%)	(kg N ha^{-1})	
Forage legumes				
Cyprus[a]				
Medicago truncatula	129	70	90	na
Vicia sativa	141	75	106	
V. villosa subsp. *varia*	191	79	151	
Philippines[b]				
Macroptilium atropurpureum	132	67	91	190 - 195d
Desmanthus vingatus	251	77	193	
Crotalaria juncea	277	80	221	
Green manures				
Philippines[c]				
Vigna radiata	58	64	37	125d
Indigofera tinctoria	113	70	79	225d
Shrubs and trees				
Australia[d]				
Codariocalyx gyroides	107	51	55	6 months[d]
Calliandra calothyrsus	210	48	101	
Gliricidia sepium	268	69	185	
Calliandra calothyrsus	79	14	11	3 months regrowth[e]
Gliricidia sepium	132	75	99	

[a]Papastylianou (1988).
[b]Ladha et al. (1995).
[c]Ladha and Peoples (unpubl.).
[d]Palmer and Peoples (unpubl.). The 6 months data represent the first cut after establishment. The 3 month regrowth was collected from 40 month old tree hedgerows that had regularly been pruned for forage every 3 months for 3 years.

to-year variation in climate regularly tips the ecological balance in favor of, or against, high legume content (Rossiter, 1966). Nevertheless, there is reliable evidence that farmers' pastures are often dominated by undesirable non-legume species.

Where pastures become a subsidiary to crop production and are primarily a means of increasing soil fertility and of providing a disease break for following crops, selective herbicides may be used to induce a high legume content. This strategy has been shown to influence greatly BNF, subsequent levels of soil nitrate (Fig. 3; Peoples et al., 1995), and crop yield (Angus et al., 1994). But there may also be some indirect or hidden costs resulting from herbicide treatment to manipulate pasture composition (Angus et al., 1994). Supplementary feeding may be required for livestock due to lost winter pasture production when grass is removed, and lime would need to be applied (100–200 kg ha^{-1}) to counteract potential soil acidification associated with leaching of nitrate down through the soil profile under the legume-dominant sward (Fig. 3; Helyar, 1991).

Where pastures are a sole source of forage for animal production, management objectives may require a different approach from the one described above. The range of options available to manage the legume content of pastures may differ between the cropping zones and permanently grazed areas. There are a number of principles that apply:

(i) Soil fertility - Pasture legumes generally have higher critical nutrient requirements for maximum

Table 13. Effect of botanical composition of pasture swards on N_2 fixation by forage legumes

Legume species and sward composition (% dry matter as legume)	Legume N yield (kg N ha^{-1})	N_2 fixation	
		P_{fix} (%)	Amount, (kg N ha^{-1})
Trifolium subterraneum[a]			
7	4	91	4
24	18	87	16
35	36	89	32
55	81	89	72
Medicago littoralis[b]			
2	86	72	62
33	116	70	81
47	159	66	105
Pueraria phaseoloides[c]			
20	12	75	9
50	24	92	22
70	44	86	38

[a]Seasonal average derived from Peoples et al. (1994d).
[b]Data derived from Butler (1988), measured over 144d.
[c]Data derived from Zaharah et al. (1986), measured over 90d.

growth than grasses (e.g. critical levels of phosphorus range from 0.36–0.52% for a range of temperate legume species, while levels for grasses are 0.17–0.26%, Pinkerton and Randall, 1994). Legumes are also less able to compete for nutrients when growing with grasses (Wilson and Simpson, 1993). There is evidence that soil fertility problems contribute to loss of legumes from pastures (e.g. Hochman et al., 1990). A key to maintaining high legume production in grass-legume systems is, therefore, to maintain high concentrations of all nutrients other than N and thus provide the N_2-fixing legume with some ecological advantage. This is illustrated in the subterranean clover data presented in Table 13. The change in pasture clover content from 7 to 55%, and increase in amounts of N_2 fixed from <4 to 72 kg N ha^{-1}, was predominantly achieved by applying phosphatic fertilizer (Table 13; Peoples et al., 1994d).

(ii) Seed reserves - It is sometimes argued that legume persistence is a consequence of the maintenance of a large seed reserve. This may be influenced by nutrient deficiencies, the length of crop-pasture rotations, disease incidence, or grazing management (Conlan et al., 1994; Dear et al., 1993; Wilson and Simpson, 1993). Therefore, periodic resowing of pastures, or undersowing of crops just prior to a pasture phase is often desirable.

(iii) Grazing management - Stocking rate and grazing system (i.e. continuous, rotational, or deferred grazing), livestock type (cattle, sheep, or goats) and selective grazing can all influence pasture composition (Ledgard and Steele, 1992; Rossiter, 1966; Thomas, 1995; Wilson and Simpson, 1993), although this may depend upon the relative growth responses of the various grass-legume associations, different growth habits and species palatability, climate, and stage during the growing season when grazing occurred. Because of the prostrate growth habit of clovers for example, strategic heavy grazing of a grassy pasture in early spring can shift botanical composition in favor of legume dominance. As a general rule, undergrazing encourages grass growth and results in pastures with depressed legume content.

(iv) Disease - Root rot by fungal diseases has been implicated in contributing to the decline in subterranean clover in pastures (Hochman et al., 1990). The persistence of a legume component in a pasture can be improved by introducing more productive, disease-resistant cultivars or tolerant species (Dear et al., 1993).

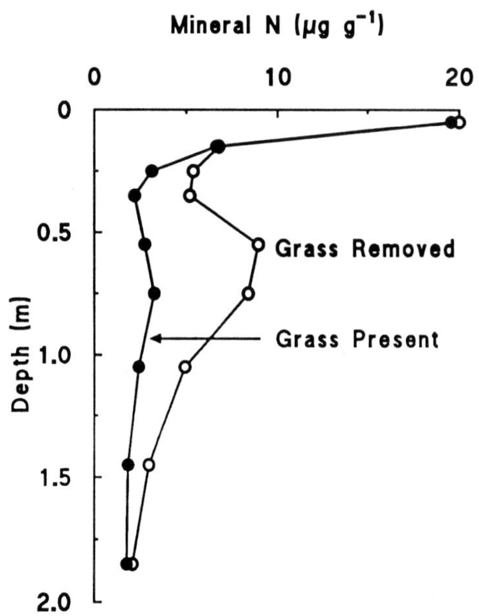

Fig. 3. Effect of manipulating pasture composition on subsequent levels of soil mineral N. The data were collected in the summer of 1993 after one-half of the pasture at a 10 ha experimental site was treated with selective herbicides to remove grasses in the winters of 1992 and 1993 (redrawn from Angus et al., 1994). The subterranean clover content of pasture was increased from 70 to 100% in 1992 and from 25 to 50% in 1993 with the application of herbicide. The difference in levels of soil nitrate beneath the herbicide-treated and untreated pasture represented around 100 kg N ha^{-1}.

Conclusions

Strategies are available for crop, forage, and tree legumes in tropical and temperate agriculture to manipulate N inputs via BNF. This can be achieved by changing P_{fix} and/or the amount of legume N accumulated. Management protocols can be imposed which either increase the numbers of effective rhizobia in soil, reduce the levels of, or legume sensitivity to, soil nitrate, or enhance the potential for legume growth, Therefore, it should be possible to manage legumes in diverse farming systems to provide a renewable source of N for agriculture.

Acknowledgements

We are indebted to many colleagues for the use of unpublished data. These data have arisen from experimentation undertaken with the support of the Australian Centre for International Agricultural Research (ACIAR), Australian International Development Assistance Bureau (AIDAB), the Wool Research and Development Corporation (WRDC), or the Grains Research and Development Corporation (GRDC) of Australia. The assistance of these funding bodies is gratefully acknowledged. We also wish to thank J Brockwell for his comments on the manuscript.

References

Allen P G 1987 Insect pests of pastures in perspective. *In* Temperate Pastures. Their Production, Use and Management. Eds. J L Wheeler, C J Pearson and G E Robards. pp 211–225. Australian Wool Corporatio/CSIRO, Melbourne.

Angus J F, Scammell G J, Peoples M B, Gault R R and Koetz E A 1994 Increasing the N contribution of pastures to crops. *In* The Role of Pastures in Cropping Systems of Southern NSW. Eds. P Orchard and K Hines. pp 41–46. NSW Agriculture, Wagga Wagga, Australia.

Armstrong E L, Pate J S and Unkovich M J 1994 Nitrogen balance of field pea crops in SW Australia, studied using the ^{15}N natural abundance technique. Aust. J. Plant Physiol. (*In press*).

Awonaike K O, Kumarasinghe K S and Danso S K A 1990 Nitrogen fixation and yield of cowpea (*Vigna unguicalata*) as influenced by cultivar and *Bradyrhizobium* strain. Field Crops Res. 24, 163–171.

Bantilan M C S and Johansen C 1995 Research evaluation and impact analysis of biological nitrogen fixation. Plant and Soil 174.

Barbetti M J and MacNish G C 1984 Effects of cultivation and cultural practices on root rot of subterranean clover. Aust. J. Exp. Agric. Anim. Husb. 24, 550–554.

Beck D P, Wery J, Saxena M C and Ayadi A 1991 Dinitrogen fixation and nitrogen balance in cool-season food legumes. Agron. J. 83, 334–341.

Becker M, Diekmann K H, Ladha J K, De Datta S K and Ottow J C G 1991 Effect of NPK on growth and nitrogen fixation of *Sesbania rostrata* as a green manure for lowland rice (*Oryza sativa L.*). Plant and Soil 132, 149–158.

Bell M J, Wright G C, Suryantini and Peoples M B 1994 The N_2-fixing capacity of peanut cultivars with differing assimilate partitioning characteristics. Aust. J. Agric. Res. 45, 1455–1468.

Bergersen F J, Turner G L, Gault R R, Chase D L and Brockwell J 1985 The natural abundance of ^{15}N in an irrigated soybean crop and its use for the calculation of nitrogen fixation. Aust. J. Agric. Res. 36, 411–423.

Bergersen F J, Brockwell J, Gault R R, Morthorpe L, Peoples M B and Turner G L 1989 Effects of available soil nitrogen and rates of inoculation on nitrogen fixation by irrigated soybeans and evaluation of δ^{15}N methods for measurement. Aust. J. Agric. Res. 40, 763–780.

Brockwell J and Bottomley P J 1994 Recent advances in inoculant technology and prospects for the future. Soil Biol. Biochem. 26 (*In press*).

Brockwell J, Bottomley P J and Thies J E 1995 Manipulation of rhizobial microflora for improving legume productivity and soil fertility: a critical assessment. Plant and Soil 174.

Brockwell J, Roughley R J and Herridge D F 1987 Population dynamics of *Rhizobium japonicum* strains used to inoculate three successive crops of soybean. Aust. J. Agric. Res. 38, 61–74.

Brockwell J, Gault R R, Herridge D F, Morthorpe L J and Roughley R J 1988 Studies on alternative means of legume inoculation:microbial and agronomic appraisals of commercial procedures for inoculating soybeans with *Bradyrhizobium japonicum*. Aust. J. Agric. Res. 39, 965–972.

Brockwell J, Gault R R, Morthorpe L J, Peoples M B, Turner G L and Bergersen F J 1989 Effects of soil nitrogen status and rate of inoculation on the establishment of populations of *Bradyrhizobium japonicum* and on the nodulation of soybeans. Aust. J. Agric. Res. 40, 753–762.

Brockwell J, Pilka A and Holiday R A 1991 Soil pH is a major determinant of the numbers of naturally occurring *Rhizobium meliloti* in non-cultivated soils in central New South Wales. Aust. J. Exp. Agric. 31, 211–219.

Brockwell J, Andrews J A, Gault R R, Gemell L G, Griffith G W, Herridge D F, Holland J F, Karsono S, Peoples M B, Roughley R J, Thompson J A, Thompson J A and Troedson R J 1991 Erratic nodulation and nitrogen fixation in field-grown pigeonpea [*Cajanus cajan* (L.) Millsp.]. Aust. J. Exp. Agric. 31, 653–661.

Butler J H A 1988 Growth and N_2 fixation by field grown *Medicago littoralis* in response to added nitrate and competition from *Lolium multiflorum*. Soil Boil. Biochem. 20, 863–868.

Carter J M, Tieman J S and Gibson A H 1994 Competitiveness and persistence of faba bean inoculants in acid and alkaline soils. Soil Biol. Biochem. 26 (*In press*).

Chandel A S, Pandey K N and Saxena S C 1989 Symbiotic nitrogen fixation and nitrogen benefits by nodulated soybean (*Glycine max* (L.) Merrill) to interplanted crops in northern India. Trop. Agric., Trinidad 66, 73–77.

Conlan D J, Dear B S and Coombes N E 1994 Effect of grazing intensity and number of grazings on herbage production and seed yields of *Trifolium subterranean*, *Medicago murex*, and *Ornithopus compressus*. Aust. J. Exp. Agric. 34, 181–184.

Corbin E J, Brockwell J and Gault R R 1977 Nodulation studies on chickpea (*Cicer arietinum* L.). Aust. J. Exp. Agric. Anim. Husb. 17, 126–134.

Craswell E T, Loneragan J F and Keerati-Kasikorn P 1987 Mineral constraints to food legume crop production in asia. *In* Food Legume Improvement for Asian Farming Systems. Eds. E S Wallis and D E Byth. pp 99–111. ACIAR Proc. No. 18, ACIAR, Canberra, Australia.

Dear B S, Murray G M, Cregan P D and Taylor P A 1993 Comparison of the performance of subterranean clover cultivars in southern New South Wales. 2. Effects of *Phytophthora clandestina* and bromoxynil on seedling survival, growth, and seed set. Aust. J. Exp. Agric. 33, 591–596.

Doughton J A, Vallis I and Saffigna P G 1993 Nitrogen fixation in chickpea 1. Influence of prior cropping or fallow, nitrogen fertilizer and tillage. Aust. J. Agric. Res. 44, 1403–1413.

Duc G, Mariotti A and Amarger N 1988 Measurement of genetic variability for symbiotic dinitrogen fixation in field-grown fababean (*Vicia faba* L.) using a low level ^{15}N-tracer technique. Plant and Soil 106, 269–276.

Eltilib A M and Ledgard S F 1988 Production and nitrogen fixation by 'Grasslands Kopu' and 'Grasslands Huia' white clovers under different nitrogen regimes. N.Z. J. Agric. Res. 31, 325–330.

Evans J, O'Connor G E, Turner G L, Coventry D R, Fettell N, Mahoney J, Armstrong E L and Walsgott D N 1989 N_2 fixation and its value to soil N increase in lupin, field pea and other legumes in south-eastern Australia. Aust. J. Agric. Res. 40, 791–805.

FAO 1992 Fertilizer use by crop: Report by the International Fertilizer Industry Association, International Fertilizer Development Center, and Food and Agriculture Organization of the United Nations. FAO, Rome.

George T and Singelton P W 1992 Nitrogen assimilation traits and dinitrogen fixation in soybean and common bean. Agron. J. 84, 1020–1028.

George T, Ladha J K, Buresh R J and Garrity D P 1992 Managing native and legume-fixed nitrogen in lowland rice-based cropping systems. Plant and Soil 141, 69–91.

Gianinazzi-Pearson V and Diem H G 1982 Endomycorrhizae in the tropics. *In* Microbiology of Tropical Soils and Plant Productivity. Eds. Y R Dommergues and H G Diem. pp 209–251. Martinus Nijhoff/ Dr W Junk Publ., the Netherlands.

Gibson A H, Dreyfus B L and Dommergues Y R 1982 Nitrogen fixation by legumes in the tropics. *In* Microbiology of Tropical Soils and Plant Productivity. Eds. Y R Dommergues and H G Diem. pp 37–73. Martinus Nijhoff/ Dr W Junk Publ., the Netherlands.

Giller K E and Cadisch G 1995 Future benefits from biological nitrogen fixation: An ecological approach to agriculture. Plant and Soil 174.

Giller K E and Wilson K J 1991 Nitrogen Fixation in Tropical Cropping Systems. CAB International, Wallingford, UK. 313p

Hardarson G, Zapata F and Danso S K A 1984 Effects of plant genotypes and nitrogen fertilizer on symbiotic nitrogen fixation by soybean cultivars. Plant and Soil 82, 397–405.

Hardarson G, Bliss F A, Cigales-Rivero M R, Henson R A, Kipe-Nolt J A, Longeri L, Manrique A, Penã-Cabriales J J, Pereira P A A, Sanabria C A and Tsai S M 1993 Genotypic variation in biological nitrogen fixation by common bean. Plant and Soil 152, 59–70.

Hardarson G 1993 Methods for enhancing symbiotic nitrogen fixation. Plant and Soil 152, 1–17.

Harper J E 1987 Nitrogen metabolism. *In* Soybean: Improvement, Production and Uses. Ed. J R Wilcox. pp 447–533. Agronomy Series No. 16, ASA Publishers, Madison, WI.

Helyar K R 1991 The management of acid soils. *In* Plant-Soil Interactions at Low pH. Eds. R J Wright, V C Baligar and R P Murrmann. pp. 365–382. Kluwer Academic Publ., Dordrecht.

Herridge D F and Bergersen F J 1988 Symbiotic nitrogen fixation. *In* Advances in Nitrogen Cycling in Agricultural Ecosystems. Ed. J R Wilson. pp 46–65. CAB International, Wallingford, UK

Herridge D F and Brockwell J 1988 Contributions of fixed nitrogen and soil nitrate to the nitrogen economy of irrigated soybean. Soil Biol. Biochem. 20, 711–717.

Herridge D F and Danso S K A 1995 Enhancing legume N_2 fixation through selection and breeding. Plant and Soil (This volume).

Herridge D F and Holland J F 1992 Production of summer crops in northern New South Wales. I. Effects of tillage and double cropping on growth, grain and N yields of six crops. Aust. J. Agric. Res. 43, 105–122.

Herridge D F and Holland J F 1993 Low nodulation and N_2 fixation limits yield of pigeonpea on alkaline vertisols of northern N S W : Effect of iron, rhizobia and plant genotype. Aust. J. Agric. Res. 44, 137–149.

Herridge D F, Bergersen F J and Peoples M B 1990 Measurement of nitrogen fixation by soybean in the field using the ureide and natural abundance methods. Plant Physiol. 93, 708–716.

Herridge D F, Rupela O P, Serraj and Beck D P 1993 Screening techniques and improved biological nitrogen fixation in cool season legumes. Euphytica 1, 1–14.

Herridge D F, Marcellos H, Felton W L, Turner G L and Peoples M B 1994 Chickpea increases soil-N fertility in cereal systems through nitrate sparing and N_2 fixation. Soil Biol. Biochem. 26 (*In press*).

Hochman Z, Osborne G J, Taylor P A and Cullis B 1990 Factors contributing to reduced productivity of subterranean clover (*Tri-*

folium subterranean L.) pastures on acidic soils. Aust. J. Agric. Res. 41, 669–682.

Hughes R M and Herridge D F 1989 Effect of tillage on yield, nodulation and nitrogen fixation of soybean in far north-coastal New South Wales. Aust. J. Exp. Agric. 29, 671–677.

Jensen E S 1987 Seasonal patterns of growth and nitrogen fixation in field-grown pea. Plant and Soil 101, 29–37.

Johnstone G R and Barbetti M J 1987 Impact of fungal and virus diseases on pasture. *In* Temperate Pastures. Their Production, Use and Management. Eds. J L Wheeler, C J Pearson and G E Robards. pp 235–248. Australian Wool Corporation/CSIRO, Melbourne.

Kumar Rao J V D K and Dart P J 1987 Nodulation, nitrogen fixation and nitrogen uptake in pigeonpea (*Cajanus cajan* (L.) Millsp) of different maturity groups. Plant and Soil 99, 255–266.

Kumar Rao J V D K and Sithanantham S 1989 Impact of nodule damage by *Rivellia angulata* on N_2 fixation, growth, and yield of pigeonpea (*Cajanus cajan L.*) growth in a vertisol. Biol. Fertil. Soils 7, 95–100.

Ladha J K, Watanabe I and Saono S 1988 Nitrogen fixation by leguminous green manure and practices for its enhancement in tropical lowland rice. *In* Green Manure in Rice Farming. pp 165–183. IRRI, Manila, Philippines.

Ladha J K, Kundu D K, Angelo-Van Coppenolle M G, Peoples M B, Carangal V R and Dart P J 1995 Legume productivity and soil nitrogen dynamics in lowland rice-based cropping systems. Soil Sci. Soc. Am. J. (*In press*).

Ledgard S F, Brier G J and Upsdell M P 1990 Effect of clover cultivar on production and nitrogen fixation in clover-ryegrass swards under dairy cow grazing. N.Z J. Agric. Res. 33, 243–249.

Ledgard S F and Steele K W 1992 Biological nitrogen fixation in mixed legume/grass pastures. Plant and Soil 141, 137–153.

Liyanage M de S, Danso S K A and Jayasundara H P S 1994 Biological nitrogen fixation in four *Gliricidia sepium* genotypes. Plant and Soil 161, 267–274.

Marshall D L, Bateman J D and Brockwell J 1993 Validation of a serial-dilution, plant-infection test for enumerating *Rhizobium leguminosarum* bv. *viciae* and its application for counting rhizobia in acid soils. Soil Biol. Biochem. 25, 261–268.

Ofori F and Stern W R 1987 Cereal-legume intercropping systems. Adv. Agron. 41, 41–90.

O'Hara G W, Boonkerd N and Dilworth M J 1988 Mineral constraints to nitrogen fixation. Plant and Soil 108, 93–110.

Papastylianou I 1988 The role of legumes in agricultural production in Cyprus. *In* Nitrogen Fixation by Legumes in Mediterranean Agriculture. Eds. D P Beck and L A Materon. pp 55–64. Martinus Nijhoff, Dordrecht.

Peña-Cabriales J J and Castellanos J Z 1993 Effects of water stress on N_2 fixation and grain yield of *Phaseolus vulgaris* L. Plant and Soil 152, 151–155.

Peoples M B and Herridge D F 1990 Nitrogen fixation by legumes in tropical and sub-troical agriculture. Adv. Agron. 44, 155–223.

Peoples M B and Craswell E T 1992 Biological nitrogen fixation: investments, expectations and actual contributions to agriculture. Plant and Soil 141, 13–39.

Peoples M B, Bell M J and Bushby H V A 1992 Effect of rotation and inoculation with *Bradyrhizobium* on nitrogen fixation and yield of peanut (*Arachis hypogaea* L., cv Virginia Bunch). Aust. J. Agric. Res. 43, 595–607.

Peoples M B, Bergersen F J, Brockwell J, Fillery I R P and Herridge D F 1994a Management of nitrogen for sustainable agricultural systems. *In* Nuclear Related Methods in Soil/Plant Aspects of Sustainable Agriculture. FAO/IAEA, Vienna, (*In press*).

Peoples M B, Gault R R, Brockwell J, Lean B and Sykes J D 1994b Nitrogen fixation by soybean in commercial irrigated crops of central and southern New South Wales. Soil Biol. Biochem. 26, (*In press*).

Peoples M B, Herridge D F, Rerkasem B, Bhromsiri A, Sampet C and Brockwell J 1994c Improving nitrogen fixation by soybean. *In* Proc. World Soybean Research Conference V (*In press*).

Peoples M B, Lilley D M, Burnett V F, Ridley A M and Garden D L 1994d The effect of surface application of lime and superphosphate to acid soils on the growth and N_2 fixation by subterranean clover in mixed pasture swards. Soil Biol. Biochem. 26 (*In press*).

Peoples M B, Herridge D F and Ladha J K 1995 Biological nitrogen fixation: an efficient source of nitrogen for sustainable agricultural production? Plant and Soil 174.

Pinkerton A and Randall P J 1994 The internal phosphorus requirements of six legumes and two grasses. Aust. J. Exp. Agric. 34, 373–379.

Ramos M L G and Ribeiro J R W Q 1993 Effect of fungicides on survival of *Rhizobium* on seeds and the nodulation of bean (*Phaseolus vulgaris* L.). Plant and Soil 152, 145–150.

Rennie R J, Rennie D A, Siripaibool C, Chaiwanakupt P, Boonkerd N and Snitwongse P 1988 N_2 fixation in Thai soybeans: effects of tillage and inoculation on ^{15}N-determined N_2 fixation in recommended cultivars and advanced breeding lines. Plant and Soil 112, 183–193.

Rerkasem B, Rerkasem K, Peoples M B, Herridge D F and Bergersen F J 1988 Measurement of N_2 fixation in maize (*Zea mays* L.)-ricebean (*Vigna umbellata* [Thumb.] Ohwi and Ohashi) intercrops. Plant and Soil 108, 125–135.

Revellin C, Leterme P and Catroux G 1993 Effect of some fungicide seed treatments on the survival of *Bradyrhizobium japonicum* and on the nodulation of soybean [*Glycine max*.(L) Merr.] Biol. Fertil. Soils 16, 211–214.

Richardson A E and Simpson R J 1988 Enumeration and distribution of *Rhizobium trifolii* under a subterranean clover-based pasture growing in a acid soil. Soil Biol. Biochem. 20, 431–438.

Rose I A, McWhirter K S and Spurway R A 1992 Identification of drought tolerance in early-maturing indeterminate soybeans [*Glycine max* (L.) Merr.]. Aust. J. Agric. Res. 43, 645–657.

Rossiter R C 1966 Ecology of the Mediterranean annual-type pasture. Adv. Agron.18, 1–56.

Sall K and Sinclair T R 1991 Soybean genotypic differences in sensitivity of symbiotic nitrogen fixation to soil dehydration. Plant and Soil 133, 31–37.

Sanginga N, Vanlauwe B and Danso S K A 1995 Management of biological N_2 fixation in alley cropping systems: estimation and contribution to N balance. Plant and Soil 174.

Saxena M C 1988 Food legumes in the Mediterranean type of environment and ICARDA's efforts in improving their productivity. *In* Nitrogen Fixation by Legumes in Mediterranean Agriculture. Eds. D P Beck and L A Materon. pp 11–23. Martinus Nijhoff, Dordrecht.

Shafiq M, Hassan A, Ahmad N and Rashid A 1994 Crop yields and nutrient uptake by rainfed wheat and mungbean as affected by tillage, fertilization, and weeding. J. Plant Nutr. 17, 561–577.

Sinclair J B 1994 Reducing losses from plant diseases. *In* Proc. World Soybean Research Conference V (*In press*).

Singelton P W, Bohlool B B and Nakao P L 1992 Legume response to rhizobial inoculation m the tropics: myths and realities. *In* Myths and Science of Soils of the Tropics. Soil Sci. Soc. Am. and Am. Soc. Agron. Spec. Publ. No. 29, 135–155.

Smith C J, Chalk P M, Noble C L, Prendergast J B and Robertson F 1993 Nitrogen fixation in a white clover-grass pasture irrigated with saline groundwater. Irrig. Sci. 13, 189–194.

Streeter J 1988 Inhibition of legume nodule formation and N_2 fixation by nitrate. CRC Critical Rev. Plant Sci. 7, 1–23.

Thies J E, Singleton P W and Bohlool B B 1991 Modelling symbiotic performance of introduced rhizobia in the field based on indicies of indigenous population size and nitrogen status of the soil. Appl. Exp. Microbiol. 57, 29–37.

Thomas R J 1995 Role of legumes in providing N for sustainable tropical pasture systems. Plant and Soil 174.

Thompson J A, Bhromsiri A, Shutsrirung A and Lillakan S 1991 Native root-nodule bacteria of traditional soybean-growing areas of northern Thailand. Plant and Soil 135, 53–65.

Troedson R J, Lawn R J, Byth D E and Wilson G L 1989 Response of field-grown soybean to saturated soil culture. 1. Patterns of biomass and nitrogen accumulation. Field Crops Res. 21, 171–187.

Wang G, Peoples M B, Herridge D F and Rerkasem B 1993 Nitrogen fixation, growth and yield of soybean grown under saturated soil culture and conventional irrigation. Field Crops Res. 32, 257–268.

Wani S P, Rupela O P and Lee K K 1995 Sustainable agriculture in the semi-arid tropics through biological nitrogen fixation in grain legumes. Plant and Soil 174.

White J W and Molano C H 1994 Production of common bean under saturated soil culture. Field Crops Res. 36, 53–58.

Wilson A D and Simpson R J 1993 The pasture resource base: status and issues. *In* Pasture Management: Technology for the 21st Century. Eds. D R Kemp and D L Michalk. pp 1–25. CSIRO, Melbourne.

Ying J, Herridge D F, Peoples M B and Rerkasem B 1992 Effects of N fertilization on N_2 fixation and N balances of soybean grown after lowland rice. Plant and Soil 147, 235–242.

Zaharah A R, Sharifuddin H A H, Razley M N and Mohd Saidi A K 1986 Measurement of nitrogen fixed by *Pueraria phaseloides* by N-15 dilution technique. Pertanika 9, 45–49.

Plant and Soil **174:** 103–118, 1995.

Role of legumes in providing N for sustainable tropical pasture systems

R. J. Thomas
Centro Internacional de Agricultura Tropical (CIAT), Apartado Aereo 6713, Cali, Colombia

Key words: agropastoral systems, farmer acceptability, forage legume, fertilization, nitrogen fixation, persistence, tropical pastures

Table of contents

Abstract
Introduction
How much N is needed to sustain the soil-plant-animal system?
How much N can be fixed
  Factors affecting BNF in pastures
  Inorganic soil N
  Phosphorus and potassium fertilization
  Efficiency of nitrogen fixation
  Estimates of BNF from biomass measurements?
Fate of fixed N
How can sustainable grass/legume pastures be achieved?
  Selection of appropriate forage legumes
  Use of fertilizers
  Alleviation of constraints
    Legume persistence
    Anti-quality factors
  Is there a need to inoculate tropical forage legumes?
  Acceptability of forage legumes by farmers
Role of forage legumes in recuperation of degraded land
The potential use of forage legumes in agropastoral systems
Conclusions
Acknowledgements
References

Abstract

Forage legumes have long been lauded for their ability to fix atmospheric nitrogen and contribute to the sustainability of agricultural production systems. However despite the benefits they bring in terms of increased herbage and animal production they are not widely used in temperate or tropical regions. In this review the amounts of biological nitrogen fixation (BNF) needed to sustain the soil-plant-animal system are discussed and related to the amounts fixed in tropical pastures. The data suggest that tropical forage legumes have the capacity to meet the requirements to balance the N cycle of grazed pastures. The actual amounts required will depend on the rate of pasture utilization and the efficiency of recycling via litter, excreta and internal remobilization. The efficiency of nitrogen fixation (% of legume N derived from fixation) is usually high in tropical pastures (> 80%) and is unlikely to be affected by inorganic soil N in the absence of N fertilizer. Thus an estimate of the amounts of N fixed could be obtained from simple estimates of legume biomass provided tissue levels of other nutrients such as phosphorus and potassium are adequate. Key factors for the achievement of sustainable grass/legume pastures include the selection of appropriate germplasm adapted to the particular environment and the judicious use of fertilizers such as phosphorus, potassium, calcium, magnesium and sulphur on acid infertile soils typical of the sub-humid and humid tropics. The main constraints to the widespread adoption of forage legumes include a lack of legume persistence, the presence of anti-quality factors such as tannins, variable *Bradyrhizobium* requirements and lack of acceptability by farmers. Strategies for the alleviation of these constraints are discussed. Forage legumes can be used to recuperate degraded soils via their ability to improve the physical, chemical and biological properties of soils and these benefits could be of particular use for small-scale resource-poor farmers. The incorporation of forage legumes into agropastoral systems is discussed as an environmentally and economically attractive means to encourage the widespread adoption of legumes in the humid tropics.

Introduction

Twenty three percent of the world's total area or 3.4 billion ha are permanent grasslands (FAO, 1993). Around 1.5 billion ha of these grasslands are in the tropics as either wild or cultivated fodder plants (Pearson and Ison, 1987; UNESCO, 1979). In most developing tropical countries animal production from pastures is low compared with developed countries, e.g. beef production is around 20 kg per animal unit per year in developing countries compared with 96 kg per animal unit per year in developed countries (Henzell, 1983). At least 700 million ha of relatively unproductive grasslands in South America, Africa, Asia and Australia are considered to be "improvable grasslands" (Pearson and Ison, 1987). Technologies for the improvement of grassland production have been considered (Breman and de Wit, 1983; Sanchez and Salinas, 1981; Teitzel, 1992; Toledo and Nores, 1986) wherein the introduction of a legume into grassland systems features prominently (e.g., Toledo, 1985). The basis of this essentially low-input (but not zero input) technology is the reliance on legume-based pastures to provide N, via biological fixation, and hence higher quality forage on offer to grazing animals. The provision of N via BNF in tropical pastures is particularly important as grasses (mainly C_4 types) frequently contain levels of N of 1.3% or less which are inadequate for animal production (Humphreys, 1991) and fertilizer N is generally less readily available to farmers for logistic and/or economic reasons.

The grasslands of Latin American tropics frequently suffer from pasture deterioration and degradation due to a number of causes including overgrazing and subsequent soil erosion, mineral deficiencies, especially N and P, and occasional attacks from pests and diseases (Thomas et al., 1994a). In Australia the expansion of pastures peaked around 1971 and since then there have also been problems of declining pasture productivity due to acidification, salinization, waterlogging and compaction, lack of phosphorus fertilization and N deficiency (Blyth and Menz, 1987; Gramshaw et al., 1989; Myers and Robbins, 1991). In Sahelian pastures N and P limitations occur along with water shortages (Breman and de Wit, 1983). In SE Asia, savannas cover around 23 million ha and are dominated by the noxious, shallow rooting grass *Imperata cylindrica* (star grass or alang-alang), where nutrient cycling is incomplete and soils degrade (Von Uexkull and Mutert, 1993). These problems are often acute leading some to suggest that these areas should perhaps be left under extensive use (Pearson and Ison, 1987). However the constraints of these marginal areas, especially in terms of soil quality, are well known (Sanchez

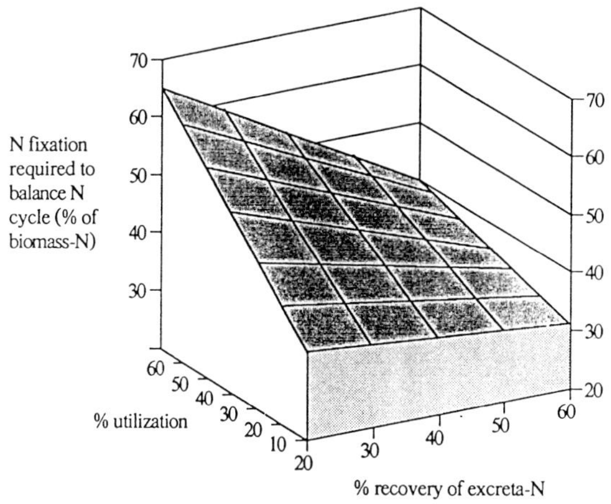

Fig. 1. Effect of variations in the % recovery of excreta-N and pasture utilization on the requirement for fixed-N.

and Logan, 1992) and it is possible to design technologies which would allow an intensification of grassland productivity using legumes (e.g., t'Mannetje, 1986; Thomas et al., 1992; Von Uexkull and Mutert, 1993). Such intensification should now consider the options for integrating more closely crop and livestock production in order to produce the additional food and fibre of animal and plant origin needed to satisfy the world's burgeoning population (Henzell, 1983; Worldwatch Institute, 1992).

The purpose of this review is to demonstrate the role that forage legumes and BNF can play in both improved pasture production and in the recuperation of degraded pastures, and to outline the main constraints that retard the adoption of legume-based pastures in regions of the developing world where they are needed the most.

The review is not comprehensive but attempts to complement several recent articles on BNF in pastures (Giller and Wilson, 1991; Ledgard and Steele, 1992; Peoples and Herridge, 1990; Peoples et al., 1995).

How much BNF is needed to sustain the soil-plant-animal system?

The amount of legume needed in a pasture is an old question that has generally been addressed from the herbage or animal production standpoint. For example, estimates of the amounts of above-ground legume biomass necessary in a pasture to maximize herbage production are in the order of 30–50% dry matter (DM) content (Harris and Thomas, 1973; Martin, 1960) with a range of 20–40% for maximum animal live weight gain (Simpson and Stobbs, 1981; Stewart, 1984; Watson and Whiteman, 1981). The question of how much legume and BNF is needed to maintain the N balance in the soil-plant-animal system has seldom been addressed. It has been estimated that to maintain the N reserves of the soil in pastures receiving no N fertilizer, a range of legume biomass of 20–31% of the pasture DM is needed for moderately grazed pastures with a 10–40% utilization (consumption by animals). This range increases to 35–45% DM in intensively used pastures with a range of utilization of 50–70% (Thomas, 1992). Others have indicated however, that as little as 10% legume could maintain the N requirements of temperate ryegrass/white clover swards (Sheehy, 1989).

The estimates by Thomas (1992) included the effects of likely variations in the recovery of N by plants via the main recycling processes, viz., excreta, plant litter decomposition and internal remobilization during senescence, on the requirement for fixed N to balance the N cycle without invoking a net drain on soil organic N (e.g. Fig. 1). Generally, as pasture utilization increases losses from the system also increase as more N passes through the animal and is excreted. Losses from excreta can be high via leaching and volatilization (e.g. 60–80%; Ball and Ryden, 1984; Simpson, 1987; Steele and Vallis, 1988) and constitute the "leaky" processes of the N cycle (in the absence of N fertilizer application). Consequently the requirement for inputs via BNF must similarly increase with increasing utilization to balance the cycle (Fig. 2).

These estimations did not include recycling via the root biomass, which can be large in N-deficient tropical pastures. Preliminary data from tropical grass/legume pastures in Colombia indicate a root:shoot ratio of approximately 1 in newly established pastures which are not excessively N-deficient. This ratio increases as deficiency increases (IM Rao, pers. commun.). Assuming a complete turnover of the root system per year and a N concentration in the roots of 0.5% N compared with 1% N in shoot tissue (Rao, unpubl.) then a 10 t ha^{-1} above ground DM will contain 100 kg N ha^{-1} shoot tissue plus 50 kg N ha^{-1} root tissue. Table 1 shows what effect the inclusion of root biomass-N would have on the requirement for legume-N to balance the cycle assuming a complete root turnover each year and a 50% recovery of root-N by growing plants. The effect varies from an additional 5% of the total plant biomass-N to 3% less depending on the % pasture utilization.

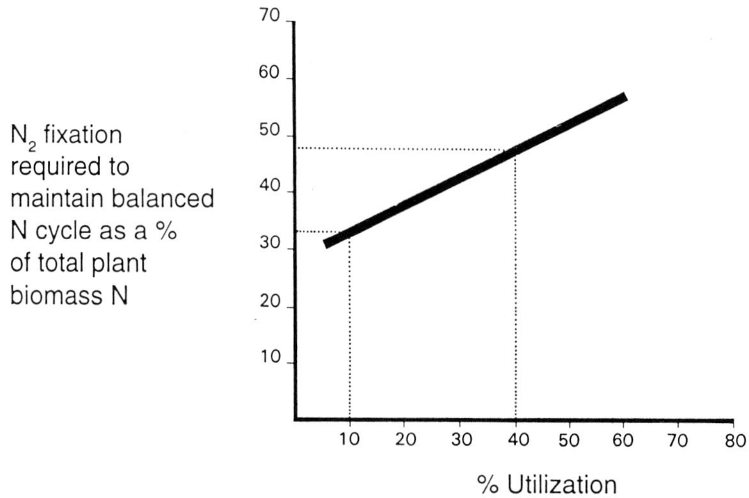

Fig. 2. Amounts of N_2 fixation required to balance the N cycle as a function of pasture utilization.

Table 1. Effect of the inclusion of roots on the requirement for legume-N to balance the N cycle of pastures grazed at different levels of utilization

	A	B[b]	C	D	E	F	G
Pasture utilization	Amount of[a] shoot-N needed to balance N cycle (% of shoot-N)	Amount of shoot-N available for recycling	Amount of shoot N available for recycling assuming complete root turnover and 50% recovery	Total amount of N (root + shoot) available for recycling	Amount of N needed to balance cycle	Amount of N needed to balance cycle as a % of total plant biomass	Difference in estimates of % biomass-N needed to balance cycle between shoot only estimates and shoot + root estimates
				(B + C)	(150–D)	E/150 × 100	(F–A)
10	34.3	65.7	25	90.7	59.3	39.5	+ 5.2
20	38.6	61.4	25	86.4	63.6	42.4	+ 3.8
30	42.9	57.1	25	82.1	67.9	45.2	+ 2.3
40	47.2	52.8	25	77.8	72.2	48.1	+ 0.9
50	51.5	48.5	25	73.5	76.5	51.0	- 0.5
60	55.8	44.2	25	69.2	80.8	53.9	- 1.9
70	60.1	39.9	25	64.9	85.1	56.7	- 3.4

[a] Data derived from Figure 4, Thomas (1992).
[b] For a 100 kg shoot N ha^{-1} pasture this value is 100–A.
Estimates are based on a 10 t DM ha^{-1} pasture with a shoot:root ratio of 1 and 1% N in shoot, 0.5% N in roots i.e. 100 kg shoot N + 50 kg root N ha^{-1} = total biomass-N is 150 kg N ha^{-1}.

Less N is needed at higher rates of utilization as the calculations assume no effect of increased grazing on root biomass-N. This may be a questionable assumption but the data illustrate that the absence of root data may not have a large effect on the estimated amounts of BNF required to balance the cycle. In addition the % N concentrations in shoot tissues of a grass/legume pasture are likely to be greater than the value of 1% used here. Values for shoot N of greater than 1% or a greater difference between %N in shoots and roots than that used in the estimations may further diminish the contribution of root N to the N balance. Data on root production and turnover of both organic matter and nutrients are urgently needed to verify these estimates particularly as it is known that root systems of tropical grasses have the capacity to immobilize substantial amounts of N resulting in pasture degradation (Bushby et al., 1992; Robbins et al., 1987).

Notwithstanding the limitations of the above analyses (Thomas, 1992) it would appear that the amounts of legume required in a pasture to balance the N cycle in terms of soil reserves are not too different from the amounts needed to maximize herbage production and individual animal performance. Thus there may not be much of a trade-off between the apparently conflicting demands for BNF in pastures for agricultural production (meat and milk) and for the replenishment of soil reserves. The key factors will be the stocking rate of animals and the rate of utilization of the pastures, i.e. factors that are controlled by the land manager and which are discussed later.

How much N can be fixed?

Reliable estimates of the amounts of N fixed in tropical pastures have appeared only relatively recently with the increasing use of ^{15}N isotope methodologies. These are summarized in Table 2 and the range reported matches the estimated annual inputs of 15–158 kg N ha^{-1} required to sustain soil-plant-animal systems producing 3–22 t forage DM ha^{-1} yr^{-1} (Thomas, 1992). There seems little doubt therefore that tropical forage legumes have the potential to sustain the N requirements of a pasture.

Noticeable in these estimations were the high proportions of legume-N derived from fixation (% Ndfa) under varying pasture conditions (average 84%). However there has been little systematic research on the effects of factors such as soil type, soil nutrients, pasture age or grazing on the % Ndfa in tropical pastures.

Factors affecting BNF in pastures

Factors affecting BNF have been extensively covered by recent reviews (Giller and Wilson, 1991; Ledgard and Steele, 1992; Sprent and Sprent, 1990) and include soil inorganic N, acidity, salinity, nutrient deficiencies (P, K, Ca, Mo, Zn, Co, Fe) or toxicities (Al, Mn), water stress, high or low temperatures, pests and diseases. The reader is referred to the cited reviews for further details and only some pertinent recent points are discussed here.

Inorganic soil N

Inorganic nitrogen in the soil is well known to be able to reduce BNF but generally in pastures and especially in the absence of N fertilizer, levels are low and are unlikely to have an inhibitory effect (Simpson, 1987). Sylvester-Bradley and Mosquera (1985) showed a reduced response to inoculation in terms of plant N in ploughed soil with subsequent higher nitrate-N levels compared with unploughed soil, but there were no differences between treatments in nodulation evaluations. In these experiments there was little competition for soil N from grasses. In rice (*Oryza sativa*)-pasture associations applications of up to 80 kg urea-N ha^{-1} (in three split applications) to rice - *Brachiaria dictyoneura-Centrosema acutifolium* mixtures sown after three different types of land preparation, had little or no effect on the nodulation of the legume (Thomas, unpubl.). In these oxisols levels of nitrate-N remain below 2 μg nitrate-N g^{-1} soil after N fertilizer applications. Thus inorganic-N may not be a significant problem for BNF in relatively infertile acid-soils.

Phosphorus and potassium fertilization

Cadisch et al. (1989) reported a marked decrease in % Ndfa from a range of 70–88 with P and K fertilization to 44–84% with no added P or K for eight tropical forage legumes that were grown in strips cleared of native savanna. The same authors showed that P fertilization had a greater effect on the % Ndfa than K with *Centrosema acutifolium* and *C. macrocarpum* (Table 3 adapted from Cadisch et al., 1993).

In field experiments on two differing soils, intimate mixtures of the grass *Brachiaria dictyoneura* and one of three forage legumes (*Arachis pintoi, Centrosema acutifolium, Stylosanthes capitata*) were given amounts of fertilizer normally used for the establishment of pastures in acid-soil savannas (kg ha^{-1}; 20 P, 20 K, 50 Ca, 20 Mg, 12 S, micronutrients and no N) and were compared with similar treatments receiving three times these amounts (Fisher et al., 1994). The % Ndfa were not significantly different between the two fertilizer treatments for any of the three legumes grown on either a sandy loam soil or a clay loam soil during the first year of establishment (Table 4). In the second year, in general, similar trends were noted with only

Table 2. Estimates of N_2 fixation and % plant N derived from fixation from tropical forage legumes using ^{15}N methodologies

Legume species	(Kg N fixed ha^{-1})	% Ndfa	Period of measurement	Reference
Calopogonium mucunoides	64	–	1 yr	Seiffert et al. (1985)
Centrosema acutifolium	43	82	17 wks	Cadisch et al. (1989)
C. macrocarpum	41	83	"	"
Desmodium ovalifolium	25	70	"	"
D. intortum	24–183	94	1 yr	Vallis et al. (1977)
Macroptilium atropurpureum	15–24	92	1 yr	"
"	23–79	83	1 yr	Shivaram et al. (1988)
Pueraria phaseoloides	9–23	82	3 mths	Zaharah et al. (1986)
"	115	88	17 wks	Cadisch et al. (1989)
Sesbania cannabina	121 – 141	80	1 season	Chapman and Myers (1987)
Stylosanthes capitata	38	87	17 wks	Cadisch et al. (1989)
S. guianensis	47	75	"	"
S. macrocephala	71	88	"	"
S. spp.	2–84	81	various	Vallis and Gardener (1985)
Zornia glabra	61	88	17 wks	Cadisch et al. (1989)

Table 3. Effect of different levels of P and K fertilizer on the % of legume-N derived from fixation in field grown *Centrosema acutifolium and C. macrocarpum*[a]

Fertilizer kg ha^{-1}		% legume shoot-N derived from fixation	
P	K	*C. acutifolium*	*C. macrocarpum*
5	60	84.9	79.2
40	60	94.6	91.6
75	60	94.5	94.0
75	30	94.8	93.0
75	0	94.3	92.9
LSD 0.05		3.1	

[a] Data from Cadisch et al. (1993), Fixation measured 14 weeks after sowing.

small differences in % Ndfa with fertilizer treatment (results not shown). The major effect of the lower level of fertilization was a reduction in legume biomass in the pastures at the two sites (Table 4).

These results suggest that the % Ndfa in forage legumes will decrease only where extreme deficiencies of nutrients such as P and K occur.

Efficiency of nitrogen fixation

The efficiency of nitrogen fixation (defined here as % Ndfa) in forage legumes has rarely been studied in grazed pastures for any length of time as most studies have dealt with the pasture establishment phase. Pasture legumes generally are poor competitors with grasses for soil N probably because of low root biomasses relative to grasses. Walker et al. (1956) and Eltilib and Ledgard (1988) showed that the proportion of nitrogen fixed by clover in a temperate pasture exceeded 80% over a wide range of mineral N supply. Vallis and Gardener (1985) showed with 10 accessions of *Stylosanthes* spp. that there were little differences in % Ndfa among accessions and little relationship between % Ndfa and total N uptake from the soil or with the age of the pasture up to 6 years old. Similarly Edmeades and Goh (1978) showed no change in the % Ndfa in grass/white clover pastures varying in age from 2 to 20 years. These results imply that the % Ndfa remains stable over time. However it should be noted that in these studies phosphate fertilizer was added either annually (Vallis and Gardener, 1985) or at the beginning of the measurement period along with other nutrients (Edmeades and Goh, 1978). Thus there remains the uncertainty of what level of efficiency (%Ndfa) can be expected from long term grass/legume pastures that do not receive maintenance levels of fertilizer as is the case for much of Latin

American pastures. The data of Cadisch et al. (1989, 1992, 1993) indicate that tissue analysis for nutrient deficiencies could be a useful guide to the likely level of % Ndfa in older pastures. Further research is needed to define the relationships between % Ndfa and mineral nutrient levels in tissues of different forage legumes.

Estimates of BNF from biomass measurements?

The bulk of the available data suggests that in tropical grass/legume pastures the %Ndfa remains relatively high (> 80%). If this is so then relatively simple estimates of legume biomass may be sufficient to estimate the amounts of nitrogen fixed using a value of around 80% for the plant N derived from fixation. In temperate grass/legume pastures Sanford et al. (1993) reported a range of % Ndfa of 0–100% with means for different species between 70–80% measured at over 200 sites. Such a comprehensive survey has not yet been reported for tropical legumes but is required to verify the use of legume biomass as a measure of N_2 fixation.

The data in Table 4 and other reports (e.g. Edmeades and Goh, 1978; Peoples et al., 1995; Vallis and Gardener, 1985) show that the amounts of nitrogen fixed will be dependent mainly on the amounts of legume present and legume productivity. Of paramount importance then is the maintenance of a legume population of 20% or greater to ensure a continued input of amounts of N that meet the requirements of both animal production and the soil N balance.

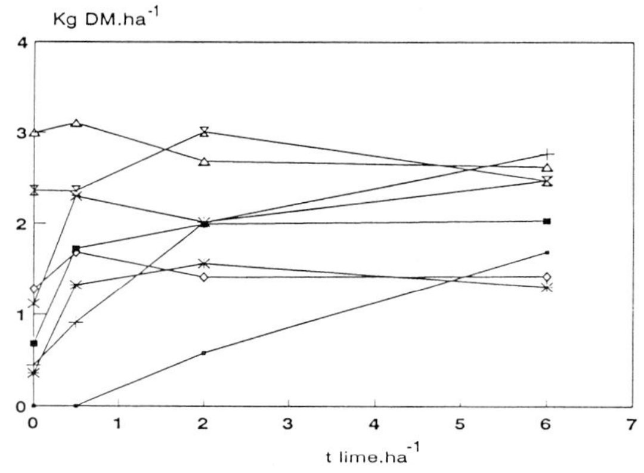

Fig. 3. Effect of lime on legume dry matter production in the field. –■– *Centrosema plumieri*, –+– *Centrosema* sp. CIAT 442, –*– *Centrosema* sp. CIAT 438, –□– *Centrosema pubescens*, –×– *Desmodium ovalifolium*, –⌑– *Pueraria phaseoloides*, –△– *Zornia* sp. – – *Stylosanthes capitata*.

Fate of fixed N

Evidence for the transfer of fixed N to companion grasses was recently discussed by Ledgard and Steele (1992) and Giller and Wilson (1991). Generally levels appear to be low with around 25% of the legume-N being transferred to the grass via the decomposition of above- and below-ground tissues, leaching from tissues into the soil and gaseous effluxes with subsequent re-uptake by grasses, and perhaps direct transfer via mycorrhizal connections. Recent evidence has shown wide variation in the short term decomposition and release of N and other nutrients from six different tropical forage legumes and it was estimated that between 2–38% of a pasture's requirement for above-ground N could be obtained from recycling via above-ground litter (Thomas and Asakawa, 1993b). Transfer via animal excreta can also be an important although seldom quantified process in tropical pastures.

The benefits of fixed N in terms of increased forage production and quality and gains in animal performance (liveweight, milk production and reproduction) have been reviewed recently (Thomas et al., 1992, 1994a; Thomas and Lascano, 1994) and were highlighted by spectacular 10-fold increases in productivity per hectare and 2-fold liveweight increases per head of cattle from grass/legume pastures compared with unimproved native savanna grasses.

How can sustainable grass/legume pastures be achieved?

Some strategies for achieving sustainable grass/legume pastures with a continuing input of N via BNF are discussed in this section along with the major constraints that need to be addressed.

Selection of appropriate forage legumes

The key to the success of forage legumes in providing N for tropical pastures is firstly the selection of germplasm adapted to the edaphic and environmental conditions and resistant to pests and diseases. In most of the tropics, soils are acid (pH <5.5) and consequently the soil constraints include acidity and often the associated toxic levels of aluminum and low availability of other plant nutrients (Sanchez and Logan,

Table 4. Amounts of N_2 fixed over 12 weeks, legume biomass and % N derived from fixation (% Ndfa)*. Values followed by same letter under each parameter are not significantly different, $p < 0.05$

Site	Fertility	*A. pintoi* (kg N fixed ha^{-1})	Biomass (kg DM ha^{-1})	% Ndfa	*C. acutifolium* (kg N fixed ha^{-1})	Biomass (kg DM ha^{-1})	% Ndfa	*S. capitata* (kg N fixed ha^{-1})	Biomass (kg DM ha^{-1})	% Ndfa
Sandy	Low	0.8a	89.9a	81.5a	1.7a	130.5a	88.9a	21.0a	1510.4a	85.6a
Loam	High	7.4b	619.7b	87.1a	2.5a	143.8a	91.7a	40.0b	2528.5b	90.2a
Clay	Low	0.9a	96.8a	71.7a	3.5a	248.0a	91.4a	14.8a	1390.1a	79.7a
Loam	High	6.8b	607.8b	85.6a	5.2a	340.9b	92.9a	31.0b	2808.6b	89.1a

*N_2 fixation measured by ^{15}N isotope dilution (Thomas and Asakawa, 1993a).

Table 5. Forage legumes for different Latin American ecosystems[a]

Legumes	Ecosystem: Savannas Colombia/Venezuela	Savannas Brazil	Humid tropics	Subhumid tropics
Arachis pintoi	+[b]	-	+	-
Centrosema acutifolium	+	+	+	-
C. brasilianum	+	+	-	+
C. macrocarpum	-	-	+	+
C. pubescens	-	-	+	-
Calopogonium mucunoides	-	+	-	-
Cratylia argentea	+	-	+	+
Desmodium ovalifolium	+	-	+	-
D. velutinum	+	-	+	-
Pueraria phaseoloides	+	-	+	-
Stylosanthes capitata	+	+	-	-
S. guianensis var. *pauciflora*	+	+	-	-
S. guianensis var. *vulgaris*	-	+	+	+

[a] Data from Miles and Lapointe (1992).
[b] + = adapted to the ecosystem; - = not adapted

1992). A large collection of over 18,000 herbaceous and woody legume accessions from over 100 genera and 600 species is maintained at CIAT, Cali, Colombia along with a collection of over 4000 strains of *Bradyrhizobium* (Franco et al., 1993). Other sources of *Rhizobium* and *Bradyrhizobium* are listed in Bushby et al. (1986). The majority of the forage legumes in the CIAT collection have been selected from acid infertile soils of Latin America and are extremely tolerant of low acidity. For example there was little response to liming when legumes were grown in an oxisol with pH 4.5 and 90% aluminum saturation (Fig. 3, adapted from Spain, 1979). Most of the legumes produced maximum growth at 0 or 0.5 t lime ha^{-1}. The promising species have been evaluated in a multi-institutional decentralized network operating throughout the Latin American region (Toledo, 1985, 1986) and a summary of their adaptation to different ecosystems in Latin America is presented in Table 5. Note that there is not a legume that is adapted to each of the four ecosystems. Thus although a wide adaptation to climate, soil and management has been advocated, especially for forage legumes introduced into native Australian pastures (Miller and Stockwell, 1991), legumes for targeted niches would seem a more appropriate objective. For the humid tropical areas of Australia *Centrose-*

ma pubsecens, C. schiedianum and *Pueraria phaseoloides* have been selected for productive and persistent grass/legume mixtures on relatively fertile soils and *Calopogonium mucunoides, Arachis pintoi* and *Centrosema* spp. have been suggested as better alternatives for pastures on poorer soils with a tendency for poorer drainage and grazing mismanagement (Teitzel, 1992).

The introduction of a new legume into an area can be very successful as exemplified by *Stylosanthes guianensis* which is grown on over 13,000 ha in tropical China 8 years after its introduction in 1982 (CIAT, 1991). In 1993 over 5,200 ha were sown in Guangdong province alone (Devendra and Sere, pers. commun.). In Australia two legumes, *Stylosanthes hamata* cv. Verano (Caribbean stylo) and *Stylosanthes scabra* cv. Seca (shrubby stylo), introduced from Venezuela and Brazil respectively, were released in 1973 and 1976. At the end of 1991 it was estimated that these legumes had been sown on 500,000 and 300,000 ha respectively (Cameron et al., 1993). These examples should encourage further selection trials.

Use of fertilizers

The second important factor in sustainable grass/legume pastures is the judicious use of fertilizers. As pointed out by Sanchez and Salinas (1981) the low-input soil management technology associated with grass/legume pastures does not imply the elimination of fertilizer use but rather a more rational and efficient use of limited amounts of fertilizer, especially phosphorus, which farmers can afford. Even with a compatible grass/legume mixture there is a need for maintenance levels of nutrients, particularly phosphorus but also in some instances, potassium, sulphur, calcium, magnesium and micronutrients. In the Australian humid tropics for example, a reapplication of 30 kg ha^{-1} of soluble P every 2 years and an application of trace elements every 4 years is recommended for productive and persistent grass/legume pastures (Teitzel, 1992). However further work is needed on maintenance fertilizers for different grass/legume pastures on different soil types based on long term evaluations.

Alleviation of constraints

The most serious constraints to the widespread use of forage legumes in pastures include problems of lack of persistence, anti-quality factors, variable *Rhizobium* requirements and poor acceptability by farmers.

Legume persistence

Lack of legume persistence is common to temperate and tropical pastures. For example in spite of early work showing that the composition of grass/white clover swards could be manipulated by timing and pressure of grazing (Jones, 1933), it is only relatively recently that guidelines have become available for the management by grazing of ryegrass-white clover pastures. These include longer resting intervals between grazing, integration of cattle and sheep grazing with conservation cuts and maintenance of sward heights around 6 cm (Evans et al., 1992; Grant and Barthram, 1991; Orr et al., 1990; Wilkins, 1982). Emphasis appears to be on the maintenance of white clover growing point numbers and avoidance of burial during wet periods (Laidlaw et al., 1992). Little or no information of this type exists for tropical species which have only been domesticated recently. Clements (1989) demonstrated the increasing susceptibility of some twining tropical legumes to loss of growing points as grazing pressure (animals per unit green DM) increased compared with more prostrate legumes which suggests that a similar strategy, i.e. maintenance of growing points under grazing, is likely to contribute to a better persistence of tropical forage legumes.

Tropical legumes show a variety of responses to grazing ranging from a rapid disappearance, e.g. *Centrosema acutifolium,* to legume dominance in *Desmodium ovalifolium* (e.g. de Santana et al., 1993). The latter contains high levels of tannins thus reducing its palatability to animals and digestibility (for discussion see Humphreys, 1991). Clements (1989) also reported that the low acceptability of the more prostrate species *Cassia rotundifolia* was a more important factor in its tolerance to grazing than the disposition of its growing points.

One of the most persistent and promising legumes to date is the forage *Arachis species, A. pintoi.* In grazing experiments in the eastern plains of Colombia this legume has persisted under heavy grazing pressure for over 6 years in association with the grass *Brachiaria humidicola* and formed good associations with three other *Brachiaria* species (Lascano, 1994). Similarly *A. glabrata* cv. Florigraze has persisted for 8 years in association with *Cynodon dactylon* and *Hemarthria altissima* in Florida (Dunavin, 1992). Possible reasons for the persistence of forage *Arachis spp.* include a prostrate stoloniferous habit (similar to white clover), an ability to flower and set seeds profusely and bury the seeds via fruiting pegs. Furthermore *A. pintoi* is easily propagated via vegetative stolons which may be detached from the mother plant by trampling. It is shade tolerant, rapidly re-establishes its leaf area index after defoliation and can survive relatively long dry periods even though it loses its leaves and appears desiccated (Fisher and Cruz, 1994). An ability to acquire aluminum-bound phosphorus from acid soils (Rao and Kerridge, 1994) may also be a factor in the superior persistence of *A. pintoi*. All the features listed above are consistent with the legume ideotype necessary for a persistent forage legume (Marten, 1989).

Grazing management is the most readily available tool to the land manager whereby a target legume content can be maintained. However a blanket recommendation cannot be made because of differences in grass and legume behaviour under grazing. Studies at CIAT have suggested that the grazing system (i.e. continuous, rotational and deferred grazing) is as important as the grazing intensity (animals/unit green biomass) with respect to maintaining an appropriate grass/legume balance (Lascano, 1991). A system of flexible grazing management has been proposed (Spain et al., 1985) for the evaluation of grass/legume mixtures which depends upon:

1. the adjustment of stocking rate (animals ha^{-1}) to maintain the amount of forage on offer between 3–6 kg DM 100 kg^{-1} liveweight day^{-1} and
2. the alteration of the grazing system to maintain the legume content between 15 and 50%.

With a high legume content an increase in the rest period from grazing is thought to increase the grass component whereas at low legume levels increased grazing via reductions in the rest period is thought to encourage the legume component at the expense of the grass component. While this experimental methodology has had some success with *Desmodium ovalifolium*, a vigorous, prostrate and unpalatable legume (de Santana et al., 1993), it has yet to be translated into management options for contrasting grass/legume pastures that are based on some simple evaluation of the state of the pasture such as the use of sward height in temperate ryegrass pastures (Hodgson et al., 1985; Parsons, 1984). Studies on the ecophysiology of tropical grass/legume pastures are required to determine if a simple indicator of sward state exists which could be used as a guideline for the persistence of a forage legume. Such an indicator will probably need to be different for associations with either the prostrate grasses (e.g., *Brachiaria*) or with the more erect bunch types (e.g., *Andropogon*).

Fisher and Thornton (1989) hypothesized that because grasses in tropical pastures are predominantly C_4 types whilst the legumes are C_3 types it is inevitable, other factors being equal, that grasses will dominate the pasture as a result of their superior rates of photosynthesis and growth. In order to obtain legume persistence the above authors argued that the legume must have some competitive or demographic advantage or that the grass must be preferentially grazed. Decision rules for grazing should therefore take these objectives into account in order to maintain the legume in the pasture even if there is a penalty in terms of animal production.

Anti-quality factors

The presence of anti-quality factors such as tannins in many tropical legumes (Humphreys, 1991; Swain, 1979) can be thought of as both advantageous and disadvantageous. Preferential grazing of the grass in a grass/legume mixture due to the unpalatability of a legume which contains anti-quality factors can result in an increase in the proportion of the legume in the pasture. It could however result in poor animal intake and production and a lower rate of litter decomposition and hence slower nutrient recycling (Thomas and Asakawa, 1993b). The latter could be a disadvantage if there is a need for a rapid release of nutrients for a subsequent crop, for example.

Is there a need to inoculate tropical forage legumes?

The previous classification of tropical forage legumes into three groups, viz., promiscuous effective, promiscuous ineffective and specific with only the latter two requiring inoculation (Date, 1977) has tended to lose its usefulness as more legumes are tested and more exceptions are reported. For example legumes previously classified as promiscuous effective such as *Pueraria phaseoloides, Centrosema macrocarpum,* and *Arachis pintoi* (Sylvester-Bradley, 1984; Sylvester-Bradley et al., 1988, 1991) responded to inoculation when grown in infertile acid soils (oxisols). Lack of an ability to generalize about the inoculation requirements of a particular forage legume is due to many factors including the wide variation in both numbers, competitiveness and effectiveness of the indigenous rhizobial population, environmental factors such as soil acidity, temperature, moisture, and microbial predators all of which can affect the survival and success of the inoculant strain. Models have been developed to predict the native rhizobial population, based on factors such as % legume cover, rainfall and extractable bases in soils (Woomer and Bohlool, 1989) and also to predict the likelihood of success of inoculation based on indigenous rhizobial populations and availability of soil mineral N (Thies et al., 1991). However there has been little or no verification of these models in tropical pastures. As suggested by Date (1977) the simplest approach, in the absence of an ability to predict an inoculation response, is to conduct simple need-to-inoculate tests. Details of these tests are available (Brockwell et al., 1988; Date, 1977; Sylvester-Bradley, 1984; Vincent, 1970) and some have been discussed further by Giller and Wilson (1991). Briefly, these tests compare the growth of uninoculated and inoculated plants with a third treatment receiving doses of N fertilizer of 30 kg ha^{-1} every 2 weeks. The uninoculated control will reveal the presence of native rhizobia and their effectiveness when compared with the other treatments. The inoculated treatment will examine the effectiveness of the inoculant strain(s) and the N fertilized treatment will indicate the growth potential in the absence of N limitation.

Acceptability of forage legumes by farmers

Farmers are generally reluctant to invest the time and resources into the establishment of legumes in pastures. There is abundant research evidence available which demonstrates the benefits of legumes to beef production e.g. in the eastern plains of Colombia on-farm trials showed an increase of between 32 to 61% in beef production from grass/legume pastures compared with grass only pastures (Ferguson, 1992). Constraints to the adoption of forage legumes include the inherent slowness of the adoption of a novel component by pastoralists, an unawareness of the relevance and benefits to beef and milk production (let alone soil improvement), low availability of commercial legume seed, limited availability of technical assistance, lack of capital and credit requirements, lack of experience in identifying "niches" for particular legumes (Ferguson, 1992). All these issues need addressing as well as a need for suitable policy incentives to improve the adoption of forage legumes by farmers (for discussion of the latter point see Bohlool et al., 1992).

One can conclude from the information presented above that the long-term success of a grass/legume pasture will be a function of the persistence of the legume and that this is a complex interaction between plant types, relative growth rates of the grass and legume, shade tolerance of the legume, relative palatability of the grass and legume or grazing preference of the animals, capacity for recruitment via legume seed or vegetative material, ability to withstand trampling and burial. In addition the socio-economic factors need to be more favourable to encourage the widespread adoption of forage legumes by farmers.

Role of forage legumes in recuperation of degraded land

Pastures can stabilize soils mainly by the complete ground cover conferred by the grass species. Legumes can also fulfill this role if they can cover the ground rapidly. Figure 4 shows that the % area covered by a number of forage legumes can be sustantial after 12 weeks growth on a hillside site in the coffee growing region of Colombia. The forage *Arachis* species, *A. pintoi*, was the most successful legume in this experiment achieving 80% ground cover.

Forage legumes have also been shown to improve the physical, chemical and biological properties of soils by increasing factors such as organic matter, cation exchange capacity, aggregate stability, infiltration rates, moisture retention, mineralisable N and P fractions and numbers of earthworms (Dalal et al., 1991; Lal et al., 1979; Mytton et al., 1993; Thomas et al., 1994b; Wilson et al., 1982). These are essentially the reversal of the degradative processes occurring in many tropical soils. The effect of forage legumes on the soil fauna may be particularly valuable for small-scale farmers as a relatively simple and cheap means to initiate remedial treatment of degraded soils and

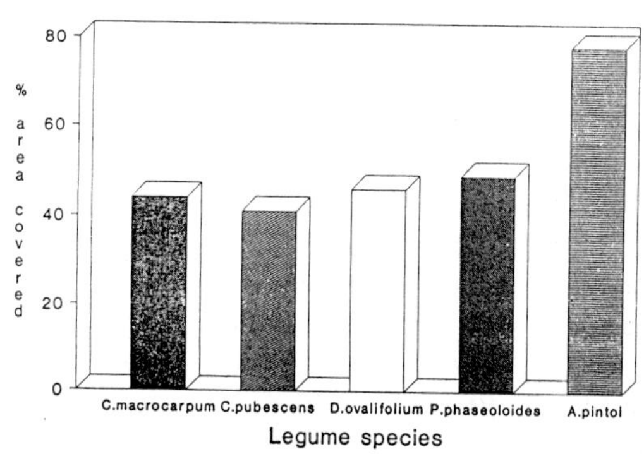

Fig. 4. Percentage soil cover with forage legumes after 12 weeks growth.

this topic merits further research (e.g., Lavelle et al., 1989).

The potential use of forage legumes in agropastoral systems

The use of agropastoral systems similar to the classic ley farming systems of the pre-fertilizer era has been discussed as an option for land management in the tropics (Saleem and Fisher, 1993). The introduction of a relatively short-term pasture phase of 3–5 years with a forage legume component is attractive from the viewpoint of the constraints noted above to the widespread use of legumes. Such a system is currently being tested in the savanna lands of Brazil, Colombia and Venezuela where a grass/legume mixture is sown simultaneously with an acid-soil tolerant upland rice variety (Sanz et al., 1994; Vera et al., 1992; Zeigler et al., 1994). The rice crop is harvested after 105–120 days and the pasture establishes at a much faster rate than the traditional methods as a result of the residual fertilizer not removed by the rice crop. Grazing of the pasture is possible after 3–5 months compared with one year using the traditional low-input pasture technology. Other advantages include a more efficient land preparation (less machinery operations), reduced soil erosion and leaching by establishing a ground cover more rapidly and completely compared with either rice or pasture alone (Thomas et al., 1994a). The pasture phase can be of short duration of 3–5 years and can be followed by another crop which can benefit from the input of N via BNF. In this system the persistence of the forage legume assumes less importance as it can be re-introduced or replaced by another forage legume or legume mixture with each crop phase. The use of rice as a pioneer crop for a grass/legume pasture is environmentally and economically attractive. An analysis of the cash flow for example indicates a net return after 3 years with rice-pasture compared with a least 5 years with pasture alone (Vera et al., 1992) and, in addition, the costs of establishing the rice-pasture association are recovered in the first year with income generated from selling the rice crop (Rivas et al., 1991).

Ley farming systems have also been estimated to be a profitable option for subtropical Australia on soils with low fertility although the lack of suitable tropical forage legumes is a current limitation (Lloyd et al., 1991). In semi-arid regions of Australia, germplasm is available but the economics of cropping and the increasing complexity of the management skills required in these areas with greater risks of crop failure due to the vagaries of climate, appear to be constraining the widespread adoption of legume-based leys (Jones et al., 1991).

The use of forage legumes in agropastoral systems holds great promise for the humid tropics and is considered to be one of the sustainable land use options currently available which can bring benefits such as improved control of pests and diseases through rotations, more efficient nutrient cycling, less loss of soil and increased productivity (National Research Council, 1993). Agropastoral systems also offer a means to overcome some of the constraints noted above for the widespread adoption of forage legumes. Further research is needed on increasing the number of crop options, on the competition between crops, grass/legume pasture and weeds in different environments, on improvements in fertilizer use efficiency and integration with the use of biologically fixed N.

Conclusions

Forage legumes can provide sufficient amounts of biologically-fixed N to increase herbage and animal production and maintain the N balance of the soil provided the legume content of the pasture is maintained at a minimum value of around 20%. As utilization of the pasture increases the requirement for legume N also increases. Careful grazing management using different grazing systems as well as grazing pressure can ensure the persistence of an adequate legume content but further research is required to define the management options for different tropical grass/legume associations.

The % Ndfa is usually greater than 80% in tropical pastures but can decline below this value if other mineral nutrient deficiencies occur. Further research is necessary to define the critical nutrient concentrations for maintaining % Ndfa above 80% especially in long-term pastures which may not receive maintenance levels of fertilization.

The forage species *Arachis pintoi* appears to be closest to the plant ideotype required for a persistent pasture legume in the tropics and an examination of the characters that confer persistence in this species is warranted.

At present the variable requirements for inoculation of forage legumes cannot be predicted with accuracy and the use of simple need-to-inoculate tests is encouraged.

Perhaps the greatest challenge for researchers is to address the issue of the poor acceptability of forage legumes by farmers and use this experience to redirect the large amount of effort currently undertaken on improving the knowledge base of the processes of biological nitrogen fixation. The latter has yet to result in any practical improvement of BNF in farmer's fields.

The integration of grass/legume pastures with cropping appears to be a promising option for increasing agricultural production, recuperating degraded soils and facilitating the wider use of forage legumes in environments where fertility is inherently low and where the use of N fertilizer is restricted by availability or cost.

Acknowledgements

I thank my colleagues from the Forage and Tropical Lowland Programs of CIAT and especially Drs M J Fisher and C E Lascano for constructive comments and suggestions on the manuscript and for the use of unpublished data.

References

Ball P R and Ryden J C 1984 Nitrogen relationships in intensively managed temperate grassland. Plant and Soil 76, 23–33.

Blyth M J and Menz K M 1987 The role of economics in pasture research evaluation. *In* Temperate Pastures, their Production, Use and Management. Eds. J L Wheeler, C J Pearson and G E Robards. pp 586–589. Australian Wool Corp/CSIRO, Australia.

Bohlool B B, Ladha J K, Garrity D P and George T 1992 Biological nitrogen fixation for sustainable agriculture: A perspective. Plant and Soil 141, 1–11.

Breman H and de Wit C T 1983 Rangeland productivity and exploitation in the Sahel. Science 221, 1341–1347.

Brockwell J, Holliday R A and Pilka A 1988 Evaluation of the symbiotic nitrogen-fixing potential of soils by direct microbiological means. Plant and Soil 108, 163–170.

Bushby H V A, Date R A and Sundram J 1986 *Rhizobium* collections and their assessment. *In* Forages in Southeast Asian and South Pacific agriculture. Eds. G J Blair, D A Ivory and T R Evans. pp 133–140. ACIAR, Canberra, Australia.

Bushby H V A, Vallis I and Myers R J K 1992 Dynamics of C in a pasture grass (*Panicum maximum* var. *trichoglume*)-soil system. Soil Biol. Biochem. 24, 381–387.

Cadisch G, Sylvester-Bradley R and Nosberger J 1989 ^{15}N-based estimation of nitrogen fixation by eight tropical forage-legumes at two levels of P:K supply. Field Crops Res. 22, 181–194.

Cadisch G, Sylvester-Bradley R and Nosberger J 1992 Variation in growth and N_2 fixation within and among *Centrosema* spp. in response to phosphorus supply. Trop. Grassl. 26, 226–234.

Cadisch G, Sylvester-Bradley R, Boller B C and Nosberger J 1993 Effects of phosphorus and potassium on N_2 fixation (^{15}N-dilution) of field-grown *Centrosema acutifolium* and *C. macrocarpum*. Field Crops Res. 31, 329–340.

Cameron D F, Miller C P, Edye L A and Miles J W 1993 Advances in research and development with Stylosanthes and other tropical pasture legumes. *In* Proc. XVII Intern. Grassl. Cong. New Zealand and Australia, Feb 8–21, 1993. pp 2109–2114. New Zealand Grassland Assoc., Palmerston North, New Zealand.

Chapman A L and Myers R J K 1987 Nitrogen contributed by grain legumes to rice grown in rotation on the Cununurra soils of the Ord irrigation area, Western Australia. Aust. J. Exp. Agric. 27, 155–163.

CIAT 1991 184-Zhuhuacao: A South American forage succeeds in China. *In* CIAT Report 1991. pp 44–48. CIAT, Cali, Colombia.

Clements R J 1989 Rates of destruction of growing points of pasture legumes by grazing cattle. pp 1027–1028. Proc. XVI International Grassland Congress, Nice, France. The French Grassland Soc., Paris, France.

Dalal R C, Strong W M, Weston E J and Gaffney J 1991 Sustaining multiple production systems 2. Soil fertility decline and restoration of cropping lands in sub-tropical Queensland. Trop. Grassl. 25, 173–180.

Date R A 1977 Inoculation of tropical pasture legumes. *In* Exploiting the legume-*Rhizobium* symbiosis in tropical agriculture. Eds. J M Vincent, A S Whitney and J Bose. pp 293–311. College of Tropical Agriculture Miscellaneous Publ. 145, Univ. of Hawaii, USA.

Dunavin L S 1992 Florigraze rhizoma peanut in association with warm-season perennial grasses. Agron. J. 32, 148–151.

Edmeades D C and Goh K M 1978 Symbiotic nitrogen fixation in a sequence of pastures of increasing age measured by a ^{15}N technique. N.Z.J. Agric. Res. 21, 623–628.

Eltilib A M and Ledgard S F 1988 Production and nitrogen fixation by "Grasslands Kopu" and "Grasslands Huia" white clovers under different nitrogen regimes. N.Z.J. Agric. Res. 31, 325–330.

Evans D R, Williams T A and Evans S A 1992 Evaluation of white clover varieties under grazing and their role in farm systems. Grass For. Sci 47, 342–352.

FAO 1993 FAO yearbook – production Vol. 46. FAO Statistics Series No. 112. FAO, Rome, Italy.

Ferguson J E 1992 Experiences at the interface of research and development with tropical pastures. *In* Pastures for the Tropical Lowlands. pp 101–120. CIAT, Cali, Colombia.

Fisher M J and Thornton P K 1989 Growth and competition as factors in the persistence of legumes in pastures. *In* Persistence of forage legumes. Eds. G C Marten, A G Matches, R F Barnes, R W Brougham, R J Clements and G W Sheath. pp 293–309. ASA/CSSA/SSSA, Madison, WI.

Fisher M J and Cruz P 1994 Some ecophysiological building blocks in the domestication of an outstanding tropical pasture legume ideotype. *In* Biology and Agronomy of Forage *Arachis*. Eds. P C Kerridge and W Hardy. CIAT, Cali, Colombia *(In press)*.

Fisher M J, Lascano C E, Thomas R J, Ayarza M A, Rao I M, Rippstein G and Thornley J H M 1994 An integrated approach to soil-plant-animal interactions on grazed legume-based pastures on tropical acid soils. *In* Proc. XVII Int. Grassl. Cong. New Zealand and Australia, Feb 8–21, 1993 *(In press)*.

Franco M A, Ocampo G I, Melo E and Thomas R J 1993 Catalogue of *Rhizobium* strains for Tropical Forage Legumes. CIAT Working Document No. 14, Cali, Colombia. 123p.

Giller K E and Wilson K J 1991 Nitrogen Fixation in Tropical Cropping Systems. CAB International, Wallingford, Oxon, UK. 313p.

Gramshaw D, Read J W, Collins W J and Carter E D 1989 Sown pastures and legume persistence: an Australian overview. *In* Persistence of Forage Legumes. Eds. G C Marten, A G Matches, R F Barnes, R W Brougham, R J Clements and G W Sheath. pp 1–21. ASA/CSSA/SSSA, Madison, WI.

Grant S A and Barthram G T 1991 The effects of contrasting cutting regimes on the components of clover and grass growth in microswards. Grass For. Sci. 46, 1–13.

Harris W and Thomas V J 1973 Competition among pasture plants. III. Effects of frequency and height of cutting on competition between white clover and two ryegrass cultivars. N.Z.J. Agric. Res. 16, 49–58.

Henzell E F 1983 Contributions of forages to worldwide food production: now and in the future. *In* Proceedings of the XIVth International Grassland Congress. pp 42–47. Westview Press, Boulder, Colorado.

Hodgson J, Mackie C K and Parker J W G 1985 Sward surface heights for efficient grazing. Grass Farmer 24, 5–10. British Grassland Society, Hurley.

Humphreys L R 1991 Tropical Pasture Utilization. Cambridge Univ. Press., Cambridge. 206p.

Jones M G 1933 Grassland management and its influence on the sward. J. R. Agric. Soc. Eng. 94, 21–42.

Jones R K, Dalgliesh N P, Dimes J P and McCown R L 1991 Sustaining multiple production systems. 4. Ley pastures in crop-livestock systems in the semi-arid tropics. Trop. Grassl. 25, 189–196.

Laidlaw A S, Teuber N G and Withers J A 1992 Out-of-season management of grass/clover swards to manipulate clover content. Grass For. Sci. 47, 220–229.

Lal R, Wilson G F and Okigbo B N 1979 Changes in properties of an alfisol produced by various crop covers. Soil Sci. 127, 377–382.

Lascano C E 1991 Harry Stobbs memorial Lecture: Managing the grazing resource for animal production in savannas of tropical America. Trop. Grassl. 25, 66–72.

Lascano C E 1994 Nutritive value and animal production. *In* Biology and Agronomy of Forage *Arachis*. Eds. P C Kerridge and W Hardy. CIAT, Cali, Colombia *(In press)*.

Lavelle P, Barois I, Martin A, Zaidi Z and Schaefer R 1989 Management of earthworm populations in agro-ecosystems: A possible way to maintain soil quality? *In* Ecology of Arable Land. Eds. M Clarholm and L Bergstrom. pp 109–122. Kluwer Academic Publ., Dordrecht.

Ledgard S F and Steele K W 1992 Biological nitrogen fixation in mixed legume/grass pastures. Plant and Soil 141, 137–153.

Lloyd D L, Smith K P, Clarkson N M, Weston E J and Johnson B 1991 Sustaining multiple production systems 3. Ley pastures in the subtropics. Trop. Grassl. 25, 181–188.

t'Mannetje L 1986 Forages in extensive grazing systems. In Forages in Southeast Asian and South Pacific agriculture. Eds. G J Blair, D A Ivory and T R Evans. pp 161–169. ACIAR, Canberra, Australia.

Marten G C 1989 Summary of the trilateral workshop on persistence of forage legumes. *In* Persistence of Forage legumes. Eds. G C Marten, A G Matches, R F Barnes, R W Brougham, R J Clements and G W Sheath. pp 569–572. ASA/CSSA/SSSA, Madison, WI.

Martin T W 1960 The role of white clover in grassland. Herb. Abstr. 30, 159–164.

Miles J W and Lapointe S L 1992 Regional germplasm evaluation: A portfolio of germplasm options for the major ecosystems of tropical America. *In* Pastures for the Tropical Lowlands. pp 9–28. CIAT, Cali, Colombia.

Myers R J K and Robbins G B 1991 Sustaining productive pastures in the tropics 5. Maintaining productive sown grass pastures. Trop. Grassl. 25, 104–110.

Mytton L R, Cresswell A and Colburn P 1993 Improvement in soil structure associated with white clover. Grass For. Sci. 48, 84–90.

Miller C P and Stockwell T G H 1991 Sustaining productive pastures in the tropics 4. Augmenting native pastures with legumes. Trop. Grassl. 25, 98–103.

National Research Council 1993 Sustainable Agriculture and the Environment in the Humid Tropics. National Academy Press, Washington DC. 702p.

Orr R J, Parsons A J, Penning P D and Treacher T T 1990 Sward composition, animal performance and the potential production of grass/white clover swards continuously stocked with sheep. Grass For. Sci. 45, 325–336.

Parsons A J 1984 Guidelines for management of continuously grazed swards. Grass Farmer 17, 5–9. British Grassland Society, Hurley.

Pearson C J and Ison R L 1987. Agronomy of Grassland Systems. Cambridge Univ. Press, Cambridge. 169p.

Peoples M B and Herridge D F 1990 Nitrogen fixation by legumes in tropical and subtropical agriculture. Adv. Agron. 44, 155–223.

Peoples M B, Herridge D F and Ladha J K 1995 Biological nitrogen fixation: An efficient source of nitrogen for sustainable agricultural production? Plant and Soil 174.

Rao I M and Kerridge P C 1994 Mineral nutrition of forage *Arachis*. *In* Biology and Agronomy of Forage *Arachis*. Eds. P C Kerridge and W Hardy. CIAT, Cali, Colombia *(In press)*.

Rivas L, Toledo J M and Sanint L R 1991 Potential impact of the use of pastures associated with crops in the tropical savannas of Latin America. *In* Trends in CIAT Commodities. pp 78–98. CIAT, Cali, Colombia.

Robbins G B, Bushel J J and Butler K L 1987 Decline in plant and animal productivity from ageing pastures of green panic (*Panicum maximum* var. *trichoglume*) J. Agric. Sci. 108, 407–417.

Saleem M M A and Fisher M J 1993 Role of ley farming in crop rotations in the tropics. *In* Proc. XVII Intern. Grassl. Cong. New Zealand and Australia, Feb 8–21, 1993. pp 2179–2187. New Zealand Grassland Assoc., Palmerston North, New Zealand

Sanchez P A and Logan T J 1992 Myths and science about the chemistry and fertility of soils in the Tropics. *In* Myths and Science of Soils in the Tropics. Eds. R Lal and P A Sanchez. pp 35–46. SSSA, Madison, WI.

Sanchez P A and Salinas J G 1981 Low-input technology for managing oxisols and ultisols in tropical America. Adv. Agron. 34, 279–406.

Sanford P, Pate J S and Unkovich M J 1993 A survey of proportional dependence of subterranean clover and other pasture legumes on N_2 fixation in south-west Australia utilizing ^{15}N natural abundance. Aust. J. Agric. Res. 45, 165–181.

Santana de J R, Pereira J M, Moreno M A and Spain J M 1993 Persistencia e qualidade proteica da consorciacao *Brachiaria humidicola-Desmodium ovalifolium* cv. Itabela sob diferentes sistemas e intensidades de pastojo. Past. Trop. 15, 2–8.

Sanz J I, Zeigler R S, Sarkarung S, Molina D L and Rivera M 1994 Rice-pasture systems for acid soil savannas in South America. I. Pasture establishment with rice. and II. Recuperation of degrading pastures with rice and rice-pasture leys. Ann. Appl. Biol. *(In press)*.

Seiffert N M, Zimmer A H, Schunke R M and Miranda C H B 1985 Reciclagem de nitrogenio em pastagem consociada de *Calopogonium mucunoides* com *Brachiaria decumbens*. Pesqui. Agropecu. Bras. 20, 529–544.

Sheehy J E 1989 How much dinitrogen fixation is required in grazed grassland? Ann. Bot. 64, 159–161.

Shivaram S, Rai P V and Hegde S V 1988 Nitrogen fixation and nitrogen balance studies in Rhizobium-VA mycorrhiza inoculated legume-grass association by ^{15}N isotope dilution technique under a field condition. Plant and Soil 111, 11–16.

Simpson J R 1987 Nitrogen nutrition of pastures. *In* Temperate Pastures, their Production, Use and Management. Eds. J L Wheeler, C J Pearson and G E Robards. pp 143–154. Australian Wool Corp/CSIRO, Australia.

Simpson J R and Stobbs T H 1981 Nitrogen supply and animal production from pastures. *In* Grazing Animals. Ed. F H W Morley. pp 261–287 World Animal Science B1, Elsevier Sci. Publs., Amsterdam.

Spain J M 1979 Pasture establishment and management in the Llanos Orientales of Colombia. *In* Pasture Production in Acid Soils of the Tropics. Eds. P A Sanchez and L E Tergas. pp 167–175. CIAT, Cali, Colombia.

Spain J M, Pereira J M and Gualdron R 1985 A flexible grazing management system proposed for the advanced evaluation of associations of tropical grasses and legumes. pp. 1153–1155. Proc XV Intern. Grassl. Cong., Kyoto, Japan.

Sprent J I and Sprent P 1990 Nitrogen Fixing Organisms. Chapman and Hall, London. 256p.

Steele K W and Vallis I 1988 The nitrogen cycle in pastures. *In* Advances in Nitrogen Cycling in Agricultural Ecosystems. Ed. J R Wilson. pp 274–291. CAB International, Wallingford, Oxon, UK.

Stewart T A 1984 Utilising white clover in grass based animal production systems. *In* Forage Legumes. Ed. D J Thomson. pp 93–103. Occasional Symposium No. 16, British Grassland Society, UK.

Swain T 1979 Tannins and lignins. *In* Herbivores – Their Interaction with Plant Secondary Metabolites. Eds. G A Rosenthal and D H Janzen. pp 657–822. Academic Press, New York.

Sylvester-Bradley R 1984 Rhizobium inoculation trials designed to support a tropical forage legume selection programme. Plant and Soil 82, 377–386.

Sylvester-Bradley R and Mosquera D 1985 Nitrification and responses to *Rhizobium* inoculation in tropical savanna as affected by land preparation. *In* Nitrogen Management in Farming Systems in Humid and Subhumid Tropics. Eds. B T Kang and J van der Heide. pp 167–183. Institute for Soil fertility, Haren, Netherlands/ IITA, Ibadan, Nigeria.

Sylvester-Bradley R, Mosquera D and Mendez J E 1988 Selection of Rhizobia for inoculation of forage legumes in savanna and rainforest soils of tropical America. *In* Nitrogen Fixation by Legumes in Mediterranean Agriculture. Eds. D P Beck and L A Materon. pp 225–234. Martinus Nijhoff Publ., Dordrecht.

Sylvester-Bradley R, Mosquera D, Asakawa N M and Sanchez G 1991 Promiscuity and responses to rhizobial inoculation of tropical kudzu (*Pueraria phaseoloides*). Field Crops Res. 27, 267–279.

Teitzel J K 1992 Sustainable pasture systems in the humid tropics of Queensland. Trop. Grassl. 26, 196–205.

Thies J E, Singleton P W and Bohlool B B 1991 Modeling symbiotic performance of introduced rhizobia in the field by use of indices of indigenous population size and nitrogen status of the soil. Appl. Environ. Microbiol. 57, 29–37.

Thomas R J 1992 The role of the legume in the nitrogen cycle of productive and sustainable pastures. Grass For. Sci. 47, 133–142.

Thomas R J and Asakawa N M 1993a Nitrogen fixation by three tropical forage legumes in an acid-soil savanna of Latin America. *In* New Horizons in Nitrogen Fixation. Eds. R Palacios, J Mora and W E Newton. p 750. Kluwer Academic Publ., Dordrecht.

Thomas R J and Asakawa N M 1993b Decomposition of leaf litter from tropical forage grasses and legumes. Soil Biol. Biochem. 25, 1351–1361.

Thomas R J, Lascano C E, Sanz J I, Ara M A, Spain J M, Vera R R and Fisher M J 1992 The role of pastures in production systems. *In* Pastures for the Tropical Lowlands. pp 121–144. CIAT, Cali, Colombia.

Thomas R J and Lascano C E 1994 The benefits of forage legumes for livestock production and nutrient cycling in pasture and agropastoral systems of acid-soil savannas of Latin America. *In* Livestock and Sustainable Nutrient Cycling in mixed Farming Systems of Sub-Saharan Africa. Eds. J M Powell, S. Fernandez-Rivera, T O Williams and C Renard. ILCA, Ethiopia, Nov 22–26, 1993 *(In press).*

Thomas R J, Fisher M J, Ayarza M A and Sanz J I 1994a The role of forage grasses and legumes in maintaining the productivity of acid soils in Latin America. Adv. Soil Sci. *(In press).*

Thomas R J, Rao I M, Ayarza M A, Sanz J I, Decaens T, Rippstein, Gijsman A J and Lavelle P 1994b Forage legumes – the means to reverse soil degradation with low inputs? *In* Proc 15th Intern. Cong. Soil Sci. Acapulco, Mexico, July 10–16, 1994 *(In press).*

Toledo J M 1985 Pasture development for cattle production in the major ecosystems of the Tropical American Lowlands. *In* Proc. XV Intern. Grass. Cong. pp 74–81. Science Council Japan/Jap Soc. Grass Sci., Kyoto, Japan.

Toledo J M 1986 Forage research networking in tropical humid and subhumid environments. *In* Forages in Southeast Asian and South Pacific Agriculture. Eds. G J Blair, D A Ivory and T R Evans. pp 69–75. ACIAR, Canberra, Australia.

Toledo J M and Nores G A 1986 Tropical pasture technology for marginal lands of Tropical America. Outlook. Agric. 15, 2–9.

UNESCO, 1979 Tropical Grazing Land Ecosystems. A state-of-knowledge report prepared by Unesco/UNEP/FAO. UNESCO, Paris. 655p.

Vallis I, Henzell E F and Evans T R 1977 Uptake of soil nitrogen by legumes in mixed swards. Aust. J. Agric. Res. 28, 413–425.

Vallis I and Gardener C J 1985 Effect of pasture age on the efficiency of nitrogen fixation by 10 accessions of *Stylosanthes* spp. Aust. J. Exp. Agric. 25, 70–75.

Vera R R, Thomas R J, Sanint L R and Sanz J I 1992 Development of sustainable ley-farming systems for the acid-soil savannas of tropical America. *In* Symposium on the ecology and sustainable agriculture in tropical biomes. An. Acad. Bras. Cienc. 64, (Supl. 1) 106–125.

Vincent J M 1970 A Manual for the Study of Root Nodule Bacteria. IBP Handbook, 15. Blackwell Scientific Publs., London. 164p.

Von Uexkull H R and Mutert E 1993 Rehabilitation of anthropic savanna. 5th regional workshop on phosphorus cycles in terrestrial and aquatic ecosystems. Budapest, Hungary. *(In press).*

Walker T W, Adams A F R and Orchiston H D 1956 Fate of labeled nitrate and ammonium nitrogen when applied to grass and clover grown separately and together. Soil Sci. 81, 339–351.

Watson S E and Whiteman P C 1981 Grazing studies on the Guadacanal Plains, Solomon Islands. 2. Effects of pasture mixtures and stocking rate on animal production and pasture components. J Agric. Sci. 97, 353–364.

Wilkins R J 1982 The effects of integration of conservation and grazing on the composition and productivity of grass/clover swards. *In* The Efficiency of N-utilization by Grass and Legume and on Grass/Clover Swards. pp 70–72. Report of EEC workshop, FAL Braunschweig, Germany.

Wilson G F, Lal R and Okigbo B N 1982 Effects of cover crops on soil structure and on yield of subsequent arable crops grown under strip tillage on an eroded alfisol. Soil Till. Res. 2, 233–250.

Woomer P and Bohlool B B 1989 Rhizobial ecology in tropical pasture systems. *In* Persistence of Forage Legumes. Eds. G C Marten, A G Matches, R F Barnes, R W Brougham, R J Clements and G W Sheath. pp 233–245. ASA/CSSA/SSSA, Madison, WI.

Worldwatch Institute 1992 Vital Signs; the Trends that are Shaping our Future. Worldwatch Institute, Norton and Co., New York. 131p.

Zaharah A R, Sharifuddin H A H, Razley M N and Mohd Saidi A K 1986 Measurement of nitrogen fixed by *Pueraria phaseoloides* by N–15 dilution technique. Pertanika 9, 45–49.

Zeigler R S, Sanz J I and Toledo J M 1994 Developing sustainable agricultural production systems for the acid savannas of Latin America. *In* International Symposium on Agroecology and Conservation in Temperate and Tropical Regions, 26–29 September 1991. Padova, Italy *(In press)*.

Plant and Soil **174:** 119–141, 1995.

Management of biological N_2 fixation in alley cropping systems: Estimation and contribution to N balance

N. Sanginga[1], B. Vanlauwe[1] and S.K.A. Danso[2]
[1]*International Institute of Tropical Agriculture (IITA), PMB 5320, Ibadan, Nigeria and* [2]*International Atomic Energy Agency (IAEA), 5 Wagramerstrasse A-1400, P.O. Box 100, Vienna, Austria*

Key words: decomposition, isotope dilution, ^{15}N, mineralization, nodulation, N use efficiency, residue management

Table of contents

Abstract

Introduction

Nodulation and N_2 fixation potential of hedgerow trees

Microbial component and occurrence of nodulation

Tree host provenance or genotypic variation in N_2 fixation

Quantification of N_2 fixation by hedgerow trees

Methods of N_2 fixation measurements

Factors affecting the validity of N_2 fixation by ^{15}N labelling techniques

Potential and actual N_2 fixation by NFTs

Factors limiting N_2 fixation by NFTs

Environmental factors

Management practices

Strategies to enhance N_2 fixation in alley cropping

Nitrogen contribution and use efficiency in alley cropping systems

Inputs of N in alley cropping

Decomposition and N-mineralization of organic residues in alley cropping systems

Above ground organic residues

Below ground organic residues

Residue N use efficiency in alley cropping systems

Definitions

Measurement of % N use efficiency

Nitrogen not taken up by the crop in alley cropping

Management practices for an increased N use efficiency in alley cropping systems

Acknowledgements

References

Abstract

Alley cropping is being widely tested in the tropics for its potential to sustain adequate food production with low agricultural inputs, while conserving the resource base. Fast growth and N yield of most trees used as hedgerows in alley cropping is due greatly to their ability to fix N_2 symbiotically with *Rhizobium*. Measurements of biological N_2 fixation (BNF) in alley cropping systems show that some tree species such as *Leucaena leucocephala*, *Gliricidia sepium* and *Acacia mangium* can derive between 100 and 300 kg N ha^{-1} yr^{-1} from atmospheric N_2, while species such as *Faidherbia albida* and *Acacia senegal* might fix less than 20 kg N ha^{-1} yr^{-1}. Other tree species such as *Senna siamea* and *S. spectabilis* are also used in alley cropping, although they do not nodulate and therefore do not fix N_2. The long-term evaluation of the potential or actual amounts of N_2 fixed in trees however, poses problems that are associated with their perennial nature and massive size, the great difficulty in obtaining representative samples and applying reliable methodologies for measuring N_2 fixed. Strategies for obtaining representative samples (as against the whole tree or destructive plant sampling), the application of ^{15}N procedures and the selection criteria for appropriate reference plants have been discussed.

Little is known about the effect of environmental factors and management practices such as tree cutting or pruning and residue management on BNF and eventually their N contribution in alley cropping. Data using the ^{15}N labelling techniques have indicated that up to 50% or more of the tree's N may be below ground after pruning. In this case, quantification of N_2 fixed that disregards roots, nodules and crowns would result in serious errors and the amount of N_2 fixed may be largely underestimated. Large quantities of N are harvested with hedgerow prunings (> 300 kg N ha^{-1} yr^{-1}) but N contribution to crops is commonly in the range of 40–70 kg N ha^{-1} season. This represents about 30% of N applied as prunings; however, N recoveries as low as 5–10% have been reported. The low N recovery in maize (*Zea mays*) is partly caused by lack of synchronization between the hedgerow trees N release and the associated food crop N demand. The N not taken up by the associated crop can be immobilized in soil organic matter or assimilated by the hedgerow trees and thus remain in the system. This N can also be lost from the system through denitrification, volatilization or is leached beyond the rooting zone. Below ground contribution (from root turnover and nodule decay) to an associated food crop in alley cropping is estimated at about 25–102 kg N ha^{-1} $season^{-1}$. Timing and severity of pruning may allow for some management of underground transfer of fixed N_2 to associated crops. However many aspects of root dynamics in alley cropping systems are poorly understood. Current research projects based on ^{15}N labelling techniques or ^{15}N natural abundance measurements are outlined. These would lead to estimates of N_2 fixation and N saving resulting from the management of N_2 fixation in alley cropping systems.

Introduction

The majority of tropical African countries are facing two major crises: (i) increasing overexploitation of wood reserves for timber or fuelwood and (ii) diminishing soil fertility of cultivated land resulting in reduced agricultural productivity. These crises are interrelated, originating from the high population growth rate (more than 2.6% per year) and a subsequent need for more food production. These have led to greatly shortened rotations, a rapid loss in productivity and the need to bring more forested land (often of marginal quality) under cultivation.

The restoration of soil fertility through fertilizer application, as practiced in most developed countries, is often too costly to be adopted by most farmers in tropical Africa, and besides, 65% of soils are fragile and therefore deteriorate rapidly under intensive cultivation. Previously, shifting cultivation with a long fallow phase maintained soil fertility, but this system is no longer often practised and in some places the fallow periods have been significantly reduced or abandoned altogether due to scarcity of new land.

For example, the reduction of fallows from 5.4 years to 1.4 years has resulted in a yield decline of cassava (*Manihot esculanta*) in some areas of humid or sub-humid areas from 10.8 t ha^{-1} to less than 2 t ha^{-1} (FAO, unpubl.). With maize and sorghum (*Sorghum vulgare*) the most commonly-grown cereals under rainfed agriculture, the soil must supply about 60 kg N (usually nitrate) and 30 kg P ha^{-1} in plant available form for each tonne of grain produce. (Weber, pers. commun.)

The problem facing farmers everywhere is often the small capacity of their soils to supply the high quantities of N required by crops and a rapid N soil decline once cropped.

The integration of trees, especially N_2-fixing trees (NFTs), into stable agroforestry systems and semi-arid silvo-pastoral systems can make a major contribution to the restoration and maintenance of soil fertility and the provision of fuelwood in addition to combating erosion and desertification. The ability to fix atmospheric N_2 allows these trees to grow well in N-impoverished soils. The nodulated roots and above-ground residues represent valuable sources for the replenishment of levels of soil organic N. Estimates of the annual value of legume N_2 fixation to the global economy have been put at US $100 billion (Burris, 1978).

The potential of such trees is already apparent in established alley cropping systems in the humid and subhumid tropics. In the "alley cropping system" rows of trees alternate with several rows of crops. The trees fix N_2 and are periodically lopped, the foliage is used as a green manure for the crops (thus giving them nutrient and building up organic matter which improves soil physical properties) and the stems are utilized for fuel.

However, much of the development of alley cropping has generally focussed on agronomy and soil fertility aspects. Relatively little attention has been paid to how the systems work and how we could improve them. For example, although N_2 fixation by trees plays a key role, almost nothing is known on what affects it, what trees actually fix N_2, and how much and how we could manage N_2 fixation more effectively.

There is little doubt that a research programme with an objective of examining N_2 fixation and restoration of soil fertility by trees would make a major contribution to maximizing the effectiveness of existing alley cropping systems, both in soil fertility and soil conservation.

The objectives of the present paper are: (i) to review the current status of BNF by trees and assess its contribution to the N balance and uptake by associated food crops in alley cropping; (ii) to show the potential and limitation of NFTs in alley cropping and identify the most significant gaps in our present knowledge and understanding.

Nodulation and N_2 fixation potential of hedgerow trees

Microbial component and occurrence of nodulation

It is not yet possible to enumerate all the NFTs belonging to the Leguminosae, since relatively few of these species have been tested for their nodulating ability. It has been known for many years that Caesalpinioid legumes nodulate less frequently than their Mimosoid and Papilionoid counterparts (Allen and Allen, 1981). This still holds, although a number of new reports of nodulated species have been made recently (Halliday, 1984). Only the nodulating NFTs currently used in alley cropping are dealt with in this section, and no attention is given to species such as *Senna siamea* which do not nodulate, and consequently do not fix N_2 but are nevertheless sometimes integrated in alley cropping systems.

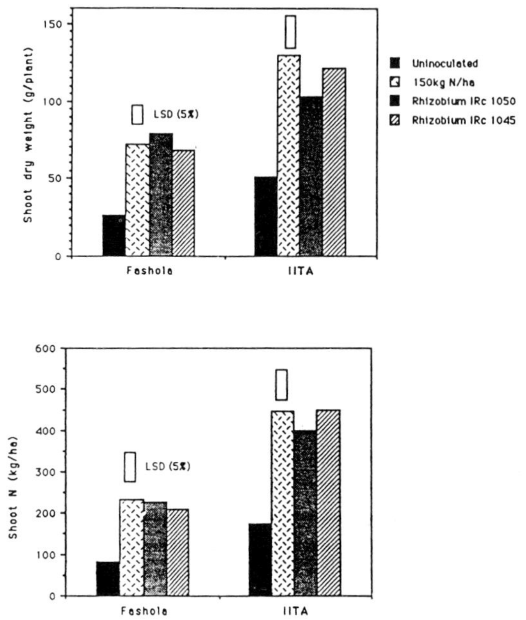

Fig. 1. Effect of urea fertilizer and inoculation with *Rhizobium* on shoot dry weight (g plant^{-1}) and N (kg ha^{-1}).

Nitrogen fixing trees used in alley cropping may nodulate with two types of *Rhizobium* (Dreyfus and Dommergues, 1981; Sanginga et al., 1989a). These are fast-growing strains belonging to the genus *Rhizobium* (sensu stricto), and slow-growing strains which form the cowpea miscellany and are now designated as *Bradyrhizobium* (Elkan, 1984).

One category of trees which nodulate with *Rhizobium* include species such as *L. leucocephala*, *Sesbania grandiflora*, *Calliandra calothyrsus*. This group is considered as specific and exhibits a narrow symbiotic range. The practical implication of the specificity of group 1 is that their establishment requires inoculation with the compatible fast-growing strains, which are generally less ubiquitous than *Bradyrhizobium*. This explains the spectacular response to inoculation of *L. leucocephala* with *Rhizobium* IRc 1045 or IRc 1050 (Sanginga et al., 1985, 1988) in two sites in Nigeria. At both places inoculated plants produced more N and dry matter than the controls (Fig. 1). This effect was statistically equivalent to the application of 150 kg N ha^{-1} as urea. Further, the strains survived and competed well in the field, as was shown in observations made on uninoculated trees 10 years after first establishment of *L. leucocephala* (Sanginga et al., 1994a).

The second category of NFTs nodulates only with *Bradyrhizobium* e.g. *Acacia mearnsii*, *Faidherbia albida* and *Tephrosia vogelii*. Inoculation of NFTs in this group rarely results in a significant yield increase since most tropical soils harbor the competent *Bradyrhizobium* (Dommergues, 1987).

A third category is more promiscuous since it nodulates both with *Rhizobium* and *Bradyrhizobium* e.g. *Acacia seyal*, *A. cyanophylla*. *G. sepium* and *Albizia lebbeck* might also belong to this group, but this needs to be further confirmed.

Turk and Keyser (1992), in a series of inoculation experiments performed to delineate rhizobial specificity of a variety of tree legumes, found that *L. leucocephala*, *C. calothyrsus* and *G. sepium* nodulated effectively with rhizobia isolated from each of the three species, presumably indicating that they belong to a common effectiveness group. But neither of *G. sepium* or *C. calothyrsus* nodulated with *Bradyrhizobium* strains. *Acacia auriculiformis*, *lebbeck* appeared to be promiscuous for nodulation and effectiveness, but *Acacia mangium* has specific requirements.

As shown above, NFTs and their rhizobia exhibit a range in their degree of specificity, but there is also a large variation for their nodulation response. The proportions and amounts of N_2 fixed by NFTs is influenced by the effectiveness of *Rhizobium* strains. In a field study conducted in Southwestern Nigeria, Sanginga et al. (1989a) indicated that estimates with the ^{15}N dilution method gave N_2 fixation of 134 kg N ha^{-1} in six months when *L. leucocephala* was inoculated with *Rhizobium* strain IRc 1045 and 98 N kg ha^{-1} when inoculated with *Rhizobium* strain IRc 1050. This N represented 34–39% of the plant total N. Sanginga et al. (1991b) have also shown that the amount of N_2 fixed in *G. sepium* provenances was influenced by the *Rhizobium* strains. Estimates of the proportion of plant N derived from N_2 fixation (% Ndfa) ranged from 36% for *Rhizobium* spp. SP14 to 71% for *Rhizobium* spp SP44. Awonaike et al. (1992) have also demonstrated that N_2 fixation and general performance of a *G. sepium* genotype was influenced by rhizobial inoculum strain. However they showed that no one strain was superior over all the host provenances and no one provenance was superior in N_2 fixation across all five bacterial strains included in the study.

Geographical restriction on nodulation occurs among NFTs used for alley cropping. Some NFTs nodulate in some areas, but not in others, and not all species within a genus e.g. *Acacia* will be nodulated by the same bacterial strain. *Leucaena leucocephala* for example, nodulated in two out of eleven sites in

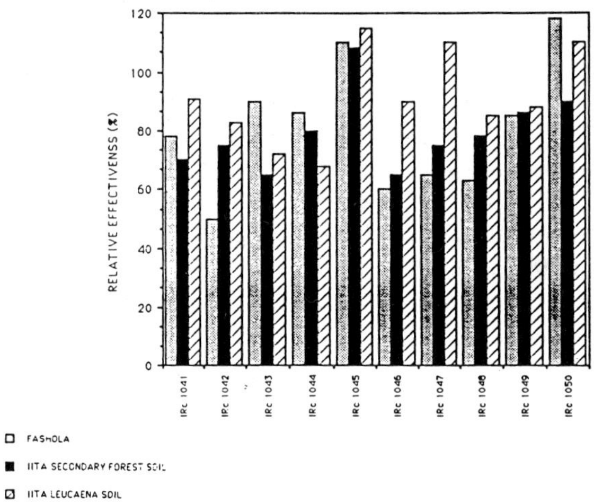

Fig. 2. Symbiotic effectiveness index (%) of different rhizobial isolates inoculated on Leucaena grown in soils from IITA and Fashola at IOWAP.

Nigeria, and in one out of four sites in Zaire but did not nodulate in Zimbabwe (Sanginga et al., 1987).

Sanginga et al. (1989a) showed that *Rhizobium* strain IRc 1047 induced high shoot dry weight on *L. leucocephala* grown in old leucaena soil according to Figure 2; however, it was less effective in other soils. Also, *Rhizobium* strain TAL 1145 from Hawaii was completely ineffective in Nigerian soils. This suggests that a particular location may not necessarily provide a better source of inoculant for that location than isolates from other environments.

It is therefore important to determine the degree of host specificity of selected NFTs to help predict the need to inoculate them at sowing, and to develop rhizobial strains that nodulate and fix N_2, with many of the useful species if possible (effective promiscuous strains). Conversely, an alternative approach might be to utilize promiscuous host NFTs or their provenances that may nodulate effectively in a soil with small rhizobial populations.

Tree host provenance or genotypic variation in N_2 fixation

The microbial partner in the NFTs/*Rhizobium* symbiosis has received considerable attention (Dommergues, 1982; Dreyfus and Dommergues, 1981; Sanginga et al., 1989a); however, much less research has been undertaken to evaluate existing variability in the NFT host species as an approach for deriving greater benefits from N_2 fixation. Attempts to select superior NFTs have often demonstrated that some leguminous trees such as *L. leucocephala* and *G. sepium*, are active in N_2 fixation, but relatively less is known about the genetic variation in N_2 fixation within and between provenances or isolines of these tree species.

Recent studies have begun examining variation in nodulation and N_2 fixation within and between leguminous tree species e.g. *G. sepium*, *L. leucocephala* and *F. albida* with *Rhizobium* spp. symbiosis. These include pot and field trials conducted in sites in Subsaharan Tropical Africa at the International Institute of Tropical Agriculture (IITA), Ibadan, Nigeria (transition forest-savanna), Yangambi, Zaire (humid forest) and at Domboshawa in Zimbabwe (savanna, Miombo woodland), while additional research was undertaken at the International Atomic Energy Agency (IAEA) laboratory in Seibersdorf, Austria. The soils used in all these experiments were low in major nutrients especially N and available P. *Rhizobium* strains which had in earlier studies proved effective on their respective hosts were used to inoculate the different tree species. The isotope dilution, the A value approach and the total N-difference methods (Danso et al., 1992) using the uninoculated NFTs treatments and non-NFT e.g. *Senna siamea*, *S. spectabilis* and *Eucalyptus* sp. as reference trees were used to measure the N_2 fixation differences between and within NFT provenance. A summary of the findings is presented below.

Many alley cropping studies at IITA have used *L. leucocephala* K8 as hedgerow tree (Kang et al., 1981a, b, 1985). Sanginga et al. (1989b) showed that when well nodulated K8 derived 40% of its N from atmospheric N_2. However, Sanginga et al. (1990a), found significant differences in N_2 fixation abilities between two isolines of *L. leucocephala* K28 and K636 and Sanginga et al. (1990c) demonstrated a variation of % Ndfa from 37 to 74 between eleven *L. leucocephala* cultivars, indicating scope for screening NFT genotypes to increase N contributions in alley cropping.

Significant differences in % Ndfa have been reported among provenances or isolines of other NFTs. For example, despite the generally poor N_2 fixation (average 20%) ascribed to *F. albida* (Sanginga et al. (1990c)), vast differences existed between provenances. One provenance derived 36% of its total N from fixation; another on the other hand, fixed only 6% of its requirements under identical conditions.

Similar variations in the proportions and amounts of N_2 fixed have been observed for *G. sepium* growing under identical conditions. Percentage Ndfa ranged from 26 to 68% between the 25 provenances examined

(Sanginga et al., 1991b). Such variations were confirmed in field conditions although values were lower than in pot experiments (Sanginga et al., 1994b).

Similarly, Liyanage et al. (1994) showed significant differences in N_2 fixation between *G. sepium* provenances in a field experiment conducted in Sri Lanka. The values of N_2 fixation ranged from 17.8 g N $tree^{-1}$ (equivalent to 86 kg N ha^{-1} at 5000 trees ha^{-1}) to 61.7 g N $tree^{-1}$ (equivalent to 300 kg N ha^{-1}). On average, % Ndfa in all *G. sepium* was about 55% equivalent to 168 kg N ha^{-1}; values also obtained by Sanginga et al. (1994) in Nigeria.

Ranking of *L. leucocephala* cultivars or *F. albida* and *G. sepium* provenances for their % Ndfa capabilities was highly dependent on the growth stage and the source of N. Growth of NFTs increased with either inoculation with *Rhizobium* spp. or N fertilizer addition. However, growth of inoculated provenances was more variable than that of plants dependent on fertilizer N suggesting large differences in the N_2-fixing abilities rather than intrinsic differences in the capacity for growth differences. The % Ndfa of selected provenances (poor, intermediate, and good fixers) increased with time, with the average values being in most cases similar within species at 36 weeks after planting. A practical implication of this is that there is a chance of reaching different conclusions on the N_2 fixation potential of species and provenances depending on the growth stage. Since trees are normally perennial, long duration studies should therefore be preferable if possible. This is a fundamental weakness of many pot studies. Except for cases where early N_2 fixation is essential for good establishment, early differences in N_2 fixation (e.g. *L. leucocephala* K636 and K28) may be of little long term consequence in the field.

The above studies have demonstrated that the genetic variability in N_2-fixing abilities of NFTs is high. No fixation may be significantly improved in any given environment by screening a large collection of different host genotypes or provenances for symbiotic performance. However, compared to the microbial symbiont, this approach has been given relatively little attention to date. Selecting superior plant genotypes or provenances should substantially increase yield of NFTs, especially in N-deficient soils. This has great implications for N_2 fixation in alley cropping and in agroforestry in general. Vegetative propagation, often relatively easy with tree species, could lead to rapid and significant farmer implementation. However, selection for high N_2 flxation should not be the sole criterion considered. In mixed ecosystems for example, it would be necessary to select for tolerance to stress conditions as well.

Quantification of N_2 fixation by hedgerow trees

Methods of N_2 fixation measurements

Until recently, few studies have been conducted on the quantification of atmospheric N_2 fixed by different *Rhizobium* NFT symbioses. To maximize the contribution of this natural and inexpensive source of N in tree-based systems such as alley cropping, reliable methods are needed for quantifying biologically N_2 fixed in trees.

Various conventional methods, such as the total N difference (Gauthier et al., 1985; Ndoye and Dreyfus, 1988; Sanginga et al., 1985, 1986), acetylene reduction assay (Högberg and Kvarnstrom, 1982; Lulandala and Hall, 1986; Roskoski, 1981) nodule number and mass (Högberg and Kvarnstrom, 1982; Roskoski, 1981) and ureide production (Herridge et al., 1994; van Kessel et al., 1988) have been used to assess N_2 fixed by NFTs. Many of these techniques, described in several critical reviews (Danso et al., 1992; Dommergues, 1987; Fried et al., 1983), are based on indirect criteria, and are either qualitative or cannot distinguish between the various sources of N in a fixing plant.

Nitrogen-15 techniques, that either involve the application of ^{15}N labelled materials to soil, or utilize natural levels of ^{15}N in soils have been suggested to be the most reliable methods for quantifying N_2-fixed (Danso et al., 1992; Peoples and Herridge, 1990). These ^{15}N techniques are particularly useful in that unlike many others, they can at a single harvest, measure the separate and integrated amounts or proportions of soil, fertilizer and atmospheric N_2 assimilated in field-grown crops.

The ^{15}N substratum labelling techniques were first used to estimate N_2 fixed in pasture and grain legumes. They have however, not been used extensively in tree N_2 fixation studies because of methodological difficulties associated with the selection of suitable reference plants, sampling protocols and the perennial nature of the trees. These factors are considered briefly below.

Factors affecting the validity of N_2 fixation estimation by ^{15}N labelling techniques

Factors that generally affect the validity of N_2 fixation measurements based on ^{15}N isotopic techniques have been reviewed in detail elsewhere (Chalk, 1985; Danso et al., 1992; Fried et al., 1983). We will only focus on some problems and possible solutions specific to trees.

Reference tree. The accurate determination of the actual amounts of N_2 fixed in the field is crucial only in some instances, such as with N-balance studies in the soil, or in the soil/tree system in alley cropping system, or for comparing N_2 fixed in different seasons, years and environments. In these cases the reference crop constitutes the main potential source of error in the ^{15}N labelling technique. The selection of an appropriate non-N_2-fixing crop is therefore an essential step in applying the ^{15}N labelling methodology to quantify N_2 fixed. However, according to Danso et al. (1986) the need for the precise quantification of N_2 fixed may not be that compelling in studies to simply compare treatment effects or rank plant genotypes or rhizobial strains for N_2 fixing abilities.

Uninoculated NFTs are among those that have been used as reference crops. If no indigenous rhizobia are present in the experimental soil, uninoculated NFTs could be suitable reference material (Gauthier et al., 1985; Sanginga et al., 1986). However, care must be taken to thoroughly examine roots of such uninoculated trees to ensure that they are not nodulated. Cross-contamination of uninoculated control from inoculated treatments has been observed in experiments involving NFTs (Sanginga et al., 1990b). Using such controls, N_2 fixation measured by isotope dilution and/or the N-difference methods will be underestimated (Sanginga et al., 1990b). It may be difficult to avoid cross-contamination in the field and the greenhouse unless special precautions are taken (Ayanaba and Bromfield, 1981). Ideally, uninoculated control plants should be compared with other potential reference crops such as known non N-fixing trees e.g. *Senna siamea*, *S. spectabilis* and *Eucalyptus* sp. These non-fixing trees, however, have to fulfil the conditions outlined by Fried et al. (1983) and Danso et al. (1992) for an appropriate reference crop.

The validity of such selections can be established by comparing the isotopic composition of N in these plants with that of non-nodulating tree provenances whenever available (Fried et al., 1983). In pot studies, Sanginga et al. (1990b) examined the suitability of *S. siamea*, a non-N_2 fixing leguminous tree and *Eucalyptus grandis* a non-leguminous tree together with the uninoculated host N_2 fixing trees, *F. albida* and *L. leucocephala* as potential reference for estimating N_2 fixed in the inoculated host N_2 fixing trees using the ^{15}N labelling techniques. Results indicated that the isotope dilution approach gave several (physiologically impossible) negative estimates of N_2 fixed in the poor N_2 fixing *F. albida* while % Ndfa determinations were positive for *L. leucocephala*. This indicated that errors due to the use of unsuitable reference plants are less critical when N_2 fixation is high. (Danso et al., 1992). For *L. leucocephala*, however the uninoculated N_2 fixing reference trees indicated significantly lower values of N_2 fixed than *S. siamea* and *E. grandis*. Such differences between the estimates given by the two non N_2 fixing reference trees may have not been real due to the nature of pot experiments. In pot experiments, errors due to reference plants are minimized because growth is restricted by physical conditions and there is a more uniform distribution of ^{15}N within the container than would normally occur in the field.

Complementary field experiments have been conducted at two sites in Nigeria, to assess the validity of *S. siamea* and *S. spectabilis* as reference plant for measuring N_2 fixation in *L. leucocephala* and *G. sepium*. Negative estimates of % Ndfa were obtained with *S. siamea* using both the ^{15}N isotope dilution and the total difference N methods at the second harvest in one site (Sanginga et al., unpubl. results). This was due to differences in pattern of N assimilation by *S. siamea* and the NFTs which coincided with a sharp decline of assimilated ^{15}N. Values of % Ndfa obtained by *S. spectabilis* on the other hand were positive, higher and less variable (CV $< 20\%$) than those of *S. siamea* (CV $> 60\%$). *Senna spectabilis* was thus considered to be a better reference for N_2 fixation in *L. leucocephala* and *G. sepium* than *S. siamea*. A potential problem with *S. siamea* could arise from the large horizontal spread of roots that will absorb N from well outside the zones within which ^{15}N was incorporated. This is particularly important if ^{15}N was applied in microplots. But data by Awonaike et al. (1993), obtained in pot experiment refute this hypothesis. *S. spectabilis* was also found to be a suitable non fixing reference species for measuring N_2 fixation in *G. sepium* in the Philippines (Ladha et al., 1993).

Awonaike et al. (1993) using the double isotope (^{15}N and ^{35}S) labelling approach, concluded that *S. siamea* did not absorb fertilizer and soil S in the same ratio as the NFTs; a criterion considered to indicate suitability of reference plants (Wagner and Zapata, 1982). However, the validity of the concept of using ^{15}N and ^{35}S was disputed by Hamilton et al. (1993) who found in a study assessing reference species for *Acacia* spp. that equal ratios of labelled to unlabelled

S between legumes and reference plants did not correspond with equal ratios of soil-derived labelled to unlabelled N.

Further search for non-nodulating NFTs, or their development (e.g. through mutation breeding) may prove valuable for N_2 quantification in NFTs. Attempts to obtain these reference trees have commenced. A simple procedure to adopt involves for example planting several hundreds of NFTs in a sand bath irrigated with heavy suspensions of *Rhizobium*. After a period of growth, the roots are examined for the absence of nodules. The identified potential non-nodulating isolines can be vegetatively propagated and again re-inoculated with *Rhizobium* to ensure that this trait is conserved. Screening NFTs for host and strain specificity could also provide suitable controls in some situations.

Genetic variation in N_2 fixation. Large plant-to-plant variation and genetic differences in nodulation and N_2-fixation of NFTs have been discussed in the previous section. The genetic heterogeneity of NFTs has hardly received the attention it deserves. As indicated by Danso et al. (1992) this genetic variation may sometimes contribute greatly to the poor precision in the estimates of N_2 fixation using the ^{15}N methodology. Duhoux and Dommergues (1985) observed that average % Ndfa based on tree size variation, varied from 16% (in the smallest trees) to 70% (in the biggest-sized trees). In such cases, it is essential to select well-matched reference crops for each plant to lessen errors due to the large genetic differences. Plant to plant variability might in many instances result from the out-crossing characteristic of number of NFTs. Where plants can be grown by vegetative propagation, this variation could certainly be reduced. This could be a practical option for several NFT species. Also, starting plants in nursery within inoculated pots or plastic bags and selecting plants of uniform size for N_2 fixation experiments could assist in reducing experimental errors.

Nitrogen-15 labelling techniques. There are several practical questions on the method of applying labelled ^{15}N materials which need to be addressed before designing ^{15}N experiments to measure N_2 fixed by NFTs. These include the N rate and ^{15}N enrichment, frequency, method and time of application, and the chemical and physical form of the fertilizer used (Chalk, 1985; Danso et al., 1986, 1992).

The amount of ^{15}N applied for labelling a tree will depend upon the N rate and the ^{15}N enrichment of the labelled material utilized. The N rates and ^{15}N enrichment used for grain and forage legumes can be readily adopted for small tree plants grown in the greenhouse or in the field. If isotope-aided experiments are performed with large-sized trees, the amount of N already present in trees may cause a significant dilution in the ^{15}N absorbed by plants. Under such conditions it is highly advisable to conduct a preliminary experiment with a few trees, to ascertain several questions, e.g. total N and its partitioning among tree organs, labelling techniques, as well as sampling procedures. The equation suggested by Danso et al. (1992) could then be used to correct for retranslocated N and also for differences in total N accumulated before ^{15}N application and harvesting.

Several workers have reported that the decline with time of in the $^{15}N/^{14}N$ ratio in soil influences the accuracy of N_2-fixation estimates; the lower the rate of decline, the less serious the errors associated with any mismatch in the N uptake between reference and fixing crops (Fried et al., 1983; Witty, 1983). This could be very important in field experiments involving NFTs because of the long period of growth. In this case, a single initial application of labelled N fertilizer may prove to be unsatisfactory for measuring N_2 fixation over a long period. Multiple additions of small amounts of ^{15}N labelled fertilizer to the soil, use of slow-release N fertilizers formulations and ^{15}N-labelled organic mater offer great promise for N_2-fixation measurements in NFTs, since the $^{15}N/^{14}N$ ratio in these cases would remain relatively stable due to the rather slow release of a small but fairly constant amount of N with time. In pot (Sanginga et al., 1992c) and field (Sanginga et al., unpubl. results) experiments comparing the effect of different amounts of ^{15}N labelled fertilizer and the frequency of application on estimates of N_2 fixed in *L. leucocephala* have shown that single application of 20 mg N kg^{-1} soil decreased N_2 fixed and that applying this amount in three and five splits gave different results, depending on whether it was repeatedly applied to the same soil or each split application was on a previously unlabelled plot. The greatest problem with these approaches is that when the applied N rates are lower than those needed to support the satisfactory growth of a non-fixing crop over several seasons in impoverished soils, poor growth (or even sometimes death) of the reference crop may severely hinder the reliability of the estimates.

In some cases the ^{15}N fertilizer has been applied to NFTs at seeding (Gauthier et al., 1985) or soon after germination (Sanginga et al., 1989b). However, it has been shown by several workers that early growth of many NFTs is slow and that nodulation and N_2-fixation is delayed sometimes for several months (Sanginga et al., 1989b). It is therefore advisable to delay ^{15}N application until both fixing and reference crops are well-matched in growth and preferably, in N uptake.

Cropping systems. Alley cropping system could in itself influence N_2 fixation in a manner similar to that shown in intercropping systems with grain and pasture legumes (Danso et al., 1986). These cropping systems can influence the suitability of the reference crop and therefore the validity of the estimates made.

Variations in N_2 fixing activity with the age of trees, or interference by different processes such as litter fall and litter decomposition, and the redistribution of N in the different compartments of the tree/soil system may present some difficulties for measuring N_2-fixation using ^{15}N isotopic methods. Significant amounts of N_2 fixed by NFTs are undoubtedly incorporated into the soil, but the magnitude has seldom been quantified. With the litter of the N_2 fixing plant being lower in ^{15}N enrichment than that of the non-N_2 fixing plant, the soil under a NFTs may in time have a lower ^{15}N enrichment than that of the non N_2-fixing plant. (Danso et al., 1992). This could be very important in agroforestry systems such as in alley cropping where NFTs and non-NFTs are grown in alternate rows. The soils under these contrasting plants have been reported to mineralize at different rates due to differences in litter quality (Vanlauwe, unpubl. data). Because NFTs may contain more N than non-NFTs they might also contribute more N to the soil. This could lead to conditions where the fixing and non-fixing crops are not absorbing N from the same pool during the growing period and thus invalidate one of the most important assumptions for the suitability of reference plants for measuring N_2 fixation (Chalk, 1985). This factor can significantly affect the validity of the ^{15}N methodology. The periodic collection and an estimation of the N in fallen leaves in a given area could be used to assess the error of measurement caused by N in falling and decomposing litter.

Transfer of N_2 fixed by NFTs to associated reference non-NFTs is another potential source of error for N_2-fixation estimates (Chalk, 1985; Danso et al., 1986; Fried et al., 1983). However, the relative importance and extent of transfer of N_2 fixed from NFTs and its impact on estimation of % Ndfa to intercropped non-NFTs is controversial and is yet to be resolved.

Sampling of plant material. Baker et al. (1992) tested different sampling strategies, and suggested that the simple procedure of collecting small numbers of leaves (20 to 60) at random from the trees was sufficient to estimate % Ndfa in *L. leucocephala* with the isotope dilution method. However, the ^{15}N enrichments in different plant parts of NFTs grown on ^{15}N enriched soils frequently differ (Sanginga et al., 1990e), and raises the question as to which plant part is most representative. Measurements of N_2 fixation, based on only one plant part may therefore not adequately represent N_2 fixed in the whole plant. Errors due to differences in ^{15}N enrichment can be minimized by sampling each of these plant parts with fairly uniform N-isotopic composition separately for N and ^{15}N determinations and using the weighted atom % ^{15}N for the whole plant. This is feasible for tree seedlings and young NFTs, but problems may occur when sampling big trees, because of their high above and below ground biomass.

NFTs can have significant proportions of their total biomass and N content below ground (Young, 1985; Smucker et al., 1994). Studies by Sanginga et al. (1990a) in a pot experiment found that *L. leucocephala* roots can contain more than 50% of the N fixed in the whole plant. However, lower values (< 10–40%) have been found in other pot studies and especially in field studies (Sanginga, unpubl. data; Liyanage et al., 1994). It is difficult to assess the recovery of roots and determine the importance of fine-root turnover. In some cases inclusion of roots may not greatly affect estimates of N_2 fixation e.g. *Gliricidia* (Liyanage et al., 1994), but in other circumstances quantification of N_2 fixed that disregards roots, nodules and crowns result in serious errors and the proportion of N_2 fixed may be largely underestimated (Danso et al., 1994). In addition, for many perennial legumes the N stored in these underground structures and unharvested portions of foliage may, after decomposition exert a significant effect on the dilution of the soil $^{15}N/^{14}N$ ratio. This is a practical problem which may not be easy to resolve, given the difficulty in recovering all roots and litter.

Some workers have looked at the feasibility of using the ^{15}N natural abundance to measure N_2 fixation in trees used in alley cropping (Peoples et al., 1991; Ladha et al., 1993). The advantages the ^{15}N natural abundance method has over the ^{15}N enrichment method are: investigation can be undertaken without the use of N fertilizers, and the uniformity of the ^{15}N level with depth in some soils and even within the different plant organs so that estimates of N_2 fixation may not be greatly affected by the choice of the non-fixing reference plant (People et al., 1991). In an alley cropping system experiment, Ladha et al. (1993) investigated if the ^{15}N natural abundance method could be applied successfully to estimate N_2 fixation by *G. sepium* using *S. spectabilis* as a reference plant. This study showed that the ^{15}N natural abundance values of total N in *S. spectabilis* were similar to those of extractable N of soil from different soil depths, suggesting that *S. spectabilis* was a suitable non-fixing reference species. The changes in ^{15}N natural abundance of total N of prunings were similar to both N_2 and non-N_2 fixing tree species and, the observed variation was considered not to have interfered with estimation of N_2 fixation. At four of the six sampling times, *G. sepium* fixed about 50% of its N requirements, whereas at the other two sampling dates N_2 fixation dropped to 30 and 35%. Comparable N_2 fixation estimates have recently been reported for field-grown *G. sepium* in alley cropping in Nigeria (Sanginga et al., 1994b) using the ^{15}N labelling technique.

However, the ^{15}N natural abundance method could sometimes be faced with difficulties such as natural variations in atom % ^{15}N in soil with depth (Steele et al., 1983) and the spatial variability in the field (Domenach, 1987). The task is complicated further by the enrichment of ^{15}N during litter decomposition and plant N uptake. This is important in alley cropping where NFTs are sequentially cut and prunings used as N source after decomposition. As suggested by Danso et al. (1992) it is advisable to estimate this stored N and make a correction for its effect on the isotopic composition of N at harvest. What is urgently needed now is an examination of procedures for obtaining representative sampling (as against whole plant sampling) that would closely reflect the overall ^{15}N enrichment of the whole plant. Linayage et al., (1994) indicated that although the atom % ^{15}N excess was lower in *G. sepium* leaves than in the other organs (all of which had similar ^{15}N enrichments), values of % Ndfa calculated for any individual organ or for the whole plant were similar. This was because the relative distribution of the ^{15}N in the different parts of the fixing plant followed the same trend as in the reference plant.

Due to the current pruning management practices in alley cropping however, ^{15}N assay in leaf samples appears to be the practical method of sampling. The validity of the technique would be based on the assumption that any defined type of leaf sample would probably provide the same answer with regard to N_2 fixation of the whole plant. Leaves of similar age and morphological position should be sampled from the tree at various intervals after application of ^{15}N. For this reason, sampling strategies advised by Baker et al., (1992) could be used in spite of difficulties described above. The natural abundance offers great promises for this type of sampling, since ^{15}N composition is often similar in all plant parts (Peoples et al., 1991; Ladha et al., 1993).

Another approach may involve the forester methods of using allometric relations to estimate the standing biomass of trees. In this case, it should be possible to estimate small portions of different organs of the tree for ^{15}N atom excess, and using the formula to estimate total N_2 fixed that is in each plant part and in the whole tree.

Sampling of NFTs for ^{15}N composition and measurement of biomass constitute the greatest problem in N_2-fixation estimation. These sampling procedure problems are further confounded by the practice of using single trees usually as experimental units to cut down cost, a practice that introduces high variability in the estimates. More research is needed to develop more reliable sampling procedures.

Potential and actual N_2 fixation by NFTs

In this paper, we have considered for N_2 fixation measurements the two parameters defined by Dommergues (1987). The N_2 fixing potential (NFP) of a species, i.e. the N_2 fixed with all environmental constraints removed, including the possible inhibitory effect of soil N. However, almost without exception the field data reported are subjected to some environmental constraints and the concept of NFP is a qualified one, and the actual N_2 fixed (ANF), is the resultant of NFP, modified by environmental constraints.

The major factors in high NFP are a high potential growth and a high percentage of N derived from the atmosphere (% Ndfa). This last parameter generally appears to be less affected by environmental conditions than total N_2 fixed (Danso et al., 1992). However, as indicated previously, there is increasing evidence of large genotype/provenance differences within NFTs species in their N_2 fixation.

Based on the above definitions, high and low NFTs have been identified. The former include such species as *L. leucocephala* and *G. sepium* for which records occur of 100 to 300 (sometimes 500) kg N fixed ha^{-1} yr^{-1} representing about 65% of their total N from atmospheric N_2. The latter include such species as *F. albida* for which the N_2 fixation values of less than 20 kg N ha^{-1} yr^{-1} i.e. 10–20% of its total N have been

Table 1. N_2 fixation estimates (proportion and actual amounts of N_2 fixed) by NFT's

Species	N_2 fixation estimates						References
	% Nitrogen derived from atmosphere			N_2 derived from atmosphere (mg plant^{-1})			
	N-Difference	Isotope D	A-value	N-Difference	Isotope	A-value	
F. albida	ND	14–44	14–37	ND	3–9	3–11	Sanginga et al. (1990c)
G. sepium	ND	19–78	28–79	ND	13–40	28–52	Sanginga et al. (1990c)
G. sepium	ND	59	ND	ND	755**	ND	Sanginga et al. (1994b)
G. sepium	ND	55	ND	ND	166*	ND	Liyanage et al. (1994)
L. leucocephala	51–61	ND	ND	224–274*	-	ND	Sanginga et al. (1985)
L. leucocephala	ND	78	ND	ND	230*	ND	Zaharah et al. (1986)
L. leucocephala	30–39	34–39	ND	76–113*	98–134*	ND	Sanginga et al. (1989b)
L. leucocephala	ND	6–52	37–72	ND	4–37	21–47	Sanginga et al. (1990c)
L. leucocephala	ND	68	ND	ND	1544**	ND	Sanginga et al. (unpubl.)
S. rostrata	35	36	ND	0.59	0.62	ND	Ndoye and Dreyfus (1988)
S. sesban	18	18	ND	0.12	0.13	ND	Ndoye and Dreyfus (1988)

ND = not determined.
* = kg ha^{-1} year^{-1}.
** = field conditions.

found. Species such as *A. lebbeck* appear to be intermediate, fixing between 60 and 120 kg N ha^{-1} (Kadiata et al., unpubl. data). It is likely that when more species are studied there will be more of a continuum. Estimates of N_2-fixed of some NFTs by different methods are summarized in Table 1.

Factors limiting N_2 fixation by NFTs

Environmental factors

Many reviews have been published dealing with the environmental factors controlling the establishment and the functioning of N_2-fixing symbioses (Danso et al., 1992; Dommergues, 1982 etc.) The present section concentrates on four limiting environmental factors which are probably among the most important for NFTs: (1) Phosphorus nutrition, (2) Levels of soil mineral N, (3) Moisture stress, and (4) Soil acidity.

Soil deficiency in phosphorus. Many tropical soils are deficient in available P, which is known to limit nodulation and N fixation and the growth of NFTs. Observations that *L. leucocephala* is not favoured by low fertility have been reported in several areas in (Ahmad and Ng, 1981; Halliday, 1981; Sanginga et al., 1985). Phosphorus supply and the absence of active *Rhizobium* have been responsible for poor nodulation and establishment of these species in some soils. Sanginga et al. 1985) observed that *L. leucocephala* needs about 80 kg P ha^{-1} for good establishment in Nigerian soils, especially when it is effectively nodulated. Similar observations have been reported by the same authors for *Acacia holocericea*, *F. albida* and *G. sepium* used in alley cropping.

Eliminating P-deficiency can be achieved by adding P fertilizer to soil and/or selecting NFTs genotypes or provenances tolerant to low P soils. For example, twenty-five provenances of *G. sepium* and eleven cultivars of *L. leucocephala* and *F. albida* were examined for growth, P uptake and P use in low P soils both in pot and field experiments (Sanginga et al., 1991a, 1994b; Sanginga, 1992). There were large differences between and within tree species in growth and P use efficiency. Results indicated that these differences were crucial at early growth stage and disappeared after one year of growth. This indicates that P is very critical initially at early growth stage and selection of provenances tolerant to low P was necessary at that stage. The same study also indicated differences between NFTs in their dependency to mycorrhizal infection. Mycorrhiza may greatly improve P supply to the host-plant by increasing the absorbing capacity of the roots thus removing some or all of the P limitation on the N_2-fixing capacity of NFTs. Work conducted at IITA showed that mycorrhizal infection increased root P and dry weight, and/or uptake of immobile elements like Zn and Cu in all woody legumes studied including exotics and indigenous species (Osonubi et al., 1991).

Tree growth is stimulated by the "tripartite" symbiosis of plant – microsymbiont – mycorrhiza. Inoculation of *L. leucocephala* by species of *Glomus* doubled plant growth and increased nodule fresh weight and nitrogenase activity each by 50% (Purcino et al., 1986). In a P-deficient soil, De la Gruz et al. (1988), obtained little increase in the N per plant in *Acacia auriculiformis*, *A. mangium* and *Albizia falcata* inoculated with only *Rhizobium* but 8- to 25-fold increases when inoculated with both *Rhizobium* and selected vesicular arbuscular mycorrhizal (VAM) fungi.

Combined N. The inhibitory effect of combined N on N_2 fixation is common in most symbiotic associations, including NFTs. Sanginga et al. (1989b), in a study conducted on an alfisol in Nigeria, showed that N_2 fixation of a well-nodulated *L. leucocephala* was reduced by 50 percent with 40 or 80 kg ha^{-1} of N fertilizer. In Senegal the absence of nodules on *Acacia senegal* is believed to be due to active nitrification of organic N (Bernard-Reversat and Poupon, 1979). Also because of the redistribution of N in the plant and the soil profile due to litter fall and its mineralization, NFTs/*Rhizobium* symbioses can be more affected by combined N than annual crops.

Two approaches may be suggested to improve N_2 fixation by trees in the presence of combined N: (i) develop specific partnerships with *Rhizobium* hosts that are more tolerant to combined N than others, or (ii) develop plants that can simultaneously absorb soil N and fix enough atmospheric N_2. Few studies have compared the relative abilities of different NFT provenances and clones to fix N_2 at different soil N levels. Variation in growth, N_2 fixation and mineral N use efficiency of provenances of *G. sepium* uninoculated, inoculated and N fertilized were studied (Sanginga et al., 1992b). N_2 fixation ranged from 56 to 74% depending on provenances and N sources. Such observations suggest that genotypic differences in physiological N use of absorbed soil N and fixed N_2 may exist. This might also result in differences in N_2 fixation inhibition in second and subsequent seasons due to the build-up in soil N, and need to be examined.

Moisture deficit. This stress is known to seriously reduce nodulation and N_2 fixation in a number of N_2-fixing systems. Soil moisture may affect N_2 fixation indirectly by its effects on plant growth, root infection and nodule formation and functioning. However, even during drought spells some perennial deep-rooted legumes (e.g. *L. leucocephala*) may still fix N_2. For example, Sanginga et al. (unpubl. data) found a high nitrogenase activity in *L. leucocephala* nodules collected at more than 2 m in an alfisol in Nigeria during the dry season. Similarly, drought resistant NFTs, such as *Prosopis* spp. (Felker et al., 1982) or *Acacia raddiana*, are probably only slightly affected in their N_2-fixing activity by moisture deficiency in the upper horizons since these plants could receive adequate moisture from below the nodule zone. By improving P nutrition in P-deficient soils or by means of other mechanisms not yet quite elucidated, mycorrhizae can also improve the tolerance of NFTs to moisture stress. Osonubi et al. (1991) showed that mycorrhiza promoted nodulation of *F. albida* and *A. nilotica* and reduced the effects of drought stress on growth of *G. sepium* and *L. leucocephala* grown in moisture-deficient soils.

Soil acidity. Soil acidity affects both plant growth and the occurrence, growth and survival of *Rhizobium* and the functioning of the symbiosis. With few exceptions, soil acidity in tropical Africa is known to be highly correlated with calcium deficiency, and with aluminium or manganese toxicity. Little information is available on the effects of soil acidity or high aluminium concentrations either on plant growth or on *Rhizobium* activity in association with NFTs. A few studies have shown that *L. leucocephala* is poorly adapted to acid soils (Ahmad and Ng, 1981; Halliday, 1981) because of its inability to establish a N_2 fixing symbiosis at low pH. However, in an experiment carried out in Hawaii, successful nodulation was not a sufficient condition to render *L. leucocephala* productive in the acid-soil locations (Halliday and Somasegaran, 1982). Despite the presence of effective nodulation, plants on plots at pH 4.5 remained stunted and yellow, yielding only 6% of the level of dry matter production measured for neighbouring plots that had been limed to pH 6.0. Thus, it appears that *L. leucocephala* rhizobia are able to multiply at a pH in which *L. leucocephala* itself can not grow.

In conclusion, the problem of soil acidity can be approached in two ways. The first and most obvious way is to apply lime. A second approach is to develop or select a NFTs symbiosis which is acid tolerant. Selection for tolerance of acidity may be especially relevant to the survival and persistence of inoculum but the tolerance of the plant genome may play the dominant overall role in some BNF systems. A combination of these two approaches offers the best hope for a speedy utilization in tropical soils.

Management practices

Cutting or pruning of tops of NFTs in alley cropping systems can cause considerable sloughing of roots and nodules and therefore affect N_2 fixation, with the extent depending on the severity of pruning.

The effect of three successive prunings on N uptake and N_2 fixation and N distribution in *L. leucocephala* was investigated in a greenhouse experiment (Sanginga et al., 1990e) and validated in the field using provenances of *G. sepium*. Two isolines, uninoculated or inoculated with three different *Rhizobium* strains and labelled with ^{15}N, were grown for 36 weeks during which they were cut every 12 weeks. The results showed that cutting affected the distribution of ^{15}N in the different plant parts. Live nodules showed the lowest atom % ^{15}N excess values (0.087), followed by leaves (0.490), branches (0.522), stems (0.591), and roots (0.857). The total N_2 fixed in the roots was about 60% of that fixed in the whole plant, while the shoots contained only 40% of the fixed N_2. It was concluded that N reserves in roots and nodules constitute a substantial N source that must be taken into account when estimating fixed N_2 and its contribution to the N balance after pruning or cutting plants. Thus, timing and severity of pruning may allow for some management of underground transfer of fixed N_2 to associated crops, as well as regulating root competition between established hedgerows of NFT's and food crops. As indicated above, the accretion of N to soil by root sloughing, and addition of above-ground litter may decrease BNF in later seasons.

Strategies to enhance N_2 fixation in alley cropping

To date the use of N_2-fixing trees in alley cropping has been largely recommended over the use of non NFTs. The success of introduction of species with a high N_2-fixing potential, e.g. *L. leucocephala* and *G. sepium*, is such that interest in N_2-fixing trees is increasing. Field experiments with other NFTs which have been carried out or are under way provide us with information about the ecology and the agronomic behavior of these plants. Only a few alley cropping studies have specifically dealt with the N_2-fixing systems. Whilst the principles and methods of study learned from work with annual and pasture legumes may apply to trees there are many special problems to be considered in studying symbiosis with NFTs. Below, we comment briefly on some research strategies in order to improve N_2 fixation in alley farming.

1. Relatively few NFTs are self-pollinated. Large plant-to-plant variation in nodulation and growth have been recorded in NFTs. Further use of NFTs requires, as a first step, screening between and within species to determine which ones exhibit the highest N_2-fixing potential. Survey of NFTs both exotic and indigenous may help to find such trees. As a second step, it appears necessary to improve our knowledge of the requirements of the selected trees with regard to their microsymbiont.

   It is possible to vegetatively propagate a number of important NFTs (e.g. *G. sepium*). The use of clones of effective host genotypes may improve N_2 fixation potential and reduce variability in host growth potentiai.

2. Some environmental and management factors affect N_2 fixation and play a critical role in the successful growth of NFTs.

   It is appropriate to further examine soil factors such as those discussed above (P, N, acidity), in order to manage N_2 fixation more effectively. Good methods must be developed for studying the effect of environment and management practices on growth of the host and the microsymbiont in the rhizosphere, the infection process, nodule genesis, nodule development and N_2 fixation processes.

3. Not all legumes fix N_2 and not all NFTs are legumes. There is a lack of knowledge on what tree species are potential N_2 fixers, e.g. leguminous trees used in established alley cropping systems are not always N_2 fixers e.g. *S. siamea*. Because some soils may contain enough N, plant growth is not a reliable indicator of N_2 fixation (Ladha et al., 1993). Nodulation is often difficult to assess with tree species because nodulation may be seasonal, because they may not occur throughout the life of the tree, and because the nodules are sometimes restricted to deep layers in the soil. Furthermore the presence or absence of nodules gives little indication of the level of N_2 fixation.

   It is critically important to develop reliable methods to survey candidate tree species in natural environments for their N_2 fixation potential. There is good evidence that with careful sampling of the test species and of reference non-fixing species, and with careful analytical procedures, the ^{15}N natural abundance method may be a useful tool for identifying NFTs (Domenach, 1987; Ladha et al., 1993; Peoples and Herridge et al., 1990, 1991). However, levels of ^{15}N natural abundance in mineral N can be low and/or variable in some ecosystems and the procedure may not always be suitable.

4. The measurement of N_2 fixation in the field, and questions such as the rate of N_2 availability to associated plants are areas requiring further study. In the management of N_2 fixation, it is essential to measure fixation in the field and how it is influenced by the strain of micro-organism, soil nutrient conditions (and their amendment, e.g. P additions), water, and management practices. Nitrogen fixation assessments based on plant growth or acetylene reduction have limited usefulness. Furthermore, a measurement of N_2 fixation is needed which distinguishes between N_2 fixed from the atmosphere and N_2 absorbed from soil. The ^{15}N isotope dilution method has proved to be a good method with herbaceous legumes and it should be developed further to provide a reliable method for measurement of N_2 fixation in trees. Applying the isotope dilution method to trees involves a unique set of problems, requiring solutions. Other methods such as ureide analysis of sap has potential for a limited number of species (Herridge et al., 1994). However, newer approaches to measurement of N_2 fixation by trees e.g. the use of ^{15}N natural abundance have a particular attraction with perennials and represent an alternative approach.

In conclusion, N_2 fixation by trees in the field should be examined (in appropriately designed plantings) to elucidate the effects of: (a) soil chemical status, especially the effect of added phosphate and trace elements; (b) season and age of the tree, and different management systems such as times and intensity of cutting of foliage or stems for green manure, fodder, and fuel; (c) soil moisture and its interactions with soil chemical status; (d) genetic variation in N_2 fixation within species.

Nitrogen contribution and use efficiency in alley cropping systems

Inputs of N in alley cropping

Total production of hedgerow trees and the fertilizer values is related to total dry matter production and also on the nutrient content of leaf material. However, limited information on annual N yield is available for some multipurpose trees (MPTs) used in alley cropping. This is the N accumulation in leafy matter, harvested several times during one year, and consisting of fixed N_2 and N absorbed from the soil.

Guevarra (1976) reported N yields between 500 and 600 kg N ha^{-1} yr^{-1} for *L. leucocephala* grown for forage in Hawaii. Observations in Southern Nigeria indicate that five annual prunings of *L. leucocephala* grown in hedgerow yielded between 150 and 560 kg N ha^{-1} yr^{-1} depending on hedgerow spacing, pruning regime, and soil type (Duguma et al., 1988; Kang et al., 1981b). Sanginga et al. (1986) reported nitrogen yields ranging between 448 and 598 kg N ha^{-1} yr^{-1} from *L. leucocephala* inoculated with an effective strain of *Rhizobium* grown on an altisol in Nigeria. Tree species such as *G. sepium*, *Flemingia macrophyla*, *Sesbania rostrata* and *S. siamea* also have large N yields as compared to the non-leguminous species e.g. *Dactyladenia barterii* and *Alchornea cardifolia*.

Despite the high N yield for species such as *L. leucocephala* and *G. sepium*, supplemental N application was still required to obtain high maize yields (Kang et al., 1981a) due in part to low use efficiency of N in the prunings, (Guevarra, 1976; Mulongoy and van der Meerch, 1988; Sanginga et al., 1988). However, the last authors indicated a 10 percent increase of the efficiency of prunings when *L. leucocephala* was effectively nodulated with *Rhizobium*. Amongst other species that have been tried in Africa is *Sesbania rostrata*. Prunings of *S. rostrata* planted as hedgerow in an alley cropping system at two different populations provided 3 and 4 t ha^{-1} of dry matter (Mulongoy, 1986). Their decomposition released an average of 70 kg N ha^{-1} during the rice cropping season, with subsequent yield increase of 20–50%. Its beneficial contribution as green manure in rotation with rice has been also reported by Rinaudo et al. (1982) in microplots, in Senegal. These reports indicate that prunings from NFTs can be used as a source of N.

Decomposition and N-mineralization of organic residues in alley cropping systems

Organically bound N, incorporated in the hedgerow through N_2 fixation, is made available to crops through the N-mineralization process. The rate at which mineralization of the fixed N_2 occurs depends on the residue quality, the activity and composition of the decomposer community and the physico-chemical environment. Management practices can alter the impact of the different N-mineralization regulating factors.

Table 2. Biochemical characterization of different hedgerow species (Kachaka et al., 1993)

	L. leucocephala	*S. siamea*	*D. barterii*	*F. macrophylla*
% N	4.33	2.79	1.41	3.26
C/N ratio	9.80	15.30	28.00	12.30
% (Hemi) cellulose	18.10	31.70	21.50	27.10
% Lignin	8.10	10.40	14.90	17.20
% Polyphenol	3.37	1.41	3.43	2.39
Lignin/N	1.90	3.70	10.60	5.30
Polyphenol/N	0.78	0.51	2.43	0.73

Above ground organic residues

Residue quality has been assessed by measuring biochemical properties which have been shown to delay or enhance the N-mineralization process. As such, high quality residues have been linked to low polyphenol content (Palm and Sanchez, 1990), a low polyphenol/N ratio (Oglesby and Fownes, 1992; Palm and Sanchez, 1991), low lignin/N ratio (Kachaka et al., 1993; Melillo et al., 1982), or low (lignin + polyphenol)/N ratio (Fox et al., 1990).

In alley cropping systems, organic residues from hedgerow trees of different quality are applied on the surface or incorporated in the soil at different times and in different amounts.

Prunings of different hedgerow species have been shown to have often very contrasting biochemical characteristics (Table 2). Plant materials of low C:N and lignin:N ratios (e.g. leaves of *G. sepium* and *L. leucocephala*) decompose faster than plant materials of relatively high C:N and lignin:N ratios (e.g. leaves of *S. siamea* and *D. barterii*). The quality of the different components of one single pruning, however, is again drastically different depending upon the plant organs and age of trees. The age of the hedgerow canopy can range in an alley cropping system with 4 prunings per year from 1.5 months (2nd and 4th pruning) over 2.5 months (3rd pruning) to 5 months (1st pruning). Kachaka et al. (1993) showed that young *Senna siamea* and *Dactyladenia barterii* leaf residues have a higher quality than old leaves. From all these observations, we can conclude that at each pruning time, qualitatively different residues are applied to the soil surface in different amounts, although produced by the same tree. This will have a pronounced impact on their N-mineralization rate.

Weeds contribute to the aboveground biomass inputs in alley cropping. The quantity depends on the weeding frequency, but it is always much lower than the hedgerow biomass. Van der Meersch et al. (1993) found a weed N content equal to 11% of the N in prunings of alley cropped *L. leucocephala* and *S. siamea* at the time of harvesting of the first season maize crop. However, weed growth has been shown to be suppressed in alley cropping by canopy closure (Yamoah et al., 1986) or by mulch applications (Lal, 1975), the latter being more effective in the presence of a slowly decomposing mulch (Kang et al., 1990). So far, weed quality has not been assessed, while this information is needed for a complete understanding of N-dynamics in alley cropping systems.

Below ground organic residues

In contrast to applications of aboveground residues, belowground tree and crop residues enter the soil continuously, due to root exudation and turnover. Nutrient fluxes to the soil from fine roots and nodules senescence and exudation might well be the same or greater than those of above ground litter fall. Smucker et al. (1994) estimated that during the 14 weeks of maize growth, up to 91 kg N ha^{-1} may have been released by decomposing maize roots in alley cropping systems while *L. leucocephala* roots could also have contributed up to 102 kg N ha^{-1} during the same period. Sanginga et al. (1988) indicated that N contribution of inoculated *L. leucocephala* roots and nodules to maize grain yield was equivalent to an average of 32 kg N h^{-1}. This is supported by observations by Sanginga et al. (1990e) showing that 50% of the N in cut *L. leucocephala* was contributed by roots and nodules which decay. These authors showed that nodule senescence and decay occurred within 3 weeks after each cutting of *L. leucocephala* shoots while new ones were being formed to continue N_2 fixation during regrowth. It is likely that fine roots might have similar turnover rate. However, it is unknown what proportions of this N "pool" are reabsorbed, leached or recycled by the plant.

Many aspects of the fine root dynamics among trees are poorly understood. Very little is known regarding the longevity of fine roots, though this is a key factor affecting C and nutrient fluxes to and from the fine roots. Compounding these problems is our inability to measure the magnitude of concurrent fine root production and mortality, especially when food crops are removed or when trees are trimmed in alley cropping systems. Because of the large amount of assimilate going below ground, it is important to understand the root "cost" to the plant, and to consider if more assimilate could be used for aboveground productivity by appropriate genotype selection. This is closely allied to an understanding of root growth and the use of soil resources.

Residue N use efficiency in alley cropping systems

Sustainable food crop production is the ultimate goal of the alley cropping systems. A high input of N in the soil through N_2 fixation will not significantly contribute to soil productivity if a substantial part of it is lost and thus not available for crop uptake. Efficient use of the fixed N is a prerequisite for recommending N_2 fixation as an environmentally sound, low cost way to improve soil fertility.

Definitions
Residue N use efficiency by the crop (% Ncrop) can be defined as the amount of crop N derived from the residues per unit of added residue N:

$$\%\text{Ncrop} = \frac{\text{(amount of crop N derived from residue)}}{\text{(amount of added residue N)}} \times 100 \quad (1)$$

The % Ncrop does not take into account any residual N from previous pruning application. In alley cropping systems, the nutrients that are lost beyond the crop rooting zone may be recovered by the deeper-rooting hedgerow trees and made available to the subsequent food crop. Residue N can also be immobilized in the soil organic matter pool, or taken up by weeds (Van der Meersch, 1992). On the other hand, residue N can be lost through denitrification, volatilization, run-off, or leaching beyond the tree rooting zone.

It is therefore necessary to extend the residue N use efficiency by the crop to the residue N use efficiency by the whole system, which can be defined at year "n" after the residue application (% Nsys_n) as in the following equation:

$$\%\text{Nsys}_n = \frac{\sum_{i=1}^{n} \text{Ngrain}_i + \text{Ncrop}_n + \text{NSOM}_n + \text{Nhedge}_n + \text{Nweed}_n}{\text{(N residue)}} \times 100 \quad (2)$$

Ncrop_n, NSOM_n, Nhedge_n and Nweed_n are the residue derived N, present in the above and belowground crop without the grains, soil organic matter, the above and belowground hedgerow and the weed biomass. N grain_i is the residue derived N present in the crop grains at year i. It is obvious that Ncrop_n, Nhedge_n and Nweed_n consist of fresh organic matter when entering the soil system, while NSOM_n consists partly of stabilized N, not more immediately available for crop growth.

Measurement of % N use efficiency
Several methods have been used by IITA scientists to measure the N use efficiency of *L. leucocephala* prunings by maize using Equation 1. The amount of crop N derived from the pruning (Ndfp) residue is calculated as follows:

$$\%\text{Ndfp} = \frac{\text{(amount of maize N derived from residues)}}{\text{(amount of total maize N)}} \times 100 \quad (3)$$

In the conventional method, % Ndfp is measured as the N uptake in an alley cropping system with added residues, as compared to a control system without added residues. In this case the choice of a suitable control is a major problem. If an alley cropping system without added residues is used as a control, the competition for nutrients other than N between the crop and the hedgerow might reduce the crop N uptake. This will obviously lead to an overestimated % Ndfp. In the case of an absolute control (plot without hedgerows and residues) omission of the competition with the hedgerow may lead to an underestimated % Ndfp, in a soil with relatively higher available N. However, in available N poor soils, % Ndfp might be overestimated. Isotope labelling techniques have also been used to measure N contribution of prunings in alley cropping systems.

In the direct isotope method, ^{15}N labelled residues are added, while in the indirect isotope method, the soil is labelled with a small amount of ^{15}N enriched fertilizer and unlabelled residues are added, and a control without prunings is included. In the indirect ^{15}N method the underlying hypothesis is that the availability of the ^{15}N to a test or subsequent crop is similar with

Table 3. Different estimates of leucaena N use efficiency by maize at the IITA main station in Ibadan, Nigeria

Methodology	Leucaena N (kg ha^{-1})			% Ndfp	% Ncrop	Reference
	1st	2nd	Sum			
Conventional methods						
Alley farming without pruning addition as control	211	159	37	62.6	6.4	Mulongoy and van der Meersch (1988)
Continuous farming as control	?	?	187	33.0	17.6	van der Meersch et al. (1993)
^{15}N methods						
Direct (^{15}N labelled residues)	78	-	78	10.1	10.0	Vanlauwe et al. (1995)
Indirect (^{15}N labelled soil)	198	164	362	19.8	6.3	Akinnifesi et al. (1994)

1st = First pruning.
2nd = Second pruning.
% Ndfp: Percentage of nitrogen derived from pruning.
% Ncrop: Percentage of nitrogen in the crop.

or without residue additions. Adding residues, however, increases the soil biological activity and may thus lead to a different soil N status as compared to the unamended microplot. It seems reasonable to state that, especially if the interest lies in drawing up a complete N balance, the direct labelling technique is the only way to get accurate figures of residue N uptake by the crop. In measuring % $Nsys_n$, ^{15}N labelled residues are indispensable. So far, no measurements of the % $Nsys_n$ in alley farming systems have been reported.

Attempts made to quantify N use in alley cropping systems have shown so far that N recovery by maize from prunings is low. Van der Meersch et al. (1993) showed that *L. leucocephala* and *S. siamea* contributed to maize 20 and 32% respectively of the N released from prunings. These percentage N recoveries are in the range of those reported for *L. leucocephala* by other studies (Guevarra, 1976; Sanginga et al., 1988). Nitrogen recoveries in alley cropped maize with *L. leucocephala* as low as 5–10% have been reported (Mulongoy and Van der Meersch, 1988; Vanlauwe, unpubl.). Similarly ^{15}N labelled mulch of *G. sepium* and *Erythrina peoppigiana* grown in alley cropping contributed about 10% of maize N (Haggar et al., 1993).

Results in Table 3 using both the indirect and direct ^{15}N labelling method also indicate low N use by maize in alley cropping. In the case of the rapidly decomposing, N rich *L. leucocephala* residues, however, a substantial amount of mineral N is available shortly after pruning application. The low % Ndfp must therefore be caused by a lack of synchronization between the *L. leucocephala* N release and the maize demand for N. This is especially relevant for the first pruning after the dry season, as this pruning contains the highest amount of N (Table 4), while at the time of its application the young crop is not able to withdraw a significant amount of the mineralized N. Van der Meersch et al. (1993) found a similarly low recovery of *S. siamea* pruning N in maize. The second *L. leucocephala* and *S. siamea* pruning has been shown to be better synchronized with the maize N needs (Van der Meersch et al., 1993). As such, N mineralized from the first pruning, recovered in the canopy regrowth might be better synchronized with crop demand. However, Haggar et al. (1993) suggested that the long-term build-up of the soil organic matter reserve of mineralizable organic N was more important than the synchrony of mulch N release and crop uptake in determining the substantially higher productivity and N uptake in the alley crop compared to the sole crop.

Nitrogen not taken up by the crop in alley cropping
Attempts are being made to quantify factors contributing to low N use in alley cropping systems. Using above low recovery figures, about 70–80% of applied pruning N is unaccounted for. Some factors are losses due to volatilization and leaching. Other N sources unaccounted for in alley cropping systems include denitrification, N retained in the soil organic matter or recovered by hedgerow trees and by weeds.

Table 4. Amounts of N added in different prunings in leucaena alley cropping systems in IITA, Ibadan, Nigeria

Species	Pruning activities (kg N ha^{-1})					Reference
	1st	2nd	3rd	4th	Total	
Leucaena[a]	1.6	55	60	NP[b]	221	van der Meersch et al. (1993)
Leucaena[a]	172	142	66	NP	380	van der Meersch et al. (1993)
Leucaena[a]	161	80	79	NP	320	van der Meersch et al. (1993)
Leucaena[a]	129[e]			96[e]	225	Kang et al. (1981a)
Leucaena[a]	112[e]			69[e]	181	Kang et al. (1981b)
Senna[a]	78	19	21	NP	118	van der Meersch et al. (1993)
Senna[a]	119	68	33	NP	292	van der Meersch et al. (1993)
Senna[a]	103		39	NP	181	van der Meersch et al. (1993)

[a]Means of fertilized (120-90-30 kg NPK ha^{-1}) and unfertilized plots.
[b]"MP" means "no pruning activity".
[c]Means of fertilized (60, 120 or 180 kg N ha^{-1}) and unfertilized plots.
[d]Means of fertilized (70 or 140 kg N ha^{-1}) and unfertilized plots.
[e]129 kg N ha^{-1} for the first 3 prunings; 96 kg N ha^{-1} for the 4th and 5th prunings; 112 kg N ha^{-1} for the first 3 prunings; 69 kg N ha^{-1} for the 4th to the 6th pruning.

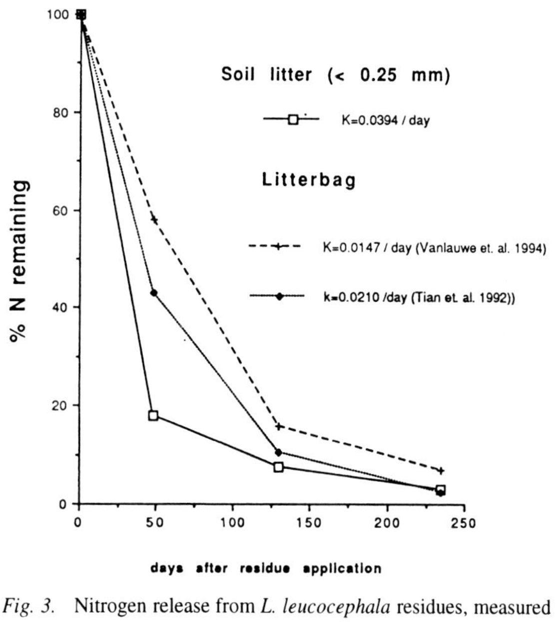

Fig. 3. Nitrogen release from *L. leucocephala* residues, measured with the floatation and the litterbag method.

N-recovery by deep-rooting trees. Hedgerow trees have been hypothesized to act as nutrient pumps by recovering mineral N_2 leached below the rooting zone of foodcrops. However, unequivocal experimental evidence with ^{15}N labelled residues has not been presented so far. Indirect evidence is based on the hypothesis that lower soil mineral N concentrations under alley cropping than under control soils are caused by tree N removal. Horst (1991) found a lower amount of nitrate in soil under *L. leucocephala* alley cropping as compared to a control soil on a Nitosol. Hauser (1990) concluded that the risk of leaching losses under the interrow space and monocrop was higher than under the *L. leucocephala* row. The main question, however, remains as to which soil layer the tree is extracting its nutrients from, because this will determine whether the tree acts as a "nutrient pump" or a competitor with the accompanying crop.

N-immobilized in microbial biomass. Some of the N released from the prunings is immobilized in the soil microbial biomass. Microbial biomass is a dynamic sink and source of nutrients. It contributes to the productivity of the system through its continuous turnover and immobilization of nutrients. This contributes to a reduction of N losses and during the remineralization of organic nutrients, are made available to crops at later times. Van der Meersch et al. (1992) found that an amount of 6 and 10 kg N ha^{-1} yr^{-1} was available in the soil microbial biomass under *L. leucocephala* and *S. siamea* alley cropping systems, respectively. Haggar et al. (1993) found that mulch N contributed only 3–5% of the microbial N pool at 40 days and this fell to zero by 105 days in alley cropping with *G. sepium* or *E. peoppigiana*.

N immobilization in soil organic matter. Fixed N, can enter the soil organic matter pool via two mechanisms. Firstly, through gradual microbial catabolism or faunal comminution, the residue particle size might be reduced to less than 0.25 mm, leaving the N in its original bond. On the other hand, mineralized N can be reimmobilized by microorganisms, associated with a particular soil organic matter fraction. In the second case, the decomposability of the N can be reduced drastically, depending on the immobilization reactions. The fraction in which the N turns up will determine its availability to crops. Mulongoy and Sanginga (1990) suggested that 34% of the *L. leucocephala* N, released during one growing season was retained in the soil organic matter. Lal (1989) observed a decline in both soil total N and organic C in *L. leucocephala* and *G. sepium* alley cropping fields, during a 4 year period. Kang et al. (1985), however, observed an increase in organic C in *L. leucocephala* alley cropping during a 6 year period, under similar conditions. Mulongoy et al. (1993) found no difference between the total N content of the control and the alley cropping plots with *L. leucocephala*, *G. sepium* and *D. barterii*, 12 years after field establishment. Van der Meersch (1992) observed no differences in soil total N under *L. leucocephala* and *S. siamea* alley cropping during a 4 year period. No consistent effect of alley cropping on the soil organic C or total N pool seems to be demonstrated. Little data, however, are available on the N dynamics of distinct soil organic matter fractions as affected by the different residue inputs.

Vanlauwe et al. (1994) indicated that N immobilized in the soil organic matter will be released, depending on the quality of the soil organic matter pool. As the soil litter pool has a very high turnover time (Fig. 3), it is necessary to quantify the contribution of the several organic inputs to this pool. Other organic matter fractions are expected to release their N at a much slower (zero order) rate and as such, may not be able to supply a substantial amount of the total N needed by the crop. Mulongoy and Gasser (1993) observed that Ibadan soil released about 45 mg N per kg dry soil in a period of 12 weeks.

As the observed % Ndfp's are very low, that means that the major part of the plant N is derived from other sources than the added prunings. This might be the residual soil litter, or the tree/crop roots. It is necessary to fully understand the origin of this non-residue derived N, as it seems to be much better synchronized with plant demand.

Losses due to volatilization, leaching and denitrification. Conditions such as favorable moisture, high temperature which occur immediately after application of residues are favorable for both N denitrification and leaching in alley cropping. The rapid release of *L. leucocephala* N and the possible inability of freshly pruned trees to absorb a lot of N due to the removal of their canopy might lead to substantial N leaching. Leaching of mineral N represented 20% of N released from *L. leucocephala* leaves at 12 weeks after pruning application (Liya, pers. commun.). Moreover, if at the time of pruning, belowground tree residues also die off, an additional amount of mineral N is expected to be released. The high amount of soluble C (Kachaka et al., 1993) and mineralized N and the possible partial anaerobic conditions in "hot spots" due to the oxygen-consuming decomposition of the soluble C might favour denitrification. No direct, independent estimates of the size of both processes, however, are available.

NH_3-volatilization might be a less important loss mechanism in alley cropping, as Sanginga and Mulongoy (1994) estimated, in greenhouse pots, that only 5% of the *L. leucocephala* and *S. siamea* N added was volatilized within 2 weeks after the residue application. Again, field estimates of volatilization are not available. Above and belowground lateral displacement of pruning N should be considered in drawing up complete N-balances, especially when ^{15}N isotopes are used.

Management practices for an increased N use efficiency in alley cropping systems

In order to develop management strategies aimed at increasing % $Nsys_n$, it is necessary to fully understand the N pathways in alley cropping systems. Nitrogen^{15} isotopes will remain necessary to unequivocally quantify the different processes. Major difficulties are experienced when trying to measure N losses through denitrification, leaching, run-off or NH_3 volatilization. Quantification of these processes, however, is vital to derive proper management strategies, as the nature of the major N loss will determine the nature of the management practices.

The advantages of alley cropping are mainly related to the presence of hedgerow trees. Both the time of residue application and the amount of residue can be altered to improve the % Ncrop. Other hedgerow management options, however, remain unexplored. Although it is known that the total amount of dry matter production depends on the pruning frequency and height (Duguma et al., 1988; Kang et al., 1990),

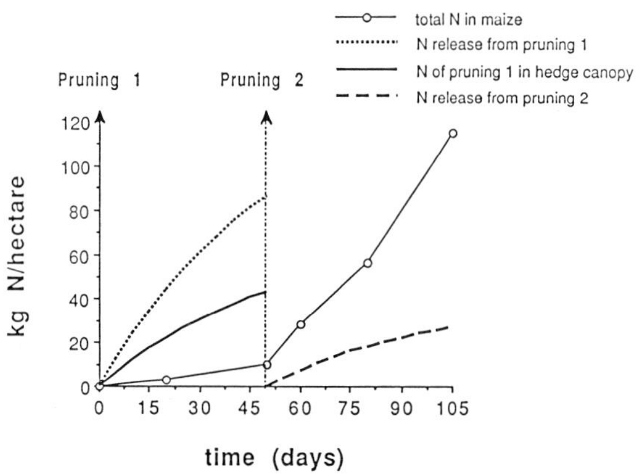

Fig. 4. Hypothesized improved synchronization between residue N, released from the first pruning and taken up the canopy regrowth and maize N uptake.

each pruning activity normally consists of a complete removal of the *L. leucocephala* canopy. This can lead to substantial N losses, as indicated above, especially for the first pruning. Moreover, before the first pruning, a high amount of surface litter is already present, accumulated through dropping of leaves during the dry season. It might be necessary to think in terms of "partial pruning" to increase the % Ncrop since pruning part of the canopy reduces the crop/tree competition for light, and leaves part of the canopy thus reducing the amount of added residue N and maintaining the ability of the trees to absorb N. The N thus recovered in the aboveground biomass could be better synchronized with maize demand for N (Fig. 4).

Synchronization of plant available N with crop demand can also be improved by manipulating the quality of the applied residues. Mixing the high quality *L. leucocephala* residues with lower quality litter might stimulate the immobilization of N in the soil organic matter pool, thus reducing its potential loss. Maize stover, returned to the soil after harvest, might be a suitable low quality material accompanying the first pruning of the second season. A prerequisite for mixing residues with different quality, however, is that both species produce a sufficient amount of biomass. Mixing *L. leucocephala* with *D. barterii* residues might only be an academic exercise, as these species do not perform well under similar soil acidity. Mulongoy et al. (1993) showed that mixing *L. leucocephala* with *S. siamea* residues (both species perform well on non acidic soils) delays N release from the *L. leucocephala* leaves. If the use efficiency of the fixed N cannot be improved, it might be necessary to rethink the inter-row spacing in alley cropping systems to adopt. A larger inter-row space will result in a lower biomass production by trees, and a lower capacity for tree roots to reabsorb mineralized N from deeper layers, but might also result in a higher crop yield (Kang et al., 1990).

Acknowledgements

Much of the published and unpublished information presented in this chapter is derived from experiments conducted at the International Atomic Energy Agency (IAEA), Seibersdorf Lab, Austria and at the International Institute of Tropical Agriculture (IITA), Ibadan, Nigeria. The support of the Belgian Government through the collaborative project between IITA and the Catholic University of Leuven on Soil Organic Matter Dynamics is also gratefully acknowledged. We also wish to thank various IITA Scientists for providing invaluable comments on the manuscript and Mrs C Tuoyo and Ms Tessy for typing the manuscript.

References

Ahmad N and Ng F S P 1981 Growth of *Leucaena leucocephala* in relation to soil pH, nutrient levels and *Rhizobium* concentration. Leucaena Research Report 2, 5–10.

Akinnifesi F K, Kang B T, Sanginga N and Tijani-Eniola H 1994 Root distribution patterns of *Leucaena* hedgerows and associated crops and their competition for N in alley cropping. Agron. J. (*Submitted*)

Allen O N and Allen E K 1981 The Leguminosae. A Source Book of Characteristics, Uses and Modulation. University of Wisconsin Press, Wisconsin. 812p.

Awonaike K O, Hardarson G and Kumarasinghe S 1992 Biological nitrogen fixation of *Gliricidia sepium/Rhizobium* symbiosis as influenced by plant genotype, bacterial strain and their interactions. Trop. Agric., Trinidad 69, 381–386.

Awonaike K O, Danso S K A and Zapata F 1993 The use of double isotope (^{15}N and ^{34}S) labelling technique to assess the suitability of various reference crops for estimating nitrogen fixation (BNF) in *Gliricidia sepium* and *Leucaena leucoccephala*. *In* Plant Nutrition: From Genetic Engineering to Field Practice. Ed. N J Barrow. pp 431–434. Kluwer Academic Publisher, Dordrecht.

Ayanaba A and Bromheld E S P 1980 The efficacy of soybean inoculation on acid soil in Tropical Africa. Plant and Soil 45, 95–106.

Baker D D, Wheeler R A and Fried M 1992 Estimation of biological nitrogen fixation by three legume tree species using the ^{15}N dilution method. *In* Biological Nitrogen Fixation and Sustainability of Tropical Agriculture. Eds. K Mulongoy, M Gueye and D S C Spencer. pp 259–265. John Wiley and Sons, New York.

Bernard-Reversal F and Poupon H 1979 Nitrogen cycling in a soil tree system in a Sahelian savanna: Example of *Acacia senegal*. *In* Nitrogen Cycling in West African Ecosystems. Ed. F

Rosswall. pp 363–369. Royal Swedish Academy of Science, Stockholm.

Burris N H 1978 The global nitrogen budget. Science or seance Nitrogen Fixation. Volume 1: Free Living Systems and Chemical Models. Eds. W E Newton and W H Orme-Johnson. pp 7–16. University of Park Press, Baltimore.

Chalk P M 1985 Estimation of N_2 fixation by isotope dilution: An appraisal of techniques involving ^{15}N enrichment and their application. Soil Biol. Biochem. 17, 389–410.

Danso S K A, Hardarson G and Zapata F 1986 Assessment of dinitrogen fixation potential of forage legumes with ^{15}N technique. *In* Potential of Forage Legumes in Farming systems of Sub-saharan Africa. Eds. I Haque, S Jutzi and P J G Neate. pp 26–58. ILCA, Ethiopia.

Danso S K A, Bowen G D and Sanginga N 1992 Biological nitrogen fixation in trees in agro-ecosystems. Plant and Soil 141, 177–196.

Danso S K A, Zapata F and Awonaike K O 1994 Biological nitrogen fixation in field grown *Robinia pseudoacacia* L. Paper presented at the Tenth Australian N_2 Fixation conference at Brisbane, September 1993.

Dela Cruz R E, Manalo M Q, Aggangan N S and Tambalo J D 1988 Growth of three legume trees inoculated with VA mycorrhizal fungi and *Rhizobium*. Plant and Soil 108, 11–115.

Dommergues V 1982 Ensuring effective symbiosis in nitrogen fixing trees. *In* Biological Nitrogen Fixation Technology for Tropical Agriculture. Eds. P H Graham and S C Harris. pp 395–413. CIAT, Cali, Colombia.

Dommergues Y R 1987 The role of biological nitrogen fixation in agroforestry. *In* Agroforestry a Decade of Development. Eds. H A Steppler and P K R Nair. pp 245–271. ICRAF, Nairobi.

Domenach A M 1987 Estimation de la fixation symbiotique chez des plantes herbacees et ligneuses; utilisation et validite de la methode basee sur la mesure des abondances isotopiques naturelles de l'azote. These de Doctorat – Lyon I. 189p.

Dreyfus B L and Dommergues Y R 1981 Nodulation of *Acacia* species by fast- and slow-growing tropical strains of *Rhizobium*. Appl. Environ. Microbiol. 41, 97–99.

Duguma B, Kang B T and Okali D U U 1988 Effect of pruning intensities of three woody leguminous species grown in alley cropping with maize and cowpea on an Alfisol. Agrofor. Syst. 6, 19–35.

Duhoux E and Dommergues Y 1985 The use of nitrogen fixing trees in forestry and soil restoration in the tropics. *In* Biological Nitrogen in Africa. Eds. H Ssali and S O Keya. pp 384–401. The Nairobi *Rhizobium* Mircen, Nairobi.

Elkan G H 1984 Taxonomy and metabolism of *Rhizobium* and its genetic relationships. *In* Biological Nitrogen Fixation, Ecology, Technology and Physiology. Ed. M Alexander. pp 1–39. Plenum Press, New York.

Felker P, Clark P R, Osborn J F and Cannel G H 1982 Biomass estimation in a young stand of mesquite (*Prosopsis* spp.). Ironwood (*Olneya tesota*), Palo Verde (*Cercidium floridium*, and *Parkinsonia aculeata*), and Leucaena (*Leucaena leucocephala*). J. Range Manage. 35, 87–89.

Fox R H, Myers R J K and Vallis 1 1990 The nitrogen mineralization rate of legume residues in soil as influenced by their polyphenol, lignin and nitrogen contents. Plant and Soil 129, 251–259.

Fried M, Danso S K A and Zapata F 1983 The methodology of measurement of N_2 fixation by non legumes as inferred from field experiment with legumes. Can. J. Microbiol. 29, 1053–1062.

Gauthier D, Diem H G, Dommergues Y R and Ganry F 1985 Assessment of N_2 fixation by *Casuarina equisetifolia* inoculated with *Frankia* ORSO2101 using ^{15}N methods. Soil Biol. Biochem. 17, 375–379.

Guevarra A B 1976 Management of *Leucaena leucocephala* (Lam.) de Wit for maximum yield and nitrogen contribution to intercropped corn. Ph.D. Thesis, University of Hawaii, Honolulu, Hawaii. 128p.

Haggar J P, Tanner E V J, Beer J W and Kass C D L 1993 Nitrogen dynamics of tropical agroforestry and annual cropping systems. Soil Biol. Biochem. 25, 1363–1378.

Halliday J 1981 Nitrogen fixation by leucaena in acid soils. Leucaena Res. Rep. 2, 71–73.

Halliday J 1984 Integrated approach to nitrogen fixing tree germplasm development. Pesqui. Agropecu. Bras. 19, 91–117.

Halliday J and Somasegaran 1982 Nodulation, nitrogen fixation, and *Rhizobium* and strai affinities in the genus *Leucaena*. *In Leucaena* research in the Asian Pacific region. pp 33–37. IDRC, Ottawa.

Hamilton S D, Hopmans P, Chalk P M and Smith C J 1993 Field estimation of N_2 fixation by *Acacia* spp. using ^{15}N isotope dilution and labelling with ^{35}S. For. Ecol. Manage. 56, 279–313.

Hauser S 1990 Water and nutrient dynamics under alley cropping versus monocropping in the humid-subhumid transition zone. Tans 14th ICSS 6, 204–209.

Hogberg P and Kvanrstrom M 1982 Nitrogen fixation by the woody legume *Leucaena leucocephala* in Tanzania. Plant and Soil 66, 21–28.

Herridge D F, Palmer B, Nurhayati D P and Peoples M B 1994 Application of the xylem solute (Ureide) method for measuring N_2 fixation in tree legumes. Soil Biol. Biochem. (*In press*).

Horst W J 1991 Cropping systems and their effect on sustained soil productivity in the tropics. *In* Nachhaltigkeit und Ressourcennutzung in Entwicklungslandern. Proceedings of a symposium 24–25/10/91 Gottingen. pp 243–268.

Kachaka S, Vanlauwe B and Merckx R 1993 Decomposition and nitrogen mineralization of prunings of different quality. *In* Soil Organic Matter Dynamics and Sustainability of Tropical Agriculture. Eds. K Mulongoy and R Merckx. pp 199–208. John Wiley and Sons, Chichester, UK.

Kang B T, Sipkens L, Wilson G F and Nangju D 1981a *Leucaena leucocephala* (Lam.) de Wit prunings as nitrogen source for maize (*Zea mays* L.) Fert. Res. 2, 279–287.

Kang B T, Wilson G F and Sipkens L 1981b Alley cropping maize (*Zea mays* L.) and leucaena (*Leucaena leucocephala* Lam de Wit) in Southern Nigeria. Plant and Soil 63, 165–179.

Kang B T, Grimme H and Lawson T L 1985 Alley cropping sequentially cropped maize and cowpea with *Leucaena* on a sandy soil in Southern Nigeria. Plant and Soil 85, 267–277.

Kang B T, Reynolds L and Atta-Krah A N 1990 Alley farming. Adv. Agron. 43, 315–359.

Kessel C van, Nakao P, Roskoski J P and Kevin K 1988 Ureide production by N_2-fixing leguminous trees. Soil Biol. Biochem. 20, 891–897.

Ladha J K, Peoples M B, Garity D P, Capuno V T and Dart P J 1993 Estimating dinitrogen fixation of hedgerow vegetation using the nitrogen-15 natural abundance method. Soil Sci. Soc. Am. J. 57, 732–737.

Lal R 1975 Role of mulching techniques in tropical soil and water management. Techn. Bull. No 1, IITA, Ibadan, Nigeria.

Lal R 1989 Agroforestry systems and soil surface management of a tropical alfisol. III. Changes in soil chemical properties. Agrofor. Syst. 8, 113–132.

Liyanage M S, Danso S K A and Yayasundara H P S 1994 Biological nitrogen fixation in four *Gliricidia sepium* genotypes. Plant and Soil (*In press*).

Lulandala L L L and Hall J B 1986 *Leucaena leucocephala*'s biological nitrogen fixation: a promising substitute for inorganic nitrogen fertilization in agroforestry systems. *In* Biotechnology of Nitrogen Fixation in the Tropics (BIOnifT). Proceedings of UNESCO Regional Symposium and Workshop, UPM, Malaysia, 25–29 August 1986.

Melillo J M, Aber J O and Muratore J F 1982 Nitrogen and lignin control of hardwood leaf litter decomposition dynamics. Ecology 63, 621–626.

Mulongoy K 1986 Potential of *Sesbania rostrata* (Brem.) as nitrogen source in alley cropping systems. Biol. Agric. Hortic. 3, 341–346.

Mulongoy K and Van der Meersch M K 1988 Nitrogen contribution by leucaena (*Leucaena leucocephala*) prunings to maize in an alley cropping system. Biol. Fertil. Soils 6, 282–285.

Mulongoy K and Sanginga N 1990 Nitrogen contribution by leucaena in alley cropping. IITA Res. 1, 14–17.

Mulongoy K and Gasser M O 1993 Nitrogen-supplying capacity of leaves of *Dactyladenia barterii* (Hook ex olw) and *Leucaena leucocephala* (Lam.) de Wit in two soils of different acidity from southern Nigeria. Biol. Fertil. Soils 16, 57–62.

Mulongoy K, Ibewiro E B, Oseni O, Kilumba N, Opara-Nadi A O and Osonubi O 1993 Effect of management practices on alley-cropped maize utilization of nitrogen derived from prunings on a degraded Alfisol in south-western Nigeria. *In* Soil Organic Matter Dynamics and Sustainability of Tropical Agriculture. Eds. K Mulongoy and R Merckx. pp 223–231. John Wiley and Sons, Chichester, UK.

Ndoye I and Dreyfus B 1988 N_2 fixation by *Sesbania rostrata* and *Sesbania sesban* estimated using ^{15}N and total N difference methods. Soil Biol. Biochem. 20, 209–213.

Oglesby K A and Fownes J H 1992 Effects of chemical composition on nitrogen mineralization from green manures of seven tropical leguminous trees. Plant and Soil 143, 127–132.

Osonubi O, Mulongoy K, Awotoye O O, Atayese M O and Okali D U U 1991 Effects of ectomycorrhizal and vesicular-arbuscular mycorrhizal fungi on drought tolerance of four leguminous woody seedlings. Plant and Soil 136, 136–143.

Palm C A and Sanchez P A 1990 Decomposition and nutrient release patterns of the leaves of three tropical legumes. Biotropica 22, 330–338.

Palm C A and Sanchez P A 1991 Nitrogen release from the leaves of some tropical legumes as affected by their lignin and polyphenolic contents. Soil Biol. Biochem. 23, 83–88.

Peoples M B and Herridge D F 1990 Nitrogen fixation by legumes in tropical and subtropical agriculture. Adv. Agron. 44, 155–223.

Peoples M B, Bergersen F J, Turner G L, Sampet C, Rerkasem B, Bhromsiri A, Nurhayati D P, Faizah A W, Sudin M N, Norhayati M and Herridge D F 1991 Use of the natural enrichment of ^{15}N in plant available soil N for the measurement of symbiotic N_2 fixation. *In* Stable Isotopes in Plant Nutrition, Soil Fertility and Environmental Studies. pp 117–129. IAEA, Vienna.

Purcino A C, Lurlarb C and Lynd J Q 1986 Mycorrhiza and soil fertility effects with growth, nodulation and nitrogen fixation of *Leucaena* grown on a typic eutrostox. Commun. Soil Sci. Plant Anal. 17, 473–469.

Rinaudo G, Dreyfus B and Dommergus Y R 1982 Influence of *Sesbania rostrata* green manure of the nitrogen content of rice crop and soil. Soil Biol. Biochem. 15, 111–113.

Roskoski J P 1981 Nodulation and N_2 fixation by *Inga jinicuil*, a woody legume in coffee plantations. I. Measurements of nodule biomass and field C_2H_2 reduction rates. Plant and Soil 59, 201–206.

Sanginga N 1992 Early growth and N_2-fixation of *Leucaena leucocephala* and *Gliricidia sepium* at different levels of P application. Fert. Res. 31, 165–173.

Sanginga N, Mulongoy K and Ayanaba A 1985 Effect of inoculation and mineral nutrients on nodulation and growth of *Leucaena leucocephala*. *In* Biological Nitrogen Fixation in Africa. Eds. H Ssali and S O Keya. pp 419–427. Nairobi MIRCEN, Nairobi.

Sanginga N, Mulongoy K and Ayanaba A 1986 Inoculation of *Leucaena leucocephala* (Lam.) de Wit with *Rhizobium* and its nitrogen contribution to a subsequent maize crop. Biol. Agric. Hortic. 3, 34–347.

Sanginga N, Mulongoy K and Ayanaba A 1987 Evaluation of indigenous strains of *Rhizobium* and *Leucaena* in Nigeria. *In* Les Arbres Fixateurs d'Azote et l'Amelioration Biologique de la Fertilite du Sol. Colloques et Seminaires. pp 416–436. ORSTOM, Paris.

Sanginga N, Mulongoy K and Ayanaba A 1988 Nitrogen contribution of *Leucaena/Rhizobium* symbiosis to soil and subsequent maize crop. Plant and Soil 112, 137–141.

Sanginga N, Mulongoy K and Ayanaba A 1989a Effectivity of indigenous rhizobia for nodulation and early nitrogen fixation with *Leucaena leucocephala* grown in Nigerian soils. Soil Biol. Biochem. 21, 231–235.

Sanginga N, Mulongoy K and Ayanaba A 1989b Nitrogen fixation of field-inoculated *Leucaena leucocephala* (Lam.) de Wit estimated by the ^{15}N and the difference methods. Plant and Soil 117, 269–274.

Sanginga N, Zapata F, Danso S K A and Bowen G D 1990a Effect of successive cutting on nodulation and nitrogen fixation of *Leucaena leucocephala* using ^{15}N dilution and the difference methods. *In* Plant Nutrition – Physiology and Applications. Ed. M L van Beusichem. pp 667–674. Kluwer Academic Publishers, Dordrecht.

Sanginga N, Danso S K A, Zapata F and Bowen G D 1990b Influence of reference trees on N_2 fixation estimates in *Leucaena leucocephala* and *Acacia albida* using ^{15}N labelling techniques. Biol. Fertil. Soils. 9, 37–42.

Sanginga N, Bowen G D and Danso S K A 1990c Assessment of genetic variability for N_2 fixation between and within provenances of *Leucaena leucocephala* and *Acacia albida* estimated by ^{15}N labelling techniques. Plant and Soil 127, 169–178.

Sanginga N, Bowen G D and Danso S K A 1990d Genetic variability in symbiotic nitrogen fixation within and between provenances of two *Casuariana* species using the ^{15}N-labelling methods. Soil Biol. Biochem. 22, 539–547.

Sanginga N, Zapata F, Danso S K A and Bowen G D 1990e Effects of successive cuttings on uptake and partitioning of ^{15}N among plant parts of *Leucaena leucocephala*. Biol. Fertil. Soils 9, 37–42.

Sanginga N, Bowen G D and Danso S K A 1991a Intra-specific variation in growth and N_2-fixation of *Leucaena leucocephala* and *Gliricidia sepium* at low levels of soil P. Plant and Soil 127, 169–178.

Sanginga N, Manrique K and Hardarson G 1991b Variation in nodulation and N_2 fixation by the *Gliricidia sepium/Rhizobium* spp. symbiosis in cacalreous. Soil Biol. Fertil. Soils 11, 273–278.

Sanginga N, Mulongoy K and Swift M J 1992a Contribution of soil organisms to the sustainability and productivity cropping systems in the tropics. Agric. Ecosyst. Environ. 41, 135–152.

Sanginga N, Danso S K A and Bowen G D 1992b Variation in growth, sources of nitrogen and N-use efficiency by provenances of *Gliricidia sepium*. Soil Biol. Biochem. 24, 1021–1026 .

Sanginga N, Zapata F, Danso S K A and Bowen G D 1992c Estimating nitrogen fixation in *Leucaena* and *Gliricidia* using different ^{15}N labelling methods. *In* Biological Nitrogen Fixation and Sustainability of Tropical Agriculture. Eds. K Mulongoy, M Gueye and D S C Spencer. pp 265–277. John Wiley and Sons, New York.

Sanginga N, Danso S K A, Mulongoy K and Ojeifo A A 1994a Persistence and recovery of introduced *Rhizobium* ten years after inoculation on *Leuceana leucocephala* grown on an Alfisol in Southwestern Nigeria. Plant and Soil 159, 199–204.

Sanginga N, Danso S K A, Zapata F and Bowen G D 1994 Field validation of intraspecific variation in phosphorus use efficiency and N_2 fixation by provenances of *Gliricidia sepium* grown in low P soils. Appl. Soil Ecol. (*In press*).

Sanginga N and Mulongoy K 1994 Increasing biological N_2 fixation and its contribution to the N cycle in alley cropping systems. *In* Proceedings of the International Conference on Alley Farming. IITA, Ibadan (*In press*).

Smucker A J M, Ellis B G and Kang B T 1994 Root, nutrient, and water dynamics in alley cropping on an Alfisol in a forest savanna transition zone. *In* Proceedings of the International Conference on Alley Farming. IITA, Ibadan (*In press*).

Steele K W 1983 Quantitative measurements of nitrogen turnover in pasture systems with particular reference to the role of ^{15}N. *In* Nuclear Techniques in Improving Pasture management. pp 17–35. IAEA, Vienna, Austria.

Tian G, Kang B T and Brussaard L 1992 Biological effects of plant residues with contrasting chemical composition under humid tropical conditions – decomposition and release. Soil Biol. Biochem. 24, 1051–1060.

Turk D and Keyser H H 1923 Rhizobia that nodulate tree legumes: Specificity of the host for nodulation and effectiveness. Can. J. Microbial. 38, 451–460.

Van der Meersch M K 1992 Soil fertility aspects of alley cropping systems in relation to sustainable agriculture. Ph.D. Thesis, K U Leuven, Belgium.

Van der Meersch M K, Mulongoy K and Merckx R 1993 Evolution of plant biomass and nutrient content in relation to soil fertility changes in two alley cropping systems. *In* Soil Organic Matter Dynamics and Sustainability of Tropical Agriculture. Eds. K Mulongoy and R Merckx. pp 143–154. John Wiley and Sons, Chichester, UK.

Vanlauwe B, Vanlangenhove G and Merckx R 1995 Impact of different rainfall regimes on the decomposition of leaf residues with contrasting quality under tropical conditions. Biol. Fertil. Soils (*In press*).

Wagner G H and Zapata F 1982 Field evaluation of reference crops in the study of nitrogen fixation using isotope techniques. Agron. J. 74, 607–612.

Witty J F 1983 Estimating N_2 fixation if the field using ^{15}N labelled fertilizer: Some problems and solutions. Soil Biol. Biochem. 15, 631–639.

Yamoah C F, Agboola A A and Mulongoy K 1986 Decomposition, nitrogen release and weed control by prunings of selected alley cropping shrubs. Agrofor. Syst. 4, 239–246.

Young A 1985 The potential of agroforestry as a practical means of sustaining soil fertility. ICRAF Working Paper 34. 28p.

Zaharah A R, Sharifuddin H A H and Subramaniam R 1986 Nitrogen fixation by *Leucaena leucocephala* as measured by N-15 dilution technique. Pertanika 91, 17–22.

Plant and Soil **174:** 143–180, 1995.

Manipulation of rhizobia microflora for improving legume productivity and soil fertility: A critical assessment

John Brockwell[1], Peter J. Bottomley[2] and Janice E. Thies[3]
[1] *CSIRO Division of Plant Industry, GPO Box 1600, Canberra, ACT 2601, Australia,* [2] *Department of Microbiology, Oregon State University, Corvallis, OR 97331–3804, USA and* [3] *University of Western Sydney (Hawkesbury), School of Agriculture and Rural Development, Bourke Street, Richmond, NSW 2753, Australia*

Key words: Azorhizobium, Bradyrhizobium, inoculation, legumes, N_2 fixation, nodulation, rhizobial ecology, *Rhizobium*, symbiosis, taxonomy

Table of contents

Abstract
Introduction
Taxonomic and nutritional considerations relating to rhizobial ecology
  Taxonomic
  Nutritional
  Summary
Indigenous and naturalized populations
Rhizobial genetics as related to ecology
Management economies
  Soil moisture
  Weed competition
  Soil temperature
  Nutrients and fertilizers
  Soil amendment
  Synergism
  Nitrate
  Crop hygiene
Specialized management systems
  Rice systems
  Stem nodulation
  Leguminous trees

Inoculation
The need for inoculation
Inoculant technology
Strains of rhizobia suitable for inoculants
Inoculant production
Application of inoculant
Preinoculation
Inoculant quality control
Measurement of success of inoculation
Manipulation of rhizobial populations
New populations (introduced rhizobia)
Naturally-occurring populations
Manipulation of naturally-occurring populations
Competition between naturally-occurring and introduced rhizobia
Competition between different components of naturally-occurring populations of rhizobia
Concluding remarks, prospects and prognostication
Acknowledgements
References

Abstract
Inputs of biologically fixed nitrogen derived from the symbiotic relationship between legumes and their root-nodule bacteria into terrestrial ecosystems amount to at least 70 million metric tons per year. It is obvious that this enormous quantity will need to be augmented as the world's population increases and as the natural resources that supply fertilizer nitrogen diminish. This objective will be achieved through the development of superior legume varieties, improvement in agronomic practice, and increased efficiency of the nitrogen fixation process itself by better management of the symbiotic relationship between plant and bacteria. This paper considers ways and means by which populations of root-nodule bacteria, established and introduced, can be manipulated ecologically, agronomically, edaphically and genetically to improve legume productivity and, as a consequence, soil fertility.

Introduction

The great family of legumes (Fabaceae J. Lindley, formerly Leguminosae) is as diverse as it is large. Representatives range from tiny herbs to huge trees and rambling lianes. They are distributed from the arctic zones to equatorial areas, from the strand to the high mountains and from the wetlands to dry deserts. Despite this diversity, perhaps less than 100 legumes have assumed any substantial agricultural importance although many others provide timber and firewood, bases for pharmaceuticals, adhesives, tannins, spices, honey, foliage, pods and seeds that are consumed by grazing or browsing domestic animals, drought reserves, shelter belts and street trees, stability for soil conservation, borrow pits and roadside batters, reclamation for degraded lands, and beautification for gardens and recreational areas.

In this contribution, we are charged with consideration of the relationship between the legumes and their symbiotic nitrogen-fixing root-nodule bacteria (the rhizobia) and how it functions to influence plant productivity and soil fertility. The invigorating effect of legume cultivation on exhausted soils is extolled in ancient texts (cit. Fred et al., 1932). Although we now know that nitrogen fixation (N_2 fixation) is largely responsible, the legume confers other more subtile benefits on the soil environment: stimulation, amplification and diversification of the microflora; breaking disease cycles inseparable from monocultures; provision of organic nitrogen (N) which interacts with soil organic carbon to enhance soil structural stability. However, the scope of our contribution is restricted to N_2 fixation and how it might be enhanced and optimized by manipulation of the soil microflora.

Not all legumes fix N. The capacity to form nodules appears to be absent from the majority of members of the segregate family Caesalpiniaceae. On the other hand nodulation, if not N_2 fixation, with root-nodule bacteria appears almost universal in the segregate family Mimosaceae and the other Fabaceae although only a relatively small proportion of the total number of species in those two groups has been properly examined (Allen and Allen, 1981).

Estimates of global N_2 fixation are essentially matters of guesswork. The value given by Burns and Hardy (1975) was 175 million metric tons per year. Legume N_2 fixation accounted for approximately 40% of that figure. Since then estimates of global N_2 fixation have been augmented because of higher levels now attributed to marine fixation (Bunt, 1988). Whatever the true figure, legume symbioses contribute at least 70 million metric tons per year, approximately one half deriving from the cool and warm temperature zones and the remainder from the tropics. N_2 fixation is an efficient source of N (Peoples et al., 1995a). Values estimated for various legume crop and pasture species are often impressive, commonly falling in the range of 200–300 kg N per hectare (Peoples et al., 1995a). Using a ^{15}N technique in Australia, Bergersen et al. (1985) calculated that a crop of soybeans fixed 234 kg N per hectare. In an experiment with soybeans in Thailand, Rennie et al. (1988) recorded 643 kg N fixed per hectare in one particular treatment. Values quoted by Burns and Hardy (1975) for other legumes include 208 kg N per hectare per year for lucerne, 105–200 kg N per hectare per year for clover, and 169 kg N per hectare for lupin.

Increased plant protein levels and reduced depletion of soil N reserves are obvious consequences of legume N_2 fixation. Less immediate benefits include diversification and improvement of the diet of animals grazing pastures and crop residues, and increased yields from rotational, non-leguminous crops. Ley farming, in which self-regenerating legume pastures are alternated with cereals in regions with a Mediterranean-type climate, combines both features (Cocks et al., 1980).

Consequences of inadequate legume nodulation are often not manifest in yield because the plant appears to compensate by increased uptake of soil N (Herridge et al., 1984). Such exploitation of the reserves of soil N is not indefinitely sustainable. Plant and soil scientists have a responsibility to devise strategies for legume cultivation that optimize N_2 fixation, conserve soil N and, indeed, augment the pool of soil N for the benefit of rotational crops. For soybeans, this objective is within reach. (Table 1 illustrates how levels of soil N were maintained for three years of double-cropping, with oats and well-nodulated, abundantly N_2-fixing soybeans, during which time >630 kg N per hectare was removed in harvesting the crops.) Data presented by Brockwell et al. (1985) suggest that optimum nodulation and N_2 fixation are functions of early colonization of the plant rhizospheres by *Bradyrhizobium japonicum*. Other information obtained in one of our laboratories (see later) indicates that the extent of rhizosphere colonization and, therefore, N_2 fixation can be enhanced by method of inoculation, manipulation of inocula, soil nitrate and soil moisture, and by choice of soybean cultivar. There is reason for confidence that, by using species with the abilities to nodulate vigorously and to fix N efficiently combined with simple, sensible agronomic strategies for cultivation of grain legumes, it is feasible to grow crops that are both highly productive of seed and contributory to reserves of soil N. We will examine this proposition within this contribution.

Taxonomic and nutritional considerations relating to rhizobial ecology

Taxonomic

For many decades the taxonomy of the Rhizobiaceae received only sporadic attention, e.g. Graham (1964). Then, the early 1980s saw the separation of the legume root-nodule bacteria into two genera, fast-growing *Rhizobium* and slow-growing *Bradyrhizobium* (Jordan, 1982; Kreig and Holt, 1984). Species formerly known as *R. leguminosarum*, *R. trifolii* and *R. phaseoli* were combined as *Rhizobium leguminosarum* and accorded the status of biovars: *viciae*, *trifolii* and *phaseoli*, respectively. The most recent listing, *Index of the Bacterial and Yeast Nomenclatural Changes* (Moore and Moore, 1992), records four genera of stem- and root-nodulating rhizobia*. They are *Rhizobium*, six species; *Bradyrhizobium*, a single species, *B. japon-*

Table 1. Effect on soil N of 3 years of double-cropping with oats and well nodulated soybeans. (Data for a grey clay soil at Trangie, New South Wales; Brockwell and Peoples, unpubl.)

	N removed in biomass of oats and in seed of soybean (kg ha^{-1})	Residual soil N	
		Total N (g kg^{-1})	Mineral N (mg kg^{-1})
Prior to oat crop 1		1.38	30.3
Oat crop 1	107.9	1.24	3.4
Soybean crop 1	174.2	1.32	14.4
Oat crop 2	20.8	1.29	4.3
Soybean crop 2	156.6	-	-
Oat crop 3	33.5	1.18	6.9
Soybean crop 3	137.2	1.34	6.5
Total for 6 crops	630.2		

icum, but with many other incompletely named phena such as *Bradyrhizobium* sp. (*Lupinus*); *Azorhizobium*, one species; and *Sinorhizobium*, two species. Since the preparation of that list, identification of significant genetic diversity in *B. japonicum* led Kuykendall et al. (1992) to propose a new species of *Bradyrhizobium* - *B. elkani*. Perhaps more significant is the proposal that phototrophic rhizobia from the stem nodules of *Aeschynomene* spp. be allocated genus status - *Photorhizobium thompsonianum* (Eaglesham et al., 1990). Now, however, an extensive study by So et al. (1994) (and changes to the rules promulgated by the International Committee responsible for the taxonomy of the rhizobia, e.g. Graham et al., 1991), has indicated that photosynthetic rhizobia belong to *Bradyrhizobium* and that the generic name *Photorhizobium* is unnecessary. Likewise, doubt has been cast on the validity of *Sinorhizobium*; by many criteria it is almost indistinguishable from *Rhizobium meliloti* (Jarvis et al., 1992) and there appears no doubt that it should be classified as *Rhizobium*.

The integrity of the traditional cross-inoculation group classification has long been questioned (e.g. Wilson, 1944) and is now in general disrepute. This is largely because of the increasing recognition of high levels of rhizobial specificity within groups of legumes once regarded as symbiotically homogeneous (e.g. Thies et al., 1991a) and because of the many anomalies in which the same host has the capacity to form nodules with strains of both *Rhizobium* and *Bradyrhizobium*. For instance, strains of both genera nodulate *Acacia longifolia* and *Kennedia prostrata* (Lawrie, 1983), other *Acacia* spp. (Dreyfus and Dommergues, 1981b) and *Prosopis glandulosa* (Jenkins et al., 1987). Soybean is nodulated by both *B. japonicum* and *R. fredii* (Dowdle and Bohlool, 1985), and Gault et al. (1994b) found both fast- and slow-growing rhizobia which nodulated and fixed N with four species of *Lotus* and with *Chamaecytisus palmenis*. There are many other examples. Trinick (1980) isolated a fast-growing strain of *Rhizobium* sp., NGR234, from *Lablab purpureus* which is symbiotically effective for many tropical legumes. A remarkable observation about NGR234 was that 29 companion isolates from *L. purpureus* were all slow-growing. This strain represents a genetic tool for investigations of host-range specificity (Bassam et al., 1986; Broughton et al., 1986; Horvath et al., 1987; Nayudu and Rolfe, 1987) and fascinating material for ecological studies of the interactions between *Rhizobium*, *Bradyrhizobium* and other members of the soil micro-community.

Nonetheless, the range of the symbiotic relationship between specific legumes and rhizobia, when used prudently, can play a role in systematic classification of both plants and bacteria. This proposition, termed symbiotaxonomy, was conceived by Norris (1959) who considered that the symbiotic characters, nodulation and N_2 fixation, expressed during the interaction between African species of *Trifolium* and

* We intend to use the collective nouns "rhizobia" and "root-nodule bacteria" to refer to strains of *Rhizobium*, *Bradyrhizobium* and *Azorhizobium*

strains of *R. leguminosarum* bv. *trifolii* could be a useful taxonomic tool. Application of the concept to the classification of biovars within *R. leguminosarum* is now a reality (Kreig and Holt, 1984). A spectacular example of the association between symbiotaxonomy and legume taxonomy was the reclassification of the complex genus formerly recognised as *Phaseolus*. Legume bacteriologists had long been aware of paradoxes within the complex: different species had fast- and slow-growing root-nodule bacteria; there were marked host plant × bacteria specificities in nodulation and N_2 fixation. These have now been partially resolved. Verdcourt (1970) recognized several new genera and species within the old *Phaseolus* complex, some nodulated by fast-growing *Rhizobium* spp., others by slow-growing *Bradyrhizobium* spp. Further resolution of the symbiotaxonomic complexities within this broad host-bacterial association has been achieved by Somasegaran et al. (1990).

A curious unresolved example of a systematic relationship, expressed as the capacity to fix N, between species of *Medicago* and strains of *R. meliloti* is shown in the matrix in Table 2. Each element of the matrix represents an association between a host species (or 2) of *Medicago* and a strain of *R. meliloti* (which itself is representative of a group of strains of similar symbiotic behaviour). The *Medicago* spp. are arranged on the matrix in (descending) order of increasing specificity in ability to fix N with the range of rhizobial strains. *M. sativa* and *M. minima* are the least specific of the *Medicago*, fixing N with 10 out of 11 groups of strains; *M. rugosa* and *M. laciniata* are the most specific and fix N with only one group of strains. Likewise, specificity amongst the strains of *R. meliloti* in their capacity to fix N with the *Medicago* hosts increases from left to right across the matrix. The phenomenon is not understood.

Taxonomic studies of rhizobia have been predominantly laboratory-oriented with little emphasis given to the behaviour of the organisms in their natural habitat, i.e. the soil and, more specifically, the rhizospheres and rhizoplanes of plants growing in the soil. Except for nodulation and N_2 fixation capabilities, very little of the abundant physiological information about the root-nodule bacteria has been useful in agriculture. We might speculate, however, whether there are discriminatory taxonomic features which induce differential behaviour of rhizobia in peat-based inoculant carriers, on legume seed surfaces, in early colonization of an emerging root, or in ability to colonize soils in the presence or absence of the rhizospheres of host or non-host legumes or non-leguminous plants. Finally, it is well worth considering whether there are physiological traits which relate to competitive nodulating attributes.

There has been a growing neglect of the ecology of legumes as scientific endeavour focuses increasingly on a restricted number of agriculturally proven species. As Bushby (1982) has emphasized, this has been accompanied in ecological studies of the root-nodule bacteria by a corresponding lack of concern for the fundamental characteristics of the soil environment whence the bacteria originated or into which they are being introduced. Further complicating the issue until recently, site-specific factors affecting the occurrence and symbiotic performance of rhizobia (Singleton et al., 1992; Woomer et al., 1988) have mostly been overlooked. Chemical, physical and morphological diversity of soils, soil temperature and soil moisture regimes, and composition of native leguminous and non-leguminous flora are often ignored. Given the impossibility of assembling truly representative collections of soil microorganisms, we submit that most ecological studies of *Rhizobium* or *Bradyrhizobium* are based on findings generated from a poorly defined fraction of the genetic diversity that exists in these genera as a whole.

For many years, there has been intensive screening of isolates of different phena of *Rhizobium* and *Bradyrhizobium* for tolerance of environmental stress. Differences have been reported in ability to tolerate acidity, excess aluminium, high temperature, low phosphate and moisture imbalance (Cassman et al., 1981; Date and Halliday, 1979; Fuhrman et al., 1986; Keyser and Munns, 1979; Keyser et al., 1979; Mahler and Wollum, 1980; Munevar and Wollum, 1981). Yet there have been few critical assessments of whether the differences correlate with nodulating and saprophytic success under the field conditions whence the isolates originated and those that have been conducted tend to be confusing. For instance, strains tolerant of low pH may (Graham et al., 1982; Hartel and Alexander, 1983) or may not (Gemell and Roughley, 1993; Howieson and Ewing, 1989) survive in acid soil better than less acid-tolerant strains. Richardson and Simpson (1989) observed that the proportion of isolates that would grow on an acid medium was unrelated to the acidity of the soil from which the rhizobia had been isolated. To the best of our knowledge, only one systematic survey of stress tolerance has been conducted on collections of root-nodule bacteria isolated from endemic legumes growing in soils of different but well-defined

Table 2. Groupings of species of *Medicago* and strains of *Rhizobium meliloti* according to the capacity of each association to fix N (+ = at least some N_2 fixation; - = nill N_2 fixation, in some cases no nodulation). Data after Brockwell and Hely (1966) and Hebb and Brockwell (unpubl.). Specific names of *Medicago* according to Heyn (1963)

Medicago species	*R. meliloti* strain group										
	1	2	3	4	5	6	7	8	9	10	11
sativa, minima	+	+	+	+	+	+	+	+	+	+	-
disciformis	+	+	+	+	+	+	+	+	+	-	-
turbinata	+	+	+	+	+	+	+	+	-	-	-
blancheana	+	+	+	+	+	+	+	-	-	-	-
truncatula, soleirolii	+	+	+	+	+	+	-	-	-	-	-
granadensis	+	+	+	+	+	-	-	-	-	-	-
murex	+	+	+	+	-	-	-	-	-	-	-
intertexta, aculeata	+	+	+	-	-	-	-	-	-	-	-
polymorpha	+	+	-	-	-	-	-	-	-	-	-
rugosa	+	-	-	-	-	-	-	-	-	-	-
laciniata	-	-	-	-	-	-	-	-	-	-	+

chemical and physical characteristics. Howieson and Ewing (1986) showed that strains of *R. meliloti*, recovered from nodules on annual medics growing in acidic soils in Sardinia, were better than strains isolated from alkaline soils in Iraq in terms of second-year nodulation of *Medicago polymorpha* and *M. murex* growing in an acidic loamy sand. A start has been made in our laboratories to evaluate the "success" of a field isolate in its site of origin in terms of the proportion of nodules it occupies relative to other resident root-nodule bacteria or of its saprophytic competence measured as population density and distribution through the soil profile relative to other rhizobia.

Some of these problems are intractable. The difficulty of obtaining a true cross-spectrum of the population of rhizobia in any soil can be illustrated by considering the limitations of the traditional "trap-host" technique. A single leguminous annual possesses a root system capable of exploring between 1 and 5 kg of soil which might contain 1×10^8 - 1×10^9 cells of rhizobia able to nodulate that particular host. Assuming that 100 nodules form on the root system it is obvious that only a minute fraction of the total population of rhizobia in the soil can be trapped by traditional methods. Moreover, host effects sometimes influence which components of populations of root-nodule bacteria in soil form nodules (e.g. Masterson and Sherwood, 1974; Robinson, 1969). Thus, conclusions about relationships between physiological characteristics of particular nodule isolates and the saprophytic competence of the soil rhizobial population as a whole will be only weakly founded. The development of selective media for direct isolation of *Rhizobium* and *Bradyrhizobium* from soil has received much attention (Bromfield et al., 1994; Bushby, 1981; Gault and Schwinghamer, 1993; Graham, 1969; Kinkle et al., 1994; Nutman, 1973; Tong and Sadowsky, 1994), but more reliable techniques are needed if we are to understand whether differences observed in physiological/metabolic traits under laboratory conditions are of any significance to saprophytic competence in the field.

Despite the differences between *Rhizobium* and *Bradyrhizobium* being sufficient to separate the genera taxonomically, few studies have focused on their comparative saprophytic behaviours under the same soil conditions. The work that is best documented has been summarized by Bushby (1982) and Parker et al. (1977). Briefly, a strain of *Bradyrhizobium* sp. for lupin persisted better in soil than strains of *R. leguminosarum* bv. *trifolii*; in the case of the latter, strain differences also occurred. Bushby and Marshall (1977) hypothesized that the higher internal water content of strains of *Rhizobium* compared with strains of *Bradyrhizobium*, when exposed to identical vapour pressures, prevented the former from reaching the level of dormancy required to remain viable. Other studies comparing the 2 genera reveal a more complex picture (Jansen van Rensburg and Strijdom, 1980; Mahler and Wollum, 1981). For example, Woomer (1990) considered there were fundamental differences between the genera in adaptation to prevalent conditions of stress: *Rhizobium* spp. were intolerant of highly weathered soils of low pH but moderately adapted to semi-arid environments; *B. japonicum* was well suited to soils of the humid tropics and intolerant of high temperature and desiccation. When desiccated rhizobia were reconstituted, Mary et al. (1994) observed loss of effectiveness in *B. japonicum* strains but not in strains of *R. meliloti*. Many aspects of these contentious issues might be clarified by conducting a systematic study of the water relations of both genera indigenous to the same soils.

Nutritional

Parker et al. (1977) raised the issue of the source of nutrients available to root-nodule bacteria in soil and its influence on symbiotic competence. Researchers characterizing carbon and N nutrition of rhizobia rarely consider the nutrient most prevalent in soil and in plant rhizospheres. For instance, legume exudates contain a higher proportion of amino acids than other small molecular weight compounds (Gaworzewska and Carlile, 1982; van Egeraat, 1975), yet metabolism of amino acids as sources of carbon and energy is rarely studied. There has been some interest in simple aromatic molecules as carbon and energy sources for rhizobia (Glenn and Dilworth, 1981) and this is germane to the discovery that exposure to flavonoid molecules of host plant origin results in *nod* gene expression in *Rhizobium* species (Firmin et al., 1986; Peters et al., 1986; Redmond et al., 1986; Zaat et al., 1987).

There is an emerging picture of the diverse capacity within *Rhizobium* and *Bradyrhizobium* to metabolize a range of aromatic compounds. Glenn and Dilworth (1981) showed that *R. leguminosarum* bv. *viciae* strain 3841 had an inducible system for catabolizing p-hydroxybenzoate and other specific aromatics, indicating the possession of a protocatechuate (3,4 dihydroxybenzoate) degradative pathway. Strain 3841 co-metabolised p-hydroxybenzoate simultaneously with glucose or succinate (Dilworth et al., 1983). Since then differences between species and between strains within species have been noted. Chen et al. (1984) found that *R. leguminosarum* bv. *trifolii* strain TA1 possessed both catechol and protocatechuate degradative pathways. Parke and Ornston (1984) showed that strains of *Bradyrhizobium* grew readily on many aromatic compounds (but not catechol); 2 out of 4 strains of *R. leguminosarum* bv. *trifolii* and both of 2 strains of *R. leguminosarum* bv. *viciae* grew on catechol. Enzymes of the protocatechuate branch of the β-ketoadipate pathway are constitutive in *Bradyrhizobium* and inducible in *Rhizobium* (Park and Ornston, 1986). Parke et al. (1985) found that *B. japonicum* strain 1–110 was constitutively chemotactic, and *R. leguminosarum* bv. *trifolii* weakly chemotactic, to aromatic compounds.

It is useful to speculate whether these differences in metabolism of aromatics might have some significance to inoculant technology, to bacterial response to aromatics of plant origin or to saprophytic competence and persistence in soil. In manufacturing inoculants, a period of "curing" (maturation) after addition of broth culture to peat carrier improves the quality of the product. Proliferation of rhizobia occurs during curing (Roughley and Vincent, 1967) and improved survival of rhizobia on seed is a consequence of curing (Burton, 1976; Thompson, 1980). Somasegaran and Halliday (1982) reported amplifications of strains of *R. leguminosarum* bv. *phaseoli* and *B. japonicum* from 1×10^4 to 1×10^{10} cells per gram of peat when the broth culture was diluted 1000 times with water prior to its incorporation into the peat. These findings pose questions about what kinds of nutrients are used by the bacteria in the peat, about the enzymatic pathways and genes expressed for nutrient utilization under these conditions, whether the strain and species differences in aromatic metabolism alluded to above have any bearing on the success of the organism in colonizing peat, and about the physiological state of the cells when they are applied to the seed or introduced into the seed bed. Were the composition of peat (or other) carrier material better characterized, it might be feasible to manipulate its content of particular aromatic compounds. Such endeavours may lead to acclimation of inoculant rhizobia prior to their introduction to the legume seed coat, soil and plant rhizosphere.

The discovery of the relationship between plant-derived compounds and the activation of nodulation (*nod*) genes in rhizobia is relevant to this theme of aromatic metabolism. Specific flavonoid compounds recovered from different legumes cause *nod* gene expression via interaction with the *nod*D gene product (Peters et al., 1986; Zaat et al., 1987). Aromatics *antagonising* induction of *nod* genes have also been identified in pea and clover root exudates (Djordjevic et al., 1987; Firmin et al., 1986). Although specific flavonoids seem to dominate in particular legume species, no host specificity appears to exist because exudates from various legumes activate common *nod* gene expression in non-nodulating rhizobia. For instance, Horvath et al. (1987) reported that *nod*D1 gene from the wide host-range *Rhizobium* strain MP1K3030, a derivative of NGR234 (Trinick, 1980), can interact with plant factors from both *Macroptilium atropurpureum* and *Medicago sativa* to cause common *nod* gene expression.

Despite technological developments that ensure that peat inoculants contain large populations of viable, symbiotically competent cells (Roughley and Vincent, 1967), we have little understanding of the state of symbiotic readiness of these cells at the time of inoculant use. The rhizobia in peat inoculants are exposed to a complex array of aromatic-like compounds, including flavonoids, some of which might be inducers or antagonizers of *nod* gene expression. This might be advantageous or deleterious depending on the relative abundance of the two classes of compounds and on the aromatic transforming capacities that the particular rhizobia possess.

Summary

There is still a need to generate ecologically-based taxonomic data which has the potential to provide leads towards understanding both the physiology of rhizobia in peat culture and those physiological characteristics which might enhance their colonization of seed, rhizosphere and soil. When comparing physiological characteristics and relating them to taxonomic and symbiotic criteria, research workers must not neglect the soil, the climate or the floral composition of the environment whence the strains of rhizobia originated.

Indigenous and naturalized populations

Studies of indigenous or naturalized (naturally-occurring, resident) populations of root-nodule bacteria in soil have been motivated by ecological and agricultural considerations. Research on rhizobia which nodulate non-agricultural legumes of significance only to the nitrogen cycle has been limited (Jenkins et al., 1987; Lange, 1961; Lawrie, 1983). In the case of agriculture, interest in naturally-occurring populations has been inspired by failure of inoculant strains to form nodules (e.g. Jenkins et al., 1954), by the displacement of inoculant strains as nodule occupants during post-establishment years by the resident rhizobia (e.g. Dudman and Brockwell, 1968; Moënne-Loccoz et al., 1994; Roughley et al., 1976), and by instances of widespread sub-optimal effectiveness of naturally-occurring rhizobia with agricultural legumes (see Bottomley, 1992; Bushby, 1982; Trinick, 1982; Vincent, 1954a, 1974).

Soils of both the temperate and tropical zones almost invariably contain populations of *Rhizobium* and/or *Bradyrhizobium*. There is sometimes semantic argument as to whether these populations should be referred to as indigenous or as naturalized or naturally-occurring. Where agriculture has been practised, there is little doubt that the population of soil rhizobia is composed partly of naturalized strains initially introduced deliberately as inoculants or accidentally as contaminants of seed or hay or by vectors such as man, animals and machinery. On the Australian continent, legumes of the tribes Vicieae and Trifolieae are not endemic and it is logical to suppose that their rhizobia are not endemic either. Therefore, naturalized is the correct term to apply to populations of the biovars of *R. leguminosarum* and *R. meliloti* which now occur in Australian soils. In the United States, *B. japonicum* falls into the same category. Mostly, however, situations are more complex with rhizobial populations being in part a legacy of the contribution of past or present endemic leguminous species. Regardless of terminology, the immediate significance of these populations relates to their impact on the productivity of leguminous crops or pastures.

Pioneering studies in Australia (see Vincent, 1954b, 1962) with medic and clover rhizobia established that strains in nodules on different plants separated by only short distances, or even within different nodules on the same plant, could be diverse. On the other hand, strains related antigenically were recovered from different locations on the Australian continent (Vincent, 1954b). The most extensive information available on the composition of populations of soil rhizobia deals with *B. japonicum* in the United States. With the benefit of a large collection of antisera (Date and Decker, 1965), several surveys showed diverse serogrouping of *B. japonicum*, with serogroup 123 dominating in nodules of soybeans growing in the north-central states (Ham, 1980; Schmidt and Robert, 1985). It is a matter of debate whether nodule dominance by strains of serogroup 123 limits N_2 fixation and restricts soybean yields. Suffice it to say that the type strain of the serogroup, USDA123, is symbiotically inferior (Caldwell and Vest, 1970) and that it has proved impossible to displace strains of serogroup 123 from the occupancy of nodules by inoculating with more effective strains at rates which can be practically and economically achieved with available technologies (Johnson et al., 1965; Weaver and Frederick, 1974)

Reasons advanced for the superior nodulating capacities of serogroup 123 are complex (Triplett and Sadowsky, 1992) and sometimes contradictory. Kamicker and Brill (1986) and Schmidt et al. (1986) confirmed the immunological heterogeneity of *B. japonicum* serogroup 123 reported previously (Gibson et al., 1971). Immunofluorescence studies provide no evidence that members of serogroup 123 are more saprophytically competent or respond more rapidly to the presence of a host rhizosphere than other serogroups (Schmidt and Robert, 1985). In soil free of *B. japonicum*, an inoculum of strain USDA123 competed poorly with an inoculum of strain USDA110 for nodulation of soybean (Kosslak and Bohlool, 1985). Streeter et al. (1991) suggested that nodular polysaccharides (NPS) produced by USDA123 confer enhanced persistence on the strain leading to its numerical dominance in soil and, subsequently, its competitive advantage in forming nodules. They later confirmed (Streeter et al., 1994) that NPS production is common among serotypes that dominate in forming nodules in the field. Numerical considerations also appear to be involved in other environments. Work in South Africa and Australia, in soils previously free of *B. japonicum*, shows that the strain first established in a soil is difficult to displace subsequently, either with a different strain or by spontaneous, antibiotic-resistant mutants of the original strain (Brockwell et al., 1987; Jansen van Rensburg and Strijdom, 1985). In addition, there is evidence for the involvement of the host genotype since certain soybean lines introduced into the United States restrict nodulation by members of *B. japonicum* serotype 123 (Triplett and Sadowsky, 1992).

The magnitude, distribution and diversity of different rhizobial components of a soil microflora are difficult to define. The difficulties are magnified by the influence of environmental factors and the host legume in determining which components of the rhizobial population dominate in forming and occupying nodules (Thies et al., 1992). The composition of a collection of naturally-occurring rhizobia can be biased by the species or cultivar of test plant used to trap the isolates or by the environmental conditions under which the test plant is grown. Findings from one of our laboratories show that amendment of an acid soil with lime or monobasic phosphate, or the use of different cultivars of *Trifolium subterraneum* or different *Trifolium* species as trap hosts, results in nodulation by different members of the soil population of *R. leguminosarum* bv. *trifolii* and/or influences the relative nodule occupancy by those rhizobia (Almendras and Bottomley, 1987; Demezas and Bottomley, 1987; Dughri and Bottomley, 1983, 1984; Valdivia et al., 1988). Based on studies at the same site over several years, at least nine major serotypes of *R. leguminosarum* bv. *trifolii* were identified and found to be representatively distributed in a pasture soil. Such antigenic diversity at a specific location has been recorded previously for *R. meliloti* (Hughes and Vincent, 1942) and *B. japonicum* (Johnson and Means, 1963). The relationships between antigenic composition and other indicators of genotypic diversity need more investigation in both local and cosmopolitan collections of rhizobia (Leung et al., 1994a).

Other data are available which indicate that the physical/chemical structure of the soil may influence which component of the population of soil rhizobia forms nodules. Bottomley (1992) remarked upon three studies which suggest a role for clay minerals in determining competitive success among rhizobia in forming nodules on three species of legumes (lentil, May and Bohlool, 1983; leucaena, Moawad and Bohlool, 1984; and chickpea, Somasegaran et al., 1988). The mechanism is improperly understood. Certain characteristics of soil such as electrostatic association between rhizobia and the mineral fraction (Marshall, 1969) and the entrapment of the bacteria within colloids (Demezas and Bottomley, 1986a) may have implications for the effective dispersal, full recovery and accurate enumeration of soil populations. Kingsley and Bohlool (1981) noted that conventional extractants were quite inadequate for recovery of rhizobia from highly weathered soils. Soil flocculation techniques must be modified for accurate enumeration by immunofluorescence of indigenous *R. leguminosarum* bv. *trifolii* from a variety of Oregon soils of similar clay content but different clay mineral composition (Almendras and Bottomley, 1987; Demezas and Bottomley, 1986a). Thompson and Vincent (1967), using intact soil cores from the field, showed that extensive nodulation of *Trifolium subterraneum* might occur even when populations of *R. leguminosarum* bv. *trifolii* were virtually undetectable with a standard soil-dilution, plant-infection procedure. A similar phenomenon was noted for *Lupinus* ssp. grown in soil containing a low-density population of *Bradyrhizobium* (Bottomley et al., 1994).

This whole area is in need of more study; the findings might have significance to soil microbiology in general. A start has been made. Already it is recognized that an understanding of the composition of indigenous soil populations requires an awareness of the legume flora that occur in the area (Woomer et al., 1988). Work from the NifTAL Center in Hawaii has made it clear that rhizobial inoculation is underutilized in tropical agriculture (Singleton et al., 1992), that the size of populations of indigenous rhizobia has a major impact on the establishment and symbiotic performance of inoculant rhizobia (Thies et al., 1991b), and that it is possible to predict the performance of introduced strains using indices of the size of populations of indigenous rhizobia and soil N status (Singleton and Tavares, 1986; Thies et al., 1991c; Turk et al., 1993).

Soil microbial biomass, its role in nutrient cycling and its magnitude as affected by tillage practice, cropping history, fertilizer regime and soil type have received much attention, e.g. Paul and Voroney (1984). In contrast, and despite the fact that the concepts of incursion and saprophytic competence have existed for many years (Chatel et al., 1968; Harris, 1954), it is only recently that rhizobial components of the soil microflora have been enumerated in relation to total soil biomass (Thies, unpubl.). Total microbial biomass follows distinct seasonal trends that mirror each other across sites. Rhizobial populations *do not* follow these same trends (Table 3) and are clearly responding to specific factors, e.g. host. Management practices have clear effects on population sizes (Gault and Brockwell, 1988; Roughley et al., 1995; Thies et al., 1995; Triplett et al., 1993) and a study of their impact on composition and distribution has commenced in one of our laboratories (JET).

We believe that an important role of the *Rhizobium* researcher is to elucidate and solve problems relating to the establishment of legumes and rhizobia in the field in a way that optimizes immediate and continuing N_2 fixation and plant productivity. Only then will the kinds of symbiotic data be generated which are of use to the practicing agronomist. We also believe that the recent literature shows an undue emphasis on controlled environment, monocultural systems where influences of other plant and environmental factors, soil effects and agricultural management practices are ignored. At the same time, agronomists often investigate aspects of legume growth in the field without consideration of the symbiotic functions. The question arises whether the nature of the populations of naturally-occurring rhizobia at a particular site may influence responses obtained from purely agronomic treatments. For instance, applications of lime and phosphate may influence which components of a mixed population of naturally-occurring rhizobia dominate in forming nodules on subterranean clover (Almendras and Bottomley, 1987; Dughri and Bottomley, 1983). Further, phosphate was found to affect the extent of bacteroid transformation by members of a naturally-occurring serogroup with the ability to nodulate extensively (Almendras and Bottomley, 1988). What impact such effects have on the response of crops to nutrient and soil amendment treatments requires further investigation.

Rhizobial genetics as related to ecology

Over the past two decades, the molecular basis of rhizobia-legume genetics has been intensively studied. For some time there has been substantial understanding of the bacterial genes involved in early nodulation (Kondorosi et al., 1985), host range (Rolfe et al., 1985), and regulation and expression of the N_2-fixing process (Ausubel et al., 1985; Hennecke et al., 1985). Moreover, detailed study of host plant genetics has identified numerous plant genes, the early and late nodulins and other plant proteins, that are specifically expressed during the initiation, development and function of mature nodules (Franssen et al., 1992). The flavonoids produced by legumes, and often enhanced in the presence of rhizobia (Recourt et al., 1991; van Brussel et al., 1990), induce transcription of nodulation genes in *Rhizobium* (Firmin et al., 1986; Peters et

Table 3. Percentage of total microbial biomass represented by populations of *Rhizobium leguminosarum* bv. *trifolii* on 3 soils in Western Australia. (Data from Thies, Sparling, Abbott and Milton, unpubl.)

Location	1993		1994		
	Aug.	Nov.	Feb.	May	Aug.
South Sterling (low rainfall)	0.017	0.056	0.002	0.002	0.010
Many Peaks (midrange rainfall)	0.049	0.013	0.004	0.054	0.030
Mount Shadforth (high rainfall)	0.140	0.021	0.011	0.011	0.011

al., 1986; Redmond et al., 1986) and *Bradyrhizobium* (Kosslak et al., 1987). Non-flavone inducers of *nod* genes of legume seed origin have also been reported (Phillips et al., 1993b). The rhizobial nodulation genes are primarily involved in the production of Nod factors which are required for initiation of root hair deformation, cortical cell division and subsequent infection. Hennecke (1990) and Long (1989) have elaborated on these matters. The way is now open for controlled, plant-mediated manipulation of field populations of invasive rhizobia (Phillips et al., 1993a).

Although the promise of enhancing legume yields as a result of genetic manipulation has thus far not materialized, there is one encouraging report - a lucerne response to inoculation with recombinant strains of *R. meliloti* (Bosworth et al., 1994). There are, however, excellent prospects for clarifying certain old, unsolved questions relating to legume agronomy from the new base of genetic knowledge. For instance, the instability of symbiotic characteristics in certain rhizobial strains has been well documented over many years (Herridge and Roughley, 1975; Kuykendall and Elkan, 1976; Labandera and Vincent, 1975; Nutman, 1946; Vincent, 1944, 1954b). Numerous field surveys have shown that many rhizobial isolates recovered from nodulated plants in the field are less effective than inoculant strains (Bottomley and Jenkins, 1983; Gibson et al., 1975; Jansen van Rensburg and Strijdom, 1985; Vincent, 1954a). There has been considerable debate about the reasons for the mediocre effectiveness of field isolates. There is the possibility that the majority of those isolates represent naturally-occurring rhizobia which are symbiotically effective with endemic legumes but which are less compatible with the host lines used for testing purposes and commercial agriculture (Gibson, 1964; Vincent, 1954b). However, it is also possible that the results reflect a genetic instability of the effectiveness trait of an inoculant strain which occurs throughout its progeny after several years in the soil. It has been reported that nodule occupancy by inoculant strains declines in the years following establishment (Dudman and Brockwell, 1968; Moënne-Loccoz et al., 1994; Roughley et al., 1976). Hypotheses to explain these phenomena have been proposed but rely on equivocal evidence. An exceptional case of poor persistence of *R. leguminosarum* bv. *trifolii* strain TA1 was linked to second-year mortality of *T. subterraneum* in Western Australia (Parker et al., 1977).

Distinguishing between inherent and environmentally-induced genetic instability of a rhizobial strain will be difficult. It has been shown often that single colony isolates from parent cultures maintained in laboratory storage display delayed nodulation, variable symbiotic effectiveness and even lack of invasiveness (Gibson et al., 1975; Jansen van Rensburg and Strijdom, 1985; Labandera and Vincent, 1975; Vincent, 1954a, 1962). The question of environmentally-induced change is difficult to address with a strain that is known to be inherently unstable. However, strains that are stable in culture do exist (e.g. *B. japonicum* strain CB1809; Gibson et al., 1990). A study of the effect of environment on strain stability will require some knowledge of the particular microsite where the experiment is to be conducted. Djordjevic et al. (1982) and Wang et al. (1986) showed that the passage of rhizobia through the host plant may influence the structure of regions of the plasmid that carry host-range determinants. In the work of Wang et al. (1986), white clover was very discriminating in this regard with strains of *R. leguminosarum* bv. *trifolii* that carried determinants for pea nodulation; 80–90% of reisolates had lost their ability to nodulate peas. Further research is required to pinpoint the exact location in the plant where these effects on the *Rhizobium* genotype are manifest.

It is interesting to speculate how gene exchange among rhizobia might relate to agronomic situations. In the field, pastures frequently contain several different legumes (species or cultivars). For instance, in North America, perennial *Lotus* spp. and *Medicago sativa* are often sown with annual clovers, such as *Trifolium subterraneum*, *T. vesiculosum* and *T. incarnatum*, and *Vicia* spp.; other legumes, such as *T. dubium*, *Lupinus* spp., *Lathyrus* spp. and *Medicago* spp. may volunteer in the pasture unpredictably. In other situations, grain legumes are grown in rotation with pastures. Under these circumstances, legumes are confronted with a diversity of strains of root-nodule bacteria and opportunities for legume-mediated genetic change among the rhizobia or genetic exchange between rhizobia and other types of bacteria surely exist. Already Kinkle and Schmidt (1991) and Kinkle et al. (1993) have reported, respectively, plasmid transfer between populations of *R. leguminosarum* bv. *viciae* and *B. japonicum* in non-sterile soil.

However, if genetic change occurs within nodules, one may ponder what proportion of the rhizobia in the nodule is represented by variant forms and whether the variants will persist sufficiently well in soil to act as successful inocula in forming new nodules. The potency of nodules as a source of genetic variants will also depend on the number and size of nodules, the proportion of the nodule content which is viable, and the extent of development and viability of bacteroids. Zhou et al. (1985) found that well-differentiated clover bacteroids were not capable of proliferation whereas soybean bacteroids were viable. Moawad et al. (1984) observed large populations of *B. japonicum* in the soil during decay of nodules on field-grown soybeans and Kuykendall et al. (1982) considered that rhizobia released from decaying nodules were a potent source of soil colonization. Bushby (1984) made similar observations on populations of rhizobia in the rhizospheres of *Vigna* spp. Thies et al. (1995) recorded significant host-specific enrichment of soil with *Bradyrhizobium* spp. in fields cropped with *Vigna unguiculata*. The question whether the nodule is a site conducive to the instigation of genetic change remains unanswered.

Naturally-occurring genetic exchange between rhizobia probably depends in part upon the development of large populations in close proximity to each other. The question arises as to the nature of the microsites which support such events. Dual (or multiple) nodule occupancy by two (or more) strains may enhance the possibility of recombination. Nodules of legumes grown in the field may be extensively co-occupied by antigenically distinct strains (Almendras and Bottomley, 1987; Demezas and Bottomley, 1986a; May and Bohlool, 1983; Renwick and Jones, 1986; Thies et al., 1992).

The role of the rhizosphere in mediating genetic exchange between rhizobia (and other bacteria) is difficult to assess. Plant rhizospheres are highly variable and strongly influenced by environmental conditions. Some plant and soil factors may be significant.

1. Root hairs: Variation has been recorded (Itoh and Barber, 1981; Nutman, 1959) in the number and length (per unit area of root) of legume root hairs. This might have impact upon the surface area of root and the mass of soil under rhizosphere influence. Soil moisture and phosphate status influence the extent of root hair development (Mackay and Barber, 1985).
2. Nutrient supply: The extent of rhizosphere exudation varies with the condition of the root environment (Bowen, 1980). This would affect the physiological activity of populations of rhizobia inhabiting the rhizosphere. It is possible to visualize rhizosphere conditions which might be apt for genetic transfer between bacteria. Even were such events to occur, their frequency might be so low that any recombinants would soon be numerically overwhelmed by other bacteria.

There is considerable debate about the extent to which genetic exchange might occur between rhizobia in soil. The concept of the biovars of *R. leguminosarum* having a similar chromosomal background was placed on a sound genetic basis by the transfer of symbiotic plasmids, and therefore host nodulating capacities, from one of those fast-growing biovars to another (Johnston and Beringer, 1977; Sadowsky and Bohlool, 1985). Young (1985) has shown with multilocus enzyme electrophoresis that isolates from nodules on clover, pea and bean growing in the same site can carry the same enzyme profile; this suggested a common chromosomal background with host nodulation determined by different *sym* plasmids. Young et al. (1987) identified the same chromosomal backgrounds at a second field site several miles distant from the first, and suggested that genetic exchange was infrequent relative to the rate of migration between populations. However, recent evidence indicates the situation to be more complex. Souza et al. (1994) present evidence that gene exchange is frequent among local soil populations of *R. etli*, and that migration between populations is limited. Schofield et al. (1984) noted that different *sym* plasmids could be identified in a collection of isolates of *R. leguminosarum* bv. *trifolii* made from plants of *Trifolium repens* growing in one small location. One isolate possessed a "recombinant" *sym* plasmid that contained *nif* genes from another family of *sym* plasmids. This particular isolate was more effective than its companions with each of the three species of *Trifolium* on which they were all tested. Broughton et al. (1987) found that transconjugation of *sym* plasmid pJB5JI from *R. leguminosarum* bv. *viciae* strain T83K3 to a non-nodulating mutant WL113 of *R. meliloti* occurred in alfalfa rhizospheres allowing nodulation to occur. Curiously, and undescribed previously, most alfalfa nodules were co-inhabited by both the unmodified *R. leguminosarum* bv. *viciae* and *R. meliloti* without evidence of transconjugation having occurred.

Research in this area may shed light on observations made about plant performance and strain behaviour.

1. Deterioration of symbiotic capabilities: Jenkins and Bottomley (1985) obtained a response to the application of nitrogenous fertilizer to alfalfa growing in a soil of limited depth and low in plant-available N and predominantly nodulated by strains of mediocre effectiveness. As soil N status is often elevated as a result of legume growth, the contribution of soil N to legume N budgets might, in the short term, alleviate the limitations of mediocre N_2 fixation. This is more likely to occur with pasture legumes, where removal of N from the ecosystem in plant and animal products is relatively slight, than with grain legumes where up to 60 kg N per hectare may be removed in every metric ton of seed harvested (Peoples et al., 1995a).
2. Loss of nodulating capacity: Brewin et al. (1983) concluded that the ability of *R. leguminosarum* bv. *viciae* to colonize rhizospheres was a chromosomal trait. Accordingly, variants of inferior effectiveness but unaffected in nodulating capacity might compete with fully effective parents for nodulation of a host. Working with a line of *Trifolium pratense* of limited root hair infectability and *R. leguminosarum* bv. *trifolii* strain A and its non-nodulating mutant Bart A, Nutman (1965) noted that the relative abundance of the two strains in the rhizospheres determined the extent of nodulation. If this were a common phenomenon, variants would need to occur at relatively high frequency to compete successfully against the parent. In this connection, Segovia et al. (1991) isolated numerous strains of non-nodulating *R. etli* (*R. leguminosarum* bv. *phaseoli*) contributing to a total soil population of *R. etli* in Mexico.
3. Adaptability of rhizobia: Many legume breeders still consider rhizobia to be of minor importance. Indeed, in Australia, N_2 fixation is not one of the six major agronomic criteria that together qualify a legume for registration as a cultivar. Choice of species or cultivar is a dynamic aspect of modern agriculture. Economic conditions and disease susceptibility persuade growers to change their legume species or cultivars on a regular basis. The questions of adaptability of rhizobia to one particular crop and the subsequent impact of this population on a new legume introduction are rarely addressed. A few instances have been documented (Ireland and Vincent, 1968; Vincent, 1954a) where populations of *R. leguminosarum* bv. *trifolii*, originally established in the presence of perennial clover species, were ineffective on annual clovers introduced later to improve seasonal pasture productivity.

Management economies

The previous sections present, at times, a somewhat sombre picture of the state of the art of *Rhizobium* and *Bradyrhizobium* ecology. While it remains true that this area of N_2 fixation research has been somewhat neglected in comparison with other fields (Jones, 1991; Postgate, 1984) there now exists enough basic information to identify those situations where manipulation of rhizobia microflora might improve crop productivity and soil fertility.

A principle of limiting factors states that "the level of crop production can be no higher than that allowed by the maximum limiting factor". A similar principle probably applies also to symbiotic N_2 fixation. We do not accept that the symbiotic relationship between legumes and bacteria is an equal one; on the contrary, we regard the plant as the dominant partner. It follows that symbiotic N_2 fixation is strongly related to the physiological state of the host plant. However effective, competitive and persistent a strain of rhizobia might be, it cannot realistically be expected to express its full capacity for N_2 fixation if limiting factors such as plant disease, nutrient deficiency, mineral toxicity (e.g. aluminium, manganese), salinity, unfavourable soil pH, weed competition, inadequate photosynthesis, temperature extremes, insufficient or excessive soil moisture, or the influences of grazing and management practices impose limitations on the vigour of the host legume (Peoples et al., 1995b; Thies et al., 1991b, c). Numerous texts deal authoritatively with the many environmental factors imposing constraints on rhizobial population dynamics, nodulation and N_2 fixation so we intend to dwell only briefly on these matters. Suffice it to say that, in the absence of applied N, the crops and pastures that yield most highly also fix the most N.

Soil moisture

Soil moisture deficiency has a pronounced effect on N_2 fixation because nodule initiation, nodule growth and nodule activity are all more sensitive to water stress than are general root and shoot metabolism (Albrecht et al., 1984; Gallacher and Sprent, 1978; Weisz et al., 1985; Zahran and Sprent, 1986). Excess soil moisture may also restrict N_2 fixation by reducing the supply of oxygen to nodulated roots. Thus, periodically irrigated legumes are prone to troughs in the level of fixation, immediately before watering when the soil is dry and immediately afterwards when the soil is saturated, and peaks between times.

Weed competition

Although weed competition might ordinarily be expected to reduce N_2 fixation as a result of a reduction in legume vigour, legume/non-legume intercropping can actually increase the level of symbiotic N_2 fixation by the legume and the total N uptake by the joint components of the system (e.g. soybean-maize, Martin et al., 1991; bean-maize, Pineda et al., 1994; ricebean-corn, Rerkasem et al., 1988). This no doubt is the result of uptake of soil N by the corn reducing the amount of soil N available to the ricebean, thereby diminishing the extent of nitrate-induced inhibition of symbiotic N_2 fixation. On the other hand, under some conditions intercropping undoubtedly suppresses legume yield when the plants compete for limited resources.

Soil temperature

Root temperature effects on legume nodulation and N_2 fixation are germane to choice of time of sowing of winter-grown legumes. Optimum temperatures for nodulation are often higher than for N_2 fixation (Gibson, 1963, 1967). Legumes grown at low temperature experience delays in nodule formation and the onset of N_2 fixation, even in the presence of adequate populations of rhizobia. For instance, with subterranean clover plants grown singly in N-free medium, nodulation was substantially delayed and growth rate restricted at root temperatures below 17°C (Gibson, 1967). The threshold root temperature below which clover growth is retarded is probably lower when clover is grown in a sward, especially when nitrate is present (Silsbury, 1984), as would occur in the field. Where soil N and soil temperature are low, seedlings may become so moribund before N_2 fixation commences that they are prone to root diseases (Hely, 1964) and physiological disorders from which they may not recover even when temperatures rise to more favourable levels. It is desirable that, soil moisture permitting, winter-grown legumes be sown before soil temperature falls to a level inhibitory to nodule formation. Further information is needed on nodule development by legumes that are indigenous to regions with cold soils.

Nutrients and fertilizers

Excellent, encapsulated reviews of the nutritional requirements of legumes, both dependent on and independent of symbiotic N_2 fixation, are provided by Munns (1977), O'Hara et al. (1988) and Smith (1982). Although the mineral requirements of the bacteria, the infection process, nodule development and nodule function are less than the requirement of the un-nodulated, N-sufficient plant, there are important exceptions: Some legumes (e.g. *Trifolium subterraneum*, Lowther and Loneragan, 1968; *Medicago sativa*, Munns, 1970; *Neonotonia wightii*, Andrew, 1976), grown in moderately acid medium, need a higher concentration of Ca during infection and nodule formation than is required for subsequent growth of the host plant. There is evidence (Munns, 1977) that the requirement for Bo for nodule development is similar to that for growth of the host. Likewise for nitrogenase (N_2-fixing) activity, there is a need for Mo and Co far in excess of other plant requirements (Evans and Russell, 1971). Nickel is required for the synthesis of functional hydrogenase (Partridge and Yates,

1982). This enzyme is involved in hydrogen cycling (Klucas et al., 1983) which in turn can improve N_2 fixation and growth of soybeans (Evans et al., 1985). It has long been known that P plays a major part in the build-up and maintenance of soil fertility through its effect on legume growth (Donald and Williams, 1954). More recently there has been an accumulation of evidence of a specific effect of P on the growth and survival of rhizobia and their capacity for nodulation and N_2 fixation (expression of differential symbiotic effectiveness - Singleton et al., 1985). This subject has been thoroughly reviewed by McLaughlin et al. (1990). A striking example of the role of adequate P nutrition in enhancing N_2 fixation by soybeans is given by Cassman et al. (1993).

Direct contact with fertilizers, especially acidic ones such as superphosphate, may drastically affect viability of inoculant (Anderson, 1970). Trace elements are sometimes applied to soil by incorporation in seed pellets. Caution is necessary in choosing non-toxic compounds when rhizobia inoculant is also a component of the pellet. For instance, Gault and Brockwell (1980) found that sodium molybdate in coated seed of alfalfa and subterranean clover was deleterious to the survival of *R. meliloti* and *R. leguminosarum* bv. *trifolii* whereas ammonium molybdate and molybdenum trioxide were not.

Soil amendment

Liming is an agricultural practice that has directly beneficial effects on legume growth by improving nodulation during pasture establishment. This results from the addition of calcium and the neutralization of acidity, the immobilization of manganese and aluminium which might otherwise depress nodulation, and the release of sorbed molybdate thereby alleviating molybdenum deficiency (Munns, 1977). Some of these benefits can often be obtained by adding small amounts of lime to inoculated seed as a pellet or a seed coating (e.g. Brockwell, 1962; Loneragan et al., 1955). This procedure protects the inoculant from contact with acid fertilizers and from inimical environmental factors and may improve rhizosphere colonization by the rhizobia and, as a consequence, nodulation.

Some species of rhizobia tolerate acidity better than others. Norris (1956) generalized that this was merely a reflection of the adaptation of some legumes, mainly tropical, to acid soils and others, mainly temperates, to less acid or alkaline soils. Indeed, the rhizobia for some species adapted to acid soils, e.g. *Lupinus* and *Ornithopus* spp., are killed by seed pelleting with calcium carbonate (Parker and Oakley, 1965). On the other hand, lime pelleting of inoculated clover seed has played a significant role in the establishment of subterranean clover pastures on very large areas of otherwise inhospitable, usually acidic soils in southern Australia (Anderson, 1970).

It is now recognized that soil acidification occurs under subterranean clover pastures (Bromfield et al., 1983; Helyar, 1976; Williams, 1980) when there is no routine top-dressing with lime. This may be associated with substantial reductions in the size of populations of *R. trifolii* (Coventry et al., 1983b). Reduction of soil acidity and associated problems of legume nodulation and N_2 fixation can be achieved by liming (e.g. Coventry et al., 1983a) which is both practicable and environmentally attractive in most circumstances. However, lime is not always readily or conveniently available and costs of transport and application may represent prohibitive impediments to its use. Furthermore, excessive liming of weakly buffered soils may create physical and nutrient disorders which are not easily rectified. Although acid-tolerant strains of rhizobia have been identified (Bromfield and Jones, 1980; Howieson and Ewing, 1986; Howieson et al., 1988; Keyser and Munns, 1979; Keyser et al., 1979; Richardson and Simpson, 1989), there is no unequivocal proof that they possess superior ability to colonize acidic soils.

Synergism

Vesicular-arbuscular (VA) mycorrhizae are able to increase plant growth by acquisition of otherwise immobile nutrients from the soil. Almost without exception, nodulated legumes are mycorrhizal. A relationship between infection by VA mycorrhizae and nodulation and N_2 fixation in legumes was first reported by Asai (1944). The relationship appears to be synergistic. Infection of legume roots by the two organisms seems to occur simultaneously (Barea and Azcon-Aguilar, 1983) without competition between them for infection sites (Smith and Bowen, 1979). Several studies (e.g. Kucey and Paul, 1982; Smith and Daft, 1977) have indicated that VA mycorrhizae may stimulate N_2 fixation independently of any influence on the growth of the host plant, but this effect may be restricted to situations where nutrient supply, in particular P, is limited. Munns and Mosse (1980) and Robson et al. (1981) consider that the responses of legumes to VA mycorrhizae and to phosphatic fertilizer are similar. It has been proposed that the symbiosis between VA mycorrhizae and plants can be managed to increase the recovery of phosphatic fertilizer (Abbott and Robson, 1982, 1987). Interactions between phosphorus uptake, mycorrhizal infection and legume N_2 fixation have been reviewed by McLaughlin et al. (1990).

The subject of dual inoculation of white clover with *Rhizobium* and mycorrhizal fungi was raised by Smith and Daft (1978). Co-inoculation of legumes with rhizobia and mycorrhizae (Mahdi and Atabani, 1992; Thiagarajan et al., 1992), may have application to commercial inoculants (Rice et al., 1995). Other microorganisms besides mycorrhiza (e.g. *Penicillium bilaji*, Downey and van Kessel, 1990; *Pseudomonas fluorescens*, Nishijima et al., 1988) in association with rhizobia have been reported to enhance legume nodulation. Soil invertebrates, collembola, increased nodule occupancy by indigenous strains of *B. japonicum* in soil growing inoculated soybeans (Lussenhop, 1993). Rice et al. (1995) suggest how preparation of co-inoculant might be effected.

Nitrate

Inhibition of nodulation and N_2 fixation processes by inorganic forms of N, mainly nitrate, are well known (e.g. Vincent, 1965). Despite a considerable literature on the subject, the mode of action of nitrate in reducing nodulation is not clearly understood. Munns (1968a, b), for instance regarded it as a complex of effects on root-hair curling, root-hair infection and the rate of development of infected root hairs. Dazzo et al. (1981), on the other hand, implicated the nitrate ion in interference with the expression of the recognition phenomena that are precursors of the infection process. The presence of nitrate in field soils has profound implications for the establishment of an effective symbiosis (e.g. Bergersen et al., 1985; Herridge et al., 1984; Thies et al., 1991c). Indeed, we believe that it is not a wise choice to plant a legume into soil containing significant amounts of available nitrate.

Crop hygiene

Legume root disease represents a serious impediment to legume establishment and stand longevity. *Pythium* spp. are the most common phytopathogens attacking food legumes but *Fusarium* and *Rhizoctonia* are important also (Matthews et al., 1988). Effectual fungicidal seed dressings exist to control these organisms. However, many fungicides are incompatible with rhizobia (Kecskés, 1970) although toxicity levels range from mild to severe (e.g. Corbin et al., 1977). Inoculant delivered directly into the seed bed in solid (e.g. Scudder, 1975) or liquid (e.g. Schiffman and Alper, 1968a, b) form provides a physical separation of rhizobia from fungicide-treated seed and, when properly performed, promotes healthy, well-nodulated seedlings (Brockwell et al., 1980).

Insect attack, foliar disease, virus infection, nematodes and weeds, including particularly the parasitic weed *Orobanche*, all have indirect influences on legume N_2 fixation through their effects in reducing host plant vigour or photosynthesis. *Sitona* weevil has a direct influence because it focusses its predation on the nodules themselves. Invasion of nodules by nematodes was reported by Robinson (1961). Pest control is essential to the maximization of symbiotic N_2 fixation.

Specialized management systems

Rice systems

Lowland rice-based cropping systems, characterized by flooding during most of the rice-growing season, comprise 125 million hectares worldwide (George et al., 1992). There is a high requirement for N in these systems, especially those producing two or more crops per year, much of which is supplied as fertilizer. There is a great potential for replacement of this fertilizer N with fixed N from rotational crops of food legumes, such as cowpea (*Vigna unguiculata*), mungbean (*V. radiata*), pigeonpea (*Cajanus cajan*), groundnut (*Arachis hypogaea*) and soybean (*Glycine max*), and by green manures such as milk vetch (*Astragalus sinicus*), *Sesbania* spp., sunnhemp (*Crotalaria juncea*), berseem (*Trifolium alexandrinum*) and indigo (*Indigofera tinctoria*) (George et al., 1992; Ladha et al., 1988). The rhizobial requirements of these legumes are frequently supplied by resident populations in the soil. Most of the information concerning survival of rhizobia under flooded conditions stems from work with soybeans. Invariably, persistence of *B. japonicum* has been little affected by flooding (Osa-Afiana and Alexander, 1979; Weaver et al., 1987), even for prolonged periods (Roughley et al., 1995). Indeed, it is obvious from the extensive work on the growth of soybean in saturated soil culture (e.g. Hartley et al., 1993; Nathansen et al., 1984; Troedson et al., 1989) that inundation per se has no drastic effect on soybean rhizobia. It is not known to what extent the evidence for tolerance of flooding in *B. japonicum* can be extrapolated to other species. For instance, Boonkerd and Weaver (1982) observed that various strains of rhizobia for cowpeas and peanuts differed in their survival in flooded soil.

Stem nodulation

The existence of legume-stem nodulation has been recognized for many decades. Stem infection by rhizobia appears to occur at root primordia which are distributed on stems of certain aquatic legumes including species of *Aeschynomene*, *Neptunia* and *Sesbania* (Ladha et al., 1990). Stem nodules have been reported also on other legume species but apparently only in circumstances of partial submergence of the stem, a situation that may trigger primordial development and allow access to rhizobia (Alazard and Duhoux, 1987). A single strain of rhizobia may be responsible for both root and stem nodulation on the same plant, but the stem nodules are greater contributors to symbiotic N_2 fixation (Becker et al., 1990; Rinaudo et al., 1988).

Genuine interest in legume-stem nodulation was stimulated by the landmark paper of Dreyfus and Dommergues (1981a). They reported stem nodules on *Sesbania rostrata* formed by an organism later classified as *Azorhizobium caulinodans* (Dreyfus et al., 1988). This bacterium, ORS571 is strongly host specific and appears to nodulate effectively with only one other species, viz. *S. paludosa* (Alazard et al., 1988). There are curious features about the ecology of *A. caulinodans*. The host, and presumably the organism itself, occurs naturally in only a few West African countries. Yet when *S. rostrata* has been introduced into Asia, and other parts of the world, it has nodulated effectively and spontaneously without inoculation (Ladha et al., 1990). It seems that the organism is seed-borne (Ladha et al., 1988). Large populations of *A. caulinodans* have been found on leaves and flowers of *S. rostrata* (Adebayo et al., 1989). Some of these bacteria may be entrapped inside the seed coat during reproductive development and later form a source of inoculum when the seed is sown (Ladha et al., 1988). *A. caulinodans* survives well, in the absence of its host, in flooded soil and in the rhizosphere of rice (Ladha et al., 1989). Strategies for inoculation, where needed, are discussed by Ladha et al. (1992).

Ladha et al. (1990) list a number of species of *Aeschynomene* that have stem nodules formed by rhizobia that produce bacteriochlorophyll and a photosynthetic reaction centre akin to that of photosynthetic bacteria (Eaglesham et al., 1990). For some time the proper classification of these organisms was in doubt but now they are regarded as species of *Bradyrhizobium* (So et al., 1994). The stem-nodulating *Neptunia* symbiont and non-photosynthetic *Aeschynomene* rhizobia are also species of *Bradyrhizobium*.

A remarkable feature of the symbiosis between *S. rostrata* and *A. caulinodans* is the large volume of N fixed (see Table 4). No higher rate of legume N_2 fixation has ever been reported. Ladha et al. (1989, 1992) reacted to this by indicating strategies for the exploitation of stem-nodulating aquatic legumes as green manures and for the capture and use the fixed N by lowland rice. Some *Aeschynomene* species also have the capacity for fixing large quantities of N and, therefore, potential as green manures (Becker et al., 1995).

Leguminous trees

Agroforestry farming systems using leguminous shrubs and trees are attracting increasing attention for alley cropping, biofertilizer, browse, cut and carry, fuelwood and shade. Mainly tropicals but also temperate species are involved. Levels of N_2 fixation vary. *Gliricidia sepium* fixed 99–185 kg N per hectare in 3–6 months (Peoples et al., 1995a); *Faidherbia albida* and *Acacia senegal* may fix less than 20 kg per hectare per year (Sanginga et al., 1995). Turk and Keyser (1992) analysed the symbiotic specificity of 10 tree legumes. They found that species of *Gliricidia*, *Calliandra* and *Leucaena* cross-nodulated effectively with strains of unclassified *Rhizobium*, that *Sesbania grandiflora* and *Robinia pseudoacacia* each required different, specific *Rhizobium* and that *Paraserianthes*, *Tephrosia* and three species of *Acacia* required *Bradyrhizobium* spp. The temperate tree *Chamaecytisus palmensis* appears to nodulate and fix N with both *Rhizobium loti* and *Bradyrhizobium* sp. (*Lotus*) (Gault et al., 1994b). Response to inoculation of tree legumes is a function of the population density of indigenous rhizobia (Turk et al., 1993).

Very often leguminous trees are established in the field by transplanting seedlings nurtured in a glasshouse or nursery. This provides a unique opportunity to apply inoculant under benign conditions and so ensure that the transplants are already vigorously and effectively nodulated at the time of field establishment. Ways and means of achieving this have been suggested by Dommergues (1987), Keyser et al. (1992) and

Table 4. Comparison of field N_2 fixation by *Sesbania rostrata* (grown in the wet season in the Philippines) and soybean (grown under irrigation in summer in Australia). *S. rostrata* data after Pareek et al. (1990); soybean data derived from Bergersen et al. (1985)

S. rostrata		Soybean	
Days after sowing	N_2 fixed (kg ha^{-1})	Days after sowing	N_2 fixed (kg ha^{-1})
25	11	50	3
45	140	78	66
55	286	114	249
65	458		

Roskoski et al. (1986). It should be noted that it is normal prudent practice to grow nursery conifers in mycorrhizal soil to obviate subsequent nutritional problems following outplanting on to production sites.

Inoculation

Used judiciously where needed and performed properly, legume inoculation is a significant agency for the manipulation of rhizobia microflora for improving crop productivity and soil fertility, particularly with soybean (Keyser and Li, 1992).

The need for inoculation

Although species of *Rhizobium*, *Bradyrhizobium* and *Azorhizobium* are, no doubt, as widely distributed as the legumes themselves, there are many soils where suitable strains for rhizobia-specific species are absent, or where the population density is so low as to pose a threat to legume establishment. "Is it necessary to inoculate?" is a question answered in different ways (see Table 5).

Field experiments have been designed to diagnose the need for inoculation (e.g. Bell and Nutman, 1971; Brockwell, 1971; Date, 1977; Thies et al., 1991b) but take at least several months to complete. Bonish (1979), using dilutions of soil samples to inoculate clover seedlings growing aseptically in test tubes, demonstrated a microbiological means for characterizing simultaneously the size and N_2-fixing capacity of soil-borne populations of rhizobia. Brockwell et al. (1988) developed this method into an expeditious assay (28 days duration) which, combined with a serial-dilution, plant-infection technique for enumerating rhizobia (Brockwell, 1963), represents a reliable guide to the need for inoculation in the field. A related procedure (Thies et al., 1991c) makes it possible to forecast the likely success of introducing inoculant rhizobia into the soil by considering indices of the size of resident rhizobial population and the N status of the soil. A unique proposal for predicting the need for inoculation on a regional basis (a case study for Northeast Thailand) using a Geographical Information System has been advanced by Thies et al. (1994).

Inoculant technology

There are a number of major texts dealing with the principles and practice of inoculant preparation and use (e.g. Brockwell, 1977; Burton, 1982; Date and Roughley, 1977; Keyser et al., 1992; Smith, 1992; Somasegaran, 1991; Somasegaran and Hoben, 1994; Thompson, 1980, 1983; Williams, 1984). In one respect, these contributions present significantly different approaches. Some methods involve the use of non-sterile peat carriers for preparation of inoculants whereas both Date and Roughley (1977) and Thompson (1980) indicate a strong preference for a sterile carrier. Indeed, Date and Roughley (1977) state that inoculants prepared with non-sterile peat may contain 100–fold fewer rhizobia than those made with sterilized peat, and that, because mortality of rhizobia is greater in unsterilized peat, this difference increases during storage. Numerous papers attest to the consistently high rhizobial cell densities, in excess of 1×10^9 per gram, achievable in sterile peat (Parker and Vincent, 1981; Somasegaran, 1985; Somasegaran and

Table 5. Diagnosis of the need to inoculate

Allen and Allen (1961) - historical indicators

1. The absence of the same or a symbiotically related legume in the immediate past history of the land;
2. Poor nodulation when the same crop was grown on the land previously;
3. When the legume follows a nonleguminous crop in a rotation;
4. In land reclamation undertakings.

Roughley and Brockwell (1987) - microbiological queries

1. How specific is the legume in its rhizobial requirements ?
2. What is the likelihood of effective rhizobia spreading from volunteer legumes ?
3. Has the legume been sown before and for how many seasons was it grown continuously ?
4. How long since it was last sown and, in the interim, were conditions likely to favour survival of the rhizobia ?

Thies et al. (1991c) - soil indices

1. How large is the resident population of competitive rhizobia ?
2. What is the level of soil nitrogen (nitrate) ?

Halliday, 1982; Strijdom and Jansen van Rensburg, 1981).

Strains of rhizobia suitable for inoculants

Based on considerations by Brockwell et al. (1982), Burton (1979), Date (1982) and Howieson and Ewing (1986), Keyser et al. (1992) list the following characters as desirable for strains for use in commercial inoculants:

1. Ability to form nodules and fix N on the target legume;
2. Ability to compete in nodule formation with populations of rhizobia already present in the soil;
3. Ability to fix N across a range of environmental conditions;
4. Ability to form nodules and fix N in the presence of soil nitrate;
5. Ability to grow well in artificial media, in inoculant carrier and in the soil;
6. Ability to persist in soil, particularly for annually regenerating legumes;
7. Ability to migrate from the initial site of inoculation;
8. Ability to colonize the soil in the absence of a legume host;
9. Ability to tolerate environmental stresses;
10. Ability to fix N with a wide range of host genotypes;
11. Genetic stability;
12. Compatibility with agrichemicals.

Inoculant manufacturers add:

13. Wide host range;
14. Low mortality on inoculated seed;
15. Ability to colonize the rhizosphere of the host plant.

Inoculant production

Commercial production of legume inoculants began 100 years ago in the United States and the United Kingdom. In 1995 they are produced in many countries on all continents. Most are prepared in powdered, organic carriers such as peat which remains the favoured base for inoculants, particularly when rendered sterile by gamma irradiation (Roughley and Vincent, 1967).

Keyser et al. (1992) regard the properties of a good inoculant carrier as:

1. High water-holding capacity;
2. Non-toxic to rhizobia;
3. Easy to sterilize by autoclaving or gamma irradiation;
4. Readily and inexpensively available;
5. Sufficiently adhesive for effectual application to seed;
6. pH buffering capacity;
7. Cation- and anion-exchange capacities.

While most peats meet these criteria, the search for alternative carrier materials continues particularly in countries that have no natural deposits of peat.

Thompson (1980) presented an imposing list of alternative inoculant carriers and it has been augmented since (see Brockwell and Bottomley, 1995). Gels are amongst the more interesting alternatives. Dommergues et al. (1979) reported that an inoculant, in which *B. japonicum* was entrapped in polyacrylamide gel, survived and nodulated soybean as well as a peat-based inoculant. The concept has been extended to the use of other polymers (Jung et al., 1982). Good survival depends upon the maintenance of moisture in the gels. Jawson et al. (1989) used several cellulose gels as inoculant carriers and reported excellent survival of *B. japonicum* and good nodulation of soybean in sand culture and in the field where there was an existing population of competing strains. Inoculation of seed with fluid gels lends itself to fluid drilling for sowing grain legumes. Kremer and Peterson (1982) prepared inoculants, by resuspending lyophilized cultures of rhizobia in vegetable oil, that survived as well as or better than peat cultures and performed well in field trials (Kremer and Peterson, 1983a). Hoben et al. (1991) reported favourably on oils as adhesives for seed inoculation. Graham-Weiss et al. (1987) fermented inoculant in a vermiculite carrier. Significantly, perhaps, none of these alternatives, except the last-mentioned has found its way into commerce.

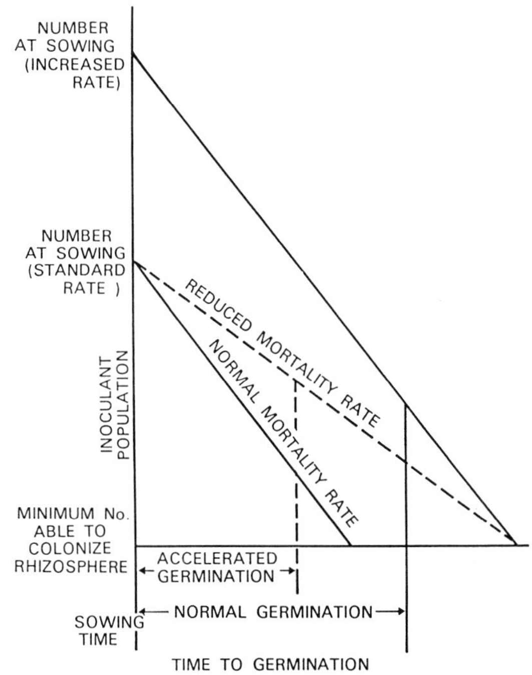

Fig. 1. A schematic illustration of basic factors involved in improving the likelihood of nodulation following rhizobial inoculation of legume seed: (i) reducing inoculant mortality rate, (ii) increasing rate of inoculant application, (iii) decreasing time to germination.

Application of inoculant

A primary aim of legume inoculation is to maximize survival of inoculant during the period between its introduction to the soil and the development of a legume rhizosphere which it can colonize. The principle is illustrated in Figure 1. The literature affirms the significance of high rates of inoculation in achieving this objective and optimizing subsequent nodulation and N_2 fixation (e.g. Berg et al., 1988; Bergersen et al., 1985; Brockwell et al., 1987; La Favre and Eaglesham, 1984; Lowther and Littlejohn, 1984; Nambiar et al., 1983; Smith et al., 1980; Somasegaran et al., 1988; Turk et al., 1993; Weaver and Frederick, 1974; Wedderburn, 1986). Application of inoculant to the seed surface prior to sowing is the traditional, most commonly used and most user-friendly means of inoculation, although viability of the rhizobia is subject to the hazards of drying (Salema et al., 1982), fertilizer contact (Kremer et al., 1982), seed coat toxicity (Materon and Weaver, 1984), incompatible pesticidal and mineral additives (Gault and Brockwell, 1980; Skipper et al., 1980) and inimical soil factors (Kremer and Peterson, 1983b; Mahler and Wollum, 1982). Proposals to extend the life expectancy of rhizobia on seed, including curing (storage at 20–27°C for up to four weeks) inoculants before use (Burton, 1976; Materon and Weaver, 1985) and suspending cultures in alginate gel rather than sucrose before application to seed (Rawsthorne and Summerfield, 1984), have not been adopted. There are numerous adhesives suitable for attaching inoculant to seed (e.g. Brockwell, 1962; Elegba and Rennie, 1984; Hoben et al., 1991). Tenacity is an important characteristic of adhesives to ensure that inoculant is not lost from the seed during handling and passage through sowing machinery. Naturally, inoculant adhesive must be free from any preservative that might diminish the viability of rhizobia.

There are some situations where seed application of rhizobia may be an inefficient means of inoculation, e.g. with seed dressed with a pesticide incompatible with rhizobia; for inoculation for broad-acre sowing of crop legumes with high seeding rates; for seeds such as peanut which are too fragile for seed-surface inoculation (Brockwell, 1982). Preparations and procedures for inoculant application directly into the seed bed are now in vogue, viz. solid inoculant ("soil implant")

(Barkdoll et al., 1983; Hegde and Brahmaprakash, 1992; Scudder, 1975) and liquid inoculant (Hely et al., 1980; Schiffman and Alper, 1968a, b). These methods are often better and never worse than conventional seed inoculation for initiating nodulation and N_2 fixation (e.g. Brockwell et al., 1980; Chamber, 1983; Danso et al., 1990a; Hale, 1981; Jensen, 1987; Muldoon et al., 1980; Rice and Olsen, 1992). Solid inoculant may be introduced into the seed bed through an insecticide attachment to the seed drill. Liquid inoculation for large scale sowings requires an inoculant tank, a pump, a manifold, and capillary tubes to deliver the liquid into the seed bed beside and beneath the seed (Brockwell, 1982; Brockwell et al., 1977) - see Figure 2. Less elaborate equipment is also effectual - see Figure 3. Australian operators prepare liquid inoculant as a suspension of peat culture in water. Frozen concentrated broth cultures have been used in the United States (Gault, 1978). Liquid inoculant prepared from peat must be free of fibre, which could block capillary tubes or nozzles, and mineral grit which might damage pumps.

Inoculant has been successfully introduced into rhizobia-free soil under irrigation water-run inoculation (Ciafardini and Barbieri, 1987). Successful nodulation and N_2 fixation appeared to depend on the infiltration characteristics of the soil (Gault et al., 1994a). Post-emergence inoculation of un-nodulated legume stands is sometimes effective (Atkin-Smith et al., 1986; Boonkerd et al., 1984; Danso et al., 1990a, b; Rogers et al., 1982) and sometimes not (Gault et al., 1984). Success apparently depends on environmental conditions at the time of treatment and is more likely in glasshouse experiments than in the field. Legume inoculant retained its viability in hydroseeding mixes (fertilizer, lime, mulch, seed, water), used for revegetation of overburden from surface mining, provided the blend had a pH >6.0 (Brown et al., 1983).

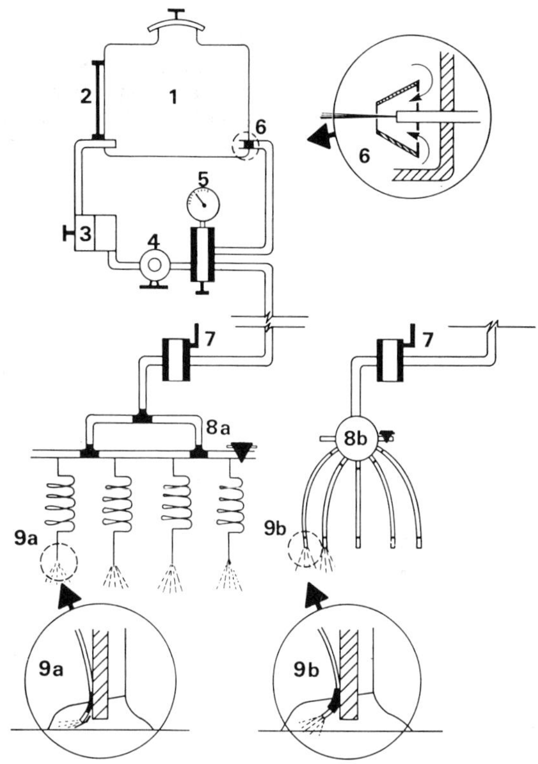

Fig. 2. Diagrammatic representation of a system for delivering liquid inoculant directly into the seed bed.

Preinoculation

Preinoculation of legume seed, i.e. inoculation with rhizobia before sale, excited the interest of inoculant manufacturers, seed merchants and research microbiologists in the 1950s, 1960s and 1970s (Thompson et al., 1975). Inoculation in conjunction with seed coating and impregnation of seed with broth inoculant were the usual processes. Generally the product was disappointing (e.g. Brockwell et al., 1975; Quackenbush et al., 1961; Schall et al., 1975) because of poor survival of rhizobia. A variant of preinoculation is "custom inoculation", i.e. inoculation of seed for sowing within, at most, ten days of treatment. The numerical quality of this product was sometimes erratic (Brockwell and Roughley, 1967) but it found favour with many farmers especially in New Zealand. Although it attracts little attention in 1995, preinoculation that is reliable has a great deal to offer and remains a tantalising concept. Smith (1992) has reviewed many aspects of inoculant application and delivery. He saw as a challenge for the future the provision of improved inoculant carriers that supported high populations of rhizobia, of inoculants with extended shelf-life which gave the rhizobia protection against environment stresses, and of systems which delivered the inoculant on to the seed and/or

Fig. 3. An Indian farmer using a basic system for delivery of liquid rhizobial inoculant directly into the seed bed. The bowl mounted on the lower part of the draw bar, nearer the operator, is the seed box from which seed is delivered into the seed bed through wide-gauge tubing. The container slung from the upper part of the draw bar, nearer the oxen, is the reservoir from which liquid inoculant is dribbled through narrow-gauge tubing into the seed bed where it is deposited alongside and beneath the seed. (Photograph compliments of Peter Dart.)

into the soil in a convenient and cost-effective manner. He made no mention of preinoculation.

Inoculant quality control

Evaluation of inoculant quality by enumeration of viable rhizobia is an accurate index of inoculating potential (Hiltbold et al., 1980). Numerical considerations are of such significance in determining quality of inoculant products and their success in the field that the necessity for quality control systems, exemplified by Roughley et al. (1984), is widely recognized. Some countries (e.g. Canada, Uruguay) have regulatory authorities supported by appropriate legislation; in others (e.g. Australia, India, New Zealand, South Africa) inoculant manufacturers participate voluntarily in quality control schemes (Thompson, 1983). In the United States, regulatory control has not been considered necessary since the 1940s. However, the results of independent tests published from time to time (e.g. Olsen et al., 1995; Schall et al., 1975; Skipper et al., 1980; Vincent and Smith, 1982) indicate that substantial proportions of the inoculants examined appear unsatisfactory for farmer use because of low populations of rhizobia and/or high numbers of microbial contaminants. Nevertheless, it is not clear to what extent legume crop failures in the United States are attributable to poor quality inoculants.

Measurement of success of inoculation

Major criteria for evaluating the success of inoculation are extent of nodulation, proportion of nodules occupied by the inoculant strain, and indices of plant response which may include the total amount of N fixed, the proportion of plant N due to N_2 fixation, and dry matter production. These criteria and their measurement are dealt with comprehensively in Bergersen (1980), Brockwell and Bottomley (1995), Keyser et al. (1992), Peoples et al. (1989), Somasegaran and Hoben (1994) and Vincent (1970). Electrophoretic and molecular techniques have attracted recent attention for characterizing introduced and existing populations of soil rhizobia. MLEE (multilocus enzyme electrophoresis), for instance, has been widely and successfully used in taxonomic, genetic and ecological studies of *Rhizobium* (Demezas et al., 1991; Eardly et al., 1990; Leung et al., 1994b; Segovia et al., 1991; Young, 1985) and *Bradyrhizobium* spp. (Bottomley et al., 1994). PCR (polymerase chain reaction) techniques, as described by Harrison et al. (1992), Welsh and McClelland (1990) and Williams et al. (1990), have been refined by Richardson et al. (1995) into a simple, accurate and expeditious tool for ecological investigations of rhizobia. It is a comfort for workers studying persistence of rhizobial strains that, by and large, most strain identification markers remain stable in soil for several years at least (Lindstrom et al., 1990).

Manipulation of rhizobial populations

Naturally-occurring populations of rhizobia are significant factors determining the establishment of inoculant strains in the field. Indeed, where there are large, competitive resident populations in the soil, inoculation is invariably futile. It is necessary to consider, therefore, how rhizobial populations in the soil can be manipulated to influence, with or without effectual inoculation, legume nodulation, N_2 fixation and plant productivity.

New populations (introduced rhizobia)

Legume bacteria are widespread as a result of the natural distribution of legumes and cultivation of leguminous crops and pastures. Nevertheless, many soils remain devoid of strains of rhizobia for particular crops; for instance, soybean rhizobia do not occur naturally in Australia (Diatloff and Brockwell, 1976). In such situations establishment of new populations is easily achieved and legume bacteriologists have the opportunity to introduce elite strains that will become permanent components of the soil microflora. They also have a paramount responsibility to ensure that such strains are genuinely elite and, once established in the soil, do not become an impediment to subsequent introduction of more desirable strains. *B. japonicum* strain CB1809 (also known as USDA136 and TAL379) exemplifies this concept. It became the recommended Australian inoculant strain for soybean in 1966 (Date, 1969). It is highly effective for soybean, aggressive in colonizing the rhizospheres of its host, is persistent in soil (Gault and Brockwell, 1988; Roughley et al., 1995), and has proved to be stable in symbiotic effectiveness (Gibson et al., 1990). Today, in 1995, it is the only strain of *B. japonicum* readily detectable in Australian soybean soils and is a valued member of the soil microcommunity.

To establish a new organism in any environment, an ecological niche is essential. For a strain of rhizobia that niche is provided by growing a host legume and the bacteria are supplied by inoculation. Provided that edaphic conditions are suitable for the healthy growth of the host and that sufficient of the inoculant survives until a rhizosphere is available for colonization (see Fig. 1), infection of the root and nodulation will occur as a matter of course. The nodule itself represents an environment akin to pure culture and, within it, there is great multiplication of the rhizobia. When, subsequently, nodule breakdown takes place, large numbers of viable cells are released into the soil (e.g. Kuykendall et al., 1982; Moawad et al., 1984; Thies et al., 1995) where they constitute a potent source of infection for subsequent crops and usually become a persistent component of the soil microflora.

Naturally-occurring populations

A knowledge of the elements comprising natural populations of legume bacteria is essential background for devising strategies for rhizobia manipulation. Naturalized populations of a species of rhizobia usually contain a number of components as described earlier. Ways and means of differentiating them are listed by Brockwell and Bottomley (1995). The characteristics of root-nodule bacteria of most practical significance are those which lead to effective N_2 fixation. The various components of mixed populations of rhizobia frequently express different levels of effectiveness. Soil acidity appears to be an important determinant of N_2-fixing capacity in field populations of *R. leguminosarum* bv. *trifolii*, ineffectiveness being more common in soil of low pH (e.g. Jones and Burrows, 1969). Considerable variation exists in N_2-fixing capacity within field populations of *R. meliloti* (e.g. Bottomley and Jenkins, 1983; see Table 1), *R. leguminosarum* bv. *trifolii* (e.g. Hagedorn et al., 1983) and *Bradyrhizobium* spp. (e.g. Thies et al., 1991a). Sometimes, sharp differences in level of effectiveness may occur between field sampling sites separated by as little as 10 cm (Gibson et al., 1975). In this context, Wollum and Cassel (1984) caution against sampling regimes biased towards one particular area of a field. This type of diversity in N_2-fixing capacity has the advantage that different rhizobial requirements of different legumes may be satisfied by different components of the same population (e.g. Brockwell and Katznelson, 1976; Bromfield, 1984; Robinson, 1969).

Great variation occurs also in the numerical size of field populations of rhizobia. The presence, continuing or periodic, of an appropriate host is a major determinant (Hiltbold et al., 1985; Rupela et al., 1987; Woomer et al., 1988) but soil acidity (Rice et al., 1977), seasonal effects and depth of sampling (Rupela et al., 1987), soil texture and density of gramineaceous herbage (Brockwell and Hely, 1962), and median annual rainfall and base status of the soil (Woomer et al., 1988) are other factors. Brockwell and Robinson (1970) considered that environmental factors had no consistent influence on *Rhizobium* occurrence except in so far as they affected components of the vegetation.

Manipulation of naturally-occurring populations

Naturally-occurring rhizobia often exist in populations of between 1.0×10^4 and 1.0×10^7 cells per gram of soil. A population of 1.0×10^4 rhizobia per gram of soil is equivalent to 1.5×10^{13} rhizobia per hectare to a depth of 10 cm given a bulk density of 1.0. With subterranean clover, inoculated at 1.0×10^4 rhizobia per seed and sown at a rate of 10 kg per hectare, the number of bacteria introduced into the soil is approximately 6.0×10^{10} cells per hectare. If the naturally-occurring population is 1.0×10^4 rhizobia per gram of soil and assuming no loss of inoculant viability, the introduced strain is outnumbered by 250:1 by naturally-occurring rhizobia. Despite advantages of its strategic placement and whatever competitive superiority the inoculant might have, it would be optimistic to expect it to be more than transiently successful in forming nodules and persisting in the soil. The situation may be exacerbated by the frequency of poor quality inoculants and by mortality of inoculant following its introduction into soil (e.g. Bowen and Kennedy, 1959; Brockwell et al., 1987). Numerous studies bear testimony to the futility of inoculation in soils containing large natural populations of rhizobia (e.g. Brockwell et al., 1987; Ham et al., 1971; Johnson et al., 1965; Thies et al., 1991b, c; Weaver and Frederick, 1974). It seems sensible, therefore, in soils where there are large numbers of naturally-occurring rhizobia, to ignore inoculation altogether and instead attempt strategically to manipulate those components of the rhizobial population which have the potential to maximize N_2 fixation.

Competition between naturally-occurring and introduced rhizobia

Where naturalized rhizobia are few or absent, introduction of new strains by inoculation of seed or soil is normally successful provided some prudence is exercised. On the other hand, where large populations of rhizobia occur, inoculation is invariably futile. It is at the intermediate levels of naturalized rhizobial populations, between 10 and 1000 per gram of soil, that competition between naturalized and introduced rhizobia for nodule formation is of practical concern, and only then if the naturalized population, or a substantial component of it, is poorly effective for the target legume.

The subject of strain competition has been ably reviewed by Dowling and Broughton (1986) and Trinick (1985). In general, it appears that the more effective strains are more competitive or, in other words, the host legume exercises a selective preference for the more effective component of a mixed population (e.g. Brockwell and Katznelson, 1976; Robinson, 1969; Singleton and Stockinger, 1983). However, there are many exceptions. Indeed, in elegant experiments with effective strains of *R. meliloti* and their ineffective mutants, Amarger (1981) demonstrated that competitive success in forming nodules was a characteristic of each parent strain, was independent of level of effectiveness, and was retained during mutation from effectiveness to ineffectiveness.

In laboratory and glasshouse experiments, successful manipulation of competitiveness has been achieved by using fungicide-resistant mutants as inocula for seed treated with fungicide active against competing strains (Jones and Giddens, 1984; Odeyemi and Alexander, 1977). The general principle lends itself to other developments for manipulating soil bacteria.

The weight of evidence indicates clearly that the best way to establish a new strain of rhizobia amongst a naturally-occurring population is to apply a heavy rate of effective, persistent inoculum placed strategically close to that point in the soil where the legume roots will first accept infections. In selecting strains suitable for inoculants, the pragmatic approach is to presume that the strain which performs best in the field is the most desirable strain for field inoculation (Brockwell et al., 1982). This presumption underlies the importance of strain testing in field situations.

Competition between different components of naturally-occurring populations of rhizobia

Selective enrichment of particular components of naturalized populations of rhizobia may be the only way to influence legume nodulation in field situations where ineffective or poorly-effective populations are so large that conventional inoculation procedures are ineffectual. This proposition has been little addressed. Renwick and Jones (1986) showed that increasing levels of lime significantly influenced the relative proportions of nodules formed on white clover by two inoculum strains applied at equal rates of inoculation. Similar observations for lime and phosphate have been made by Almendras and Bottomley (1987). Work by Demezas and Bottomley (1986a, 1987) suggested that different selective preferences may be exercised by different cultivars or species of clover. They cautioned, however, that nodule formation by competing strains is not necessarily an index of proportional representation in the rhizosphere and that the incidence of dual (or multiple) occupancy of nodules by more than one strain makes results difficult to interpret (Demezas and Bottomley, 1986b).

Absolute specificity is defined as the circumstance where a legume will accept infection from, and nodulate and fix abundant N with, only a single species or a small group of strains of root-nodule bacteria, e.g. *Medicago laciniata* - see Table 1. It provides the host plant with a symbiotic bacterium not subject to the vagaries of competition from other, less effective nodule bacteria. The phenomenon of absolute specificity is rare in nature. More common is "absolute" promiscuity where the legume nodulates and fixes N abundantly with a wide range of organisms, e.g. cowpea (Fred et al., 1932; Thies et al., 1991a). The N economy of such legumes is independent of rhizobial strain competition for infection sites because of the overwhelming likelihood that whatever strain(s) is successful in nodule formation, its nodules will provide an adequate supply of N to the host. Unfortunately for maximization of symbiotic N_2 fixation, the majority of legumes occupy a central position between specificity and promiscuity. A strategy that legume breeders might employ to improve N_2 fixation by alleviating problems of competition between inoculant strain and naturalized rhizobia, e.g. in soybean breeding, is to select for symbiotic promiscuity. Appropriate germplasm of species of both host and microsymbiont, *Glycine* and *Bradyrhizobium*, occur in wild populations (Thompson et al., 1991). Plant breeding to improve levels of legume N_2 fixation is also feasible (Heichel, 1982; Herridge and Danso, 1995; Mytton et al., 1984).

Concluding remarks, prospects and prognostication

Today, in 1995, the urgency of seeking alternatives to inorganic forms of fertilizer N is perceived to be less than at the time of the energy crises of the 1970s. Yet much of the nitrogenous fertilizer in current use is a subsidized by-product of fossil fuels. On the other hand, the volume of biological N_2 fixation is not constrained by a finite resource. Plant and soil scientists have a responsibility to make biological N_2 fixation, in particular N_2 fixation by legumes, an economically efficient substitute for fertilization of crops and pastures with inorganic N. It is not enough merely to maximize N_2 fixation itself; effectual means of efficiently using fixed N are just as important. For instance, it is essential to contain losses due to leaching of N made soluble by mineralization and subsequent nitrification of organic N of legume origin. Potential loss from soil of the products of fixed N by denitrification to nitrogen oxides and dinitrogen also needs to be considered.

One of the multiple benefits of legume growth is the potential capacity of the legume to conserve and augment the pool of soil N. Augmentation occurs when the amount of soil N removed as animal product, hay and seed is less than the amount of fixed atmospheric N that remains behind in legume residues. Pasture legumes used for forage fulfil this role effectively because only meat, milk and/or fibre are removed and much of the fixed N is returned to the ecosystem where it is available for rotational crops. This concept is the basis of ley farming where leguminous pastures and cereal crops alternate. Rather more fixed N is removed from the system when the legume is harvested for hay and more still by grain legumes grown for their seed. We believe it is feasible to meet the challenge of devising simple, sensible strategies for cultivating legumes that are both high yielding and significantly contributory to reserves of soil N. Some of the information we have presented suggests that N_2 fixation can be enhanced by choice of cultivar, control of soil nitrate and soil moisture and, in particular, by manipulation of rhizobial inocula, either supplied as inoculant or already naturalized in the soil. The principle of limiting factors is especially applicable to legume inoculation, nodulation and N_2 fixation. Whatever advances are possible through improvements in inoculants, delivery systems and manipulation of rhizobial populations in the soil, they will not be realised in the absence of sound agronomic practice. If a leguminous crop or pasture has a low yield potential because of disease, nutrient deficiency, weed competition, insect infestation and/or hostile edaphic factors, high-quality inoculants or better means of handling rhizobia in the soil will be of little help. Only healthy legumes free of environmental stress can express fully their potential for N_2 fixation - see also Robson and Bottomley (1991) and Thies et al. (1991c).

The legume inoculant industry has made, and continues to make an enormous contribution to the Earth's capacity to feed and clothe its peoples. It is a paradox that, despite nearly 100 years of experience, most of the inoculant produced in the world today is of relatively poor quality (e.g. Olsen et al., 1995) and that frankly some of it is extremely bad. Even good quality inoculants are often not used to best advantage. We venture to say that 90% of all inoculant has no practical impact whatsoever on the productivity of the legumes for which it is used and/or on the conservation of N in the soils in which they are grown. Of course, some good inoculants are produced and some of those are used properly in situations where they are needed for N_2 fixation and the conservation of soil N. In these circumstances, legume inoculation may be the most cost-effective of all agricultural practices. Too often, unfortunately, the reverse is true and many farmers, even scientists, in the developed and developing worlds, see little value in the practice of inoculation. We are pessimistic about the prospects for the inoculant industry and its capacity for large-scale production of high quality inoculants. Existing inoculants of highest quality tend to be those produced by small factories under the umbrella of a quality control authority. Within this context, Hoben and Somasegaran (1992) and Somasegaran et al. (1992) have described small and medium-scale fermenters for production of legume inoculants. It is our opinion that the future of the legume inoculant industry, and its potential benefits for food and fibre production worldwide, depends on improving inoculant quality - both numerical and competitive, on targeting specific legumes including new crops and on becoming more user-friendly.

The last major advance in inoculant quality was generated by the work of Roughley and Vincent (1967). They demonstrated a 10-fold improvement in numerical quality from the use of peat carrier sterilized by irradiation. Commercial inoculants prepared in this way contain ca. 2.0×10^9 rhizobia per gram of peat. A population of this magnitude occupies <0.13% of the total volume of the inoculant. Surely it is feasible to aim for a further 10-fold increase. The desirability of this is illustrated by upward responses in nodulation and seed yield to increasing rates of inoculation (Hume and Blair, 1992; Roughley et al., 1993). Griffith and Roughley (1992) and Griffith et al. (1992) have suggested how some advance might be made through control of moisture levels in peat culture and correct use of packaging films. The oxygen and nutrient requirements of rhizobia in peat culture remain to be properly addressed. The stage of growth of the fermenter-grown culture of rhizobia when it is added to the peat carrier may be a significant factor determining the final population of the peat inoculant, and its responsiveness to plant signals that lead to the induction of *nod* genes (Djordjevic et al., 1987).

For the future, strain selection will have an important role in maintaining and enhancing inoculant quality. There is scope to improve the N_2-fixing capacity of inoculant strains by mutagenesis and genetic engineering (Paau, 1991). Previously neglected sources in nature will be explored for superior rhizobia exemplified by the successful discovery of acid tolerant *Medicago* spp. and *R. meliloti* strains by Howieson and Ewing (1986, 1989) and Bounejmate and Robson (1992). In this context East Asia and the tropics, for instance, may well yield useful strains for North American-type soybeans (La Favre et al., 1991; Ravuri and Hume, 1992; Thompson et al., 1991). Studies will continue on strain competition and the ability of inoculant strains to compete successfully with naturalized rhizobia for nodule sites. Some innovative options for achieving these objectives, as indicated by recent work, include the use of soil fumigation (Kishinevsky et al., 1992), use of inhibitors of *nod*Y expression in the host legume (Cunningham et al., 1991) to manipulate competition between strains, and the application of phage to seed (Basit et al., 1992) or soil (Hashem and Angle, 1988, 1990) to influence the outcome of strain competition. The practicalities of these propositions have been considered by Brockwell and Bottomley (1995).

Where inoculant is unnecessary or no good, its application is a waste of time and money. To claim otherwise will ultimately affect the industry's credibility. Legume inoculants must become more user-friendly. Shelf life of inoculants must be extended (e.g. Griffith and Roughley, 1992) and they must become easier to apply. The wider use of multi-strain inoculants will reduce the total number of inoculants available as well as the scope for user error. Somasegaran and Bohlool (1990) made an extensive comparison of multistrain and single-strain inoculants. They found that in almost all cases the effectiveness of a multi-strain inoculant exceeded or equalled the performance of the best strain in that inoculant. Paau (1989) also obtained excellent results with a multi-strain soybean inoculant fermented directly in the point-of-use container in a vermiculite carrier. Excellent extension programmes promoting rhizobial inoculants are conducted in Thailand (Chanaseni and Kongngoen, 1992) and worldwide by the NifTAL Center, Paia, Hawaii. The credibility of the inoculant industry will be well served by also publicizing those situations where inoculation is not required.

Dual inoculation of legumes may have the potential to exploit synergistic effects on N_2 fixation by rhizobia accompanied by other organisms. The subject of co-inoculation of white clover with *Rhizobium* and VA mycorrhizal fungi was raised by Smith and Daft (1978). For a period, it was considered that additional benefit from mycorrhizas accrued only under conditions of low fertility where the fungal hyphae scavenged plant nutrients otherwise unavailable to the host legume. Now it is recognized that such co-inoculation may have wider application (Badr El-Din and Moawad, 1988; Mahdi and Atabani, 1992; Thiagarajan et al., 1992). Other microorganisms as well have been reported to enhance nodulation directly, e.g. *Penicillium bilaji* (Downey and van Kessel, 1990), *Pseudomonas fluorescens* (Nishijima et al., 1988), or indirectly by improving the health and vigour of the host legume, e.g. *Bacillus thuringiensis* (Keyser et al., 1992). Rice et al. (1995) suggest how commercial preparation of a co-inoculant might be effected.

It may be that the best way to increase N_2 fixation in many situations is to ignore legume inoculation altogether and instead exploit the characteristics of rhizobial populations already established in the soil. The breeding of symbiotically promiscuous lines of legumes is a matter for serious consideration, e.g. soybeans (Herridge and Danso, 1995). These plants would not require inoculation because they would nodulate and fix N vigorously with the large diverse populations of *B. japonicum* that occur in many tropical soils (e.g. Thompson et al., 1991). Alternatively, it may be possible to muzzle the competitiveness of naturalized rhizobia by exploiting requirements for specific rhizobia possessed by certain legumes. Examples of strain-specific legumes are well known and their potential for plant breeding well understood (Lie, 1978). Fobert et al. (1991), working in a glasshouse, exploited the specificity between a symbiotic gene in pea and a nodulation gene in *R. leguminosarum* bv. *viciae* to preempt competition from soil rhizobia against inoculant strains. The authors considered their results sufficiently encouraging to justify field experimentation. The work of Kipe-Nolt et al. (1992) suggests that certain wild accessions of *Phaseolus* bean have sufficient nodulation specificity to exploit in a breeding programme to produce a strain-specific plant. A fascinating possibility for the use of a selective host to give competitive advantage to particular rhizobial strains has been proposed by Cregan and Keyser (1988). Yet another possibility for contriving better means for a legume to use naturalized populations of rhizobia (or inoculant strains) or to prevail over environmental constraints to symbiotic N_2 fixation lies in supernodulation. Plant lines with greatly enhanced capacity for nodulation and N_2 fixation (supernodulaters) have already been identified from collections of germplasm (Herridge and Betts, 1988) and developed by mutagenesis: soybean - Carroll et al. (1985), Gremaud and Harper (1989); *Phaseolus* bean - Park and Buttery (1988); pea - Duc and Messager (1989).

Our speculations about manipulation of rhizobia may have wider application. As agricultural biotechnology evolves, much greater use will be made of soil microorganisms both as inoculants and by manipulating them in the soil or the plant rhizosphere. Mycorrhizal fungi, *Azolla* and *Anabaena* in rice culture, benign organisms that compete for rhizosphere space with root pathogens thereby reducing the incidence of plant diseases, organisms that dissolve forms of phosphate otherwise unavailable to plants, free-living diazotrophs and cellulolytic organisms are all presently under utilized. Ecological principles and practices that are appropriate for the manipulation of the rhizobia will quite likely prove suitable models for these and other soil microorganisms as well.

Acknowledgments

We thank Lyn Abbott, Nui Milton and Graham Sparling for allowing us to refer to their unpublished research results relating to rhizobial populations and microbial biomass. We are indebted to Gary Bullard and David Pulsford for presenting us with an inoculant manufacturer's viewpoint on strain selection. We are most grateful to Pat French, Helen Kaminski, Inara Licis and Diane Hebb who assisted greatly with the preparation and presentation of the manuscript.

References

Abbott L K and Robson A D 1982 The role of vesicular-arbuscular mycorrhiza in agriculture and the selection of fungi for inoculation. Aust. J. Agric. Res. 33, 389–408.

Abbott L K and Robson A D 1987 Managing VA mycorrrhizal fungi in pastures. *In* Proc. 7th NACOM Conference. Eds. D M Sylvia, L L Hung and J H Graham. pp 10–12. Inst. Forest Agric. Sci., Florida.

Adebayo A, Watanabe I and Ladha J K 1989 Epiphytic occurrence of *Azorhizobium caulinodans* and other rhizobia on host and nonhost legumes. Appl. Environ. Microbiol. 55, 2407–2409.

Alazard D and Duhoux E 1987 Diversity of stem nodulation sites in *Aeschynomene* spp. J. Plant Physiol.132, 123–125.

Alazard D, Ndoye I and Dreyfus B 1988 *Sesbania rostrata* and other stem–nodulated legumes. *In* Nitrogen Fixation: Hundred Years After. Eds. H Bothe, F J de Bruijn and W E Newton. pp 765–769. Gustav Fischer, Stuttgart.

Albrecht S L, Bennett J M and Boote K J 1984 Relationship of nitrogenase activity to plant water stress in field-grown soybeans. Field Crops Res. 8, 61–71.

Allen E K and Allen O N 1961 The scope of nodulation in the Leguminosae. *In* Recent Advances in Botany. Vol. 1. Proc. 9th Int. Bot. Congr. pp 585–588. Univ. Toronto Press, Toronto.

Allen O N and Allen E K 1981 The Leguminosae. A Source Book of Characteristics, Uses and Nodulation. Univ. Wisconsin Press, Madison, WI. 812p.

Almendras A S and Bottomley P J 1987 Influence of lime and phosphate on nodulation of soil-grown *Trifolium subterraneum* L. by indigenous *Rhizobium trifolii*. Appl. Environ. Microbiol. 55, 2090–2097.

Almendras A S and Bottomley P J 1988 Cation and phosphate influences on the nodulating characteristics of indigenous serogroups of *Rhizobium trifolii* on soil grown *Trifolium subterraneum*. Soil Biol. Biochem. 20, 345–351.

Amarger N 1981 Competition for nodule formation between effective and ineffective strains of *Rhizobium meliloti*. Soil Biol. Biochem.13, 475–480.

Anderson A J 1970 Trace elements for sheep pastures and fodder crops in Australia. J. Aust. Inst. Agric. Sci. 36, 15–29.

Andrew C S 1976 Effect of calcium, pH and nitrogen on the growth and chemical composition of some tropical and temperate pasture legumes. 1. Nodulation and growth. Aust. J. Agric. Res. 27, 611–623.

Asai T 1944 Uber die Mykorrhizenbildung der leguminosen pflanzen. Jap. J. Bot. 13, 463–484.

Atkin–Smith R E, Gault R R and Brockwell J 1986 Effective rhizobia can be introduced, by surface application, into poorly-nodulated subterranean clover pasture. *In* Proc. 8th Aust. Nitrogen Fixation Conference. AIAS Occ. Publ. No. 25. Eds. W Wallace and S E Smith. pp 151–152. Aust. Inst. Agric. Sci., Adelaide.

Ausubel F M, Buikema W J, Earl D C, Klingensmith J A, Nixon B and Szeto W W 1985 Organization and regulation of *Rhizobium meliloti* and *Parasponia Bradyrhizobium* nitrogen fixation genes. *In* Nitrogen Fixation Research Progress. Eds. H J Evans, P J Bottomley and W E Newton. pp 165–171. Martinus Nijhoff, Dordrecht.

Badr El-Din S M S and Moawad H 1988 Enhancement of nitrogen fixation in lentil, faba bean, and soybean by dual inoculation with Rhizobia and mycorrhizae. Plant and Soil 108, 117–124.

Barea J M and Azcon-Aguilar C 1983 Mycorrhizas and their significance in nodulating nitrogen-fixing plants. Adv. Agron. 36, 1–54.

Barkdoll A W, Sartain J B and Hubbell D H 1983 Effect of soil implanted granular and pellet *Rhizobium* inoculant in *Phaseolus vulgaris* L. in Honduras. Soil Crop Sci. Soc. Fla. Proc. 42, 184–189.

Basit H A, Angle J S, Salem S and Gewailly E M 1992 Phage coating of soybean seed reduces nodulation by indigenous soil bradyrhizobia. Can. J. Microbiol. 38, 1264–1269.

Bassam B J, Rolfe B G and Djordjevic M A 1986 *Macroptilium atropurpureum* (siratro) host specificity genes are linked to a *nod*D-like gene in the broad host range *Rhizobium* strain NGR234. Mol. Gen. Genet. 203, 49–57.

Becker M, Ladha J K and Ali M 1995 Green manure technology: Potential, usage, and limitations - a case history for lowland rice. Plant and Soil 174.

Becker M, Ladha J K and Ottow J C 1990 Growth and N_2 fixation of two stem-nodulating legumes and their effect as green manure on lowland rice. Soil Biol. Biochem. 22, 1109–1119.

Bell F and Nutman P S 1971 Experiments on nitrogen fixation by nodulated lucerne. Plant and Soil, special vol., 231–264.

Berg R K, Jr, Loynachan T E, Zablotowicz R M and Lieberman M T 1988 Nodule occupancy by introduced *Bradyrhizobium japonicum* in Iowa soils. Agron. J. 80, 876–881.

Bergersen F J (Ed.) 1980 Methods for Evaluating Biological Nitrogen Fixation. John Wiley and Sons, Chichester, UK. 702p.

Bergersen F J, Turner G L, Chase D L, Gault R R and Brockwell J 1985 The natural abundance of ^{15}N in an irrigated soybean crop and its use for the calculation of nitrogen fixation. Aust. J. Agric. Res. 36, 411–423.

Bonish P M 1979 Clover rhizobia in soils: assessment of effectiveness using the plant infection count method. N. Z. J. Agric. Res. 22, 89–93.

Boonkerd N, Arunsri C, Rungrattanakasin W and Vasurat Y 1984 Effect of postemergence inoculation on field grown soybeans. *In* Advances in Nitrogen Fixation Research. Eds. C Veeger and W E Newton. p 327. Martinus Nijhoff / Dr. Junk Publ., The Hague.

Boonkerd N and Weaver R W 1982 Survival of cowpea rhizobia in soil as affected by soil temperature and moisture. Appl. Environ. Microbiol. 43, 585–589.

Bosworth A H, Williams M K, Albrecht K A, Kwiatkowski R K, Beynon J, Hankinson T R, Ronson C W, Cannon F, Wacek T C and Triplett E W 1994 Alfalfa yield response in inoculation with recombinant strains of *Rhizobium meliloti* with an extra copy of *dctABD* and/or modified *nifA* expression. Appl. Environ. Microbiol. 60, 3815–3832.

Bottomley P J 1992 Ecology of *Bradyrhizobium* and *Rhizobium*. *In* Biological Nitrogen Fixation. Eds. G Stacey, R H Burris and H J Evans. pp 293–348. Chapman and Hall, New York.

Bottomley P J, Cheng H-H and Strain S R 1994 Genetic structure and symbiotic characteristics of a *Bradyrhizobium* population recovered from a pasture soil. Appl. Environ. Microbiol. 60, 1754–1761.

Bottomley P J and Jenkins M B 1983 Some characteristics of *Rhizobium meliloti* isolates from alfalfa fields in Oregon. Soil Sci. Soc. Am. J. 47, 1153–1157.

Bounejmate M and Robson A D 1992 Differential tolerance of genotypes of *Medicago truncatula* to low pH. Aust. J. Agric. Res. 43, 731–737.

Bowen G D 1980 Misconceptions, concepts and approaches in rhizosphere biology. *In* Contemporary Microbial Ecology. Eds. D C Ellwood, J N Hedger, M J Latham, J M Lynch and J H Slater. pp 283–305. Academic Press, London.

Bowen G D and Kennedy M 1959 Effect of high soil temperature on *Rhizobium* spp. Q. J. Agric. Sci.16, 177–197.

Brewin N J, Wood E A and Young J P W 1983 Contribution of the symbiotic plasmid to the competitiveness of *Rhizobium leguminosarum*. J. Gen. Microbiol. 129, 2973–2977.

Brockwell J 1962 Studies on seed pelleting as an aid to legume seed inoculation. 1. Coating materials, adhesives, and methods of inoculation. Aust. J. Agric. Res. 13, 638–649.

Brockwell J 1963 Accuracy of a plant-infection test for counting populations of *Rhizobium trifolii*. Appl. Microbiol. 11, 377–383.

Brockwell J 1971 An appraisal of an IBP experiment on nitrogen fixation by nodulated legumes. Plant and Soil, special vol., 265–272.

Brockwell J 1977 Application of legume seed inoculants. *In* A Treatise on Dinitrogen Fixation. Section IV. Agronomy and Ecology. Eds. R W F Hardy and A H Gibson. pp 277–309. John Wiley and Sons, New York.

Brockwell J 1982 Inoculation methods for field experimenters and farmers. *In* Nitrogen Fixation in Legumes. Ed. J M Vincent. pp 211–227. Academic Press, Sydney.

Brockwell J and Bottomley P J 1995 Recent advances in inoculant technology and prospects for the future. Soil Biol. Biochem. 27 (*In press*).

Brockwell J, Diatloff A, Roughley R J and Date R A 1982 Selection of rhizobia for inoculants. *In* Nitrogen Fixation in Legumes. Ed. J M Vincent. pp 173–191. Academic Press, Sydney.

Brockwell J, Gault R R and Chase D L 1977 Inoculating soybeans with root-nodule bacteria. Farmers' Newsl. 104, 7–12. Irrigation Research and Extension Committee, Griffith, NSW.

Brockwell J, Gault R R, Chase D L, Hely F W, Zorin M and Corbin E J 1980 An appraisal of practical alternatives to legume seed inoculation: field experiments on seed bed inoculation with solid and liquid inoculants. Aust. J. Agric. Res. 31, 47–60.

Brockwell J, Gault R R, Chase D L, Turner G L and Bergersen F J 1985 Establishment and expression of soybean symbiosis in a soil previously free of *Rhizobium japonicum*. Aust. J. Agric. Res. 36, 397–409.

Brockwell J and Hely F W 1962 Some aspects of the ecology of *Rhizobium meliloti* in the brown acid soils of the Macquarie Region of New South Wales. Fld Stn Rec., Div. Plant Ind., CSIRO (Aust.) 1(2), 1–11.

Brockwell J and Hely F W 1966 Symbiotic characteristics of *Rhizobium meliloti*: an appraisal of the systematic treatment of nodulation and nitrogen fixation interactions between hosts and rhizobia of diverse origins. Aust. J. Agric. Res. 17, 885–899.

Brockwell J, Herridge D F, Roughley R J, Thompson J A and Gault R R 1975 Studies on seed pelleting as an aid to legume seed inoculation. 4. Examination of preinoculated seed. Aust. J. Exp. Agric. Anim. Husb. 15, 780–787.

Brockwell J, Holliday R A and Pilka A 1988 Evaluation of the symbiotic nitrogen–fixing potential of soils by direct microbiological means. Plant and Soil 108, 163–170.

Brockwell J and Katznelson J 1976 Symbiotic characteristics of *Rhizobium trifolii* from Israel in association with ten species of *Trifolium*. Aust. J. Agric. Res. 27, 799–810.

Brockwell J and Robinson A C 1970 Observations on the natural distribution of *Rhizobium* spp. relative to physical features of the landscape. Proc. 11th Int. Grassl. Congr. pp 438–441. Univ. Queensland Press, Brisbane, Australia.

Brockwell J and Roughley R J 1967 An examination of the numbers of nodule bacteria associated with legume seed following commercial multiple inoculation. J. Aust. Inst. Agric. Sci. 33, 204–207.

Brockwell J, Roughley R J and Herridge D F 1987 Population dynamics of *Rhizobium japonicum* strains used to inoculate three successive crops of soybean. Aust. J. Agric. Res. 38, 61–74.

Bromfield E S P 1984 The preference for strains of *Rhizobium meliloti* by cultivars of *Medicago sativa* grown on agar. Can. J. Microbiol. 30, 1179–1183.

Bromfield E S P and Jones D G 1980 Studies on acid tolerance of *Rhizobium trifolii* in culture and in soil. J. Appl. Bacteriol. 48, 253–264.

Bromfield E S P, Wheatcroft R and Barran L R 1994 Medium for direct isolation of *Rhizobium meliloti* from soil. Soil Biol. Biochem. 26, 423–428.

Bromfield S M, Cumming R W, David D J and Williams C H 1983 Change in soil pH, manganese and aluminium under subterranean clover pasture. Aust. J. Exp. Agric. Anim. Husb. 23, 181–191.

Broughton W J, Samrey U and Stanley J 1987 Ecological genetics of *Rhizobium meliloti* symbiotic plasmid transfer in the *Medicago sativa* rhizosphere. FEMS Microbiol. Lett. 40, 251–255.

Broughton W J, Wong C H, Lewin A, Samrey U, Myint H, Meyer H, Dowling D N and Simon R 1986 Identification of *Rhizobium* plasmid sequences involved in recognition of *Psophocarpus*, *Vigna*, and other legumes. J. Cell Biol. 102, 1173–1182.

Brown M R, Wolf D D, Morse R D and Neal J L 1983 Viability of *Rhizobium* in fertilizer slurries used for hydroseeding. J. Environ. Qual. 12, 388–390.

Bunt J S 1988 Nitrogen fixation in the sea. *In* Microbiology in Action. Eds. W G Murrell and I R Kennedy. pp 209–219. Research Studies Press, Letchworth, UK.

Burns R C and Hardy R W F 1975 Nitrogen Fixation in Bacteria and Higher Plants. Springer Verlag, New York. 189p.

Burton J C 1976 Methods of inoculating seeds and their effect on survival of rhizobia. *In* Symbiotic Nitrogen Fixation in Plants. International Biological Programme No. 7. Ed. P S Nutman. pp 175–189. Univ. Press, Cambridge, UK.

Burton J C 1979 *Rhizobium* species. *In* Microbial Technology, 2nd ed., Vol. 1. Eds. H J Peppler and D Perlman. pp 29–58. Academic Press, New York.

Burton J C 1982 Modern concepts in legume inoculation. *In* Biological Nitrogen Fixation Technology for Tropical Agriculture. Eds. P H Graham and S C Harris. pp 105–114. CIAT, Cali, Colombia.

Bushby H V A 1981 Quantitative estimation of rhizobia in non-sterile soil using antibodies and fungicides. Soil Biol. Biochem. 13, 237–239.

Bushby H V A 1982 Ecology. *In* Nitrogen Fixation. Volume 2: *Rhizobium*. Ed. W J Broughton. pp 35–75. Clarendon Press, Oxford.

Bushby H V A 1984 Colonization of rhizospheres and nodulation of two *Vigna* species by rhizobia inoculated onto seed: influence of soil. Soil Biol. Biochem. 16, 635–641.

Bushby H V A and Marshall K C 1977 Water status of rhizobia in relation to their susceptibility to desiccation and to their protection by montmorillonite. J. Gen. Microbiol. 99, 19–27.

Caldwell B E and Vest G 1970 Effects of *Rhizobium japonicum* strains on soybean yield. Crop Sci. 10, 19–21.

Carroll B J, McNeil D L and Gresshoff P M 1985 A supernodulation and nitrate-tolerant symbiotic (*nts*) soybean mutant. Plant Physiol. 78, 34–40.

Cassman K G, Singleton P W and Linquist B A 1993 Input/output analysis of the cumulative soybean response to phosphorus on an Ultisol. Field Crops Res. 34, 23–36.

Cassman K G, Munns D N and Beck D P 1981 Growth of *Rhizobium* strains at low concentrations of phosphate. Soil Sci. Soc. Am. J. 45, 520–523.

Chamber M A 1983 Influence of several methods for rhizobial inoculation on nodulation and yield of soybeans. Plant and Soil 74, 203–209.

Chanaseni C and Kongngoen S 1992 Extension programs to promote rhizobial inoculants for soybean and groundnut in Thailand. Can. J. Microbiol. 38, 594–597.

Chatel D L, Greenwood R M and Parker C A 1968 Saprophytic competence as an important character in the selection of *Rhizobium* for inoculation. Trans. 9th Int. Congr. Soil Sci., Vol 2. pp 65–73. Augus and Robertson, Sydney, Australia.

Chen Y P, Glenn A R and Dilworth M J 1984 Uptake and oxidation of aromatic substrates by *Rhizobium leguminosarum* MNF3841 and *Rhizobium trifolii* TA1. FEMS Microbiol. Lett. 21, 201–205.

Ciafardini G and Barbieri C 1987 Effects of cover inoculation of soybean on nodulation, nitrogen fixation, and yield. Agron. J. 79, 645–648.

Cocks P S, Mathison M J and Crawford E J 1980 From wild plants to pasture cultivars: annual medics and subterranean clover in southern Australia. *In* Advances in Legume Science. Eds. R J Summerfield and A H Bunting. pp 569–596. Royal Botanic Gardens, Kew.

Corbin E J, Brockwell J and Gault R R 1977 Nodulation studies on chickpea (*Cicer arietinum*). Aust. J. Exp. Agric. Anim. Husb. 17, 126–134.

Coventry D R, Hirth J R, Reeves T G and Burnett V F 1983a Growth and nitrogen fixation by subterranean clover in response to inoculation, molybdenum application and soil amendment with lime. Soil Biol. Biochem. 17, 791–796.

Coventry D R, Hirth J R, Reeves T G and Jones H R 1983b Development of populations of *Rhizobium trifolii* and nodulation of subterranean clover following the cropping phase in crop-pasture rotations in southern Australia. Soil Biol. Biochem. 17, 17–22.

Cregan P B and Keyser H H 1988 Influence of *Glycine* spp. on competitiveness of *Bradyrhizobium japonicum* and *Rhizobium fredii*. Appl. Environ. Microbiol. 54, 803–808.

Cunningham S, Kollmeyer W D and Stacey G 1991 Chemical control of interstrain competition for soybean nodulation by *Bradyrhizobium japonicum*. Appl. Environ. Microbiol. 57, 1886–1892.

Danso S K A, Kapuya J and Hardarson G 1990a Nitrogen fixation and growth of soybean as influenced by varying the methods of inoculation with *Bradyrhizobium japonicum*. Plant and Soil 125, 81–86.

Danso S K A, Zapata F and Awonaike K O 1990b Effect of postemergence, supplemental inoculation on nodulation and symbiotic performance of soybean (*Glycine max* (L.) Merrill) at three levels of soil nitrogen. Appl. Environ. Microbiol. 56, 1793–1798.

Date R A 1969 A decade of legume inoculant control in Australia. J. Aust. Inst. Agric. Sci. 35, 27–37.

Date R A 1977 Inoculation of tropical pasture legumes. *In* Exploiting the Legume-*Rhizobium* Symbiosis in Tropical Agriculture. Eds. J M Vincent, A S Whitney and J Bose. pp 293–311. Univ. Hawaii, Honolulu.

Date R A 1982 Collection, isolation, characterization and conservation of *Rhizobium*. *In* Nitrogen Fixation in Legumes. Ed. J M Vincent. pp 95–109. Academic Press, Sydney.

Date R A and Decker A M 1965 Minimal antigenic constitution of 28 strains of *Rhizobium japonicum*. Can. J. Microbiol. 11, 1–8.

Date R A and Halliday J 1979 Selecting *Rhizobium* for acid, infertile soils of the tropics. Nature, London 277, 62–64.

Date R A and Roughley R J 1977 Preparation of legume seed inoculants. *In* A Treatise on Dinitrogen Fixation. Section IV. Agronomy and Ecology. Eds. R W F Hardy and A H Gibson. pp 243–275. John Wiley and Sons, New York.

Dazzo F B, Hrabek E M, Urbano M R, Sherwood J E and Truchet G 1981 Regulation of recognition in the *Rhizobium*-clover symbiosis. *In* Current Perspectives in Nitrogen Fixation. Eds. A H Gibson and W E Newton. pp 292–295. Aust. Acad. Sci., Canberra.

Demezas D H and Bottomley P J 1986a Autecology in rhizospheres and nodulating behaviour of indigenous *Rhizobium trifolii*. Appl. Environ. Microbiol. 52, 1014–1019.

Demezas D H and Bottomley P J 1986b Interstrain competition between representatives of indigenous serotypes of *Rhizobium trifolii*. Appl. Environ. Microbiol. 52, 1020–1025.

Demezas D H and Bottomley P J 1987 Influence of soil and nonsoil environments on nodulation by *Rhizobium trifolii*. Appl. Environ. Microbiol. 53, 596–597.

Demezas D H, Reardon T B, Watson J M and Gibson A H 1991 Genetic diversity among *Rhizobium leguminosarum* bv. Trifolii strains revealed by allozyme and restriction fragment length polymorphism analyses. Appl. Environ. Microbiol. 57, 3489–3495.

Diatloff A and Brockwell J 1976 Ecological studies of root-nodule bacteria introduced into field environments. 4. Symbiotic properties of *Rhizobium japonicum* and competitive success in nodulation of two *Glycine max* cultivars by effective and ineffective strains. Aust. J. Exp. Agric. Anim. Husb. 16, 514–521.

Dilworth M J, McKay I, Franklin M and Glenn A R 1983 Catabolite effects on enzyme induction and substrate utilization in *Rhizobium leguminosarum*. J. Gen. Microbiol. 129, 359–366.

Djordjevic M A, Redmond J W, Batley M and Rolfe B G 1987 Clovers secrete specific phenolic compounds which either stimulate or repress *nod* gene expression in *Rhizobium trifolii*. EMBO J. 6, 1173–1179.

Djordjevic M A, Zurkowski W and Rolfe B G 1982 Plasmids and stability of symbiotic properties of *Rhizobium trifolii*. J. Bacteriol. 151, 560–568.

Dommergues Y R 1987 The role of biological nitrogen fixation in agroforestry. *In* Agroforestry: A Decade of Development. Eds. H A Steppler and P K R Nair. pp 245–271. International Council for Research in Agroforestry, Nairobi.

Dommergues Y R, Diem H G and Divies C 1979 Polyacrylamide-entrapped *Rhizobium* as an inoculant for legumes. Appl. Environ. Microbiol. 37, 779–781.

Donald C M and Williams C H 1954 Fertility and productivity of a podzolic soil as influenced by subterranean clover (*Trifolium subterranean* L.) and superphosphate. Aust. J. Agric. Res. 5, 664–687.

Dowdle S F and Bohlool B B 1985 Predominance of fast-growing *Rhizobium japonicum* in a soybean field in the People's Republic of China. Appl. Environ. Microbiol. 54, 1171–1176.

Dowling D N and Broughton W J 1986 Competition for nodulation of legumes. Annu. Rev. Microbiol. 40, 131–157.

Downey J and van Kessel C 1990 Dual inoculation of *Pisum sativum* with *Rhizobium leguminosarum* and *Penicillium bilaji*. Biol. Fertil. Soils 10, 194–196.

Dreyfus B, Garcia J L and Gillis M 1988 Characterization of *Azorhizobium caulinodans* gen. nov., sp. nov., a stem-nodulating nitrogen-fixing bacteria isolated from *Sesbania rostrata*. Int. J. Syst. Bacteriol. 38, 89–98.

Dreyfus B L and Dommergues Y R 1981a Nitrogen-fixing nodules induced by *Rhizobium* on the stem of the tropical legume *Sesbania rostrata*. FEMS Microbiol. Lett. 10, 313–317.

Dreyfus B L and Dommergues Y R 1981b Nodulation of *Acacia* species by fast- and slow- growing tropical strains of *Rhizobium*. Appl. Environ. Microbiol. 41, 97–99.

Duc G and Messager A 1989 Mutagenesis of pea (*Pisum sativum* L.) and the isolation of mutants for nodulation and nitrogen fixation. Plant Sci. 60, 207–213.

Dudman W F and Brockwell J 1968 Ecological studies of root-nodule bacteria introduced into field environments. I. A survey of field performance of clover inoculants by gel immune diffusion serology. Aust. J. Agric. Res. 19, 739–747.

Dughri M H and Bottomley P J 1983 Effect of acidity on the composition of an indigenous soil population of *Rhizobium trifolii* found in nodules of *Trifolium subterraneum* L. Appl. Environ. Microbiol. 46, 1207–1213.

Dughri M H and Bottomley P J 1984 Soil acidity and the composition of an indigenous population of *Rhizobium trifolii* in nodules of different cultivars of *Trifolium subterraneum* L. Soil Biol. Biochem. 16, 405–411.

Eaglesham A R J, Ellis J M, Evans W R, Fleishman D E, Hungria M and Hardy R W F 1990 The first photosynthetic N_2-fixing *Rhizobium*: characteristics. *In* Nitrogen Fixation: Achievements and Objectives. Eds. P M Gresshoff, L E Roth, G Stacey and W E Newton. pp 805–811. Chapman and Hall, New York.

Eardly B D, Materon L A, Smith N H, Johnson D A, Rumbaugh M D and Selander R K 1990 Genetic structure of natural populations of the nitrogen-fixing bacterium *Rhizobium meliloti*. Appl. Environ. Microbiol. 56, 187–194.

Elegba M S and Rennie R J 1984 Effect of different inoculant adhesive agents on rhizobial survival, nodulation, and nitrogenase (acetylene-reducing) activity of soybeans (*Glycine max* (L.) Merrill). Can. J. Soil Sci. 64, 631–636.

Evans H J and Russell S A 1971 Physiological chemistry of symbiotic nitrogen fixation by legumes. *In* Chemistry and Biochemistry of Nitrogen Fixation. Ed. J R Postgate. pp 191–244. Plenum Press, London.

Evans H J, Hanus F J, Haugland R A, Cantrell M A, Xu L-S, Russell S A, Lambert G R and Harker A R 1985 Hydrogen recycling in nodules affects nitrogen fixation and growth of soybeans. *In* World Soybean Research Conference III. Proceedings. Ed. R Shibles. pp 935–942. Westcourt Press, Boulder, CO.

Firmin J L, Wilson K E, Rossen L and Johnston A W B 1986 Flavonoid activation of nodulation genes in *Rhizobium* reversed by other compounds present in plants. Nature, London 324, 90–92.

Fobert P R, Roy N, Nash J H E and lyer V N 1991 Procedure for obtaining efficient root nodulation of a pea cultivar by a desired *Rhizobium* strain and preempting nodulation by other strains. Appl. Environ. Microbiol. 57, 1590–1594.

Franssen H J, Vijn I, Yang W C and Bisseling T 1992 Developmental aspects of the *Rhizobium*-legume symbiosis. Plant Mol. Biol. 19, 89–107.

Fred E B, Baldwin I L and McCoy E 1932 Root Nodule Bacteria and Leguminous Plants. Stud. Sci. Univ. Wisconsin No. 5. 343p.

Fuhrmann J, Davey C B and Wollum A G, II 1986 Desiccation tolerance of clover rhizobia in sterile soils. Soil Sci. Soc. Am. J. 50, 639–644.

Gallacher A E and Sprent J I 1978 The effect of different water regimes on growth and nodule development ofgreenhouse grown *V. faba*. J. Exp. Bot. 29, 413–423.

Gault R R 1978 A study of developments and trends, in New Zealand, the U.S.A., and Canada, in the technology associated with the exploitation of the nitrogen–fixing legume root nodule bacteria, *Rhizobium* spp., for use in legume crops new to Australian agriculture. Final Report to the Winston Churchill Memorial Trust. CSIRO, Canberra. 143 p.

Gault R R, Banks L W, Chase D L and Brockwell J 1984 Remedial measures to salvage unnodulated soybean crops. J. Aust. Inst. Agric. Sci. 50, 244–246.

Gault R R, Bernardi A L, Thompson J A, Andrews J A, Banks L W, Hebb D M and Brockwell J 1994a Studies on alternative means of legume inoculation: appraisal of application of inoculant suspended in irrigation water (water-run inoculation). Aust. J. Exp. Agric. 34, 401–409.

Gault R R and Brockwell J 1980 Studies on seed pelleting as an aid to legume inoculation. 5. Effects of incorporation of molybdenum compounds in the seed pellet on inoculant survival, seedling nodulation and plant growth of lucerne and subterranean clover. Aust. J. Exp. Agric. Anim. Husb. 20, 63–70.

Gault R R and Brockwell J 1988 The autecology of a strain of *Bradyrhizobium japonicum* in a soybean rice rotation. Proc. Sth Aust. Soybean Research Workshop. pp 82–84. NSW Agriculture and Fisheries, Sydney.

Gault R R, Pilka A, Hebb D M and Brockwell J 1994b Nodulation studies on legumes exotic to Australia: symbiotic relationships between *Chamaecytisus palmensis* (tagasaste) and *Lotus* spp. Aust. J. Exp. Agric. 35, 385–394.

Gault R R and Schwinghamer E A 1993 Direct isolation of *Bradyrhizobium japonicum* from soil. Soil Biol. Biochem. 25, 1161–1166.

Gaworzewska E T and Carlile M J 1982 Positve chemotaxis of *Rhizobium leguminosarum* and other bacteria towards root exudates from legumes and other plants. J. Gen. Microbiol. 128, 1179–1188.

Gemell L G and Roughley R J 1993 Field evaluation in acid soils of *Rhizobium legurninosarum* bv. *trifolii* selected for their tolerance or sensitivity to acid soil factors in agar medium. Soil Biol. Biochem. 25, 1447–1452.

George T, Ladha J K, Buresh R J and Garrity D P 1992 Managing native and legume-fixed nitrogen in lowland rice-based cropping systems. Plant and Soil 141, 69–91.

Gibson A H 1963 Physical environment and symbiotic nitrogen fixation. I. The effect of root temperature on recently nodulated *Trifolium subterraneum* L. plants. Aust. J. Biol. Sci. 16, 28–42.

Gibson A H 1964 Genetic control of strain-specific ineffective nodulation in *Trifolium subterraneum* L. Aust. J. Agric. Res. 15, 37–49.

Gibson A H 1967 Physical environment and symbiotic nitrogen fixation. IV. Factors affecting the early stages of nodulation. Aust. J. Biol. Sci. 20, 1087–1104.

Gibson A H, Curnow B C, Bergersen F J, Brockwell J and Robinson A C 1975 Studies of field populations of *Rhizobium*: effectiveness of strains of *Rhizobium trifolii* associated with *Trifolium subterraneum* L. pastures in south-eastern Australia. Soil Biol. Biochem. 7, 95–102.

Gibson A H, Demezas D H, Gault R R, Bhuvaneswari T V and Brockwell J 1990 Genetic stability in rhizobia in the field. Plant and Soil 129, 37–44.

Gibson A H, Dudman W F, Weaver R W, Horton J C and Anderson I C 1971 Variation within serogroup 123 of *Rhizobium japonicum*. Plant and Soil, special vol., 33–37.

Glenn A R and Dilworth M J 1981 Oxidation of substrates by isolated bacteroids and free living cells of *Rhizobium leguminosarum* 3841. J. Gen. Microbiol. 126, 243–247.

Graham P H 1964 The application of computer techniques to the taxonomy of the root-nodule bacteria of legumes. J. Gen. Microbiol. 35, 511–517.

Graham P H 1969 Selective medium for growth of *Rhizobium*. Appl. Microbiol. 17, 769–770.

Graham P H, Sadowsky M J, Keyser H H, Barnet Y M, Bradley R S, Cooper J E, De Ley D J, Jarvis B D W, Roslycky E B, Strijdom B W and Young J P W 1991 Proposed minimal standards for the description of new genera and species of root- and stem-nodulating bacteria. Int. J. Syst. Bacteriol. 41, 582–587.

Graham P H, Viteri S E, Mackie F, Vargas A T and Palacios A 1982 Variation in acid soil tolerance among strains of *Rhizobium phaseoli*. Field Crops Res. 5, 121–128.

Graham-Weiss L, Bennett M L and Paau A S 1987 Production of bacterial inoculants by direct fermentation on nutrient-supplemented vermicilite. Appl. Environ. Microbiol. 53, 2138–2140.

Gremaud M F and Harper J E 1989 Selection and initial characterization of partially nitrate tolerant mutants of soybean. Plant Physiol. 89, 169–173.

Griffith G W and Roughley R J 1992 The effect of moisture potential on growth and survival of root nodule bacteria in peat culture and on seed. J. Appl. Bacteriol. 73, 1–13.

Griffith G W, Roughley R J and Brown J F 1992 The effect of packaging films on growth and survival of rhizobia in peat culture. Lett. Appl. Microbiol. 14, 241–243.

Hagedorn C, Ardahl A H and Materon L A 1983 Characteristics of *Rhizobium trifolii* populations associated with subclover in Mississippi soils. Soil Sci. Soc. Am. J. 47, 1148–1152.

Hale C N 1981 Methods of white clover inoculation - their effect on competition for nodule formation between naturalised and inoculated strains of *Rhizobium trifolii.* N.Z. J. Exp. Agric. 9, 169–172.

Ham G E 1980 Inoculation of legumes with *Rhizobium* in competition with naturalized strains. *In* Nitrogen Fixation, Vol. 2. Eds. W E Newton and W H Orme-Johnson. pp 131–138. Univ. Park Press, Baltimore.

Ham G E, Frederick L R and Anderson I C 1971 Serogroups of *Rhizobium japonicum* in soybean nodules sampled in Iowa. Agron. J. 63, 69–72.

Harris J R 1954 Rhizosphere relationships of subterranean clover. Aust. J. Agric. Res. 5, 247–270.

Harrison S P, Mytton L R, Skot L, Dye M and Cresswell A 1992 Characterisation of *Rhizobium* isolates by amplification of DNA polymorphisms using random primers. Can. J. Microbiol. 38, 1009–1015.

Hartel P G and Alexander M 1983 Growth and survival of cowpea rhizobia in acid, aluminium-rich soils. Soil Sci. Soc. Am. J. 47, 502–506.

Hartley R A, Lawn R J and Byth D E 1993 Genotypic variation in growth and seed yield of soybean [*Glycine max* (L.) Merr.] in saturated soil culture. Aust. J. Agric. Res. 44, 689–702.

Hashem F M and Angle J S 1988 Rhizobiophage effects on *Bradyrhizobium japonicum* nodulation and soybean growth. Soil Biol. Biochem. 20, 69–73.

Hashem F M and Angle J S 1990 Rhizobiophage effects on nodulation, nitrogen fixation, and yield of field-grown soybeans (*Glycine max* L. Merr.). Biol. Fertil. Soils 9, 330–334.

Hegde S V and Brahmaprakash G P 1992 A dry granular inoculant of *Rhizobium* for soil application. Plant and Soil 144, 309–311.

Heichel G H 1982 Breeding alfalfa for improved nitrogen fixation: a physiological perspective. Iowa State J. Res. 56, 255–280.

Hely F W 1964 Calcium carbonate as an aid to clover establishment in soils with low winter temperatures. Fld Stn Rec., Div. Plant Ind., CSIRO (Aust.) 3(1), 63–68.

Hely F W, Hutchings R J and Zorin M 1980 Methods of rhizobial inoculation and sowing techniques for *Trifolium subterranean* L. establishment in a harsh winter environment. Aust. J. Agric. Res. 31, 703–712.

Helyar K R 1976 Nitrogen cycling and soil acidification. J. Aust. Inst. Agric. Sci. 42, 217–221.

Hennecke H 1990 Nitrogen fixation genes involved in the *Bradyrhizobium japonicum*-soybean symbiosis. FEBS Lett. 268, 422–426.

Hennecke H, Alvarez-Morales A, Betancourt-Alvarez M, Ebeling S, Filser M, Fischer H-M, Gubler M, Hahn M, Kaluza K, Lamb J W, Meyer L, Regensburger B, Struder D and Weber J 1985 Organization and regulation of symbiotic nitrogen fixation genes from *Bradyrhizobium japonicum. In* Nitrogen Fixation Research Progress. Eds. H J Evans, P J Bottomley and W E Newton. pp 157–163. Martinus Nijhoff, Dordrecht.

Herridge D F and Betts J H 1988 Field evaluation of soybean genotypes selected for enhanced capacity to nodulate and fix nitrogen in the presence of nitrate. Plant and Soil 110, 129–135.

Herridge D F and Danso S K A 1995 Enhancing crop legume N_2 fixation through selection and breeding. Plant and Soil 174.

Herridge D F and Roughley R J 1975 Variation in colony characteristics and symbiotic effectiveness of *Rhizobium*. J. Appl. Bacteriol. 38, 19–27.

Herridge D F, Roughley R J and Brockwell J 1984 Effect of rhizobia and soil nitrate on the establishment and functioning of the soybean symbiosis in the field. Aust. J. Agric. Res. 35, 149–161.

Heyn C C 1963 The Annual Species of *Medicago*. Scripta Hierosolymitana, Vol. 12. pp 1–154. Hebrew Univ., Jerusalem.

Hiltbold A E, Patterson R M and Reed R B 1985 Soil populations of *Rhizobium japonicum* in a cotton-corn-soybean rotation. Soil Sci. Soc. Am. J. 49, 343–348.

Hiltbold A E, Thurlow D L and Skipper H D 1980 Evaluation of commercial soybean inoculants by various techniques. Agron. J. 72, 675–681.

Hoben H J, Aung N N, Somasegaran P and Kang U G 1991 Oils as adhesives for seed inoculation and their influence on the survival of *Rhizobium* spp. and *Bradyrhizobium* spp. on inoculated seeds. World J. Microbiol. Biotechnol. 7, 324–330.

Hoben H J and Somasegaran P 1992 A small glass fermentor for the production of *Rhizobium* inoculum. World J. Microbiol. Biotechnol. 8, 333–334.

Horvath B, Bachem C W B, Schell J and Kondorosi A 1987 Host-specific regulation of nodulation genes in *Rhizobium* is mediated by a plant-signal, interacting with the *nod*D gene product. EMBO J. 6, 841–848.

Howieson J G and Ewing M A 1986 Acid tolerance in the *Rhizobium meliloti-Medicago* symbiosis. Aust. J. Agric. Res. 37, 55–64.

Howieson J G and Ewing M A 1989 Annual species of *Medicago* differ greatly in their ability to nodulate on acid soils. Aust. J. Agric. Res. 40, 843–850.

Howieson J G, Ewing M A and d'Antuono M F 1988 Selection for acid tolerance in *Rhizobium meliloti*. Plant and Soil 105, 179–188.

Hughes D Q and Vincent J M 1942 Serological studies of the root-nodule bacteria. III. Tests of neighbouring strains of the same species. Proc. Linn. Soc. NSW 67, 141–152.

Hume D J and Blair D H 1992 Effect of numbers of *Bradyrhizobium japonicum* applied in commercial inoculant on soybean yield in Ontario. Can. J. Microbiol. 38, 588–593.

Ireland J A and Vincent J M 1968 A quantitative study of competition for nodule formation. Trans 9th Int. Congr. Soil Sci., Vol. 2. pp 85–93. Augus and Robertson, Sydney, Australia.

Itoh S and Barber S A 1983 Phosphorus uptake by six plant species as related to root hairs. Agron. J. 75, 457–461.

Jansen van Rensberg H and Strijdom B W 1980 Survival of fast and slow-growing *Rhizobium* spp. under conditions of relatively mild desiccation. Soil Biol. Biochem. 12, 353–356.

Jansen van Rensberg H and Strijdom B W 1985 Effectiveness of *Rhizobium* strains used in inoculants after their introduction into soil. Appl. Environ. Microbiol. 49, 127–131.

Jarvis B D W, Downer H L and Young J P W 1992 Phylogeny of fast-growing soybean-nodulating rhizobia supports synonymy of *Sinorhizobium* and *Rhizobium* and assignment to *Rhizobium fredii*. Int. J. Syst. Bacteriol. 42, 93–96.

Jawson M D, Franzluebbers A J and Berg R K 1989 *Bradyrhizobium japonicum* survival in and soybean inoculation with fluid gels. Appl. Environ. Microbiol. 55, 617–622.

Jenkins H V, Vincent J M and Waters L M 1954 The root-nodule bacteria as factors in clover establishment in the red basaltic soils of the Lismore district, New South Wales. III. Field inoculation trials. Aust. J. Agric. Res. 5, 77–89.

Jenkins M B and Bottomley P J 1985 Seasonal response of uninoculated alfalfa to N fertilizer: soil N, nodule turnover, and symbiotic effectiveness of *Rhizobium meliloti*. Agron. J. 76, 959–963.

Jenkins M B, Virginia R A and Jarrell W M 1987 Rhizobial ecology of the woody legume mesquite (*Prosopis glandulosa*) in the Sonoran desert. Appl. Environ. Microbiol. 53, 36–40.

Jensen E S 1987 Inoculation of pea by application of *Rhizobium* in the planting furrow. Plant and Soil 97, 63–70.

Johnson H W and Means U M 1963 Serological groups of *Rhizobium japonicum* recovered from nodules of soybeans (*Glycine max*) in field soils. Agron. J. 55, 269–271.

Johnson H W, Means U M and Weber C R 1965 Competition for nodule sites between strains of *Rhizobium japonicum* applied as inoculum and strains in the soil. Agron. J. 57, 179–185.

Johnston A W B and Beringer J E 1977 Chromosomal recombination between *Rhizobium* species. Nature, London 267, 611–613.

Jones D G 1991 Symbiotic nitrogen fixation - exploitation and unachieved potential. Ann. Appl. Biol. 118, 249–259.

Jones D G and Burrows A C 1969 Acid production and symbiotic effectiveness in *Rhizobium trifolii*. Soil Biol. Biochem. 1, 57–61.

Jones R and Giddens J 1984 Introduction of effective N_2-fixing rhizobial strains into the soybean plant by use of fungicide resistance. Agron. J. 76, 599–602.

Jordan D C 1982 Transfer of *Rhizobium japonicum* Buchanan 1980 to *Bradyrhizobium* gen. nov., a genus of slow-growing, root nodule bacteria from leguminous plants. Int. J. Syst. Bacteriol. 32, 136–139.

Jung G, Mugnier J, Diem H G and Dommergues Y R 1982 Polymer-entrapped rhizobium as an inoculant for legumes. Plant and Soil 65, 219–231.

Kamicker B J and Brill W J 1986 Identification of *Bradyrhizobium japonicum* nodule isolates from Wisconsin soybean farms. Appl. Environ. Microbiol. 51, 487–492.

Kecskés M 1970 Comparative investigations of the action of fungicides on *Rhizobium leguminosarum* F. and its symbiosis with *Vicia sativa*. Meded. Fac. Landbouwwet. Rijksuniv. Gent 35, 505–514.

Keyser H H and Li F 1992 Potential for increasing biological nitrogen fixation in soybean. Plant and Soil 141, 119–135.

Keyser H H and Munns D N 1979 Tolerance of rhizobia to acidity, aluminium and phosphate. Soil Sci. Soc. Am. J. 43, 519–523.

Keyser H H, Munns D N and Hohenberg J S 1979 Acid tolerance of rhizobia in culture and in symbiosis with cowpea. Soil Sci. Soc. Am. J. 43, 719–722.

Keyser H H, Somasegaran P and Bohlool B B 1992 Rhizobial ecology and technology. *In* Soil Microbial Ecology. Applications in Agricultural and Environmental Management. Ed. F B Metting Jr. pp 205–226. Marcel Dekker, New York.

Kingsley M T and Bohlool B B 1981 Release of *Rhizobium* spp. from tropical soils and recovery for immunofluorescence enumeration. Appl. Environ. Microbiol. 42, 241–248.

Kinkle B K, Sadowsky M J, Johnstone K and Koskinen W C 1994 Tellurium and selenium resistance in rhizobia and its potential use for direct isolation of *Rhizobium meliloti* from soil. Appl. Environ. Microbiol. 60, 1674–1677.

Kinkle B K, Sadowsky M J, Schmidt E L and Koskinen W C 1993 Plasmids pJP4 and r68.45 can be transferred between populations of bradyrhizobia in nonsterile soil. Appl. Environ. Microbiol. 59, 1762–1766.

Kinkle B K and Schmidt E L 1991 Transfer of the pea symbiotic plasmid pJB5JI in nonsterile soil. Appl. Environ. Microbiol. 57, 3264–3269.

Kipe-Nolt J A, Montealegre M and Tohme J 1992 Restriction of nodulation by the broad host range *Rhizobium tropici* strain CIAT899 in wild accessions of *Phaseolus vulgaris* L. New Phytol. 120, 489–494.

Kishinevsky B D, Lobel R, Gurfel D and Nemas C 1992 Soil fumigation with methyl bromide as a means of increasing the occurrence of the inoculum strain in peanut nodules. Soil Biol. Biochem. 24, 845–848.

Klucas R V, Hanus F J, Russell S A and Evans H J 1983 Nickel: A micronutrient element for hydrogen-dependent growth of *Rhizobium japonicum* and for expression of urease activity in soybean leaves. Proc. Natl. Acad. Sci. USA 80, 2253–2257.

Kondorosi A, Horvath B, Gottfert M, Putnoky P, Rastas K, Gyorgypal Z, Kondorosi E, Torok I, Bachem C, John M, Schmidt J and Schell J 1985 Identification and organization of *Rhizobium meliloti* genes relevant to the initiation and development of nodules. *In* Nitrogen Fixation Research Progress. Eds. H J Evans, P J Bottomley and W E Newton. pp 73–78. Martinus Nijhoff, Dordrecht.

Kosslak R M and Bohlool B B 1985 Influence of environmental factors on interstrain comptition in *Rhizobium japonicum*. Appl. Environ. Microbiol. 49, 1128–1153.

Kosslak R M, Bookland R, Barkei J, Paaren H E and Appelbaum E R 1987 Induction of *Bradyrhizobium japonicum* common *nod* genes by isoflavones isolated from *Glycine max*. Proc. Natl. Acad. Sci. USA 84, 7428–7432.

Kreig N R and Holt J G 1984 Bergey's Manual of Determinative Bacteriology, Vol. 1. Williams and Wilkins, Baltimore. 964p.

Kremer R J and Peterson H L 1982 Effect of inoculant carrier on survival of *Rhizobium* on inoculated seed. Soil Sci. 134, 117–125.

Kremer R J and Peterson H L 1983a Field evaluation of selected *Rhizobium* in an improved legume inoculant. Agron. J. 75, 139–143.

Kremer R J and Peterson H L 1983b Effects of carrier and temperature on survival of *Rhizobium* spp. in legume inocula: development of an improved type of inoculant. Appl. Environ. Microbiol. 45, 1790–1794.

Kremer R J, Polo J and Peterson H L 1982 Effect of suspending agent and temperature on survival of *Rhizobium* in fertilizer. Soil Sci. Soc. Am. J. 46, 539–542.

Kucey R M N and Paul E A 1982 Carbon flow, photosynthesis and N_2 fixation in mycorrhizal and nodulated faba beans (*Vicia faba* L.). Soil Biol. Biochem. 14, 407–412.

Kuykendall L D, Devine T E and Cregan P B 1982 Positive role of nodulation on the establishment of *Rhizobium* in subsequent crops of soybean. Can. J. Microbiol. 7, 79–81.

Kuykendall L D and Elkan G H 1976 *Rhizobium japonicum* derivatives differing in nitrogen fixing efficiency and carbohydrate ulitization. Appl. Environ. Microbiol. 32, 511–519.

Kuykendall L D, Saxena B, Devine T E and Udell S E 1992 Genetic diversity in *Bradyrhizobium japonicum* Jordan 1982 and a proposal for *Bradyrhizobium elkanii* sp. nov. Can. J. Microbiol. 38, 501–503.

Labandera C A and Vincent J M 1975 Loss of symbiotic capacity in commercially useful strains of *Rhizobium trifolii*. J. Appl. Bacteriol. 39, 209–211.

Ladha J K, Garcia M, Miyan S, Padre A T and Watanabe I 1989 Survival of *Azorhizobium caulinodans* in the soil and rhizosphere of wetland rice under *Sesbania rostrata*-rice rotation. Appl. Environ. Microbiol. 55, 454–460.

Ladha J K, Pareek R P and Becker M 1992 Stem-nodulating legume-*Rhizobium* symbiosis and its agronomic use in lowland rice. *In* Advances in Soil Science, Vol. 20. pp 147–192. Springer-Verlag, New York.

Ladha J K, Pareek R P, So R and Becker M 1990 Stem-nodule symbiosis and its unusual properties. *In* Nitrogen Fixation: Achievement and Objectives. Eds. P M Gresshoff, L E Roth, G Stacey and W E Newton. pp 633–640. Chapman and Hall, New York.

Ladha J K, Watanabe I and Saono S 1988 Nitrogen fixation by leguminous green manure and practices for its enhancement in tropical lowland rice. *In* Sustainable Agriculture: Green Manure in Rice Farming. pp 165–183. IRRI, Los Banos, Philippines.

La Favre A K, Sinclair M J, La Favre J S and Eaglesham A R J 1991 *Bradyrhizobium japonicum* native to tropical soils: novel sources of strains for inoculants for US-type soya bean. Trop. Agric. 68, 243–248.

La Favre J S and Eaglesham A R J 1984 Increased nodulation of "non-nodulating" (rj_1rj_1) soybean by high dose inoculation. Plant and Soil 80, 297–300.

Lange R T 1961 Nodule bacteria associated with the indigenous Leguminosae of south-west Australia. J. Gen. Microbiol. 61, 351–359.

Lawrie A C 1983 Relationships among rhizobia from native Australian legumes. Appl. Environ. Microbiol. 45, 1822–1828.

Leung K, Strain S R, de Bruijn F J and Bottomley P J 1994a Genotypic and phenotypic comparisons of chromosomal types within an indigenous soil population of *Rhizobium leguminosarum* bv. *trifolii*. Appl. Environ. Microbiol. 60, 416–426.

Leung K, Yap K, Dashti N and Bottomley P J 1994b Serological and ecological characteristics of a nodule-dominent serotype from an indigenous soil population of *Rhizobium leguminosarum* bv. *trifolii*. Appl. Environ. Microbiol. 60, 408–415.

Lie T A 1978 Symbiotic specialization in pea plants: the requirement of specific *Rhizobium* strains for peas from Afghanistan. Ann. Appl. Biol. 88, 462–465.

Lindstrom K, Lipsanen P and Kaijalainen S 1990 Stability of markers used for identification of two *Rhizobium galegae* inoculant strains after five years in the field. Appl. Environ. Microbiol. 56, 444–450.

Loneragan J F, Meyer D, Fawcett R G and Anderson A J 1955 Lime-pelleted clover seeds for nodulation on acid soils. J. Aust. Inst. Agric. Sci. 21, 264–265.

Long S R 1989 *Rhizobium*-legume nodulation: life together in the underground. Cell 56, 203–214.

Lowther W L and Littlejohn R P 1984 Effect of strain of rhizobia, inoculation level, and pelleting on the establishment of oversown *Lotus pedunculatus* "Grasslands Maku". N.Z. J. Exp. Agric. 12, 287–294.

Lowther W L and Loneragan J F 1968 Calcium and nodulation in subterranean clover (*Trifolium subterraneum* L.). Plant Physiol. 43, 1362–1366.

Lussenhop J 1993 Effects of two collembola species on nodule occupancy by two *Bradyrhizobium japonicum* strains. Soil Biol. Biochem. 25, 775–780.

McLaughlin M J, Malik K A, Memon K S and Idris M 1990 The role of phosphorus in nitrogen fixation in upland crops. Proc. Workshop Phosphorus Requirements for Sustainable Agriculture in Asia and Oceania. pp 295–305. IRRI, Los Banos, Philippines.

Mackay A D and Barber S A 1985 Effect of soil moisture and phosphate level on root hair growth of corn roots. Plant and Soil 86, 321–331.

Mahdi A A and Atabani I M A 1992 Response of *Bradyrhizobium*-inoculated soyabean and lablab bean to inoculation with vesicular-arbuscular mycorrhizae. Exp. Agric. 28, 399–407.

Mahler R L and Wollum A G, II 1980 Influence of water potential on the survival of rhizobia in a Goldsboro loamy sand. Soil Sci. Soc. Am. J. 44, 988–992.

Mahler R L and Wollum A G, II 1981 The influence of soil water potential and soil texture on the survival of *Rhizobium japonicum* and *Rhizobium leguminosarum* isolates in the soil. Soil Sci. Soc. Am. J. 45, 761–766.

Mahler R L and Wollum A G, II 1982 Seasonal fluctuation of *Rhizobium japonicum* under a variety of field conditions in North Carolina. Soil Sci. 134, 317–324.

Marshall K C 1969 Studies by microelectrophoretic and microscopic techniques of the sorption of illite and montmorillonite to rhizobia. J. Gen. Microbiol. 56, 301–306.

Martin R C, Voldeng H D and Smith D L 1991 Nitrogen transfer from nodulating soybean to maize or to nonnodulating soybean in intercrops: the ^{15}N dilution method. Plant and Soil 132, 53–63.

Mary P, Dupuy N, Dolhem–Biremon C, Defives C and Tailliez R 1994 Differences among *Rhizobium meliloti* and *Bradyrhizobium japonicum* strains in tolerance to desiccation and storage at different relative humidities. Soil Biol. Biochem. 26, 1125–1132.

Masterson C L and Sherwood M T 1974 Selection of *Rhizobium trifolii* strains by white and subterranean clovers. Ir. J. Agric. Res. 13, 91–99.

Materon L A and Weaver R W 1984 Toxicity of arrowleaf clover seed to *Rhizobium trifolii*. Agron. J. 76, 471–473.

Materon L A and Weaver R W 1985 Inoculant maturity influences survival of rhizobia on seed. Appl. Environ. Microbiol. 49, 465–467.

Matthews S, Powell A A and Spaeth S C 1988 Seedling vigour and susceptibility to diseases and pests. *In* World Crops: Cool Season Food Legumes. Ed. R J Summerfield. pp 619–625. Kluwer, Dordrecht.

May S N and Bohlool B B 1983 Competition among *Rhizobium leguminosarum* strains for nodulation of lentils (*Lens esculenta*). Appl. Environ. Microbiol. 45, 960–965.

Moawad H and Bohlool B B 1984 Competition among *Rhizobium* spp. for nodulation of *Leucaena leucocephala* in two tropical soils. Appl. Environ. Microbiol. 48, 5–9.

Moawad H A, Ellis W R and Schmidt E L 1984 Rhizosphere response as a factor in competition among three serogroups of indigenous *Rhizobium japonicum* for nodulation of field-grown soybeans. Appl. Environ. Microbiol. 47, 607–612.

Moënne-Loccoz Y, Sen D, Krause E S and Weaver R W 1994 Plasmid profiles of rhizobia used in inoculants and isolated from clover fields. Agron. J. 86, 117–121.

Moore W E C and Moore L V H 1992 Index of the Bacterial and Yeast Nomenclatural Changes. American Society for Microbiology, Washington DC.

Muldoon J F, Hume D J and Beversdorf W D 1980 Effects of seed- and soil-applied *Rhizobium japonicum* inoculants on soybeans in Ontario. Can. J. Plant Sci. 60, 399–409.

Munevar F and Wollum A G, II 1981 Growth of *Rhizobium japonicum* strains at temperatures above 27°C. Appl. Environ. Microbiol. 42, 272–276.

Munns D N 1968a Nodulation of *Medicago sativa* in solution culture. II. Compensating effects of nitrate and of prior nodulation. Plant and Soil 28, 246–257.

Munns D N 1968b Nodulation of *Medicago sativa* in solution culture. III. Effects of nitrate on root hairs and infection. Plant and Soil 29, 33–47.

Munns D N 1970 Nodulation of *Medicago sativa* in solution culture. V. Calcium and pH requirements during infection. Plant and Soil 32, 90–107.

Munns D N 1977 Mineral nutrition and the legume symbiosis. *In* A Treatise on Dinitrogen Fixation. Section IV. Agronomy and Ecology. Eds. R W F Hardy and A H Gibson. pp 353–391. John Wiley and Sons, New York.

Munns D N and Mosse B 1980 Mineral nutrition of legume crops. *In* Advances in Legume Science. Eds. R J Summerfield and A H Bunting. pp 115–125. HMSO, London.

Mytton L R, Brockwell J and Gibson A H 1984 The potential for breeding an improved lucerne-*Rhizobium* symbiosis.1. Assessment of genetic variability. Euphytica 33, 401–410.

Nambiar P T C, Ravishanker H N and Dart P J 1983 Effect of *Rhizobium* numbers on nodulation and dinitrogen fixation in groundnut. Exp. Agric. 19, 243–250.

Nathanson K, Lawn R J, de Jabrun P L M and Byth D E 1984 Growth, nodulation and nitrogen accumulation by soybean in saturated soil culture. Field Crops Res. 8, 73–92.

Nayudu M and Rolfe B G 1987 Analysis of R-primes demonstrates that genes for broad host range nodulation of *Rhizobium* strain NGR234 are dispersed on the Sym plasmid. Mol. Gen. Genet. 206, 326–337.

Nishijima F, Evans W R and Vesper S J 1988 Enhanced nodulation of soybean by *Bradyrhizobium* in the presence of *Pseudomonas fluorescens*. Plant and Soil 111, 149–150.

Norris D O 1956 Legumes and the *Rhizobium* symbiosis. Emp. J. Exp. Agric. 27, 247–270.

Norris D O 1959 *Rhizobium* affinities of African species of *Trifolium*. Emp. J. Exp. Agric. 27, 87–97.

Nutman P S 1946 Variation within strains of clover nodule bacteria in the size of nodule produced and in the effectivity of the symbiosis. J. Bacteriol. 51, 411–432.

Nutman P S 1959 Some observations on root-hair infection by nodule bacteria. J. Exp. Bot. 10, 250–263.

Nutman P S 1965 The relationship between nodule bacteria and the legume host in the rhizosphere and in the process of infection. *In* Ecology of Soil-Borne Pathogens. Eds. K F Baker and W C Snyder. pp 231–247. Univ. California Press, Berkeley.

Nutman P S 1973 Effects of antibacterial substances on *Rhizobium*. *In* Rothamsted Experimental Station Report for 1972, part 1. pp 83–84. Rothamsted Experimental Station, Harpenden, UK.

O'Hara G W, Boonkerd N and Dilworth M J 1988 Mineral constraints to nitrogen fixation. Plant and Soil 108, 93–110.

Odeyemi O and Alexander M 1977 Use of fungicide-resistant rhizobia for legume inoculation. Soil Biol. Biochem. 9, 247–251.

Olsen P E, Rice W A and Collins M M 1995 Biological contaminants in North American rhizobial legume inoculants. Soil Biol. Biochem. 27 (*In press*).

Osa-Afiana L O and Alexander M 1979 Effect of moisture on the survival of *Rhizobium* in soil. Soil Sci. Soc. Am. J. 43, 925–930.

Paau A S 1989 Improvement of *Rhizobium* inoculants. Appl. Environ. Microbiol. 55, 862–865.

Paau A S 1991 Improvement of *Rhizobium* inoculants by mutation, genetic engineering and formulation. Biotechnol. Adv. 9, 173–184.

Pareek R P, Ladha J K and Watanabe I 1990 Estimating N_2 fixation by *Sesbania rostrata* and *S. cannabina* (syn. *S. aculeata*) in lowland rice by the ^{15}N dilution method. Biol. Fertil. Soils 10, 77–88.

Park S J and Buttery B R 1988 Nodulation mutants of white bean (*Phaseolus vulgaris* L.) induced by ethyl-methane sulphonate. Can. J. Plant Sci. 68, 199–202.

Parke D and Ornston L N 1984 Nutritional diversity of Rhizobiaceae revealed by auxanography. J. Gen. Microbiol. 130, 1743–1750.

Parke D and Ornston L N 1986 Enzymes of the β-ketoadipate pathway are inducible in *Rhizobium* and *Agrobacterium* spp. and constitutive in *Bradyrhizobium* spp. J. Bacteriol. 165, 288–292.

Parke D, Rivelli M and Ornston L N 1985 Chemotaxis to aromatic and hydroaromatic acids: comparison of *Bradyrhizobium japonicum* and *Rhizobium trifolii*. J. Bacteriol. 113, 417–422.

Parker C A and Oakley A E 1965 Reduced nodulation of lupins and serradella due to lime pelleting. Aust. J. Exp. Agric. Anim. Husb. 5, 144–146.

Parker C A, Trinick M J and Chatel D L 1977 Rhizobia as soil and rhizosphere inhabitants. *In* A Treatise on Dinitrogen Fixation. Section IV. Agronomy and Ecology. Eds. R W F Hardy and A H Gibson. pp 311–352. John Wiley and Sons, New York.

Parker F E and Vincent J M 1981 Sterilization of peat by gamma irradiation. Plant and Soil 61, 285–293.

Partridge C D P and Yates M G 1982 Effect of chelating agents on hydrogenase in *Azotobacter chroococcum*. Biochem. J. 204, 339–344.

Paul E A and Voroney R P 1984 Field interpretation of microbial biomass activity measurements. *In* Current Perspectives in Microbial Ecology. Eds. M J Klug and C A Reddy. pp 509–514. American Society for Microbiology, Washington DC.

Peoples M B, Faizah A W, Rerkasem B and Herridge D F 1989 Methods for Evaluating Nitrogen Fixation by Nodulated Legumes in the Field. ACIAR, Canberra. 76 p.

Peoples M B, Herridge D F and Ladha J K 1995a Biological nitrogen fixation: an efficient source of nitrogen for sustainable agricultural production? Plant and Soil 174.

Peoples M B, Ladha J K and Herridge D F 1995b Enhancing legume N_2 fixation through plant and soil management. Plant and Soil 174.

Peters N K, Frost J W and Long S R 1986 A plant flavone, luteolin, induces expression of *Rhizobium meliloti* nodulation genes. Science 233, 977–980.

Phillips D A, Dakora F D, Leon–Barrios M, Sande E and Joseph C M 1993a Signals released from alfalfa regulate microbial activities in the rhizosphere. *In* New Horizons in Nitrogen Fixation. Eds. J Palacios, J Mora and W E Newton. pp 197–202. Kluwer Acad. Publ., Dordrecht.

Phillips D A, Joseph C M and Maxwell C A 1993b Non-flavonoid inducers of *nod* genes in *Rhizobium meliloti*: apparent *nod*D2 activities released naturally from alfalfa seeds add new dimensions to rhizosphere biology. *In* Advances in Molecular Genetics of Plant-Microbe Interactions, Vol. 2. Eds. E W Nester and D P S Verma. pp 169–173. Kluwer Acad. Publ., Dordrecht.

Pineda J A, Kipe-Nolt J A and Rojas E 1994 *Rhizobium* inoculation increases of bean and maize yields in intercrops on farms in the Peruvian sierra. Exp. Agric. 30, 311–318.

Postgate J R 1984 Concluding remarks. *In* Advances in Nitrogen Fixation Research. Eds. C Veeger and W E Newton. pp ix-x. Martinus Nijhoff / Dr. Junk Publ., The Hague.

Quackenbush F W, Carter A S and Shenberger L C 1961 Inspection of legume inoculants and growth substances. Inspection Rep. No. 28. Purdue Univ. Agric. Exp. Stn, Lafayette, IN. 8p.

Ravuri V and Hume D J 1992 Performance of a superior *Bradyrhizobium japonicum* and a selected *Sinorhizobium fredii* strain with soybean cultivars. Agron. J. 84, 1051–1056.

Rawsthorne S and Summerfeld R J 1984 An assessment of different techniques for inoculating *Phaseolus vulgaris* with *Rhizobium*. Exp. Agric. 20, 119–127.

Recourt K, Schripsema J, Kijne J W, van Brussel A A N and Lugtenberg B J J 1991 Inoculation of *Vicia sativa* subsp. *nigra* roots with *Rhizobium leguminosarum* biovar *viciae* results in release of *nod* gene activating flavanones and chalcones. Plant Mol. Biol. 16, 841–852.

Redmond J W, Batley M, Djordjevic M A, Innes R W, Kuempel P L and Rolfe B G 1986 Flavones induce expression of nodulation genes in *Rhizobium*. Nature, London 323, 632–635.

Rennie R J, Rennie D A, Siripaibool C, Chaiwanakupt P, Boonkerd N and Snitwongse P 1988 N_2 fixation in Thai soybeans: Effects of tillage and inoculation on ^{15}N-determined N_2 fixation in recommended cultivars and advanced breeding lines. Plant and Soil 112, 183–193.

Renwick A and Jones D G 1986 The manipulation of white clover "host preference" for strains of *Rhizobium trifolii* in an upland soil. Ann. Appl. Biol. 108, 291–302.

Rerkasem B, Rerkasem K, Peoples M B, Herridge D F and Bergersen F J 1988 Measurement of N_2 fixation in maize (*Zea mays* L.) - ricebean (*Vigna umbellata* [Thunb.] Ohwi and Ohashi) intercrops. Plant and Soil 108, 125–135.

Rice W A and Olsen P E 1992 Effects of inoculation method and size of *Rhizobium meliloti* population in the soil on nodulation of alfalfa. Can. J. Soil Sci. 72, 57–67.

Rice W A, Olsen P E and Leggett M E 1995 Co-culture of *Rhizobium* and phosphate–solubilizing fungus in sterile peat. Soil Biol. Biochem. 27 (*In press*).

Rice W A, Penney D C and Nyborg M 1977 Effects of soil acidity on rhizobia numbers, nodulation and nitrogen fxation by alfalfa and red clover. Can. J. Soil Sci. 57, 197–203.

Richardson A E and Simpson R J 1989 Acid-tolerance and symbiotic effectiveness of *Rhizobium trifolii* associated with a *Trifolium subterraneum* L.-based pasture growing in an acid soil. Soil Biol. Biochem. 21, 87–95.

Richardson A E, Viccars L A, Watson J M and Gibson A H 1995 Differentiation of *Rhizobium* strains using the polymerase chain reaction with random and directed primers. Soil Biol. Biochem. 27 (*In press*).

Rinaudo G, Dreyfus B and Dommergues Y 1988 Stem–nodulating legumes as green manure for rice in West Africa. *In* Green Manure in Rice Farming. pp 97–109. IRRI, Los Banos, Philippines.

Robert F M and Schmidt E L 1985 Response of three indigenous serogroups of *Rhizobium japonicum* to the rhizosphere of pre-emergent seedlings of soybeans. Soil Biol. Biochem. 17, 579–580.

Robinson A C 1969 Host selection for effective *Rhizobium trifolii* by red clover and subterranean clover in the field. Aust. J. Agric. Res. 20, 1053–1060.

Robinson P E 1961 Root-knot nematodes and legume nodules. Nature, London 189, 506–507.

Robson A D and Bottomley P J 1991 Limitations in the use of legumes in agriculture and forestry. *In* Biology and Biochemistry of Nitrogen Fixation. Eds. M J Dilworth and A R Glenn. pp 320–349. Elsevier, Amsterdam.

Robson A D, O'Hara G W and Abbott L K 1981 Involvement of phosphorus in nitrogen fixation by subterranean clover (*Trifolium subterraneum* L.). Aust. J. Plant Physiol. 8, 427–436.

Rogers D D, Warren R D Jr and Chamblee D S 1982 Remedial postemergence inoculation with *Rhizobium*. Agron. J. 74, 613–619.

Rolfe B G, Innes R W, Schofield P R, Watson J M, Sargent C L, Kuempel P L, Plazinski J, Canter-Cremers H and Djordjevic M A 1985 Plant secreted factors induce the expression of *R. trifolii* nodulation and host-range genes. *In* Nitrogen Fixation Research Progress. Eds. H J Evans, P J Bottomley and W E Newton. pp 79–85. Martinus Nijhoff, Dordrecht.

Roskoski J P, Pepper I and Pardo E 1986 Inoculation of leguminous trees with rhizobia and VA mycorrhizal fungi. For. Ecol. Manage. 16, 57–68.

Roughley R J, Blowes W M and Herridge D F 1976 Nodulation of *Trifolium subterraneum* L. by introduced rhizobia in competition with naturalized strains. Soil Biol. Biochem. 8, 403–407.

Roughley R J and Brockwell J 1987 Grain legumes and soil micro-organisms. *In* Grain Legumes 1987 Research and Production Seminar. AIAS Occ. Publ. No. 28. Eds. D R de Kantzow and M G May. pp 66–69. Aust. Inst. Agric. Sci., Sydney.

Roughley R J, Bullard G K and Gemell L G 1984 The Australian Inoculants Research and Control Service. N.S.W. Dept. Agric., Gosford, NSW. 21p.

Roughley R J, Gault R R, Gemell L G, Andrews J A, Brockwell J, Dunn B W, Griffith G W, Hartley E J, Hebb D M, Peoples M B and Thompson J A 1995 Autecology of *Bradyrhizobium japonicum* in soybean-rice rotations. Plant and Soil (*In press*).

Roughley R J, Gemell L G, Thompson J A and Brockwell J 1993 The number of *Bradyrhizobium* sp. (*Lupinus*) applied to seed and its effect on rhizosphere colonization, nodulation and yield of lupin. Soil Biol. Biochem. 25, 1453–1458.

Roughley R J and Vincent J M 1967 Growth and survival of *Rhizobium* spp. in peat culture. J. Appl. Bacteriol. 30, 362–376.

Rupela O P, Toomsan B, Mittal S, Dart P J and Thompson J A 1987 Chickpea *Rhizobium* populations: survey of influence of season, soil depth and cropping pattern. Soil Biol. Biochem. 19, 247–252.

Sadowsky M J and Bohlool B B 1985 Differential expression of the pea symbiotic plasmid pJB5JI in genetically dissimilar backgrounds. Symbiosis 1, 125–138.

Salema M P, Parker C A, Kidby D K and Chatel D L 1982 Death of rhizobia on inoculated seed. Soil Biol. Biochem. 14, 13–14.

Sanginga N, Vanlauwe B and Danso S K A 1995 Management of biological N_2 fixation in alley cropping systems: estimation and contribution to N balance. Plant and Soil 174.

Schall E D, Shenberger L C and Swope A 1975 Inspection of legume inoculants and pre-inoculated seeds. Inspection Rep. No. 106. Purdue Univ. Agric. Exp. Stn Lafayette, IN. 12p.

Schiffman J and Alper Y 1968a Effects of *Rhizobium*-inoculum placement on peanut inoculation. Exp. Agric. 4, 203–208.

Schiffman J and Alper Y 1968b Inoculation of peanuts by application of *Rhizobium* suspensions into the planting furrows. Exp. Agric. 4, 219–226.

Schmidt E L and Robert F M 1985 Recent advances in the ecology of *Rhizobium*. *In* Nitrogen Fixation Research Progress. Eds. H J Evans, P J Bottomley and W E Newton. pp 379–385. Martinus Nijhoff, Dordrecht.

Schmidt E L, Zidwick M J and Abebe H M 1986 *Bradyrhizobium japonicum* serocluster 123 and diversity among member isolates. Appl. Environ. Microbiol. 51, 1212–1215.

Schofield P R, Ridge R W, Rolfe B G, Shine J and Watson J M 1984 Host-specific nodulation is encoded on a 14 kb fragment in *Rhizobium trifolii*. Plant Mol. Biol. 3, 3–11.

Scudder W T 1975 *Rhizobium* inoculation of soybeans for subtropical and tropical soils. 1. Initial field trials. Soil Crop Sci. Soc. Fla Proc. 34, 79–82.

Segovia L, Pinero D, Palacios R and Martinez-Romero E 1991 Genetic structure of a soil population of nonsymbiotic *Rhizobium leguminosarum*. Appl. Environ. Microbiol. 57, 426–433.

Silsbury J H 1984 Comparison of the growth rates of dinitrogen fixing subterranean clover swards with those assimilating nitrate ions. Plant and Soil 80, 201–213.

Singleton P W, AbdelMagid H M and Tavares J W 1985 Effect of phosphorus on the effectiveness of strains of *Rhizobium japonicum*. Soil Sci. Soc. Am. J. 49, 613–616.

Singleton P W, Bohlool B B and Nakao P L 1992 Legume response to rhizobial inoculation in the tropics: myths and realities. *In* Myths and Science of Soils in the Tropics. SSSA Special Publ. No. 29. pp 135–155. Soil Sci. Soc. Am., Madison, WI.

Singleton P W and Stockinger K R 1983 Compensation against ineffective nodulation in soybean. Crop Sci. 23, 69–72.

Singleton P W and Tavares J W 1986 Inoculation response of legumes in relation to the number and effectiveness of indigenous *Rhizobium* populations. Appl. Environ. Microbiol. 51, 1013–1018.

Skipper H D, Palmer J H, Giddens J E and Woodruff J M 1980 Evaluation of commercial soybean inoculants from South Carolina and Georgia. Agron. J. 72, 673–674.

Smith F W 1982 Mineral nutrition of legumes. *In* Nitrogen Fixation in Legumes. Ed. J M Vincent. pp 155–172. Academic Press, Sydney.

Smith R S 1992 Legume inoculant formulation and application. Can. J. Microbiol. 38, 485–492.

Smith R S, Ellis M A and Smith R E 1980 Effect of *Rhizobium japonicum* inoculant rates on soybean nodulation in a tropical soil. Agron. J. 72, 505–508.

Smith S E and Bowen G D 1979 Soil temperature, mycorrhizal infection and nodulation of *Medicago truncatula* and *Trifolium subterraneum*. Soil Biol. Biochem. 11, 469–473.

Smith S E and Daft M J 1977 Interactions between growth, phosphate content and N_2 fixation in mycorrhizal *Medicago sativa*. Aust. J. Plant Physiol. 4, 403–413.

Smith S E and Daft M J 1978 The effect of mycorrhizas on the phosphate content, nitrogen fixation and growth of *Medicago sativa*. *In* Microbial Ecology. Eds. M W Loutit and J A R Miles. pp 314–319. Springer-Verlag, Heidelberg.

So R B, Ladha J K and Young J P W 1994 Photosynthetic symbionts of *Aeschynomene* spp. form a cluster with bradyrhizobia on the basis of fatty acid and ribosomal rRNA analyses. Int. J. Syst. Bacteriol. 44, 392–403.

Somasegaran P 1985 Inoculant production with diluted liquid cultures of *Rhizobium* spp. and autoclaved peat: evaluation of diluents, *Rhizobium* spp., peats, sterility requirements, storage, and plant effectiveness. Appl. Environ. Microbiol. 50, 398–405.

Somasegaran P 1991 Inoculant production with emphasis on choice of carriers, methods of production, and reliability testing/quality assurance guidelines. *In* Expert Consultation on Legume Inoculant Production and Quality Control. Ed. J A Thompson. pp 87–106. FAO, Rome.

Somasegaran P and Bohlool B B 1990 Single-strain versus multistrain inoculation: effect of soil mineral N availability on rhizobial strain effectiveness and competition for nodulation on chick-pea, soybean and dry bean. Appl. Environ. Microbiol. 56, 3298–3303.

Somasegaran P and Halliday J 1982 Dilution of liquid *Rhizobium* cultures to increase production capacity of inoculant plants. Appl. Environ. Microbiol. 44, 330–333.

Somasegaran P and Hoben H J 1994 The Handbook for Rhizobia: Methods in Legume-Rhizobia Technology. Springer Verlag, New York. 450p.

Somasegaran P, Hoben H J and Burton J C 1992 A medium–scale fermentor for mass culture of rhizobia. World J. Microbiol. Biotechnol. 8, 335–336.

Somasegaran P, Hoben H J and Gurgun V 1988 Effect of inoculation rate, rhizobial strain competition and nitrogen fixation in chickpea. Agron. J. 80, 68–73.

Somasegaran P, Hoben H J and Lewinson L 1990 Symbiotic interactions of *Phaseolus acutifolius* and *P. acutifolius* × *P. vulgaris* hybrid progenies in symbiosis with *Bradyrhizobium* spp. and *Rhizobium leguminosarum* bv. *phaseoli*. Can. J. Microbiol. 37, 497–503.

Souza V, Eguiarte L, Avila G, Cappello R, Gallardo C, Montoya J and Pinero D 1994 Genetic structure of *Rhizobium etli* biovar *phaseoli* associated with wild and cultivated bean plants (*Phaseolus vulgaris* and *Phaseolus coccineus*) in Morelos, Mexico. Appl. Environ. Microbiol. 60, 1260–1268.

Streeter J G, Salminen S O, Beuerlein J E and Schmidt W H 1994 Factors influencing the synthesis of polysaccharide by *Bradyrhizobium japonicum* bacteroids in field-grown soybean nodules. Appl. Environ. Microbiol. 60, 2939–2943.

Streeter J G, Salminen S O and Carslon R W 1991 Synthesis of polysaccharide in soybean nodules by *Bradyrhizobium japonicum* strains which are highly competitive for nodule formation. Proc. 13th North American Symbiotic Nitrogen Fixation Conf. p 12.

Strijdom B W and Jansen van Rensburg H 1981 Effect of steam sterilization and gamma irradiation of peat on quality of *Rhizobium* inoculants. Appl. Environ. Microbiol. 41, 1344–1347.

Thiagarajan T R, Ames R N and Ahmed M H 1992 Response of cowpea (*Vigna unguiculata*) to inoculation with co-selected vesicular-arbuscular mycorrhizal fungi and *Rhizobium* strains in field trials. Can. J. Microbiol. 38, 573–576.

Thies J E, Bohlool B B and Singleton P W 1991a Subgroups of the cowpea miscellany: symbiotic specificity within *Bradyrhizobium* spp. for *Vigna unguiculata, Phaseolus iunatus, Arachis hypogaea* and *Macroptilium atropurpureum*. Appl. Environ. Microbiol. 57, 1540–1545.

Thies J E, Bohlool B B and Singleton P W 1992 Environmental effects on competition for nodule occupancy between introduced and indigenous rhizobia and among introduced strains. Can. J. Microbiol. 38, 493–500.

Thies J E, Cook S E and Corner R J 1994 Use of Bayesian influence in a Geographical Information System to determine regional legume inoculation requirements. *In* Proceedings of Resource Technology '94. New Opportunities. Best Practice. pp 475–488. Australian Department of Resources, Melbourne.

Thies J E, Singleton P W and Bohlool B B 1991b Influence of the size of indigenous rhizobial populations on establishment and symbiotic performance of introduced rhizobia on field-grown legumes. Appl. Environ. Microbiol. 57, 19–28.

Thies J E, Singleton P W and Bohlool B B 1991c Modeling symbiotic performance of introduced rhizobia in the field by use of indices of indigenous population size and nitrogen status of the soil. Appl. Environ. Microbiol. 57, 29–37.

Thies J E, Woomer P L and Singleton P W 1995 Enrichment of *Bradyrhizobium* spp. populations in soil due to cropping of the homologous host legume. Soil Biol. Biochem. 27 (*In press*).

Thompson J A 1980 Production and quality control of legume inoculants. *In* Methods for Evaluating Biological Nitrogen Fixation. Ed. F J Bergersen. pp 489–533. John Wiley and Sons, Chichester, UK.

Thompson J A 1983 Production and quality control of carrier-based legume inoculants. Inform. Bull. No. 17. ICRISAT, Patancheru, India.

Thompson J A, Bhromsiri A, Shutsrirung A and Lillakan S 1991 Native root-nodule bacteria of traditional soybean-growing areas of northern Thailand. Plant and Soil 135, 53–65.

Thompson J A, Brockwell J and Roughley R J 1975 Preinoculation of legume seed. J. Aust. Inst. Agric. Sci. 41, 253–254.

Thompsom J A and Vincent J M 1967 Methods of detection and estimation of rhizobia in soil. Plant and Soil 26, 72–84.

Tong Z and Sadowsky M J 1994 A selective medium for the isolation and quantifcation of *Bradyrhizobium japonicum* and *Bradyrhizobium elkanii* from soils and inoculants. Appl. Environ. Microbiol. 60, 581–586.

Trinick M J 1980 Relationships amongst the fast growing rhizobia of *Lablab purpureus, Leucaena leucocephala, Mimosa* spp., *Acacia farnesiana* and *Sesbania grandiflora* and their affinities with other rhizobial groups. J. Appl. Bacteriol. 49, 39–53.

Trinick M J 1982 Biology. *In* Nitrogen Fixation. Vol. 2: *Rhizobium*. Ed. W J Broughton. pp 76–146. Clarendon Press, Oxford.

Trinick M J 1985 *Rhizobium* strain competition for host nodulation. *In* World Soybean Research Conference III: Proceedings. Ed. R Shibles. pp 911–917. Westview Press, Boulder, CO.

Triplett E W, Albrecht K A and Oplinger E S 1993 Crop rotation effects on populations of *Bradyrhizobium japonicum* and *Rhizobium meliloti*. Soil Biol. Biochem. 25, 781–784.

Triplett E W and Sadowsky M J 1992 Genetics of competition for nodulation in legumes. Annu. Rev. Microbiol. 46, 399–428.

Troedson R J, Lawn R J, Byth D E and Wilson G L 1989 Response of field-grown soybean to saturated soil culture. 1. Patterns of biomass and nitrogen accumulation. Field Crops Res. 21, 171–187.

Turk D and Keyser H H 1992 Rhizobia that nodulate tree legumes: specificity of the host for nodulation and effectiveness. Can. J. Microbiol. 38, 451–460.

Turk D, Keyser H H and Singleton P W 1993 Response of tree legumes to rhizbial inoculation in relation to the population density of indigenous rhizobia. Soil Biol. Biochem. 25, 75–81.

Valdivia B, Dughri M H and Bottomley P J 1988 Antigenic and symbiotic characterization of indigenous *Rhizobium leguminosarum* bv. *trifolii* recovered from root nodules of *Trifolium pratense* L. sown into subterranean clover pasture soils. Soil Biol. Biochem. 20, 267–274.

van Brussel A A N, Recourt K, Pees E, Spaink H P, Tak T, Wijffelman C A, Kijne J W and Lugtenberg B J J 1990 A biovar-specific signal of *Rhizobium leguminosarum* bv. *viciae* induces increased nodulation gene-inducing activity in root exudate of *Vicia sativa* subsp. *nigra*. J. Bacteriol. 172, 5394–5401.

van Egeraat A W S M 1975 The possible role of homoserine in the development of *Rhizobium leguminosarum* in the rhizosphere of pea seedlings. Plant and Soil 42, 381–386.

Verdcourt B 1970 Studies in the *Leguminosae-Papilionoïdeae* for the 'Flora of Tropical East Africa': IV. Kew Bull. 24, 507–569.

Vincent J E and Smith M S 1982 Evaluation of inoculant viability on commercially inoculated legume seed. Agron. J. 74, 921–922.

Vincent J M 1944 Variation in the nitrogen fixing property of *Rhizobium trifolii*. Nature, London 153, 496–497.

Vincent J M 1954a The root-nodule bacteria as factors in clover establishment in the red basaltic soils of the Lismore district, New South Wales. Aust. J. Agric. Res. 5, 55–60.

Vincent J M 1954b The root-nodule bacteria of pasture legumes. Proc. Linn. Soc. NSW 79, iv-xxxii.

Vincent J M 1962 Australian studies of the root-nodule bacteria. A review. Proc. Linn. Soc. NSW 87, 8–38.

Vincent J M 1965 Environmental factors in the fxation of nitrogen by the legume. *In* Soil Nitrogen. Eds. W V Bartholomew and F C Clark. pp 384–435. Am. Soc. Agron., Madison, WI.

Vincent J M 1970 A Manual for the Practical Study of Root-Nodule Bacteria. IBP Handbook No. 15. Blackwell, Oxford. 164p.

Vincent J M 1974 Root-nodule symbiosis with *Rhizobium*. *In* The Biology of Nitrogen Fixation. Ed. A Quispel. pp 265–341. North-Holland, Amsterdam.

Wang C L, Beringer J E and Hirsch P R 1986 Host plant effects on hybrids of *Rhizobium leguminosarum* biovars *viceae* and *trifolii*. J. Gen. Microbiol. 132, 2063–2070.

Weaver R W and Frederick L R 1974 Effect of inoculum rate on competitive nodulation of *Glycine max* L. Merrill. II. Field studies. Agron. J. 66, 233–236.

Weaver R W, Morris D R, Boonkerd N and Sij J 1987 Populations of *Bradyrhizobium japonicum* in fields cropped with soybean-rice rotations. Soil Sci. Soc. Am. J. 51, 90–92.

Wedderburn M E 1986 Effect of applied nitrogen, increased inoculation, broadcast lime, and seed pelleting on establishment of *Lotus pedunculatus* cv. "Grassland Maku" in tussock grasslands. N.Z. J. Exp. Agric. 14, 31–46.

Weisz P R, Denison R F and Sinclair T R 1985 Response to drought stress of nitrogen fixation (acetylene reduction) rates by field-grown soybeans. Plant Physiol. 78, 525–530.

Welsh J and McClelland M 1990 Fingerprinting genomes using PCR with arbitrary primers. Nucl. Acids Res. 18, 7213–7218.

Williams C H 1980 Soil acidifcation under clover pasture. Aust. J. Exp. Agric. Anim. Husb. 20, 561–567.

Williams J G K, Kubelik A R, Livak K J, Rifalski J A and Tingey S V 1990 DNA polymorphisms amplified by arbitrary primers are useful as genetic markers. Nucl. Acids Res. 18, 6531–6535.

Williams P M 1984 Current use of legume inoculant technology. *In* Biological Nitrogen Fixation, Ecology, Technology and Physiology. Ed. M Alexander. pp 173–200. Plenum Press, New York.

Wilson J K 1944 Over five hundred reasons for abandoning the cross-inoculation groups of the legumes. Soil Sci. 58, 61–69.

Wollum A G, II and Cassel D K 1984 Spatial variability of *Rhizobium japonicum* in two North Carolina soils. Soil Sci. Soc. Am. J. 48, 1082–1086.

Woomer P, Singleton P W and Bohlool B B 1988 Ecological indicators of native rhizobia in tropical soils. Appl. Environ. Microbiol. 54, 1112–1116.

Woomer P L 1990 Predicting the abundance of indigenous and the persistence of introduced rhizobia in tropical soils. Ph.D. thesis, University of Hawaii.

Young J P W 1985 *Rhizobium* population genetics: enzyme polymorphism in isolates from peas, clover, bean and lucerne grown at the same site. J. Gen. Microbiol. 131, 2399–2408.

Young J P W, Demetriou L and Apte R G 1987 *Rhizobium* population genetics: enzyme polymorphism in *Rhizobium leguminosarum* from plants and soil in a pea crop. Appl. Environ. Microbiol. 53, 397–402.

Zaat S A J, Wijffelman C A, Spaink H P, van Brussel A A N, Okker R J H and Lugtenberg B J J 1987 Induction of the *nod*A promoter of *Rhizobium leguminosarum* sym plasmid pRL1JI by plant flavanones and flavones. J. Bacteriol. 169, 198–204.

Zahran H H and Sprent J I 1986 Effects of sodium chloride and polyethylene glycol on root-hair infection and nodulation of *Vicia faba* L. plants by *Rhizobium leguminosarum*. Planta 167, 303–309.

Zhou J C, Tchan Y T and Vincent J M 1985 Reproduction capacity of bacteroids in nodules of *Trifolium repens* L. and *Glycine max* (L.) Merr. Planta 163, 473–482.

Plant and Soil **174:** 181–194, 1995.

Green manure technology: Potential, usage, and limitations. A case study for lowland rice

M. Becker[1], J.K. Ladha[2] and M. Ali[3]
[1]*West Africa Rice Development Association (WARDA), BP 2551, Bouaké, Côte d'Ivoire,* [2]*International Rice Research Institute (IRRI), PO Box 933, Manila, Philippines and* [3]*Asian Vegetable Research and Development Center (AVRDC), PO Box 42, Shanhua, Tainan, Taiwan 74199*

Key words: green manure, lowland rice, nitrogen fertilizer, nitrogen fixation, *Oryza sativa*

Table of contents

Abstract
Introduction
Potential benefits from green manure
Factors limiting adoption of green manures
  Agronomic factors
  Economic factors
The suitable environment for green maure crops
  Socio-economic niches for green manure use
  Bio-physical niches for green manure use
Prognosis and research needs
References

Abstract

The growing concern about the sustainability of tropical agricultural systems stands in striking contrast to a world-wide decline in the use of soil-improving legumes. It is timely to assess the future role that soil-improving legumes may play in agricultural systems. This paper reviews recent progress, potential, and limitations of green manure technology, using lowland rice cropping systems as the example.

Only a few legume species are currently used as green manures in lowland rice. *Sesbania cannabina* is the most widely used pre-rice green manure for rice in the humid tropics of Africa and Asia. *Astragalus sinicus* is the prototype post-rice green manure species for the cool tropics. Stem-nodulating *S. rostrata* has been most prominent in recent research. Many green manure legumes show a high N accumulation (80–100 kg N ha^{-1} in 45–60 days of growth) of which the major portion (about 80%) is derived from biological N_2 fixation. The average amounts of N accumulated by green manures can entirely substitute for mineral fertilizer N at current average application rates.

With similar N use efficiencies, green manure N is less prone to loss mechanisms than mineral N fertilizers and may therefore contribute to long-term residual effects on soil productivity.

Despite a high N_2-fixing potential and positive effects on soil physical and chemical parameters, the use of green manure legumes for lowland rice production has declined dramatically world-wide over the last 30 years. Land scarcity due to increasing demographic pressure and a relatively low price of urea N are probably the main determining factors for the long-term reduction in pre-rice green manure use. Post-rice green manures were largely substituted for by high-yielding early-maturing grain legumes. Unreliability of green manure performance, non-availability of seeds, and labor intensive operations are the major agronomic constraints. The recognition and extrapolation of niches where green manures have a comparative advantage may improve an often unfavorable economic comparison of green manure with cash crop or fertilizer N. Socio-economic factors like the cost of land, labor, and mineral N fertilizer are seen to determine the cost-effectiveness and thereby farmers' adoption of sustainable pre-rice green manure technology. Hydrology and soil texture determine the agronomic competitiveness of a green manure with N fertilizers and with alternative cash crops. In general, the niches for pre-rice green manure are characterized by a relatively short time span available for green manure growth and a soil moisture regime that is unfavorable for cash crops (flood-prone rainfed lowlands with coarse-textured soils).

Given the numerous agronomic and socio-economic constraints, green manure use is not seen to become a relevant feature of favourable rice-growing environments in the foreseeable future. However, in environments where soil properties and hydrology are marginal for food crop production, but which farmers may be compelled to cultivate in order to meet their subsistence food requirements, green manures may have a realistic and applicable potential.

Introduction

The use of soil-improving legumes in agriculture has been a topic of debate for many years. Opinions range from the conclusion that green manures will never be a significant factor (Norman, 1982), to the viewpoint that agronomic exploitation of biological nitrogen fixation, including green manure use, will become much more important in the future (Postgate, 1989). With growing concerns of sustainability, it is time to thoroughly assess the role that soil-improving legumes may play in agricultural systems. While this paper focuses on green manure use in lowland rice-based cropping systems, the general conclusions are applicable to a wider range of crops and ecosystems.

Throughout tropical rice farming systems, there are concerns about sustainability as most countries move into a post-green revolution phase (Cassman and Pingali, 1994; Pingali et al., 1990). Intensive cultivation has been shown to potentially degrade the resource base (Pingali et al., 1990). When faced with declining crop yield farmers and researchers have opted for crop diversification, crop rotation, and the insertion of green manures into the farming system (King, 1911). Legumes can in many instances be used as green manures in rice. Green manures can be fitted into rice farming systems in either the pre-rice or post-rice phase (Garrity and Flinn, 1988). Pre-rice green manures are cultivated for 6 to 10 weeks before the establishment of the rice crop and incorporated into the soil during land preparation. Post-rice green manures are relay-established into the growing rice (Liu, 1988) or sown after rice harvest (Chen, 1988), when the period available for growth is usually longer than that of pre-rice legumes.

Post-rice green manures have historically played a greater role than pre-rice species and are exemplified by *Astragalus sinicus* for the cool tropics (Chen, 1988) and *Indigofera tinctoria, Calopogonium* spp. (Ali and Narciso, 1993), *Crotalaria juncea, Clitoria ternatea, Desmanthus virgatus* and *Macroptilium atropurpureum* (Ladha et al., 1994) in the warm tropics. Prototype species for the pre-rice niche include flood-tolerant legume species such as *Sesbania* and *Aeschynomene* spp. (Ladha et al., 1992) or the more or less drought-tolerant legumes *Crotalaria* and *Tephrosia* spp. (Abrol and Palaniappan, 1988). Flood-tolerant legumes have been most prominent in recent research, a development that was triggered by the recognition of their high N_2-fixing capacity and the world-wide spread of the stem-nodulating *Sesbania rostrata*.

Numerous studies have examined growth and nitrogen fixation of legumes and the effects of green manure on soils and on the succeeding rice crop (Becker et al., 1990; Buresh and De Datta, 1991; Ladha et al., 1992; Singh et al., 1991), and have concluded that incorporation of green manure legumes can add large quantities

Table 1. Leguminous green manure crops in lowland rice-based cropping systems

Genus	Region
Pre-rice niche (short growing season)	
Flood-tolerant legumes	
Sesbania	Asia
	Africa
Aeschynomene	Asia
	Africa
Drought-tolerant legumes	
Crotalaria	S Asia
Tephrosia	S Asia
Post-rice niche (long growing season)	
Cool tropics	
Astragalus	E Asia
Warm tropics	
Indigofera	SE Asia
Calopogonium	W Africa
Forage legumes	Asia
	S America
	Africa

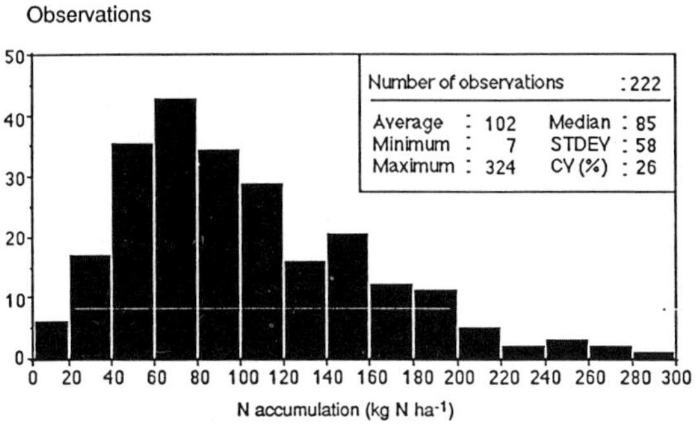

Fig. 1. N accumulation by leguminous green manures in rice-based cropping systems (compiled from Alazard and Becker, 1987; Becker et al., 1988; Becker, 1990, 1993; Bhatti et al., 1985; Biswas, 1988; Buresh and De Datta, 1991; Diekmann, 1992; Furoc et al., 1989; Ladha et al., 1992; Meelu and Morris, 1988; Morris et al., 1989; Roger and Watanabe, 1986; Ventura et al., 1987).

of biologically fixed N to lowland rice cropping systems and improve the soil productivity. However, the adoption of sustainable green manure technology in tropical lowland rice farming systems has been limited in the past, and their use is currently declining. Has the potential of green manures loomed larger in the agronomists mind than in the farmers?

To answer this question, a critical analysis of the potential and the limiting factors of green manure technology is needed in order to assess the future scope for soil-improving legumes in tropical agriculture. Such analysis may help the scientific community to target future green manure research activities and funding agencies to justify further investments in research on soil-improving legumes.

Potential benefits from green manure use

Relatively few species of green manure legumes are encountered in lowland rice-based cropping systems (Table 1). In the cool tropics of Asia (China, Japan) the post-rice niche is dominated by *Astragalus sinicus* (Liu, 1988) and in the Americas by vetches and clovers (Schultze-Kraft, 1988; Westcott and Mikkelsen, 1988). In the warm tropical post-rice situation, legumes are seldom grown for the sole purpose of soil improvement; these legumes are exemplified by *Indigofera tinctoria* in Asia (Ali and Narciso, 1995) and *Calopogonium* spp. in Africa (Evans and Rotar, 1987; Roy et al., 1988). More common are forage legumes like *Stylosanthes, Desmodium*, and *Clitoria* spp. (Ladha et al., 1994). This is particularly true for the South American post-rice situation.

Among the pre-rice green manures, *Sesbania cannabina* has been most widely used in the warm tropics of Africa and Asia. The stem-nodulating green manure *S. rostrata* (origin: West Africa) has been subjected to intense research since its introduction into Asia some 10 years ago. It is recommended as a green manure in both Asia and Africa and is currently grown on about 500,000 hectares in the Delta and the Central Plain of Myanmar (Mar et al., 1995). *S. sesban, S. speciosa* (Arunin et al., 1988; Liu, 1988), and *A. indica* (Crozat and Sangchyo-sawat, 1985; Ladha et al., 1992) are less common but can be locally important. *Sesbania speciosa*, for example, is grown in parts of Thailand, where it provides edible flowers in addition to an N-rich biomass (Evans and Rotar, 1987). Stem-nodulating *A. afraspera* and *A. nilotica* have only recently been domesticated (Alazard and Becker, 1987) and their use is largely limited to research farms and extension demonstration trials.

Nitrogen accumulation by green manure legumes in tropical rice-based cropping systems is influenced by

Table 2. Nitrogen fixation (%N derived from the atmosphere % Ndfa) in 50- to 70-day-old flood-tolerant legumes

Species	Season	Age	% Ndfa	Method	Reference
S. rostrata	SD	60	45	Difference	N'Doye and Dreyfus (1988)
	SD	55–65	63–72	Difference	Pareek et al. (1990)
	LD	55–65	85–87	Difference	Pareek et al. (1990)
	LD	53	44	Difference	Rinaud et al. (1988)
	LD	56	88	ARA / $^{15}N_2$	Becker et al. (1990)
	SD	56	83	ARA / $^{15}N_2$	Becker et al. (1990)
	SD	60	51	^{15}N dilution	N'Doye and Dreyfus (1988)
	LD	53	39	^{15}N dilution	Rinaud et al. (1988)
	LD	56	85	^{15}N dilution	Becker (1993)
	SD	56	86	^{15}N dilution	Becker (1993)
	LD	50	78	$\partial^{15}N$	Yoneyama et al. (1991)
S. cannabina	SD	55–65	59–62	Difference	Pareek et al. (1990)
	LD	55–65	62–88	Difference	Pareek et al. (1990)
	SD	55–65	83–86	^{15}N dilution	Pareek et al. (1990)
	LD	56	78	^{15}N dilution	Becker (1993)
	SD	56	93	^{15}N dilution	Becker (1993)
S. sesban	SD	60	11	Difference	N'Doye and Dreyfus (1988)
	SD	60	18	^{15}N dilution	N'Doye and Dreyfus (1988)
	LD	56	79	^{15}N dilution	Pareek et al. (1990)
	SD	56	83	^{15}N dilution	Pareek et al. (1990)
A. afraspera	LD	56	77	ARA / $^{15}N_2$	Becker et al. (1990)
	SD	56	68	ARA /$^{15}N_2$	Becker et al. (1990)
	LD	56	84	^{15}N dilution	Becker (1993)
	SD	56	76	^{15}N dilution	Becker (1993)
A. nilotica	LD	56	86	^{15}N dilution	Becker et al. (1990)
	SD	56	82	^{15}N dilution	Becker et al. (1990)
A. indica	LD	56	80	^{15}N dilution	Becker (1993)
	SD	56	89	^{15}N dilution	Becker (1993)
	LD	70	93	$\partial^{15}N$	Yoneyama et al. (1991)
Ndfa average:	78 %				

water regime, soil fertility, photo-period, inoculation, and growth duration (Buresh and De Datta, 1991). Very few papers report N accumulation and N_2 fixation data from upland or post-rice green manures (reviewed in Yost and Evans, 1988). On the other hand, numerous papers (more than 200 observations) report N yield and N_2 fixation in flood-tolerant legumes. The amount of N accumulated in a 45–60-day-old green manure crop varies with species, season, and site. Values range from as low as 7 kg N ha^{-1} to more than 300 kg N ha^{-1} (Fig. 1). An average N accumulation of 80–100 kg N ha^{-1} corresponds to the average amount of mineral fertilizer N applied to lowland rice in Asia (IRRI, 1990).

For legumes to maintain or increase soil N as desired, they must fix large amounts of atmospheric N_2. A survey of the literature indicates that on average, 75–80% of green manure legume N is derived from the atmosphere (Table 2). Difference and isotopic

Table 3. Percent nitrogen losses from mineral and organic N sources in lowland rice

Urea	Green manure	Loss mechanism	Reference
23.2	20	NO_3 leaching	Buresh and De Datta (1991)
4.5	1	NH_3 volatilization	Buresh and De Datta (1991)
28.9	5	NH_3 volatilization	Buresh and De Datta (1991)
40	4	NH_3 volatilization	De Datta and Buresh (1989)
42	11	NH_3 volatilization	De Datta and Buresh (1989)
23	16	^{15}N balance	Buresh and De Datta (1991)
59	19	^{15}N balance	Buresh and De Datta (1991)
35	16	^{15}N balance	Biswas (1988)
62	64	^{15}N balance	Westcott and Mikkelsen (1988)
19	12	^{15}N balance	Buresh and De Datta (1991)
39	7	^{15}N balance	Diekmann (1992)
39	9	^{15}N balance	Diekmann (1992)
31	13	^{15}N balance	Diekmann (1992)
31	16	^{15}N balance	Diekmann (1992)
42	12	^{15}N balance	Becker (1993)
42	8	^{15}N balance	Becker (1993)
31	11	^{15}N balance	Becker et al. (1994)
35	5	^{15}N balance	Becker et al. (1994)
Mean: 35	Mean: 14	18 observations	10 reported studies

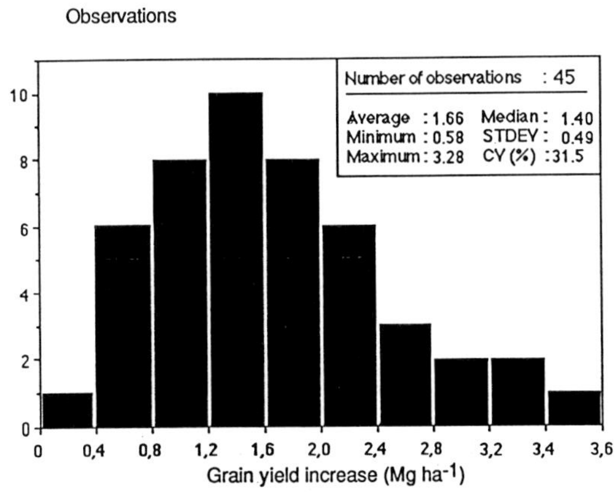

Fig. 2. Lowland rice grain yield increases over unfertilized contol treatment, due to leguminous green manure incorporation (compiled from Ali and Narciso, 1993; Becker et al., 1988; Becker, 1993; Bhuiyan et al., 1988; Buresh and De Datta, 1991; Ladha et al., 1992; Roger and Watanabe, 1986; Singh et al., 1991).

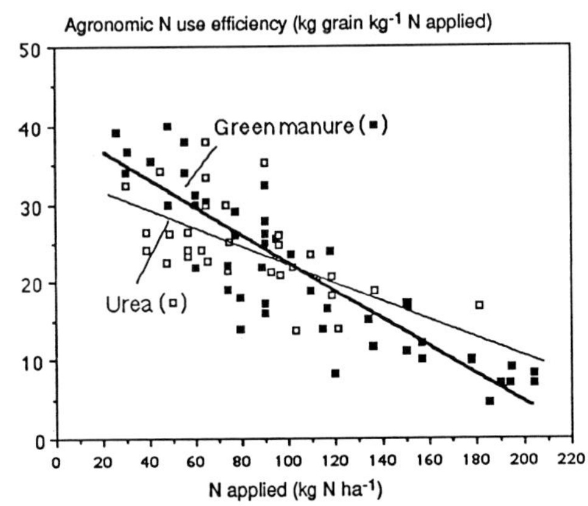

Fig. 3. Comparison of agronomic N use efficiency by lowland rice of mineral fertilizer (split-applied urea) and green manure N (compiled from Becker et al., 1988, 1990; Becker, 1990, 1993; Bhatti et al., 1985; Biswas, 1988; Buresh and De Datta, 1991; Diekmann, 1992; Furoc et al., 1989; Ladha et al., 1992; Meelu, 1988; Morris et al., 1989; Roger and Watanabe, 1986; Ventura et al., 1987).

methods were used in the reported studies, and the choice of reference datum (soil exchangeable ammonium, uninoculated legumes, rice, or weeds) seem to influence the estimates. The lowest percentages of N derived from the atmosphere (% Ndfa) were obtained when uninoculated legumes were used as reference plants (N'Doye and Dreyfus, 1988). The possibility that spontaneous nodulation occurred in these cases cannot be discounted, and could account for the low %

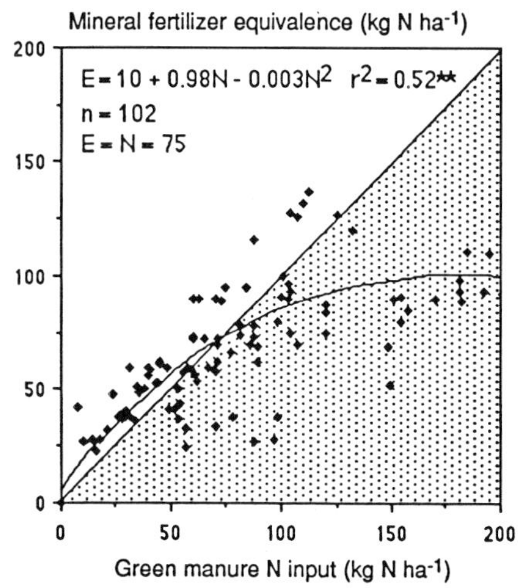

Fig. 4. Mineral fertilizer N equivalence of green manure N in flooded lowland rice (compiled from Becker et al., 1988; Buresh and De Datta, 1991; Ladha et al., 1992; Singh et al., 1991).

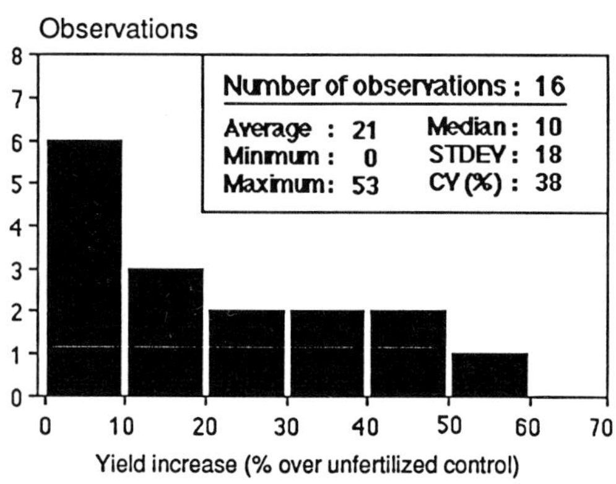

Fig. 5. Residual effect of legume green manures on grain yield of lowland rice. (compiled from Ali and Narciso, 1993; Becker, 1990, 1993; Becker et al., 1994; Biswas, 1988; Ladha et al., 1992; Morris et al., 1989; Singh et al., 1991).

Ndfa values. Using soil N or rice as a reference, more than 80% of the total N was determined to be derived from biological N_2 fixation. This figure is higher than the reliance upon N_2 fixation commonly reported for grain legumes (reviewed by Garrity and Becker, 1995; Norman, 1982; Peoples and Herridge, 1990; Peoples et al., 1995), underlining the high potential of green manure legumes to not only accumulate substantial amounts of N in their biomass, but also to add large quantities of atmospheric N to the soil.

Many studies have shown that the use of green manure legumes can increase the yield of a subsequent lowland rice crop, and reduce the requirements for inorganic N fertilizer (Buresh and De Datta, 1991). Rice grain yield increases due to green manuring, over unfertilized control treatments, range from 0.5 to 3.3 Mg grain ha^{-1} with an average of 1.7 Mg ha^{-1} (Fig. 2). Assuming an average application rate of 80 kg N ha^{-1} (IRRI, 1990), green manure N shows an agronomic N use efficiency (kg rice grain increase over unfertilized control kg^{-1} N applied) of about 20. This figure is similar to the average N use efficiency of mineral N fertilizer (Becker, 1993). Similar results were reported by Ali and Narciso (1993), who compared N efficiency in long-term experiments in India, Indonesia, and the Philippines conducted by the International Network on Soil Fertility and Fertilizer Evaluation (INSURF). Lowland rice seems to be using organic N as efficiently as mineral fertilizer N (see also Peoples et al., 1995). However, the nature of this relationship seems to change slightly according to the amount of N input (Fig. 3). Overall, linear curves of N use efficiency against N application do not differ significantly among N sources, but tend to decrease more with green manure at higher application rates.

This trend becomes clearer when using the mineral fertilizer equivalence of organic N sources. The mineral fertilizer equivalence of green manure is the amount of split-applied urea N required to obtain equivalent yield. Figure 4 indicate that mineral fertilizer equivalence increases with applied N rate though follows a quadratic response function. At application rates of less than 80 kg N ha^{-1}, lowland rice uses green manure N more efficiently than urea N. However, when green manure is applied in excess of 100 kg N ha^{-1} organic fertilizer use efficiency declines more rapidly than that of urea. To target green manure N accumulation in excess of 100 kg N ha^{-1} may not be useful where a short-term increase in rice grain yield is the objective.

In addition to the quantity of added green manure N and its mineralization rate, the N contribution of green manure (as for mineral fertilizer N) to rice depends on the magnitude of N losses. Mineral N fertilizers are generally not efficiently used by rice and are prone to high losses as N gases (Buresh and De Datta, 1991). In flooded rice soils, ammonia volatilization is widely recognized as an important mechanism for loss of mineral fertilizer N applied to tropical lowland rice. High floodwater ammoniacal N concentrations following application of N fertilizer, high temperature, and elevated floodwater pH resulting from photosynthetic activity create a favorable environment for ammonia loss (De Datta and Buresh, 1989; Roger and Watanabe, 1986). In non-flooded soils, nitrate leaching and denitrification during the dry-wet transition period are the major N loss mechanisms (George et al., 1992). If fertilizer applications were well synchronized with plant N uptake patterns, N losses from lowland rice fields would be greatly reduced. Organic materials, acting as slow release source of N, are expected to more closely match N supply and rice N demand, and this could reduce N losses (Becker et al., 1994; McGill and Myers, 1987). Nitrogen losses from applied mineral and organic fertilizers have been studied by a number of researchers, using different methods (Table 3). Average N losses in flooded soils from applied green manure are considerably lower (14%) than those from split-applied urea (35%), thus resulting in less pollution to the environment.

The facts that (a) the N input from mineral and organic N sources and their use efficiency by lowland rice are similar, and that (b) N losses are less from green manure than from urea raises the question of the fate of the portion of the organic N that is neither lost nor taken up by the rice plant. It has been suggested that large portions of organically applied N remain undecomposed under flooded soil conditions and may be available for a subsequent crop (Becker et al., 1994; Bouldin, 1988), and leguminous green manures incorporated before cropping with lowland rice were shown to increase grain yield of a following wheat or rice crop (Buresh and De Datta, 1991; Singh et al., 1991). Other authors, however, detected little or no residual response to green manure N (Ladha et al., 1992; Ventura et al., 1987). A review of several published studies (Fig. 5) indicates an average residual effect of green manuring on the grain yield of a subsequent unfertilized rice crop of 10–20%. This may appear small but might in the long-run contribute to the restoration of degraded soils and the sustainability of rice-based cropping systems. However, no conclusive research data on such long-term effects of green manuring are currently available.

Green manures frequently have positive effects on a subsequent rice crop, resulting from factors other than supplying N (rotation effects). Other nutritional effects include the mobilization of P, Si, Zn, Cu, Mn, and other nutrient elements as a result of increased microbial activity (CO_2 formation), a decrease in Redox potential, and the microbial reduction of pedogenus iron (Fe^{3+}) oxides and hydroxides (Becker, 1990). Incorporation of an easily decomposable biomass with relatively high N content can lead to an increased mineralization of the organic soil fraction (priming effect) in rice soils (Broadbent, 1979) and possibly enhance heterotrophic N_2 fixation (Ladha et al., 1989). The addition of organic material significantly increases microbial biomass in the soil and conserves N, P, K, and micronutrients in a "biological pool" with a high turnover (Ottow, 1978). Incorporation of legume biomass can also improve soil physical parameters (soft puddle). Green manure can increase the cation exchange capacity, improve soil buffering capacity, decrease porosity in light-textured soils, and help reclaim alkaline and acid soils (Arunin et al., 1988). A green manure can conserve nitrate that is mineralized during the dry fallow between two rice crops and that is prone to nitrate leaching and denitrification in the dry-wet transition before rice (George et al., 1992). Increased soil biological activity as a result of green manuring can accelerate the mineralization of persistent pesticides and their metabolites (Ferreira and Rughu, 1981). Green manures also have been shown to reduce the incidence of bacterial diseases in rice as a result of an enhanced biological barrier in the rhizosphere (Premalata Dath, 1981). Planting a legume instead of leaving the land fallow can help control weeds, particularly through the use of "creeping" cover crops in the post-rice niche (Yost and Evans, 1988). Furthermore, green manure legumes may offer the potential to reduce insect pest and nematode populations in the succeeding rice crop (Germani et al., 1983). Finally, many green manure species have additional uses such as food, feed, and fuel, or have medicinal properties (Evans and Rotar, 1987).

In summary, many green manure legumes show a high potential for N accumulation of which the major portion is derived from biological N_2 fixation. At average N accumulation rates, green manures can entirely substitute for mineral fertilizer N at current average application rates. With similar N use efficiency, green manure N is less prone to loss than mineral fertilizer N, and continuous green manure use may improve physical and chemical soil properties. In the long-run, green manure use may help restore degraded soils and enhance the production sustainability of rice-based systems.

Factors limiting adoption of green manures

Despite high N_2 fixation (Ladha et al., 1992), reduced N losses (Buresh and De Datta, 1991), increased rice grain yields (Singh et al., 1991), and various positive effects on soil physical and chemical parameters (Becker et al., 1988), the use of green manure legumes in lowland rice production systems has been dramatically declining over the last 30 years (Roger and Watanabe, 1986; Rosegrant and Roumassett, 1988). Today, green manures are estimated to cover less than 5 million hectares in tropical Asia, confined mostly to China, Myanmar and Vietnam (Garrity and Becker, 1995). In some countries like Nepal, Pakistan, and the Philippines where soil-improving legumes were once extensively used, these are now rarely encountered (Ali and Narciso, 1995). In Japan where green manures formerly played an important role in lowland rice culture, they have now literally disappeared (Ishikawa, 1988). In the United States, green manure crops have been widely grown in rotation with rice, but their use has declined to less than 5% of the planted rice area (Westcott and Mikkelsen, 1988). What are the possible reasons for this long-term trend, which includes the Peoples Republic of China (Ali and Narciso, 1993; Chen, 1988)?

Several reviews (Ladha et al., 1992; Roger and Watanabe, 1986) and surveys conducted by IRRI (Ali and Narciso, 1993; Garrity and Flinn, 1988) identified the following major constraints to green manure use in lowland rice production systems:

- the unavailability of appropriate green manure legume seeds;
- a high variability in the green manure performance, particularly in problem environments;
- a costly establishment and incorporation of green manure;
- a high price of land and labor, and
- a relatively low mineral fertilizer price.

Agronomic factors

Seed

The availability of seeds is a limiting factor for any crop, and where seeds are not commercially available (most green manure species), farmers have to assure their own stock of quality seeds. However, seed production is one of the most costly and labor intensive aspects of green manure production (Garrity and Becker, 1995). An unfavorable plant architecture, shattering of seeds, asynchronous flowering, and in many instances the need for seed scarification are some typical constraints related to the semi-domesticated nature of most green manure species. Furthermore, the degree of photo-period sensitivity has major implications on the seasonal adaptation and seed production potential of legumes (Becker et al., 1990). Seed production methods have been studied for *S. rostrata* and *A. afraspera*, and some promising concepts include the planting of seed legumes on bunds or widely spaced in the rice field (Herrera et al., unpubl.). Ideally, a soil-improving legume should produce sufficient seed before the time of green manure incorporation (Garrity and Becker, 1995).

Variability

Figures 1 and 3 indicate not only a high potential of green manure use in rice, but also highlight the tremendous variability in the possible benefit of green manure practices. A survey of 18 long-term experiments showed that the coefficient of variation of rice grain yields was consistently higher in green manure than in mineral fertilizer treatments (Ali and Narciso, 1993). Year-to-year and site-to-site variability of the green manure performance and the resulting rice grain yield are tied to an insufficient adaptation of the legumes to problem environments, including unfavorable photo-period (Becker et al., 1990) and hydrology (Garrity and Becker, 1995), marginal soils (Arunin et al., 1988) and high pest pressure. Pest management in green manures is a problematic issue, as these crops are often desirable food sources to pests and pathogens (Garrity and Flinn, 1988; Germani et al., 1983; Liu, 1988), yet, farmers may be reluctant to invest in protection measures for a crop that does not yield direct cash income. The resulting uncertainty of the outcome of his investment in a green manure crop is certainly a serious disincentive for farmers to adopt soil-improving legumes.

Cultivation

As with any crop, intensive tillage tends to ensure favorable early growth conditions for a rapid green manure crop establishment. However, such practices may not be justified in a crop that bears no economic yield. Currently, tillage and seeding operations are major components of the labor and cash outlay by farmers who grow green manure crops (Garrity and Flinn, 1988). However, in the long-run, the only viable option is to move toward minimum or zero tillage methods for green manure crop establishment. Effective, low cost stand establishment is essential to the economic viability of the system. Successful cultivation of a green manure crop results in a dense stand which is subsequently difficult to incorporate into the soil. The additional time required for green manure incorporation in land preparation for transplanted rice in Southeast

Asia is estimated to be 25% higher than in weedy fallow plots (Garrity and Flinn, 1988). In many countries the absence of appropriate machinery and/or animal traction, combined with a shortage and the high cost of labor are major green manure production constraints.

After incorporation of fresh biomass, the anaerobic decomposition of the organic material forms large amounts of volatile fatty acids, gases (e.g. CO_2 and methane), alcohol, and phenolic substances (Roger and Watanabe, 1986). The unpleasant smell of some of these substances may prompt farmers to favor mineral N against green manure. Further, organic acids and phenolic compounds, which may accumulate during anaerobic decomposition of organic matter, can retard root elongation (Chou and Chiou, 1979), shoot growth and nutrient uptake of lowland rice (Cannell and Lynch, 1984). The adverse effects on rice seedlings increase with decreasing soil pH.

Economic factors

Economic feasibility determines the ultimate fate of any technology. A number of economic analyses indicate an unfavorable comparison of green manure vs inorganic fertilizer use (Ali and Narciso, 1995; Garrity and Flinn, 1988). The most important economic factors affecting green manure use are the input prices for land, labor and fertilizers (Ali and Narciso, 1993). The land price is one of the dominant factors limiting green manure use. Wherever land can also be used for the production of cash crops, strictly soil-improving legumes can not compete. The place of green manure crops in well-drained irrigated fields is, therefore, very limited, since the period after cultivating major economic crops (rice and wheat) is preferred for cash crop production (mainly vegetables). A comparison of grain and soil-improving legumes at existing international grain and fertilizer prices indicates that a legume grain yield of 200 kg ha^{-1} is sufficient to make grain and green manure legumes equally beneficial.

The ready availability and relatively low cost of mineral N fertilizer (about 380 US$\$$$Mg^{-1}$ of urea N) also contributes to the current unfavorable economic comparison of green manures with mineral N (Ali and Narciso, 1993). Since the manufacturing of N fertilizers requires energy, their cost is closely related to oil prices and the future exploitation of new energy sources. In many instances, however, the low fertilizer price is brought about by direct and indirect subsidies. In fact, the period between 1960 and 1990, when green manure use declined to less than 5% of the lowland rice area, the consumption rate of mineral fertilizer increased by about 10% per annum (IRRI, 1990). One should, however, keep in mind that mineral N fertilizer production uses large quantities of fossil fuel that may not always be available at current prices.

It may be concluded that in most instances, non-adoption of green manures at farm level is related to the riskiness of green manures (performance variability), time-consuming and labor-intensive crop establishment, lost opportunity costs, the high price of land, and a low mineral fertilizer price. Progress in the areas of species selection and improvement, seed production, pest management and crop establishment will considerably expand the options for soil-improving legumes in all niches and may improve the currently unfavorable comparison of green manures with other non-rice crops and mineral N fertilizers. Economic evaluations of green manure production systems show an unfavorable comparison of green manure vs. mineral fertilizer N at current input prices. Such studies, however, consider the legumes solely as a N source and fail to account for residual or long-term effects and the numerous additional (non-N) benefits.

The suitable environment for green manure crops

Increased green manure adoption at the farm-level requires the recognition and identification of economic and biophysical scenarios within which green manures have a comparative advantage over other non-rice crops and mineral fertilizers.

Socio-economic niches for green manure use

The economic environment for green manure cultivation is determined in the first place by the input prices for land and labor, and the ready availability and price of N fertilizer. At the country or regional level, national policy decisions vis-a-vis green manures (or for that matter mineral fertilizers) will affect the environment for green manure adoption. Policy environments differ drastically among countries. In China and Myanmar, for example, there is a strong government support for green manure. In Bhutan and a number of African countries, green manure use may be stimulated by the limited availability of inorganic fertilizers, while Taiwan is promoting green manure as a means to reduce the rice area (Garrity and Becker, 1995).

Socio-economic conditions at the farm-level further refine this general pattern for potential green

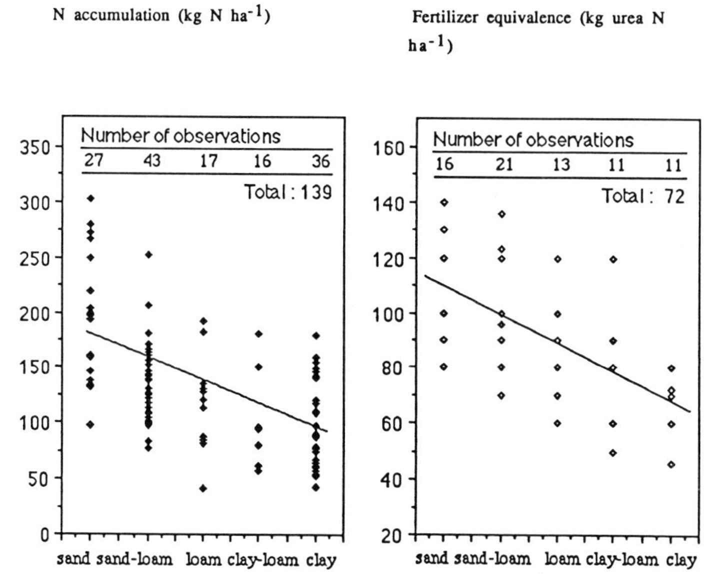

Fig. 6. Mineral N fertilizer substitution in lowland rice by leguminous green manures as influenced by soil texture (compiled from Becker et al., 1988; Becker, 1990, 1993; Buresh and De Datta, 1991; Garrity and Becker, 1995; Ghai et al., 1985, 1988; Ladha et al., 1992; Morris et al., 1989; Rinaudo et al., 1988; Singh et al., 1991; Ventura et al., 1987).

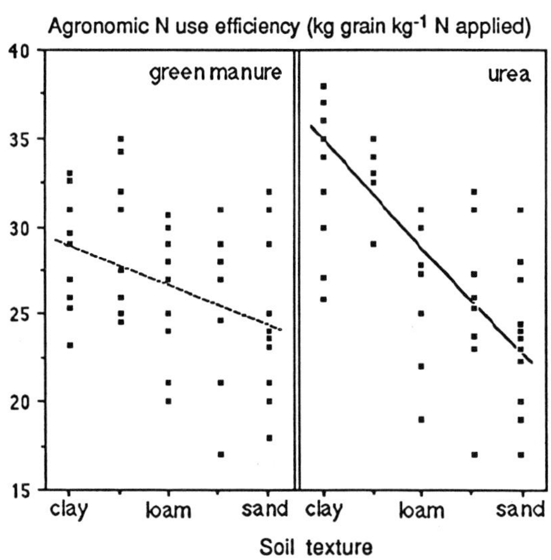

Fig. 7. Influence of soil texture on N use efficiency of mineral and organic amended N by lowland rice (compiled from Becker et al., 1988; Becker, 1990, 1993; Buresh and De Datta, 1991; Garrity and Becker, 1995; Ghai et al., 1985, 1988; Ladha et al., 1992; Morris et al., 1989; Rinaudo et al., 1988; Singh et al., 1991; Ventura et al., 1987).

manure adoption. Distance from the local market may determine farmers' access to mineral fertilizers. Sufficient high-quality seeds of appropriate legume species is crucial for green manure use. Labor availability and time allocation during the transition between two main crops have to be adequate for green manure seeding and incorporation operations. Finally, the incorporation of a relatively large biomass into the soil requires suitable implements and/or animal or mechanical traction.

Biophysical niches for green manure use

A critical analysis of 38 published references attempts a holistic evaluation and extrapolation of the biophysical niches where green manures have a comparative advantage over mineral N fertilizer use. Major agroecological zones set the general framework for the suitability of a given green manure measure. Within these agroecological zones, soil type and hydrology seem to be the dominant spatial factors determining niches for green manure use.

Legume N accumulation and mineral fertilizer equivalence of green manure seem to be generally higher in light-textured than in heavy soils (Fig. 6). The use efficiency of green manure and mineral N by lowland rice indicates an interesting relationship. As the percentage sand content increases among experimental sites from various studies, there is a distinct tendency for N use efficiency to decrease. The nature of this relationship contrasts between the two N sources, as the

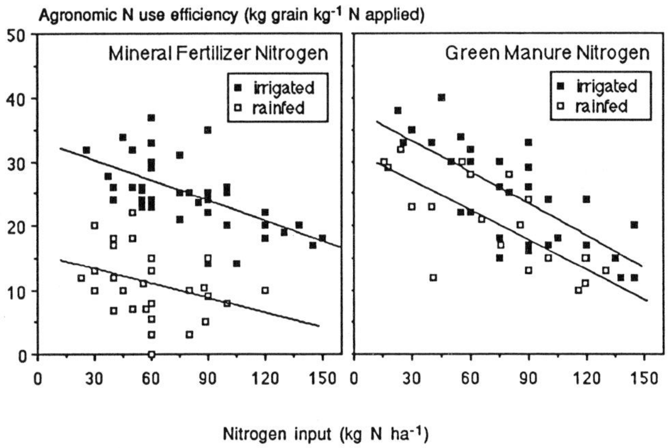

Fig. 8. Fertilizer use efficiency of mineral and organic amended N by irrigated and rainfed lowland rice. (compiled from Becker, 1990, 1993; Bhuiyan et al., 1988; Buresh and De Datta, 1991; Diekmann, 1992; Garrity and Becker, 1995; Huysman, 1983; Singh et al., 1991).

N use efficiency of urea declines more drastically than that of green manure. The efficiency of both N sources is similar on clay soils, but tends to be substantially greater for green manures in sandy soils (Fig. 7). This may be associated with an increased green manure N accumulation and higher N leaching losses from mineral compared to organic N in coarse-textured soils.

Hydrology is a dominant spatial factor that controls the crop sequence and determines the adaptation of green manures into rice farming systems. Farmers prefer cash crops over green manures. However, where severe soil moisture deficit or water excess increases the risks in cash crop production, the likelihood increases for inclusion of pre-rice green manures into the cropping patterns. Severe flooding in the pre-rice situation eliminates most alternative food crops and consequently favors a flood-tolerant green manure legume. However, unreliable water availability associated with many rainfed situations also increases the riskiness and variability of green manure production. Hydrology further appears to determine the competitiveness of green manures with N fertilizers. Repeated soil drying and wetting cycles during rice growth in rainfed environments increase mineral N losses by denitrification and leaching (George et al., 1992). N use efficiency tends to be higher at all levels of urea application under irrigated compared to rainfed conditions. However, when nitrogen is supplied by green manure, N use efficiency is very similar between irrigated and rainfed conditions (Fig. 8). Consequently, green manure use may be preferable over mineral N fertilizers in many unfavorable rainfed situations.

A step-wise identification of the above economic and agronomic factors will allow the definition of extrapolation domains where a given green manure species or technology may be considered. According to this analysis, the socio-economic environment is recognized to determine the impact that any attempt to introduce changes in existing cropping patterns may have. In the biophysical environment, the suitability of a green manure measure seems to be principally determined by a relatively short time span available for green manure growth and a moisture regime that is unfavorable for cash crop growth.

Prognosis and research needs

What is the future of green manures? The relative costs of labor, mineral N fertilizer prices, and alternative economical uses of soil-improving legumes will ultimately determine the cost-effectiveness and thereby the large-scale adoption of sustainable pre-rice green manure technology. Unless there are major breakthroughs in green manure technology and/or a significant increase in mineral fertilizer prices, the use of soil-improving legumes will continue to play only a limited role in tropical agriculture. Two factors may determine the scale of farmers' future acceptance of soil improving legumes:

– Green manure species and systems will have to provide multiple use solutions to be acceptable, since farmers are looking for direct returns to their investments. In these situations the green manure performs additional functions beside being an N source: It must provide food, fodder, fuel, or industrial products. Green manures with direct economic functions will be the key in most situations. However, very little is known about the potential additional uses of current green manure species.

– A reduction in the performance variability of soil-improving legumes requires targeting particular green manure species to well defined niches within cropping systems. The niches where single-purpose green manures fit will be limited to conditions where alternative crops are not competitive or where green manure provides more than just N (e.g. increased soil water-holding capacity, alleviation of toxicities, weed suppression). The target environments for green manure use appear in many instances to be marginal for cash crop production. However, with growing pressure on the tropical world's land for food production, marginal areas are increasingly being cultivated. Green manure legumes may be a vital component in the sustainable agricultural use of such unfavorable environments. In this context soil-improving legumes may well become an increasingly important research area in the future.

Research in the following areas may help soil-improving legumes to prove their potential in a wider range of environments than used at present and thereby add to the sustainability of tropical agricultural systems.

– Species selection for crop adaptation, seed production, pest management, and stress resistance. Most current and prospective green manure legumes have not received any formal plant improvement and many of the recently introduced species are actually undomesticated. Other than ad-hoc efforts by farmers who have cultivated green manures, plant improvement has been totally neglected. The future involvement of plant breeders in species development is highly desirable.

– The importance of agronomic management issues (like green manure crop establishment and incorporation) contrasts with the large gaps in knowledge and technology. In particular, alternative crop establishment practices, including zero tillage, may improve the economics of green manure production and needs further evaluation.

– Since green manure use is not economical in many instances at the current low fertilizer prices and relatively high costs of labor, multi-purpose legumes need to be identified and targeted for environments where they have a comparative advantage over other non-rice crops and N fertilizer use. Recognition of such niches and the use of appropriate legumes will improve both the short- and the long-term benefits of green manure legumes, and possibly reverse the current trend of a declining area of soil-improving legumes towards an accelerated acceptance of sustainable green manure technology. These environments are in many instances marginal for food crop production. Farmers, however, may be compelled to cultivate such marginal lands to meet their subsistence food requirements. Development of appropriate green manure technology packages for unfavorable environments is seen to be a rewarding area of research, wherein breakthroughs could help the resource-poor subsistence farmers to sustain and improve their food production.

References

Abrol I P and Palaniappan S P 1988 Green manure crops in irrigated and rainfed lowland rice-based cropping systems. *In* Sustainable Agriculture: Green Manure in Rice Farming. pp 71–82. The International Rice Research Institute, Los Baños, IRRI, Philippines.

Alazard D and Becker M 1987 *Aeschynomene* as green manure for rice. Plant and Soil 101, 141–143.

Ali M and Narciso J H 1995 Comparative evaluation and farmer's perceptions of green manure use in the rice-based farming systems. *In* Proceedings of the International Rice Research Conference. The International Rice Research Institute, Los Baños, Philippines (*In press*).

Ali M and Narciso J H 1993 The perception and reality of GM use in rice: An economic evaluation. IRRI Social Science Division Paper Series. The International Rice Research Institute, Los Baños, Philippines. 42p.

Arunin S, Dissataporn C, Anuluxtipan Y and Nana D 1988 Potential of *Sesbania* as green manure in saline rice soils in Thailand. *In* Sustainable Agriculture: Green Manure in Rice Farming. pp 83–95. The International Rice Research Institute, Los Baños, IRRI, Philippines.

Becker M, Ladha J K and Ottow J C G 1988 Stem-nodulating legumes as green manure for lowland rice. Philipp. J. Crop Sci. 13, 121–127.

Becker M 1990 Potential use of the stem-nodulating legumes *Sesbania rostrata* and *Aeschynomene afraspera* as green manure for lowland rice (*Oryza sativa* L.). Ph.D. Thesis, University of Giessen, Germany. 113p.

Becker M 1993 Flood-tolerant legumes as green manures for lowland rice-based cropping systems. Terminal Report. The International Rice Research Institute, Los Baños, Philippines. 130p.

Becker M, Ladha J K and Ottow J C G 1990 Growth and nitrogen fixation of two stem-nodulating legumes and their effect as green manure on lowland rice. Soil Biol. Biochem. 22, 1109–1119.

Becker M, Ladha J K and Ottow J C G 1994 Nitrogen losses and lowland rice yield as affected by residue N release. Soil Sci. Soc Am. J. 58, 1660–1665.

Bhatti H M, Yasin M and Rashid M 1985 Evaluation of *Sesbania* green manuring in rice-wheat rotation. *In* Nitrogen and the Environment. Eds. K A Malik, S H M Naqvi and M I H Aleem. pp 275–284. Nuclear Institute for Agriculture and Biology, Faisalabad, Pakistan.

Bhuiyan N I, Zaman S K and Panaullah G M 1988 Dhaincha green manure: a potential nitrogen source for rainfed lowland rice. *In* Proceedings of the Workshop on Experiences with Modern Rice Cultivation in Bangledesh, 7 April 1988. pp 108–125. Bangladesh Rice Research Institute, Dhaka, Bangladesh.

Biswas T K 1988 Nitrogen dynamics and nitrogen-15 balance in lowland rice as affected by green manure and urea application. Ph.D. thesis, Indian Agric. Res. Inst., New Delhi, India.

Bouldin D R 1988 Effect of green manure on soil organic matter content and nitrogen availability. *In* Sustainable Agriculture: Green Manure in Rice Farming. pp 151–164. The International Rice Research Institute, Los Baños, Philippines.

Broadbent F E 1979 Losses of nitrogen from some flooded soils in tracer experiments. *In* Nitrogen and Rice. pp 105–117. The International Rice Research Institute, Los Baños, Philippines.

Buresh R J and De Datta S K 1991 Nitrogen dynamics and management of rice-legume cropping systems. Adv. Agron. 45, 1–59.

Cannell R Q and Lynch J M 1984 Possible adverse effects of decomposing crop residues on plant growth. *In* Organic Matter and Rice. pp 455–475. The International Rice Research Institute, Los Baños, Philippines.

Cassman K G and Pingali P L 1994 Extrapolating trends from long-term experiments to farmers fields: the case of irrigated rice systems in Asia. *In* Agricultural Sustainability in Economic, Environmental and Statistical Terms. Eds. V Varnett, P Payne and R Steiner. John Wiley and Sons, Ltd., London, UK (*In press*).

Chen L 1988 Green manure cultivation and use for rice in China. *In* Sustainable Agriculture: Green Manure in Rice Farming. pp 63–70. The International Rice Research Institute, Los Baños, Philippines.

Chou C H and Chiou S J 1979 Auto intoxification mechanism of *Oryza sativa*. J. Chem Ecol. 5, 839–859.

Crozat I and Sangchyo-sawat C 1985 Evaluation of different green manures on rice yield in Sonkla Laka Basin. J. Sci. Technol. 7, 391–397.

De Datta S K and Buresh R J 1989 Integrated nitrogen management in irrigated rice. Adv. Soil Sci. 10, 143–169.

Diekmann K H 1992 Einfluβ der Gründüngung mit stengelknöllchen-bildenden Leguminosen auf die Stickstoffdynamic im Reisboden und die Ertragsbildung von Reis (*Oryza sativa* L.). Ph.D. Thesis, University of Giessen, Germany. 189p.

Evans D O and Rotar P P 1987 *Sesbania* in agriculture. Westview Tropical Agriculture Series, No. 8. Westview Press, Boulder, Colorado, USA.

Ferreira J and Rughu K 1981 Decontamination of hexochlorocyclohexane isomers in soil by green manure application. Sci. Technol. Lett. 2, 357–364.

Furoc R E, Dixon M A, Meelu O P, Morris R A and Pandey R K 1989 Rice response to intersown and pre-rice *Sesbania* green manure and rates and methods of nitrogen application. Philipp. J. Crop Sci. 13, 77–80.

Garrity D P and Flinn J C 1988 Farm-level management systems for green manure crops in Asian rice environments. *In* Sustainable Agriculture: Green Manure in Rice Farming. pp 111–129. The International Rice Research Institute, Los Baños, IRRI, Philippines.

Garrity D P and Becker M 1995 Where do green manures fit in Asian rice farming systems? *In* Proceedings of the International Rice Research Conference, The International Rice Research Institute, Los Baños, Philippines (*In press*).

Germani G, Reversat G and Luc M 1983 Effect of *Sesbania rostrata* on *Hirschaniella oryzae* in flooded rice. J. Nematol. 15, 269–271.

George T, Ladha J K, Buresh R J and Garrity D P 1992 Managing native and legume-fixed nitrogen in lowland rice-based cropping systems. Plant and Soil 141, 69–91.

Ghai S K, Rao D L and Batra L 1985 Comparative study of the potential of *Sesbania* for green manuring. Trop. Agric. 62, 52–56.

Ghai S K, Rao D L N and Batra L 1988 Nitrogen contribution to wetland rice by green manuring with *Sesbania* spp. in an alkaline soils. Biol. Fertil. Soils 6, 22–25.

Huysman A 1983 Estimating risk of fertilizer use in rainfed rice production. Int. Rice Res. Paper Series, No. 93.

IRRI 1990 World Rice Statistics. The International Rice Research Institute, Los Baños, Philippines. 320p.

Ishikawa M 1988 Green manure in rice: the Japan experience. *In* Sustainable Agriculture: Green Manure in Rice Farming. pp 45–61. The International Rice Research Institute, Los Baños, Philippines.

King F H 1911 Farmers of Forty Centuries or Permanent Agriculture in China, Korea and Japan. Rodale Press, Inc. Book Division, Pennsylvania.

Ladha J K, Tirol-Padre A, Punzalan G C and Watanabe I 1989 Effect of inorganic N and organic fertilizers on nitrogen-fixing (acetylene-reducing) activity associated with wetland rice plants. *In* Nitrogen Fixation with Non-legumes. Eds. F A Skinner, R M Boddey and I Fendrik. pp 263–272. Kluwer Academic Publishers, Dordrecht.

Ladha J K, Kundu D K, Angelo-Van Coppenolle M G, Peoples M B, Carangal V R and Dart P J 1994 Grain and forage legume effects on soil nitrogen dynamics and nitrogen availability to following rice crops. Soil Sci. Soc. Am. J. (Submitted)

Ladha J K, Pareek R P and Becker M 1992, Stem-nodulating legume-Rhizobium symbiosis and its agronomic use in lowland rice. Adv. Soil Sci. 20, 147–192.

Liu C 1988 Integrated use of green manure in rice fields in South China. *In* Sustainable Agriculture: Green Manure in Rice Farming. pp 319–331. The International Rice Research Institute, Los Baños, Philippines.

Mar M, Saing T, Thein T and Palis R K 1995 *Sesbania* green manure program for rice farming in Myanmar. *In* Proceedings of the International Rice Research Conference, 21–25 April 1992. The International Rice Research Institute, Los Baños, Philippines (*In press*).

Mc Gill W B and Myers R J K 1987 Controls on dynamics of soil and fertilizer nitrogen. Soil fertility and organic matter as critical components of production systems. Soil Sci. Soc. Am. Special Publ. No. 19, 73–99.

Meelu O and Morris R 1988 Green manure management in rice based cropping systems. *In* Sustainable Agriculture: Green Manure in Rice Farming. pp 209–222. The International Rice Research Institute, Los Baños, Philippines.

Morris R A, Furoc R E, Rajbhandari N K, Marqueses E P and Dizon M A 1989 Rice response to waterlog-tolerant green manures. Agron. J. 81, 803–809.

N'Doye I and Dreyfus B L 1988 N_2 fixation by *Sesbania rostrata* and *Sesbania sesban* estimated using ^{15}N and total N difference methods. Soil Biol. Biochem. 20, 209–213.

Norman M J T 1982 A role for legumes in tropical agriculture. *In* Biological Nitrogen Fixation in Tropical Agriculture. Eds. P H Graham and S C Harris. pp 9–25. Centro Internacional de Agricultura Tropical, Cali, Columbia

Ottow J C G 1978 Chemie und Biochemie des Humuskörpers unserer Böden. Naturwiss. 65, 413–420.

Pareek R P, Ladha J K and Watanabe I 1990 Estimation of N_2 fixation by *Sesbania rostrata* and *S. cannabina* in lowland rice soil by ^{15}N dilution method. Biol. Fertil. Soils 10, 77–88.

Peoples M B and Herridge D F 1990 Nitrogen fixation by legumes in tropical and sub-tropical agriculture. Adv. Agron. 44, 152–223.

Peoples M B, Herridge D F and Ladha J K 1995 Biological nitrogen fixation: an efficient source of nitrogen for sustainable agricultural production. Plant and Soil 174.

Pingali P L, Moya P F and Velasco L E 1990 The post-green revolution blues in Asian rice production. IRRI Social Science Division Paper Series No 90–01. The International Rice Research Institute, Los Baños, Philippines.

Postgate J 1989 Fixing the nitrogen fixers. New Sci. 3, 57–61.

Premalata Dath 1981 Effect of soil amendment with green manure on sheath blight of rice. Int. Rice Res. Newsl. 21, 4–6.

Rinaudo G, Alazard D and Moudiongui A 1988 Stem nodulating legumes as green manures for rice in West Africa. *In* Sustainable Agriculture: Green Manure in Rice Farming. pp 97–109. The International Rice Research Institute, Los Baños, Philippines.

Roger P A and Watanabe I 1986 Technologies for utilizing biological nitrogen fixation in wetland rice: Potentialities, current usage and limiting factors. Fert. Res. 9, 39–77.

Rosegrant M W and Roumassett J H 1988 Economic feasibility of green manure in rice-based cropping systems. *In* Sustainable Agriculture: Green Manure in Rice Farming. pp 13–27. The International Rice Research Institute, Los Baños, Philippines.

Roy A C, Wanki S B C and Takow J A 1988 Use of green manure in rice farming systems in West and Northwest Cameroon. *In* Sustainable Agriculture: Green Manure in Rice Farming. pp 333–342. The International Rice Research Institute, Los Baños, Philippines.

Schultze-Kraft R 1988 Collection and evaluation of tropical legume germplasm. *In* Sustainable Agriculture: Green Manure in Rice Farming. pp 343–358. The International Rice Research Institute, Los Baños, Philippines.

Singh Y, Khind C S and Singh B 1991 Efficient management of leguminous green manures in wetland rice. Adv. Agron. 45, 135–189.

Ventura W, Mascarina G B, Furoc R E and Watanabe I 1987 *Azolla* and *Sesbania* as biofertilizers for lowland rice. Philipp. J. Crop Sci. 12, 61–69.

Westcott M P and Mikkelsen D S 1988 Effect of green manure on rice soil fertility in the United States. *In* Sustainable Agriculture: Green Manure in Rice Farming. pp 257–274. The International Rice Research Institute, Los Baños, Philippines.

Yoneyama T, Uchiama T, Sasakawa H, Gamo T, Ladha J K and Watanabe I 1991 Nitrogen accumulation and changes in natural ^{15}N abundance in the tissue of legumes with emphasis on N_2 fixation by stem-nodulating plants in upland and paddy fields. Soil Sci. Plant Nutr. 37, 75–82.

Yost R and Evans D 1988 Green manure and legume covers in the tropics. HITAHR, University of Hawaii. Research Series 055. 43p.

Plant and Soil **174:** 195–209, 1995.

Biological nitrogen fixation associated with sugar cane and rice: Contributions and prospects for improvement

R.M. Boddey, O.C. de Oliveira, S. Urquiaga, V.M. Reis, F.L. de Olivares, V.L.D. Baldani and J. Döbereiner
EMBRAPA - Centro Nacional de Pesquisa de Agrobiologia, Itaguaí, 23851–970, Rio de Janeiro, Brazil

Key words: Acetobacter diazotrophicus, biological nitrogen fixation, endophytic bacteria, *Herbaspirillum* spp., sugar cane, wetland rice

Table of contents

Abstract

Introduction

Quantification of biological nitrogen fixation

  Sugar cane

  Wetland rice

Plant-associated N_2-fixing bacteria

  Sugar cane

  Wetland rice

Prospects for the future

Acknowledgements

References

Abstract

^{15}N isotope and N balance studies performed over the last few years have shown that several Brazilian varieties of sugarcane are capable of obtaining over 60% of their nitrogen (>150 kg N ha^{-1} year^{-1}) from biological nitrogen fixation (BNF). This may be due to the fact that this crop in Brazil has been systematically bred for high yields with low fertilizer N inputs. In the case of wetland rice, N balance experiments performed both in the field and in pots suggest that 30 to 60 N ha^{-1} crop^{-1} may be obtained from plant-associated BNF and that different varieties have different capacities to obtain N from this source. $^{15}N_2$ incorporation studies have proved that wetland rice can obtain at least some N from BNF and acetylene reduction (AR) assays also indicate differences in N_2-fixing ability between different rice varieties. However in situ AR field estimates suggest plant-associated BNF inputs to be less than 8 kg N ha^{-1} crop^{-1}. The problems associated with the use of the ^{15}N dilution technique for BNF quantification are discussed and illustrated with data from a recent study performed at EMBRAPA–CNPAB. Although many species of diazotrophs have been isolated from the rhizosphere of both sugarcane and wetland rice, the recent discovery of endophytic N_2-fixing bacteria within roots, shoots and leaves of both crops suggests, at least in the case of sugarcane, that these bacteria may be the most important contributors to the observed BNF contributions. In sugarcane both *Acetobacter diazotrophicus* and *Herbaspirillum* spp. have been found within roots and aerial tissues and these microorganisms, unlike *Azospirillum* spp. and other rhizospheric diazotrophs, have been shown to survive poorly in soil. *Herbaspirillum* spp. are found in many graminaceous crops, including rice (in roots and aerial tissue), and are able to survive and pass from crop to crop in the seeds. The physiology, ecology and infection of plants by these endophytes are fully discussed in this paper. The sugarcane/endophytic diazotroph association is the first efficient N_2-fixing system to be discovered associated with any member of the gramineae. As yet the individual roles of the different diazotrophs in this system have not been elucidated and far more work on the physiology and anatomy of this system is required. However, the understanding gained in these studies should serve as a foundation for the improvement/development of similar N_2-fixing systems in wetland rice and other cereal crops.

Introduction

The "green revolution" in agriculture of the developing world which resulted in large increases in cereal grain production since the 1960s, has been a result of the development of plant genotypes highly responsive to chemical fertilizers, particularly nitrogen. It requires approximately 18.5 Mcal of fossil energy to produce one kg of fertilizer nitrogen and even though, unlike other fertilizers, there is an unlimited supply of this element in the air, this is more than 6 times the energy required to produce either phosphate or potassium fertilizers (Da Silva et al., 1978). With the inevitable price rises of fossil fuels (not to mention proposed carbon taxes) that must occur over the next few decades due to the depletion of petroleum reserves and increased production costs of other fuels, now is the time that alternative strategies for nitrogen supply should be developed before these increased costs force farmers to cut N inputs which will result in drastic yield reductions in the staple cereal crops which feed the burgeoning human population of the Third World.

In traditional wetland rice culture yields of 2 to 3 t grain ha^{-1} (either one or two crops year^{-1}) seem to be sustainable indefinitely, even where no N fertilizer is applied, if flood water is well controlled. For such yields to be sustained between 60 and 80 kg of nitrogen are required for each crop (Bennett and Ladha, 1992) and while some of this input may be supplied in rainfall and irrigation water, several field N balance studies suggest that N is supplied in part by nitrogen-fixing organisms (Firth et al., 1973; Koyama and App, 1979; Walcott et al., 1977).

Virtually all of the varieties of sugar cane planted in Brazil were bred under conditions of low N fertilizer inputs. Probably for this reason, the plant-crop rarely responds to nitrogen fertilizer (Azeredo et al., 1981), and while ratoon crops do often respond to N application, quantities applied rarely exceed 100 kg N ha^{-1} and fertilizer use efficency is usually less than 35% (Oliveira et al., 1994; Sampaio et al., 1984). A sugar cane crop yielding 100 t cane ha^{-1} accumulates between 180 and 250 kg N ha^{-1} (Orlando-Filho et al., 1980; Stanford and Ayres, 1964). The mean Brazilian yield is 65 to 70 t cane ha^{-1} and average whole crop N accumulation is between 100 to 120 kg N. Of this approximately two thirds is transported to the mill in the cane stems and a further 25% is in the senescent leaves (trash), which in Brazil, as in most countries, is burned off before harvest (Oliveira et al., 1994). Less than 10% of the N in the form of flag leaves remains in the field. It is apparent from these data that continuous cropping of sugar cane should deplete soil N reserves such that cane yields eventually decline. However, such decline in yields or soil N reserves are not normally observed even after many decades, or even centuries, of cane cropping. Such observations have led several authors to suggest that sugar cane may benefit significantly from inputs from biological nitrogen fixation (BNF) (Döbereiner, 1961; Purchase, 1980; Ruschel et al., 1978).

Quantification of biological nitrogen fixation

Sugar cane

Only a few studies have been published on the quantification of the BNF contribution to sugar cane and all of them were performed in Brazil. Experiments using ^{15}N-labelled N_2 gas conducted at the Centro de Energia Nuclear na Agricultura (CENA) in Piracicaba (São Paulo) showed that 90 day-old sugar cane plants obtained considerable N from BNF (Ruschel et al., 1975). However, because of the difficulties of exposing plants grown in the field to controlled atmospheres, the agronomic significance of these N inputs could not be evaluated (Matsui et al., 1981). In a subsequent

^{15}N-aided N balance study performed at CNPAB, sugar cane was grown in pots containing 64 kg soil (Lima et al., 1987). Both the N balance and ^{15}N enrichment data indicated that between 40 and 60% of plant N was derived from plant-associated BNF and extrapolation to the field (15,000 plants ha^{-1}) suggested inputs of over 150 kg N ha^{-1} $year^{-1}$.

Our group has recently completed a three-year ^{15}N isotope dilution and N balance study on 10 sugar cane varieties grown in a concrete tank ($20 \times 6 \times 0.8$m) filled with soil amended with ^{15}N-labelled organic matter, and using *Brachiaria arrecta* as a non-N_2-fixing control plant (Urquiaga et al., 1992). The soil had a low N content (0.108% N) and was fertilized with phosphorus, potassium and micronutrients and well irrigated throughout the experiment, but no N fertilizer was added. In the first year yields of fresh cane of the commercial varieties were high, ranging from the equivalent of 175 to 230 t ha^{-1}, and in the varieties CB 45–3 and SP 70–1143 these high yields were maintained during the subsequent two ratoon crops. In these same varieties and the *Saccharum spontaneum* variety, Krakatau, the nitrogen accumulation also continued to be high and stable over the three years. However, other varieties (e.g., CB 47–89, NA 56–79, SP 71–799, Chunee) showed a decline in total N content after the first year as would be expected from the observed decline in the availability of soil N. Over the whole three years, the weighted mean ^{15}N enrichments of all of the sugar cane varieties were much lower than that of the non-N_2-fixing *B. arrecta* control, indicating large contributions of plant associated BNF (Table 1).

At the second and third annual harvests (first and second ratoon crops) there were only small difference in the ^{15}N enrichments between the different varieties and that of the control crop, which was due to the carry-over of labelled nitrogen from one harvest to the next in the stem bases and roots of cane varieties, which did not occur in the case of the *B. arrecta*. The interpretation of the ^{15}N data was further complicated by the fact that the uptake of soil N by the *B. arrecta* was almost certainly inhibited towards the end of each growing season due to shading of this crop by the tall sugar cane plants, and this probably resulted in a somewhat higher ^{15}N enrichment in the control crop than otherwise would have occurred.

These difficulties are fully discussed in the original paper (Urquiaga et al., 1992), and because of them it was decided to perform a total N balance on the whole tank by the careful analysis of the N content of soil samples taken at plant emergence in the first year in comparison with samples taken at the final harvest. These data showed that there were significantly ($p<0.05$) positive N balances associated with the varieties CB 45–3, SP 70–1143, SP 79–2312 and Krakatau, and that there was a good agreement between the ^{15}N dilution and the total N balance estimates of the contributions of BNF to the sugar cane varieties (Table 1).

These results were recently confirmed in a long-term nitrogen balance experiment conducted on a sugar cane plantation in Pernambuco, NE Brazil (Oliveira et al., 1994). In this experiment the effect of pre-harvest burning of the cane (to remove the senescent leaves) on the yield and N accumulation of the crop, and N balance of the cropping system, were investigated. At the end of the 9 year study the total N accumulated in the system was found to be between 300 and 620 kg ha^{-1} greater than the initial N (Table 2). This extra N was attributed to a mean annual BNF input to the crop of between 38 and 77 kg N ha^{-1}, this being a minimum estimate as gaseous or leaching losses were not quantified.

Wetland rice

Several field N balance studies on lowland rice have been reported from studies in Thailand (Firth et al., 1973; Walcott et al., 1977), Japan (Koyama and App, 1979) and at the experimental fields of the International Rice Research Institute (IRRI) in the Philippines (App et al., 1984; Ventura et al., 1986). All studies report a positive balance even when N from rainfall and irrigation water were discounted indicating inputs of between 30 and 60 kg N ha^{-1} $crop^{-1}$, but in these studies no data are available to determine what proportion of this N may be derived from free-living N_2-fixing cyanobacteria in the flood water, heterotrophic N_2 fixers in the soil or those associated with the plant.

Various nitrogen balance experiments have been performed in pots which indicate that the plant/soil system can benefit from biological N_2 fixation (BNF) even when the activity of cyanobacteria on the soil surface is inhibited by shading (De and Sulaiman, 1950; Willis and Green, 1948). In a very careful N balance study performed in pots by App et al. (1980) on 4 to 6 consecutive crops, the contribution of plant-associated BNF was estimated to be equivalent to 18% of plant N. In a further N balance study on 83 wild and cultivated rice cultivars (in 6 separate experiments each with 3 consecutive crops) reported by App et al. (1986), large and significant differences between cultivars were found.

Table 1. ^{15}N enrichment and total nitrogen accumulation of sugar cane and *Brachiaria arrecta* and estimates of nitrogen derived from BNF using N balance and ^{15}N isotope dilution techniques (g N m^{-2}). Means of 4 replicates. After Urquiaga et al. (1992)

Variety / Species	Weighted mean atom % ^{15}N excess	Final N content of soil	N accum. whole plant 3 years	Estimates of BNF contribution: All three years N balance[z]	All three years ^{15}N[y]	Annual mean N balance	Annual mean ^{15}N
		(g N m^{-2})					
CB 47–89	0.191bcd	835	61.4bc	39.7	34.8c	13.2	11.6
CB 45–3	0.166cde	864	84.3ab	62.6	52.6b	20.9	17.5
NA 56–79	0.198bc	884	57.8c	36.1	32.6c	12.0	10.9
IAC 52–150	0.188bcd	924	59.6bc	37.9	33.8c	12.6	11.3
SP 70–1143	0.146de	852	77.5bc	55.8	51.9b	18.6	17.3
SP 71–799	0.183bcd	860	56.9c	35.2	33.3c	11.7	11.1
SP 79–2312	0.198bc	845	63.6c	41.9	35.4c	14.0	11.8
Chunee	0.227b	826	33.0d	11.3	16.9d	3.8	5.6
Caiana	0.190bcd	857	11.6d	-10.1	6.7d	- 3.4	2.2
Krakatau	0.133e	857	102.8a	81.1	71.8a	27.0	23.9
B. arrecta	0.443a	830	24.9	3.2	-	1.1	-
CV (%)	13.6	5.1	25.0	29.2		29.2	

[z] N balance estimate of BNF contribution = total N accumulated by crop + mean total N content of soil in tank at final harvest - mean total N content of soil in tank at emergence. Mean change in soil N content from emergence until final harvest = 27.1 g N m^{-2} with a standard error of the difference between the means of 22.0 g N m^{-2}. N balances greater than 37.7 g N m^{-2} (12.4 g N m^{-2} $year^{-1}$) were significantly greater than zero (p=0.05, Student t test).
[y] ^{15}N isotope dilution estimate of BNF contribution = (total N accumulated by the crop) × (1 - (weighted mean atom % ^{15}N excess of sugar cane)/(weighted mean atom % ^{15}N excess of *B. arrecta*).

The positive N balances were equivalent to between 16 and 70 kg N ha^{-1} $crop^{-1}$ assuming 25 plants m^{-2} and, although in all 6 experiments there were significant correlations between N balance and plant N uptake, because of the nature of this technique it cannot necessarily be assumed that the N was fixed and immediately incorporated into the plants.

Direct evidence that heterotrophic diazotrophs can contribute significant quantities of N to rice plants was obtained by the short-term exposure of individual plants to ^{15}N enriched N_2 gas (Ito et al., 1980; Yoshida and Yoneyama, 1980; Eskew et al., 1981; Nayak et al., 1986), but most of the labelled nitrogen fixed remained in the rhizosphere soil. However, these data do not permit estimation of BNF contributions over the entire plant growth cycle.

There are many studies which have used the acetylene reduction (AR) assay to study BNF associated with rice. The early studies (e.g. Rinaudo and Dommergues, 1971; Yoshida and Ancajas, 1970, 1973) utilized an assay on excised roots. Later studies on rice and other grasses and cereals suggested that these techniques were unreliable and perhaps overestimated actual N_2 fixing activity (Barber et al., 1976; Koch, 1977; Tjepkema and Van Berkum, 1977), and subsequently in situ assays were developed (Balandreau and Dommergues, 1971; Boddey et al., 1978; Lee et al., 1977). The use of these in situ techniques in the field showed considerable AR activity associated with field grown plants (Watanabe et al., 1978a, 1981) but this technique suffers from several disadvantages for the estimation of actual BNF contributions to the plants: Firstly, the AR technique measures nitrogenase activity and not incorporation of fixed N into the plant, secondly much of the evolved ethylene may be retained in the waterlogged soil and not diffuse into the atmosphere which is sampled, and finally the measure is instantaneous and requires many assays thoughout the growing season if overall contributions of BNF to the crop are to be assessed (Boddey, 1987; Roger and Watan-

Table 2. Effect of pre-harvest burning on total nitrogen balance (g N m^{-2}) of the soil/plant system of field grown sugar cane over a sequence of the plant crop followed by 7 ratoon crops. Means of 16 replicates

Treatment	N accumulated by crop in over 8 cuts 1983–1992	Total N in soil/plant system Considering soil N content in the layer: 0–20cm			0–60cm		
		N[a] Initial	N[b] Final	Balance	N[a] Initial	N[b] Final	Balance
		(g N m^{-2})					
Burned	58.3	365.9	354.1	-11.8	789.0	744.6	-44.4
Unburned	73.6	369.1	400.6	+31.5	774.3	828.7	+54.4
HSD[c] p=0.05	7.0	24.2	29.7	30.6	74.5	64.1	61.9
CV (%)[d]	14.2	8.9	10.6	*(10.3)*[e]	12.8	11.0	*(20.9)*[e]

[a] Initial N in soil plant/system = total N in soil at planting + added fertilizer N.
[b] Final N in soil/plant system = total N in soil at final harvest + N accumulated by crop over 8 harvests (1983 to 1992).
[c] Honest significant difference (Tukey).
[d] Coefficient of variation.
[e] Value in *italics* = Standard error of mean.

abe, 1986). In studies where many in situ AR assays were taken, the estimates of total "acetylene reduced" thoughout the whole crop cycle were approximately 40 to 60 m mol ethylene m^{-2} (Boddey and Ahmad, 1981; Watanabe et al., 1978b) which extrapolate to only 5 to 8 kg N_2 fixed ha^{-1}. Results from the excised root and in situ AR assays on wetland rice were of similar magnitude (Boddey et al., 1978; Boddey, 1981) and it is generally considered that this technique over-estimates N_2-fixing activity (Berkum and Bohlool, 1980; Giller, 1987).

It seems therefore that there is a considerable disparity between the N balance and AR estimates of plant-associated BNF to wetland rice. Some of the field and pot N balance studies suggest contributions of more than 30 kg N ha^{-1} $crop^{-1}$ whereas the acetylene reduction studies suggest inputs not higher than 8 kg ha^{-1}.

The ^{15}N isotope dilution technique has the potential to estimate contributions of BNF to the plants over the whole growth season and unlike the N balance and acetylene reduction techniques, it estimates fixed N actually incorporated into the plant tissue (Chalk, 1985; McAuliffe et al., 1958). The main problem with this technique lies in labelling the soil with ^{15}N. If the enrichment varies with area, depth or time, different plants (the control and different rice varieties) may have different N uptake patterns and do not obtain the same ^{15}N enrichment in the soil derived N, an assumption essential to the application of the technique (Boddey, 1987; Witty, 1983). In the studies reported so far the soil N was not stable with time and no suitable non-N_2-fixing control plant was found that would grow in waterlogged soil (Nayak et al., 1986; Ventura and Watanabe, 1983).

A recent study was conducted at our institute (CNPAB) near Rio de Janeiro (Oliveira, 1994) and at the first planting 40 rice varieties were planted in a tank (20 × 6 × 0.6m) filled with waterlogged soil amended with ^{15}N-labelled compost (Urquiaga et al., 1992) and inoculated with soil taken from a long-established rice paddy in the Paraíba valley of São Paulo State. Analyses of leaf samples showed that there was a considerable decline in ^{15}N enrichment in the plant tissue during plant growth and earlier maturing varieties showed higher ^{15}N enrichments than later maturing varieties (Table 3). There were considerable differences in total N accumulation and ^{15}N enrichment between different varieties but higher N accumulation was not well correlated with lower ^{15}N enrichment even within each maturity group (Table 4).

Subsequently 20 of these rice cultivars were replanted in the same tank. Again ^{15}N enrichment in plant tissue decreased with time and the varieties

Table 3. Grain production, N accumulation and ^{15}N enrichment of leaf samples at 40 days after emergence (DAE) and of whole plant at final harvest of 5 rice varieties from each of 3 maturity groups. Plants grown in tank of soil labelled with ^{15}N. Means of 4 replicates. After Oliveira (1994)

Rice variety	Grain yield[z] (g m^{-2})	N accumulation (g N m^{-2})	^{15}N enrichment Atom % ^{15}N exc. Final[y] harvest
Maturity group 1 (60–85 DAE)			
Labelle	396 d	7.37 d	0.2074a
CNA 6837	814 ab	9.28 bc	0.2305a
Bluebelle	633 c	8.55 cd	0.1984a
BR-IRGA-410	766 ab	9.25 bc	0.2301a
BR-IRGA-409	759 ab	10.54 ab	0.2134a
C.V. (%)	10.3	9.0	11.7
Mean for whole group 7 varieties	711	94.9	0.2160
Maturity group 2 (80–110 DAE)			
IR 4432–28–5	942 b	17.82 a	0.1475 c
MG-1	1097 a	15.83 ab	0.1559 bc
IR-841	701 c	11.46 c	0.1586 bc
CICA-9	930 b	14.43 b	0.1618 bc
CNA 4215	698 c	8.87 d	0.1822 ab
C.V. (%)	11.0	11.0	11.2
Mean for whole group 18 varieties	906	13.3	0.1608
Maturity group 3 (110–140 DAE)			
Metica-1	1130 ab	16.19 b	0.1557 cd
De-Abril	1070 ab	22.18 a	0.1421 d
IAC-4440	1100 ab	15.30 b	0.1973 a
CICA-8	1070 ab	14.32 b	0.1758 abc
IR-42	799 c	15.68 b	0.1388 d
C.V. (%)	10.8	12.0	10.5
Mean for whole group 15 varieties	1060	15.0	0.1621

[z] Grain at 14% humidity.

[y] Weighted mean ^{15}N enrichment of whole plant.

Table 4. Regressions of total nitrogen accumulation and ^{15}N enrichment at final harvest of 40 rice varieties divided into 3 maturity groups planted in waterlogged ^{15}N-labelled soil. First crop (1989/90). After Oliveira (1994)

Maturity group	Days after emergence	Correlation coefficient	Probability	No. of data points
1	60–85	+ 0.281	0.147	28
2	85–110	- 0.320	0.006	72
3	110–140	- 0. 201	0.124	60

IR 42 and IR 4432–28–5 showed significantly lower ^{15}N enrichment and higher N accumulation than the variety IAC 4440 and the non-N_2-fixing control plant, *Brachiaria arrecta* (data not shown). Data from the N balance study of App et al. (1986) as well as acetylene reduction assays and a natural abundance (delta) ^{15}N study both performed at IRRI in the Philippines also suggest that the variety IR 42 is able to obtain significant contributions from plant associated BNF (Barraquio et al., 1986; Ladha et al., 1987a, b; Watanabe et al., 1987a).

Results from the third planting of this ^{15}N experiment were lost due to a fire in the drying oven but at the fourth planting just these 3 varieties were planted with the same control plant and harvests were taken at six times during plant growth (Table 5). The acetylene reduction activity of the 4 crops was evaluated by incubating the plant/soil system at constant temperature in the dark as described by Barraquio et al,. (1986). No significant differences were found between varieties but the rice varieties were far higher in AR activity than the *B. arrecta* control (data not shown). After the 3rd harvest (86 DAE) the ^{15}N enrichment of the *B. arrecta* control was lower than that of the rice varieties (significantly so at the final harvest) but this result could not be due to a soil N uptake pattern different from the rice varieties as the data indicate that the ^{15}N enrichment of the soil mineral N was virtually stable during crop growth. Furthermore, while the variety IR 4432–28–5 had a lower ^{15}N enrichment than the other two rice varieties the total N accumulation of this cultivar showed a tendency to be lower.

Hence, the data obtained in this study do not confirm significant BNF contributions to any of the 3 varieties of wetland rice even though two of them were pre-selected for high N accumulation and low ^{15}N enrichment. Whether this is due to adverse soil fertility factors or indicates that BNF inputs are generally very low requires further investigation. The results illustrate the difficulties involved in the application of this technique for quantifying BNF contributions to wetland rice and the necessity to use soil with a uniform and stable ^{15}N enrichment.

Plant-associated N_2-fixing bacteria

Sugar cane

In the 1950s Döbereiner (1961) found N_2-fixing bacteria of the genus *Beijerinckia* in high numbers in sugar cane fields, with selective enrichment in the rhizosphere and especially on the root surface. At the same time a new species of *Beijerinckia* was discovered (*B. fluminense*) associated with this crop (Döbereiner and Ruschel, 1958). Subsequently, other authors (Graciolli et al., 1983; Purchase, 1980) isolated a wide range of N_2-fixing bacteria from the roots, stems and even leaves of sugar cane including species of *Erwinia, Azotobacter, Derxia, Azospirillum* and *Enterobacter*. None of these bacteria seemed to occur in large enough numbers to account for the extremely high rates of N_2 fixation reported above.

More recently, a new species of N_2-fixing bacteria, *Acetobacter diazotrophicus*, was found to occur in large numbers in the roots and stems of sugar cane (Cavalcante and Döbereiner, 1988; Gillis et al., 1989). This most extraordinary diazotroph was originally isolated from semi-solid sugar cane juice inoculated with dilutions of sugar cane roots and stems which showed acetylene reduction (nitrogenase) activity in dilutions up to 10^{-6} to 10^{-7} (fresh weight). A more specific medium (LGIP) has now been developed (Reis et al., 1994).

The bacteria is a small, Gram-negative, aerobic rod showing pellicle formation in N-free semi-solid medi-

Table 5. Total nitrogen accumulation and ^{15}N enrichment of 3 rice varieties and *Brachiaria arrecta* planted in waterlogged ^{15}N labelled soil during the plant growth cycle. Fourth crop (1992/3). Harvested area 0.5 m^2. Means of 4 replicates. After Oliveira (1994)

Variety	Days after emergence of rice					
	36	52	86	94	108	130
Total N accumulation (g N m^{-2})						
IR 42	0.728ab	0.692ab	1.951ab	2.389a	3.837a	4.449a
IAC 4440	0.902a	0.767a	3.101a	3.299a	4.093a	4.799a
IR 4432–28–5	0.792a	0.774a	2.013ab	2.476a	3.757a	4.055a
B. arrecta	0.166b	0.321b	0.867b	0.601b	1.609b	1.009b
C.V. (%)	41.37	27.61	44.61	54.5	27.01	18.63
^{15}N enrichment (Atom % ^{15}N excess)						
IR 42	0.0549c	0.0558a	0.0552a	0.0527a	0.0536a	0.0553ab
IAC 4440	0.0680a	0.0558a	0.0482a	0.0553a	0.0497a	0.0606a
IR 4432–28–5	0.0643ab	0.0561a	0.0552a	0.0553a	0.0484a	0.0484bc
B. arrecta	0.0582bc	0.0549a	0.0517a	0.0482a	0.0428a	0.0419c
C.V. (%)	5.69	9.36	10.9	8.38	16.11	8.65

Means in the same column followed by the same letter are not significantly different at p=0.05 (Tukey).

um with 100 g L^{-1} sucrose but without cane juice, forming a thick surface pellicle after 7 to 10 days. Best growth occurs with high sucrose or glucose concentrations (100 g L^{-1}) and strong acid production results in a final pH of 3.0 or less. Growth and N_2 fixation (more than 100 n moles C_2H_2 mL^{-1} h^{-1}) continues at this pH for several days (Stephan et al., 1991). Ethanol is also used as a C source for growth and is oxidized to CO_2 and H_2O. Dark brown colonies form on potato agar with 100 g L^{-1} sucrose, and dark orange colonies on N-poor (0.02 g L^{-1} yeast extract) mineral agar medium with 100 g L^{-1} sucrose and bromothymol blue. The bacterium possesses no nitrate reductase and N_2 fixation is not affected by high levels (25m*M*) of NO_3^-. Also NH_4^+ causes only partial inhibition of nitrogenase, especially when grown on 100 g L^{-1} sucrose (Boddey et al., 1991; Teixeira et al., 1987).

Another interesting aspect is that *A. diazotrophicus* growing in 10% sucrose showed an optimum dissolved oxygen concentration for acetylene reduction in equilibrium with 0.2 kPa O_2 in the atmosphere, but continued to fix N_2 up to 4.0 kPa, showing a much higher O_2 tolerance than *Azospirillum* spp. (Reis et al., 1990).

Experiments on mixed cultures of *A. diazotrophicus* with an amylolytic yeast (*Lypomyces kononenkoae*), used as a model system for plant/bacteria interactions, showed that 48% of the total nitrogen fixed by the bacteria was transferred to the yeast, starting right from the beginning of the culture (Cojho et al., 1993). These results are important in that until now the lack of evidence for efficient transfer of fixed N from diazotrophs to plants has been a source of scepticism that such associations could be of agronomic importance.

This bacterium has been found in many sugar cane varieties in several regions of Brazil as well as in Mexico, Cuba and Australia (Fuentes-Martinez et al., 1993, Li and Macrae, 1992) and numbers were in the range of 10^3 to 10^7 in roots, basal and apical stems, leaves and in sugar cane trash (Döbereiner et al., 1988). It was not found in soil between rows of sugar cane plants or roots from 12 different weed species taken from cane fields. It was also not found in grain or sugar sorghum, but was isolated from a few samples of washed roots and aerial parts of *Pennisetum purpureum* cv Cameroon, and from sweet potatoes (Döbereiner et al., 1988, 1994; Paula et al., 1989).

Sterile micropropagated sugar cane seedlings were not infected by *A. diazotrophicus* by traditional root inoculation methods, and generally infection of cane plants by this bacterium is rare except when inoculated "in vitro". However, under these conditions *A. diazotrophicus* was found to colonize extensively the exterior and interior of the shoot and root (James et al., 1994). This study was performed using immuno-gold labelling with both optical and electron microscopic techniques. On the root surface the bacteria was found especially in cavities in lateral root junctions and these junctions and the root tips appeared to be preferred sites of bacterial entry. Within the roots *A. diazotrophicus* was observed in apparently intact, enlarged epidermal cells, and at the base of the stem within xylem vessels through which the bacteria appear to migrate upwards in the transpiration stream so that all shoot tissues become infected. The difficulty of infection of plants grown in soil or vermiculite can be overcome by co-inoculation with VA mycorrhizal fungi, especially originating from fungal spores infected by the bacteria (Paula et al., 1991). This technique of introduction of a N_2-fixing bacteria into sugar cane plants may be important for introducing selected, or genetically improved, strains into plants for further propagation in the field via stem cuttings.

Bacterial taxonomists working in Belgium found that the bacteria known as *Pseudomonas rubrisubalbicans*, a sugar cane endophyte which causes mottled stripe disease in some varieties from the USA and other countries, but not in Brazilian varieties, was closely related genetically to a N_2-fixing bacterium called *Herbaspirillum seropedicae* (Gillis et al., 1991). *Herbaspirillum* was first isolated from the roots of maize and other cereals at our Centre (Baldani et al., 1986). Most of the isolates of *P. rubrisubalbicans* were found to be able to fix nitrogen and were identical in most other respects to *Herbaspirillum* (Pimentel et al., 1991). Recently results from DNA/rRNA hybridization and computer-assisted auxanographic tests have established that this generically-misnamed plant endophyte, *"Pseudomonas" rubrisubalbicans*, must now be included in the genus *Herbaspirillum* (Gillis et al., 1991).

Recently a more specific culture medium (JNFb) for this organism has been developed and $^{15}N_2$ gas incorporation confirmed, not only in strains of the original *H. seropedicae*, but also in isolates from different culture collections of *H. rubrisubalbicans* identified as the causitive organism of mottled stripe disease (Table 6). *Herbaspirillum* spp. have been isolated from sugarcane leaves, stems and roots and are other N_2-fixing bacteria which do not survive well in the soil but only within plants (Baldani et al., 1992a).

Table 6. $^{15}N_2$ incorporation into cells of *P. rubrisubalbicans* and *H. seropedicae* strains grown in semi-solid JNFb medium. Means of three replicates. After Baldani et al. (1992)

Strains	Atom % ^{15}N excess
P. rubrisubalbicans	
M1 (LMG[a] 1278)	0.5271
M4 (ATCC[b] 19308)	0.4891
M5 (LMG 6415)	0.5681
M6 (LMG 6420)	0.5172
IBSBF 175 (LMG 10462)	0.3881
H. seropedicae	
Z67 (ATCC 35892)	0.5881
Z78 (ATCC 35893)	0.4405
ZM 176	0.4891
Controls	
M4 + 20m*M* NH_4^+	0.0002
Z67 + 20m*M* NH_4^+	0.0000

[a] LMG Belgian type culture collection.
[b] ATCC American type culture collection.

When non-sterile soil was inoculated with 10^8 cells g^{-1} of either species of *Herbaspirillum*, the number of viable cells decreased until the bacteria was undetectable after 21 days with *H. rubrisubalbicans* and 28 days with *H. seropedicae* (Olivares et al., 1993). However, 50 days after *Herbaspirillum* spp. were undetectable, surface-sterilized, sorghum seeds were planted in these pots and *Herbaspirillum* spp. were detected in the roots and rhizosphere soil when the plants were 30 days old.

In both monoxenic sugarcane and sorghum plants inoculated with *Herbaspirillum* spp. the bacteria have been localized, using the immunogold technique and both electron and optical microscopy, within the meta and protoxylem (Olivares et al., in preparation). In the case of a sugar cane variety (B–4362), susceptible to mottled stripe disease, *H. rubrisubalbicans* was found to completely block some of the xylem vessels, whereas in a resistant variety the bacteria were encapsulated by membranes probably of plant origin.

Wetland rice

As long ago as 1929, an Indian research worker suggested that wetland rice plants were able to obtain some contribution of nitrogen from N_2-fixing bacteria associated with the plant roots (Sen, 1929). His evidence was based on the isolation of *Azotobacter* spp. from rice roots. Since this time many diazotrophs have been isolated from the rhizosphere and roots of rice including species of *Beijerinckia* (Dobereiner and Ruschel, 1962), *Azospirillum* (Baldani and Dobereiner, 1980; Baldani et al., 1981; Ladha et al., 1982), *Alicagenes* (Qui et al., 1981), *Pseudomonas*, (Barraquio et al., 1983), *Klebsiella* and *Enterobacter* (Ladha et al., 1983); *Flavobacterium* (Bally et al., 1983), and *Agromonas* (Ohta and Hattori, 1983). However, the presence of N_2 fixing bacteria associated with rice roots does not necessarily mean that the plants obtain significant contributions from biological nitrogen fixation (BNF). For example, in a study of the inoculation of wheat plant grown in ^{15}N-labelled soil numbers of *Azospirillum brasilense* above 10^6 cells g fresh root^{-1} were counted on washed/surface sterilized roots and plant N uptake was significantly increased by *Azospirillum* inoculation, but ^{15}N enrichment data showed that the response was not due to BNF inputs (Boddey et al., 1986).

Azospirillum spp. have been isolated in considerable numbers from the rhizosphere and histosphere of wetland rice (Baldani et al., 1981; Ladha et al., 1982, 1987b; Omar et al., 1989) and a new species of *Pseudomonas* (*P. diazotrophicus*) was reported to dominate the rhizosphere bacterial population (Barraquio et al., 1982; Watanabe et al., 1987b). However, as has been pointed out by several authors, N_2-fixing bacteria are distant from the main sources of carbon substrates in the root (the vascular tissue) and are in competition with other soil microorganisms for these substrates (Barber and Lynch, 1977; Berkum and Bohlool, 1980; Kennedy and Tchan, 1992). On the other hand N_2-fixing bacteria found within rice roots or aerial tissue are unlikely to suffer from these disadvantages, and in view of the discovery of endophytic diazotrophs in sugar cane, research at our Centre has focussed on the search for such bacteria in lowland rice.

In the first report of the discovery of *Herbaspirillum seropedicae*, this N_2-fixing bacteria was isolated from washed roots of upland rice as well as from those of wheat, maize and sorghum (Baldani et al., 1986). Further studies have shown that this bacteria can be found in seeds, stems and leaves of rice as well as roots. Roots, stems and leaves of rice plants grown from seeds which were surface sterilized using hydrogen peroxide followed by acidified hypochlorite, were found to be infected with *H. seropedicae*, and only careful surface sterilization of dehulled seeds prevented this (Baldani et al., 1992b; V L D Baldani, unpubl. data). In the experiment described above to quantify BNF contributions to rice plants grown in the tank of ^{15}N-labelled soil (Oliveira, 1994) counts of *Herbaspirillum* spp. were made using the selective medium described by Baldani et al. (1992). The results showed that numbers of *Herbaspirillum* in washed roots, shoots and leaves were as high as 10^7, 10^5 and 10^4 cells g fresh weight^{-1}, respectively, and the ontogenic variation in numbers varied in a similar manner to the acetylene reduction activity associated with the plants (Fig. 1).

A further N_2-fixing bacteria has been found to be present in rice roots, shoots and leaves in numbers similar to those reported in Figure 1. As was suggested before, for the first attempts to isolate N_2-fixing bacteria from plants it is desirable to base isolation media on the carbon substrates known to be available within the plants (Boddey and Döbereiner, 1988). This was why malate was chosen for the semi-solid media first used to isolate *Azospirillum* as it was known to be an important constituent of maize sap (Döbereiner and Day, 1976). For the same reason cane juice was used for the first attempts to isolate diazotrophs from sugar cane (Cavalcante and Döbereiner, 1988). Boreau (1977) investigated the root exudates of 20 day-old sterile rice plants and discovered that glucose was the single most important carbon source and that in the organic acid fraction oxalate and citrate were quantitatively most important. Based on these results, a N-free semi-solid medium containing glucose, oxalate and citrate (medium 'M') was inoculated with dilutions of washed rice roots and rice stems (Oliveira, 1992). Slow but significant growth with initial acid production was observed, indicating the consumption of glucose. The medium was later alkalinized, indicating subsequent use of the dicarboxylic acids. Maximal AR activity was observed after 10 days incubation in N-free medium and AR activity continued until the 18th day of growth. The bacteria are small motile rods, but have not yet been identified as any of the known diazotrophs. The isolates grow best at pH between 5 and 6 and growth is very slow at pH 7. They use glucose, mannitol, celobiose, maltose, sucrose or trehalose as sole carbon sources and will hydrolyse Tween 80. This bacteria is most closely related phenotypically to *Herbaspirillum seropedicae* and *H. rubrisubalbicans* but whether it is a member of this genus awaits further investigation using DNA/rRNA homology tests etc.

A further possible candidate for an endophytic diazotroph which will infect rice plants are bacteria of the newly denominated genus *Azoarcus* (Reinhold-Hurek

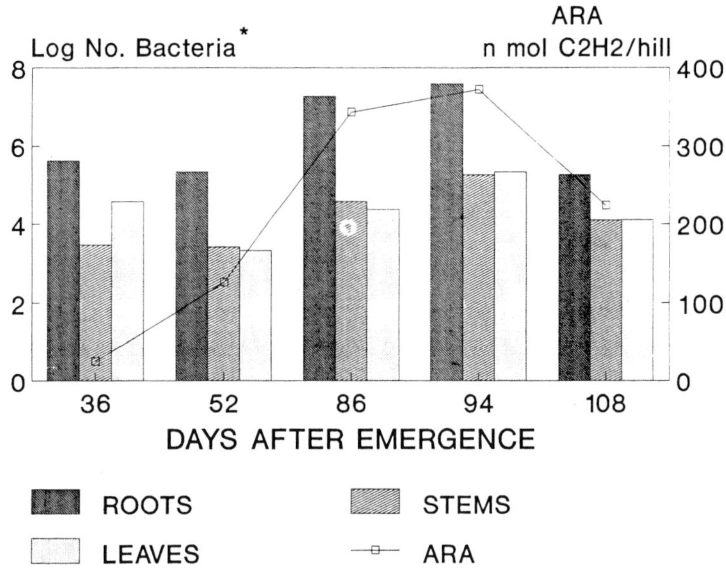

Fig. 1. Counts of *Herbaspirillum* spp. in roots, stems and leaves, and acetylene reduction activity (plant soil system, Barraquio et al., 1986 of the wetland rice variety IR 42 grown in a tank of ^{15}N-labelled waterlogged soil. * Bacterial numbers expressed per g fresh weight plant tissue (after Oliveira et al., 1994).

et al., 1993). The bacteria (labelled with the beta-glucuronidase reporter gene) were found to be able to penetrate rice roots, forming large inter- and intracellular colonies in the root cortex and just occasionally within the stele and were also found within the stem bases and shoots (Desomer et al., 1992; Hurek et al., 1991).

Prospects for the future

Brazilian sugar cane varieties are known to be capable of obtaining very considerable contributions of biologically fixed N under field conditions. Recent data suggest that water supply is critical to the maintenance of high BNF activity. A recent trial (16 areas totalling 900 ha) at a sugar cane plantation in Campos (NE Rio de Janeiro State) showed that where year round irrigation was used there was no response of ratoon cane to 200 kg ha^{-1} of urea fertilizer and yields of ratoon crop cane averaged 95 t ha^{-1}. As a result of this trial the plantation managers abandoned N fertilization on 4000 ha of irrigated cane making an annual economy of US $ 250,000 (Boddey, 1995). All attempts to isolate *Acetobacter diazotrophicus* from sugar cane from anywhere in the world have been successful except where high N fertilizer additions have been made (J Caballero Mellado, pers. commun.). Apart from Brazil no data are yet available for the occurrence of *Herbaspirillum* spp. in this crop.

The complete absence of *A. diazotrophicus* in soil and the restricted occurrence of *Herbaspirillum* spp., suggest that once selected (or even genetically manipulated) strains of these bacteria are established in cane plants in the field, the chances are slight that wild type strains will contaminate the plants to compete with them. For phytosanitary reasons the use of direct planting of monoxenic micropropagated cane plantlets is now being tested at several cane plantations in São Paulo state and this may soon offer an economically viable opportunity to propagate cane plants infected by superior strains of endophytic diazotrophs.

With regard to wetland rice it is evident that for BNF to contribute to high rice yields a great improvement in its efficiency is required. A meeting held at IRRI (Philippines) in 1992 was dedicated solely to this subject. Three possible strategies to increase BNF contributions to wetland rice were discussed (Bennett and Ladha, 1992):

1. Induction of "nodulation" of rice using hydrolytic enzymes (Al-Mallah et al., 1989), 2,4-D (Kennedy and Tchan, 1992) or other means (Rolfe and Bender, 1990) and subsequent infection with *Rhizobium, Azospirillum* or other diazotrophs. True

N_2-fixing legume nodules are complicated structures equipped with vascular tissue to supply C substrate and export fixed N. They possess a sophisticated oxygen protection mechanism with leghaemoglobin and both fixed and variable physical barriers to O_2 diffusion, and an array of specific enzyme systems and feedback controls. The induction of deformations on the root to house bacteria only constitutes a tiny fraction of the symbiotic system and the remaining parameters are dictated pricipally by the plant genome, the *Rhizobium* being mainly responsible for "switching on" the plant nodulation program (Dénarié and Roche, 1991). It thus seems that the induced nodulation strategy has little chance of success especially as true legume nodules serve to protect the nitrogenase system from external oxygen flux from the soil and in wetland rice the soil is anaerobic and oxygen flow to the root is via the aerenchyma (expanded cortex) of the root.

2. Direct integration of *nif* genes into the plant genome. Attempts to introduce just 2 of these genes into tobacco chloroplasts has met with some success although expression was found to be at extremely low levels (Dowson-Day et al., 1991). So far it is not known exactly how many, or which, *Rhizobium* genes will be necessary to make an active N_2-fixing system nor what levels of activity could be achieved.
3. Improvement/modification of existing associations of N_2-fixing bacteria with rice plants. Little enthusiasm has been expressed for this strategy as almost all attention has been focussed on diazotrophs found in the rice rhizosphere (Kennedy and Tchan, 1992). However, the recent discovery that some sugarcane varieties can obtain very large contributions of BNF under field conditions, and the existence of abundant populations of endophytic diazotrophs (*A. diazotrophicus* and *Herbaspirillum* spp.) in this crop which are probably responsible for this activity, opens up entirely new avenues for developing a similar system for rice or other cereal crops. Already one of these endophytic diazotrophs, *Herbaspirillum* spp., has been isolated in moderately high numbers from within roots and aerial tissue of rice, although evidence is lacking that these organisms contribute any significant quantities of fixed N to the plants. However, when more knowledge is accumulated concerning how the N_2-fixing system in sugarcane functions, it should be a much smaller step to try to introduce this into a plant which already can be infected by similar diazotrophs than trying to build a whole N_2-fixing system from scratch.

Acknowledgements

The authors thank Sra Dorimar de Santos Felix for her help in arranging the literature.

References

Al-Mallah M K, Davey M R and Cocking E C 1989 Formation of nodular structures on rice seedlings by rhizobia. J. Exp. Bot. 40, 473–478.

App A A, Watanabe I, Alexander M, Ventura W, Daez C, Santiago T and De Datta S K 1980 Nonsymbiotic nitrogen fixation associated with the rice plant in flooded soils. Soil Sci. 130, 283–289.

App A A, Watanabe I, Ventura T S, Bravo M and Jurey C D 1986 The effect of cultivated and wild rice varieties on the nitrogen balance of flooded soil. Soil Sci. 141, 448–452.

App A A, Santiago T, Daez C, Menguito C, Ventura W, Tirol A, Po J, Watanabe I, De Datta S K and Roger P 1984 Estimation of the nitrogen balance for irrigated rice and the contribution of phototrophic nitrogen fixation. Field Crops Res. 9, 17–27.

Azeredo D F, Bolsanello J, Weber M and Vieira J R 1986 Nitrogênio em cana-planta, doses e fracionamento. STAB 4, 26–32.

Baldani J I, Baldani V L D, Seldin L and Döbereiner J 1986 Characterization of *Herbaspirillum seropedicae* gen. nov., sp. nov., a root-associated nitrogen-fixing bacterium. Int. J. Syst. Bacteriol. 36, 86–93.

Baldani J I, Pereira P A A, Rocha R E M and Döbereiner J 1981 Especificidade na infecçÁo de raízes por *Azospirillum* spp. em plantas com via fotossintetica C3 e C4. Pesq. Agropec. Bras. 16, 325–330.

Baldani V L D and Döbereiner J 1980 Host-plant specificity in the infection of cereals with *Azospirillum* spp. Soil Biol. Biochem. 12, 433–439.

Baldani V L D, Baldani J I, Olivares F L and Döbereiner J 1992a Identification and ecology of *Herbaspirillum seropedicae* and the closely related *Pseudomonas rubrisubalbicans*. Symbiosis 13, 65–73.

Baldani V L D, James E, Baldani J I and Döbereiner J 1992b Localization of the N_2-fixing bacteria *Herbaspirillum seropedicae* within root cells of rice. An. Acad. Bras. Cienc. 64–431.

Bally R, Thomas-Bauzon D, Heulin T and Balandreau J 1983 Determination of the most frequent N_2-fixing bacteria in a rice rhizosphere. Can. J. Microbiol. 29, 881–887.

Barber D A and Lynch J M 1977 Microbial growth in the rhizosphere. Soil Biol. Biochem. 9, 305–308.

Barber L E, Tjepkema J D, Russell S A and Evans H J 1976 Acetylene reduction (nitrogen fixation) associated with corn inoculated with *Spirillum*. Appl. Environ. Microbiol. 32, 108–113.

Barraquio W L, Daroy M L G, Tirol A C, Ladha J K and Watanabe I 1986 Laboratory acetylene reduction assay for relative measurement of N_2-fixing activities associated with field-grown wetland rice plants. Plant and Soil 90, 359–372.

Barraquio W L, Guzman M R de, Barrion M and Watanabe I 1982 Population of aerobic heterotrophic nitrogen fixing bacteria associated with wetland and dryland rice. Appl. Environ. Microbiol. 43, 124–128.

Barraquio W L, Ladha J K and Watanabe I 1983 Isolation and identification of N_2-fixing *Pseudomonas* associated with wetland rice. Can. J. Microbiol. 29, 867–873.

Bennett J and Ladha J K 1992 Introduction: Feasibility of nodulation and nitrogen fixation in rice. *In* Nodulation and Nitrogen Fixation in Rice. Eds. G S Khush and B Bennett. pp 1–14. IRRI, Manila.

Berkum P van and Bohlool B B 1980 Evaluation of nitrogen fixation by bacteria in association with roots of tropical grasses. Microbiol. Rev. 44, 491–517.

Boddey R M 1981 Biological nitrogen fixation in the rhizosphere of lowland rice. PhD thesis, University of the West Indies, St Augustine, Trinidad. 275p.

Boddey R M 1987 Methods for quantification of nitrogen fixation associated with gramineae. CRC Crit. Rev. Plant Sci. 6, 209–266.

Boddey R M 1995 Biological nitrogen fixation in sugar cane: A key to energetically viable bio-fuel production. CRC Crit. Rev. Plant Sci. (*In press*).

Boddey R M and Ahmad N 1981 Seasonal variations in nitrogenase activity of various rice varieties measured with an in situ acetylene reduction technique in the field. *In* Associative N_2 Fixation. Eds. P B Vose and A P Ruschel. pp 219–229. CRC press, Boca Raton, Florida.

Boddey R M and Döbereiner J 1988 Nitrogen fixation associated with grasses and cereals: recent results and perspectives for future and research. Plant and Soil 108, 53–65.

Boddey R M, Quilt P and Ahmad N 1978 Acetylene reduction in the rhizosphere of rice: methods of assay. Plant and Soil 50, 567–574.

Boddey R M, Baldani V L D, Baldani J I and Döbereiner J 1986 Effect of inoculation of *Azospirillum* spp. on the nitrogen assimilation of field grown wheat. Plant and Soil 95, 109–121.

Boddey R M, Urquiaga S, Reis V and Döbereiner J 1991 Biological nitrogen fixation associated with sugar cane. Plant and Soil 137, 111–117.

Boreau M 1977 Application de la cromatographie en phase gazeuse á l'etude de l'exudation racinaire du riz. Cah. ORSTOM Sér. Biol. 12, 75–81.

Cavalcante V A and Döbereiner J 1988 A new acid-tolerant nitrogen-fixing bacterium associated with sugarcane. Plant and Soil 108, 23–31.

Chalk P M 1985 Estimation of N_2 fixation by isotope dilution: An appraisal of techniques involving ^{15}N enrichment and their application. Soil Biol. Biochem. 17, 389–410.

Cojho E H, Reis V M, Schenberg A C G and Döbereiner J 1993 Interactions of *Acetobacter diazotrophicus* with an amylolytic yeast in nitrogen-free batch culture. FEMS Microbiol. Lett. 106, 341–346.

Da Silva J G, Serra G E, Moreira J R, Gonçalves J C, Goldemberg J 1978 Energy balance for ethyl alcohol production from crops. Science 210, 903–906.

De P K and Sulaiman M 1950 Influence of algal growth in the rice fields on the yield of crop. Indian J. Agric. Sci. 20, 327–342.

Dénarié J F and Roche P 1991 *Rhizobium* nodulation signals. *In* Molecular signals in plant-microbe communications. Ed. D P S Verma. pp. 295–324. CRC Press, Boca Raton, Florida.

Desomer J, Holsters B, Reinhold-Hurek B and Van Monatgu M 1992 Exploring new soil bacteria. *In* Nodulation and Nitrogen Fixation in Rice. Eds. G S Khush and B Bennett. pp 77–82. IRRI, Manila.

Döbereiner J 1961 Nitrogen-fixing bacteria of the genus *Beijerinckia* Derx in the rhizosphere of sugar cane. Plant and Soil 15, 211–216.

Döbereiner J and Day J M 1976 Associative symbiosis in tropical grasses: Characterization of microorganisms and dinitrogen fixing sites. *In* Nitrogen Fixation. Eds. W E Newton and C J N Nyman. pp 518–538. Washington State Univ. Press, Pullman, Washington.

Döbereiner J and Ruschel A P 1958 Uma nova espécie de *Beijerinckia*. R. Biol. 1, 261–272.

Döbereiner J and Ruschel A P 1961 Inoculação do arroz com bactérias fixadoras de nitrogênio do gênero *Beijerinckia* Derx. R. Bras. Biol. 21, 397–407.

Döbereiner J, Baldani V L D, Olivares F and de Reis V M 1994 Endophytic diazotrophs: The key to BNF in gramineous plants. *In* Nitrogen Fixation with Non-Legumes. Eds. N A Hegazi, M Fayez and M Monib. pp 395–408. Am. Univ. in Cairo Press, Cairo, Egypt.

Döbereiner J, Reis V and Lazarine A C 1988 A new N_2 fixing bacteria in association with cereals and sugarcane. *In* Nitrogen Fixaton: Hundred years after. Eds. H Bothe, F J De Bruijn and W E Newton. pp 717–722. Gustav Fischer, Stuttgart.

Dowson-Day M, Ashurst J L, Watts J, Dixon R A and Merrick M J 1991 Studies of the potential for expression of nitrogenase Fe-protein in cells of higher plants. *In* Nitrogen Fixation. Eds. M Polsinelli, R Materassi and M Vincenzini. pp 659–669. Kluwer Academic Publ., Dordrecht.

Eskew D L, Eaglesham A R J and App A A 1981 Heterotrophic N_2 fixation and distribution of newly fixed nitrogen in a rice-flooded soil system. Plant Physiol. 68, 48–52.

Firth P, Thitipoca H, Suthipradit S, Wetselaar R and Beech D F 1973 Nitrogen balance studies in the Central Plain of Thailand. Soil Biol. Biochem. 5, 41–46.

Fuentes-Ramirez L E, Jiminez-Salgado T, Abarca-Ocampo I R and Caballero Mellado J 1993 *Acetobacter diazotrophicus*, an indolacetic acid-producing bacterium isolated from sugarcane cultivars in Mexico. Plant and Soil 154, 145–150.

Giller K E 1987 Use and abuse of the acetylene reduction assay for measurement of 'associative' nitrogen fixation. Soil Biol. Biochem. 19, 783–784.

Gillis M, Döbereiner J, Pot B, Goor M, Falsen E, Hoste B, Reinhold B and Kersters K 1991 Taxonomic relationships between [*Pseudomonas*] *rubrisubalbicans*, some clinical isolates (EF group 1), *Herbaspirillum seropedicae* and [*Aquaspirillum*] *autotrophicum*. *In* Nitrogen Fixation. Eds. M Polsinelli, R Materassi and M Vincenzini. pp 292–294. Kluwer Academic Publ., Dordrecht.

Gillis M, Kerters B, Hoste D J, Kroppenstedt R M, Stephan M P, Teixeira K R S, Döbereiner J and De Ley J 1989 *Acetobacter diazotrophicus* sp. nov. a nitrogen fixing acetic acid bacterium associated with sugar cane. Int. J. Syst. Bacteriol. 39, 361–364.

Graciolli L A, Freitas J R de and Ruschel A P 1983 Bactérias fixadoras de nitrogênio nas raízes, caules e folhas de cana-de-açúcar (*Saccharum* sp.). R. Microbiol. (São Paulo) 14, 191–196.

Hurek T, Reinhold-Hurek B, Van Montagu M and Kellenberger E 1991 Infection of intact roots of Kallar grass and rice seedlings by 'Azoarcus'. Dev. Plant Soil Sci. 48, 235–242.

Ito O, Cabrera D and Watanabe I 1980 Fixation of dinitrogen-15 associated with rice plants. Appl. Environ. Microbiol. 39, 554–558.

James E K, Reis V M, Olivares F L, Baldani J I and Döbereiner J 1994 Infection of sugar cane by the nitrogen-fixing bacterium *Acetobacter diazotrophicus*. J. Exp. Bot. 45, 757–766.

Kennedy I R and Tchan Y T 1992 Biological nitrogen fixation in non-leguminous field crops: Recent advances. Plant and Soil 141, 93–118.

Koch B L 1977 Associative nitrogenase activity by some Hawaiian grass roots. Plant and Soil 47, 703–706.

Koyama T and App A A 1979 Nitrogen balance in flooded rice soils. *In* Nitrogen and Rice. pp 95–104. IRRI, Manila.

Ladha J K, Barraquio W L and Watanabe I 1982 Immunological techniques to identify *Azospirillum* associated with wetland rice. Can. J. Microbiol. 28, 478–485.

Ladha J K, Barraquio W L and Watanabe I 1983 Isolation and identification of nitrogen-fixing *Enterobacter clocae* and *Klebsiella panticola* associated with rice plants. Can. J. Microbiol. 29, 1301–1308.

Ladha J K, Tirol A C, Daroy M L G, Caldo G and Watanabe I 1987a Plant-associated N_2 fixation (C_2H_2-reduction) by five rice varieties, and relationship with plant growth characters as affected by straw incorporation. Soil Sci. Plant Nutr. 32, 91–96.

Ladha J K, Tirol Padre A, Punzalan G and Watanabe I 1987b Nitrogen fixing (C_2H_2-reducing) activity and plant growth characters of 16 wetland rice varieties. Soil Sci. Plant Nutr. 33, 187–200.

Lee K K, Castro T and Yoshida T 1977 Nitrogen fixation throughout growth, and varietal differences in nitrogen fixation by the rhizosphere of rice planted in pots. Plant and Soil 48, 613–619.

Li R and MacRae I C 1992 Specific identification and enumeration of *Acetobacter diazotrophicus* in sugarcane. Soil Biol. Biochem. 24, 413–419.

Lima E, Boddey R M and Döbereiner J 1987 Quantification of biological nitrogen fixation associated with sugar cane using a ^{15}N aided nitrogen balance. Soil Biol. Biochem. 19, 165–170.

Matsui E, Vose P B, Rodrigues N S R and Ruschel A P 1981 Use of ^{15}N enriched gas to determine N_2-fixation by undisturbed sugar cane plant in the field. *In* Associative N_2 Fixation. Vol. 2. Eds. P B Vose and A P Ruschel. pp 153–161. CRC press, Boca Raton, Florida.

Mc Auliffe C, Chamblee D S, Uribe Arango H and Woodhouse W W 1958 Influence of inorganic nitrogen on nitrogen fixation by legumes as revealed by N^{15}. Agron. J. 50, 334–337.

Nayak D N, Ladha J K and Watanabe I 1986 The fate of marker *Azospirillum lipoferum* inoculated into rice and its effect on growth, yield and N_2 fixation of plants studies by acetylene reduction, $^{15}N_2$ feeding and ^{15}N dilution techniques. Biol. Fertil. Soils 2, 7–14.

Ohta H and Hattori T 1983 *Agromonas oligotrophica* gen. nov., sp. nov., a nitrogen-fixing oligotrophic bacterium. Antonie Leeuwenhoek 49, 429–446.

Olivares F L, Baldani V L D, Baldani J I and Döbereiner J 1993 Ecology of *Herbaspirillum* spp. and ways of infection and colonization of cereals with these endophytic diazotrophs. *In* Poster at 6th Int. Symp. Nitrogen Fixation with Non-legumes. Ismailia, Egypt. 6–10 Sept. Program and Abstracts, 118p.

Oliveira E 1992 Estudo da associação entre bactérias diazotróficas e arroz. MSc Thesis, Univ. Fed. Rural Rio de Janeiro, Itaguaí, RJ.

Oliveira O C de 1994 Quantificação da fixaçao biológica de nitrogênio em arroz (*Oryza sativa*, L.) inundado. MSc Thesis, Univ. Fed. Rural Rio de Janeiro, Itaguaí, RJ.

Oliveira O C de, Urquiaga S and Boddey R M 1994 Burning cane: the long term effects. Int. Sugar J. 96, 272–275.

Omar A M N, Richard C, Weinhard P and Balandreau J 1989 Using the spermosphere model technique to describe the dominant nitrogen-fixing microflora associated with wetland rice in two Egyptian soils. Biol. Fertil. Soils 7, 158–163.

Orlando Filho J, Haag H P and Zambello E Jr 1980 Crescimento e absorção de macronutrientes pela cana-de-açúcar, variedade CB41–76, em função da idade, em solos do Estado de São Paulo. Bol. Tec. Planalsucar, Piracicaba, São Paulo 2, 1–128.

Paula M A de, Döbereiner J and Siqueira J O 1989 Efeito da inoculaçao com fungo micorrízico VA e bactérias diazotróficas no crescimento e produçao de batata-doce. *In* 22nd Congresso Brasileiro de Ciência do Solo, Recife, SBCS. pp 109. Programa e resumos.

Paula M A de, Reis V M and Döbereiner J 1991 Interactions of *Glomus clarum* with *Acetobacter diazotrophicus* in infection of sweet potato (*Ipomoea batatas*), sugarcane (*Saccharum* spp.) and sweet sorghum (*Sorghum vulgare*). Biol. Fertil. Soils 11, 111–115.

Pimentel J P, Olivares F, Pitard R M, Urquiaga S, Akiba F and Döbereiner J 1991 Dinitrogen fixation and infection of grass leaves by *Pseudomonas rubrisubalbicans* and *Herbaspirillum seropedicae*. Plant and Soil 137, 61–65.

Purchase B S 1980 Nitrogen fixation associated with sugarcane, Proc. S. Afr. Sugar Technol. Assoc., June. pp 173–176.

Qui Y S, Zhou S P and Mo X Z 1981 Study of nitrogen fixing bacteria associated with rice root. 1. Isolation and identification of organisms. Acta Microbiol. Sinica 21, 468–472.

Reinhold-Hurek B, Hurek T, Gillis M, Hoste B, Vancanneyt M, Kersters K and De Ley J 1993 *Azoarcus* gen. nov., a nitrogen fixing Proteobacteria associated with roots of Kallar grass (*Leptochloa fusca* (L.) Kunth), and description of two species *Azoarcus indigens* sp. nov. and *Azoarcus communis* sp. nov. Int. J. Syst. Bacteriol. 43, 574–588.

Reis V M, Olivares F L and Döbereiner J 1994 Improved methodology for isolation of *Acetobacter diazotrophicus* and confirmation of its endophytic habitat. World J. Microbiol. Biotechnol. 10, 101–104.

Reis V M, Zang Y and Burris R H 1990 Regulation of nitrogenase activity by ammonium and oxygen in *Acetobacter diazotrophicus*, An. Acad. Bras. Cienc. 62, 317.

Rinaudo G and Dommergues Y 1971 Validité de l'estimation de la fixation biologique de l'azote dans la rhizosphere par la methode de reduction de l'acetylene. Ann. Inst. Pasteur 121, 93–99.

Roger P A and Watanabe I 1986 Technologies for utilizing biological nitrogen fixation in wetland rice: potentialities, current usage, and limiting factors. Fert. Res. 9, 39–77.

Rolfe B G and Bender G L 1990 Evolving a *Rhizobium* for non-legume nodulation. *In* Nitrogen Fixation: Achievements and Objectives. Eds. P M Gresshof, L E Roth, G Stacey and W L Newton. pp 779–780. Chapman and Hall, New York.

Ruschel A P, Henis Y and Salati E 1975 Nitrogen-15 tracing of N-fixation with soil-grown sugar cane seedlings. Soil Biol. Biochem. 7, 181–182.

Ruschel A P, Victoria R L, Salati E and Henis Y 1978 Nitrogen fixation in sugar cane (*Saccharum officinarum*). *In* Environmental Role of Nitrogen-fixing Blue-green Algae and Asymbiotic Bacteria. Ed. U Granhall. pp 297–303. Ecological Bulletins, Swedish Natural, Stockholm. (Ecological Bulletins/NFR 26).

Sampaio E V S B, Salgado I H and Bettany J 1984 Dinâmica de nutrientes em cana-de-açúcar. I. Eficiência da utilizaçao de uréia (^{15}N) em aplicaçao única ou parcelada. Pesq. Agropec. Bras. 19, 943–949.

Sen J 1929 Is bacterial association a factor in nitrogen assimilation by rice plants ? Agric. J. India 24, 229–231.

Stanford G and Ayres A S 1964 The internal nitrogen requirement of sugarcane. Soil Sci. 98, 338–344.

Stephan M P, Oliveira M, Teixeira K R S, Martinez Dretz G and Döbereiner J 1991 Physiology and dinitrogen fixation of *Acetobacter diazotrophicus*. FEMS Microbiol. Lett. 77, 67–72.

Teixeira K R S, Stephan M P and Döbereiner J 1987 Physiological studies of *Saccharobacter nitrocaptans* a new acid tolerant N_2-fixing bacterium. *In* Poster presented at 4th International Symposium on Nitrogen Fixation with Non-legumes, Rio de Janeiro. Final program abstracts.

Tjepkema J and Van Berkum P 1977 Acetylene reduction by soil cores of maize and sorghum in Brazil. Appl. Environ. Microbiol. 33, 626–629.

Urquiaga S, Cruz K H S and Boddey R M 1992 Contribution of nitrogen fixation to sugar cane: Nitrogen-15 and nitrogen balance estimates. Soil Sci. Soc. Am. J. 56, 105–114.

Ventura T S, Bravo M, Daez C, Ventura V, Watanabe I and App A 1986 Effects of N-fertilizers, straw, and dry fallow on the nitrogen balance of a flooded soil planted with rice. Plant and Soil 93, 405–411.

Ventura W and Watanabe I 1983 ^{15}N dilution technique of assessing the contribution of nitrogen fixation to rice plant. Soil Sci. Plant Nutr. 29, 123–130.

Walcott J J, Chauviroj M, Chinchest A, Choticheuy P, Ferraris R and Norman B W 1977 Long-term productivity of intensive rice cropping systems on the central plain of Thailand. Exp. Agric. 13, 305–316.

Watanabe I, Deguzman M R and Cabrera D A 1981 The effect of nitrogen fertilizer on N_2 fixation in the paddy field measured by in situ acetylene reduction assay. Plant and Soil 59, 135–140.

Watanabe I, Lee K and Guzman M de 1978a Seasonal change of N_2 fixing rate in rice field assayed by in situ acetylene reduction technique. II. Estimate of nitrogen fixation associated with rice plants. Soil Sci. Plant Nutr. 24, 465–471.

Watanabe I, Lee K K and Alimagno B V 1978b Seasonal change of N_2-fixing rate in rice field assayed by in situ acetylene reduction technique. I. Experiments in long-term fertility plots. Soil Sci. Plant Nutr. 24, 1–13.

Watanabe I, Yoneyama T, Padre B and Ladha J K 1987a Difference in natural abundance of ^{15}N in several rice (*Oryza sativa* L.) varieties: Applications for evaluating N_2 fixation. Soil Sci. Plant Nutr. 33, 407–415.

Watanabe I, So R, Ladha J K, Fujimura Y, Katayama and Kuraishi H 1987b A new nitrogen-fixing species of pseudomonad: *Pseudomonas diazotrophicus* sp. nov. isolated from the root of wetland rice. Can. J. Microbiol. 33, 670–678.

Willis W H and Green V E Jr 1948 Movement of nitrogen in flooded soil planted to rice. Soil Sci. Soc. Am. Proc. 13, 229–237.

Witty J F 1983 Estimating N_2 fixation in the field using ^{15}N-labelled fertilizer: some problems and solutions. Soil Biol. Biochem. 15, 631–639.

Yoshida T and Ancajas R R 1970 Application of the acetylene reduction method in nitrogen fixation studies. Soil Sci. Plant Nutr. 16, 234–237.

Yoshida T and Ancajas R R 1973 Nitrogen-fixing activity in upland and flooded rice fields. Soil Sci. Soc. Am. Proc. 37, 42–46.

Yoshida T and Yoneyama T 1980 Atmospheric dinitrogen fixation in the flooded rice rhizosphere as determined by the N-15 isotope technique. Soil Sci. Plant Nutr. 26, 551–560.

Plant and Soil **174:** 211–224, 1995.

Biological N_2 fixation by heterotrophic and phototrophic bacteria in association with straw

Margaret M. Roper and J. K. Ladha
CSIRO Centre for Mediterranean Agricultural Research, Private Bag, P.O. Wembley, Western Australia 6014; International Rice Research Institute, P.O. Box 933, 1099 Manila, Philippines

Key words: decomposition, heterotrophic/phototrophic bacteria, N_2 fixation, straw

Table of contents

Abstract

Introduction

Asymbiotic diazotrophs

Straw components and their direct utilisation by diazotrophs

Aerobic and anaerobic cellulose degradation

N_2 fixation in soils by bacteria using products of straw decomposition

  N_2 fixation in flooded soils

  N_2 fixation in dryland soils

Factors affecting N_2 fixation associated with straw

Ways to enhance straw-associated N_2 fixation

Conclusions

References

Abstract

Much of the crop residues, including cereal straw, that are produced worldwide are lost by burning. Plant residues, and in particular straw, contain large amounts of carbon (cellulose and hemicellulose) which can serve as substrates for the production of microbial biomass and for biological N_2 fixation by a range of free-living, diazotrophic bacteria. Microorganisms with the dual ability to utilise cellulose and fix N_2 are rare, but some strains that utilize hemicellulose and fix N_2 have been found. Generally, cellulolysis and diazotrophy are carried out by a mixed microbial community in which N_2-fixing bacteria utilise cellobiose and glucose produced from straw by cellulolytic microorganisms. N_2-fixing bacteria include heterotrophic and phototrophic organisms and the latter are apparently more prominent in flooded soils such as rice paddies than in dryland soils. The relative contributions of N_2 fixed by heterotrophic diazotrophic bacteria compared with cyanobacteria and other phototrophic bacteria depend on the availability of substrates from straw decomposition and on environmental pressures. Measurements of asymbiotic N_2 fixation are limited and variable but, in rice paddy systems, rates of 25 kg N ha^{-1} over 30 days have been found, whereas in dryland systems with wheat straw, in situ measurements have indicated up to 12 kg N ha^{-1} over 22 days. Straw-associated N_2 fixation is directly affected by environmental factors such as temperature, moisture, oxygen concentration, soil pH and clay content as well as farm management practices. Modification of managements and use of inoculants offer ways of improving asymbiotic N_2 fixation.

In laboratory culture systems, inoculation of straws with cellulolytic and diazotrophic microorganisms has resulted in significant increases in N_2 fixation in comparison to uninoculated controls and gains of N of up to 72 mg N fixed g^{-1} straw consumed have been obtained, indicating the potential of inoculation to improve N gains in composts that can then be used as biofertilisers. Soils, on the other hand, contain established, indigenous microbial populations which tend to exclude inoculant microorganisms by competition. As a consequence, improvements in straw-associated N_2 fixation in soils have been achieved mostly by specific straw-management practices which encourage microbial activity by straw-decomposing and N_2-fixing microorganisms.

Further research is needed to quantify more accurately the contribution of asymbiotic N_2 fixation to cropping systems. New strains of inoculants, including those capable of both cellulolytic and N_2-fixing activity, are needed to improve the N content of biofertilisers produced from composts. Developments of management practices in farming systems may result in further improvements in N_2 fixation in the field.

Introduction

Maintenance of adequate levels of soil organic matter is essential for a sustainable, high production of crops. Several countries traditionally have utilised crop residues such as straw to maintain or improve the organic matter content of soil (Ayanaba and Okigbo, 1975; Egawa, 1975). However, intensive agriculture systems, with high inorganic fertiliser inputs, limit the return of crop residues to the soil. Worldwide, most crop residues (including cereal straw, rice straw and sugar cane trash) are lost by burning (Flinn and Marciano, 1984; Ponnamperuma, 1984). Straw contains useful macronutrients such as nitrogen (N), phosphorus and potassium, but in particular, it consists of a large amount of carbon which serves as a substrate for the production of microbial biomass and for biological N_2 fixation. It appears that there are two ways in which crop residues are used for energy by asymbiotic N_2-fixing bacteria, i.e. directly through the use of some hemicellulose components (Halsall et al.,1985; Ladha et al.,1986a) or indirectly through the use of products of straw decomposition (Jensen and Swaby, 1941; Lynch and Harper, 1983; Roper and Halsall, 1986). This paper reviews published work on N_2 fixation associated with crop residues under controlled laboratory conditions and in the field. Strategies to increase the efficiency of use of crop residues for N_2 fixation are presented.

Asymbiotic diazotrophs

Several groups of asymbiotic N_2-fixing bacteria have been identified in soils and flooded systems and those genera which include N_2-fixing species are listed in Table 1. The heterotrophic diazotrophs depend on carbon, e.g. from straw, for energy and the most common isolates from soils are (*Azotobacter*, *Azomonas*, *Beijerinckia* and *Derxia*, *Clostridium* and *Bacillus*, *Klebsiella* and *Enterobacter*, and *Azospirillum*, *Desulfovibrio* and *Desulfotomaculum*) (Havelka et al., 1982; Roper and Halsall, 1986). Autotrophic bacteria generally derive their energy from photosynthesis (Havelka et al., 1982).

Straw components and their direct utilisation by diazotrophs

Cellulose and hemicellulose are major structural polysaccharides in straw and comprise between 50–70% of its dry weight (Harper and Lynch, 1981; Theander and Åman, 1978). The low molecular weight carbohydrates range from 0.3 to 1.3% of the dry weight and are mostly fructose, glucose, sucrose and sugar alcohols (arabinitol and mannitol) (Theander and Åman, 1984). A wide range of low molecular weight carbon compounds can be utilised as carbon and energy substrates by diazotrophs. However, very few organ-

Table 1. Genera which include asymbiotic N_2-fixing organisms from soils and flooded systems (from Dalton, 1980; Havelka et al., 1982; Staley, 1989; Stewart, 1980)

Heterotrophic bacteria	Autotrophic bacteria
Aerobic	Cyanobacteria
Azotobacter	Heterocystous filamentous forms
Azomonas	*Anabaena*
Beijerinckia	*Anabaenopsis*
Derxia	*Aulosira*
	Cylindrospermum
Microaerobic	*Nostoc*
Azospirillum	*Calothrix*
Aquaspirillum	*Fischerella*
Thiobacillus	*Hapalosiphon*
Pseudomonas	*Scytonema*
Xanthobacter	*Tolypothrix*
Rhizobium	*Stigonema*
Methylosinus	*Westiella*
Methylococcus	
Mycobacterium	Non-heterocystous filamentous forms
	Lyngbya
Facultative anaerobic	*Phormidium*
Klebsiella	*Plectonema*
Erwinia	*Oscillatoria*
Enterobacter	*Pseudanabaena*
Citrobacter	
Escherichia	Unicellular forms
Bacillus	*Gloeothece*
	Cyanothece
Anaerobic	*Synechococcus*
Desulfovibrio	
Desulfotomaculum	Other phototrophic bacteria
Clostridium	*Rhodospirillum*
	Rhodopseudomonas
	Rhodomicrobium
	Chromatium
	Thiocystis
	Chlorobium

isms have the dual ability to utilise cellulose and fix N_2. The only reports of cellulolytic, N_2-fixing bacteria are a marine bacterium isolated from a shipworm (Greene and Freer, 1986; Waterbury et al., 1983) and four strains of anaerobic bacteria isolated from forest soil and fresh-water mud (Leschine et al., 1988).

Xylan, which is the predominant component of hemicellulose, can be utilised by *Azospirillum* (*A. lipoferum* and *A. brasilense*) as the sole carbon source for N_2 fixation (Halsall et al., 1985; Ladha et al., 1986a). Halsall et al. (1985) also measured nitrogenase activity when xylan was replaced by powdered, gamma-irradiated straw, suggesting the use by diazotrophs, of xylan, pectins, and other easily available carbohydrates of low molecular weight in the straw. Ladha et al. (1986a) reported N_2 fixation by a range of N_2-fixing bacteria isolated from the rice rhizosphere (*Azospirillum lipoferum*, *Pseudomonas diazotrophicus*, *Enterobacter cloacae* and *Klebsiella planticola*) in the presence of hydrogen peroxide-treated rice straw, again suggesting the utilisation of xylan as a sole carbon source. The pretreatment of straw with hydrogen peroxide frees its components and partly solubilises lignin, hemicellulose and the alcohol extractable fraction. Ladha et al. (1986a) found several hundred times higher nitrogenase activity by *Azospirillum lipoferum* in the presence of hydrogen peroxide-treated straw than in untreated straw. These reports suggest that the enzyme xylanase is much more widespread among diazotrophs than has been assumed previously.

Aerobic and anaerobic cellulose degradation

Since most N_2-fixing bacteria are unable to utilise cellulose directly as a substrate for N_2 fixation, cellulose must first be degraded to simpler intermediates before being utilised by diazotrophs. In nature, most cellulose is degraded aerobically and the final product is CO_2. The degradation occurs in two steps: i) degradation by cellulolytic microorganisms (primary microorganisms) to cellobiose and glucose, and ii) utilisation of cellobiose, glucose and other free sugars by secondary microorganisms which are unable to hydrolyse cellulose. In anaerobic systems, such as in flooded soils, methane is produced from decomposing organic matter and, the cellulose is oxidised to proprionate, butyrate, lactate, acetate, molecular hydrogen, and CO_2. Anaerobic decomposition of straw in flooded soil is reviewed in detail by Yoshida (1975), Neue and Scharpenseel (1984) and Watanabe (1984).

Organisms involved in the aerobic decomposition of straw include a broad range of bacteria, actinomycetes, fungi, protozoa, nematodes and worms (Chatterjee and Nandi, 1981; Dickinson, 1974; Imshenetsky, 1967; Veal and Lynch, 1984; Zeikus, 1981), and the populations of these organisms increase in response to the addition of straw (Doran, 1980). Generally, there is a succession of microorganisms involved in decomposition (Swift, 1982) with the primary decomposition being attributed largely to the microflora, which in turn, are consumed by the soil fauna (Forbes, 1974).

Characterisation of cellulolytic microorganisms in flooded soil is limited (Watanabe and Furusaka, 1980). However, there are reports of the occurrence of aerobic fungi and bacteria, and anaerobic bacteria that decompose cellulose (Araragi and Tangcham, 1979; Saito et al., 1977a, b; Vostrov and Dolgikh, 1970). Among the anaerobic cellulolytic bacteria, *Clostridium dissolvens* has been found in flooded soil sown to rice (Saito et al., 1977a, b).

N_2 fixation in soils by bacteria using products of straw decomposition

Virtually all diazotrophic heterotrophs from dryland and flooded soils can utilise the products of cellulose decomposition including carbohydrates and some organic acids and alcohols (Jensen, 1981; La Rue, 1977; Rao, 1978; Roper and Halsall, 1986). Phototrophic nonsulphur purple bacteria which occur in fairly high numbers in flooded soil (Kobayashi et al., 1967; Ladha et al., 1987; Watanabe et al., 1978), also can utilise products of cellulose degradation (Kobayashi, 1982; Stanier et al., 1974).

N_2 fixation in flooded soils

Although the flooded rice ecosystem remains continuously flooded during crop growth, and is referred to as anaerobic, it actually consists of aerobic and anaerobic phases in close proximity to each other. This ecosystem has two major zones – the rhizosphere and the nonrhizosphere. The rhizosphere, the zone in close proximity to the roots, has intense microbial activity which is supported mostly by the release of carbon compounds from the root system. Based on physicochemical properties, the nonrhizosphere zone can be divided into four environments – the floodwater, the oxidised soil layer, the reduced soil layer and the subsoil. The floodwater and the oxidized soil layer are photic environments with a positive redox potential, while the reduced soil layer is nonphotic with a redox potential that is generally negative. The soil below the plough pan layer is either aerobic or anaerobic depending on the drainage.

N_2 fixation has been reported in flooded soil amended with straw (Barrow and Jenkinson, 1962; Brouzes et al., 1969; Magdoff and Bouldin, 1970; Rice and Paul, 1972). The application of straw to the surface or subsurface layers of flooded soil results in intense microbial activity which leads to high oxygen demand and consumption. As a result, an anaerobic to microaerobic environment, conducive to N_2 fixation by heterotrophic and phototrophic bacteria, is created. It has been suggested that either i) the low molecular weight products of aerobic degradation of polysaccharides in the aerobic zone diffuse into the anaerobic zone where they support N_2 fixation by anaerobic bacteria such as *Clostridium*, or ii) the products of anaerobic degra-

Table 2. Biological N_2 fixation or N gain by straw-amended flooded soil (modified from Roger and Watanabe, 1986)

Amount of soil used (g)	Straw			Incubation time (days)	Quantification method[b]	Efficiency: mg N fixed per g of straw		Reference
	Type	Nature	Amount (g)			Added	Consumed	
5	Barley	Powdered	0.5 (10)[a]	28	Kj	2.2–2.5		Barrow and Jenkinson (1962)
			1.0 (20)	28	Kj	0.8–1.0		
			2.0 (40)	28	Kj	1–2		
			0.72 (15)	14	Kj	1		
				28	Kj	2		
				56	Kj	5		
0.6	Wheat	Powdered	0.006 (1)	28	^{15}N	6.7		Rice et al. (1967)
			0.03 (5)	28	^{15}N and Kj	4.4		
			0.12 (20)	28	Kj	2.3		
0.1 (+ 2 g sand-clay)	Wheat	Powdered	0.12 (20)	28	Kj	2.2	16.1	Rice and Paul (1972)
5	Rice	Powdered	0.05 (1)	30	^{15}N	1.7		Rao (1976)
			0.1 (2)	30	^{15}N	7.07		

[a]Figures in parentheses are straw concentration in percent.
[b]Kj – Kjeldahl, ^{15}N – ^{15}N incorporation.

dation of cellulose in the anaerobic zone diffuse into the overlying aerobic zone to support N_2 fixation by aerobic bacteria (Knowles, 1976).

N_2 fixation, supported by straw in flooded soil, has been evaluated in two types of experiments, i) under laboratory conditions with rather high amounts of straw (equivalent to 1 to 100% of soil, dry weight basis), and ii) under greenhouse or field conditions with lower amounts of straw (equivalent to < 1% of soil, dry weight basis). The first type of experiment probably simulated a composting environment.

Laboratory studies have been reviewed in a paper by Roger and Watanabe (1986). A wide range of values for N_2 fixation and N gain have been obtained by researchers due to differences in the form and amount of straw, the time of incubation, and the methods used for quantification (Table 2). Since the amount of straw used in most of these experiments was very high and practically impossible to apply in the field, data on the amount of N_2 fixation should not be extrapolated on a hectare basis. The most practical way of expressing N_2 fixed could be based on the amount of organic matter added and consumed (Roger and Watanabe, 1986); however, the amount of straw consumed was frequently unrecorded. Based on some published data, N_2 fixed varied from 1 to 7 mg N per g of straw added in 14 to 56 days of incubation (Table 2).

Only a few quantitative data on the amounts of N_2 fixed or N gained following straw application in greenhouse or field conditions are available. Rao (1980) quantified N_2 fixation in flooded soil amended with straw and planted to rice. Chopped straw (5 and 10 t ha^{-1}) was incorporated one month before rice transplanting. Straw amended soil (5 g) was sampled from 0–10 cm depth at tillering and at harvest in the first rice crop and again at tillering in the succeeding rice crop; the soil samples were incubated with $^{15}N_2$ for 30 days. N_2 fixation in the straw-amended soil was 2 to 4 times that of the control without straw. Fixation was highest at the tillering stage of the first crop after which there was a gradual decline, with negligible amounts measured in the control and in the treatment with 5 t ha^{-1} straw in the second crop. The treatment with 10 t ha^{-1} straw still showed appreciable N_2 fixation at the tillering stage of the second rice crop, suggesting a residual effect. Extrapolation of the values of ^{15}N incorporation in straw-amended soil for the 30-day duration, indicated rates of N_2 fixation of about 7 kg ha^{-1} in the unamended soil and an average of 25 kg ha^{-1} in soil with straw amendment, based on a per hectare furrow slice of 0.7×10^6 kg dry soil ha^{-1}.

The figures may have been higher for the whole crop duration. However, these rates of N_2 fixation should be used with caution because the ^{15}N measurements were made under laboratory conditions.

Santiago-Ventura et al. (1986) conducted a N balance pot experiment for three consecutive rice crops and reported about twice the N gain following straw incorporation (equivalent to 10 t ha^{-1}) compared with the control. When the N gain was expressed per g of straw added, it ranged from 2 to 4 mg N fixed g^{-1} straw added. Apart from these quantitative data on N gain and N_2 fixation, several reports of semiquantitative measurements are available (Kimura et al., 1979; Kobayashi, 1982; Ladha et al., 1986b, 1987; Matsuguchi, 1979; Rao, 1976, 1978; Reddy and Patrick, 1979; Rice, 1979; Wada et al., 1979; Yoneyama et al., 1977; Yoo et al., 1982). Time-course measurements of N_2 fixation using acetylene reduction methods (ARA) generally showed stimulation from a few days to about 40 days after straw application, followed by a gradual decline (Ladha et al., 1986b, 1987). Activity was higher during the dry season than during the wet season. The trend of ARA in field studies was more or less similar to that in the laboratory studies (Rao, 1976; Rice and Paul, 1972; Rice, 1979; Yoneyama et al., 1977).

Most laboratory experiments were incubated in the dark permitting only heterotrophic N_2 fixation. On the other hand, greenhouse and field experiments included N_2 fixation by both heterotrophic and phototrophic microorganisms. The relative contributions of N_2 fixation by heterotrophic and phototrophic microorganisms in straw-amended soil are not known. Furthermore, among phototrophic microorganisms, the relative contributions of cyanobacteria (blue-green algae) and phototrophic (photosynthetic or photoorganotrophic) bacteria are not clear. Diverse types of cyanobacteria (Roger and Kulasooriya, 1980) and phototrophic purple nonsulphur bacteria of the family *Rhodospirillaceae* (Kobayashi et al., 1967) are ubiquitous in rice fields. Their growth is stimulated by straw application, although to a lesser extent with cyanobacteria (Matsuguchi, 1979; Yoo et al., 1982), and to a greater extent with the phototrophic bacteria (Kobayashi and Haque, 1971; Ladha et al., 1986a, 1987; Matsugushi and Yoo, 1981; Yoo et al., 1982).

In order to determine the relative contributions of cyanobacteria and phototrophic bacteria to N_2 fixation in flooded soil, Habte and Alexander (1980) used a herbicide (propanil 3', 4'-dichloropropionanilide) to inhibit nitrogenase activity by cyanobacteria. The herbicide had no effect on other N_2-fixing bacteria including phototrophic bacteria. They found significantly higher N_2-fixing activity in the herbicide-treated soil than in the untreated soil after 30 days of rice growth but no difference after 55 days. Their study concluded that the contribution of N_2 fixation by phototrophic bacteria may be as high as that by cyanobacteria; furthermore, it may be higher if cyanobacterial growth is poor. Reddy and Patrick (1979) recorded 3- to 4-fold higher nitrogenase activity in straw-amended soil incubated in the light compared with the dark, and suggested that increased activity was due to phototrophic bacteria. Similarly, Yoo et al. (1982) and Ladha et al. (1987) suspected that phototrophic bacteria are more important than the cyanobacteria in flooded soils. Yoo et al. (1982) suggested that cyanobacterial growth may be inhibited by the increased population of molluscs resulting from straw application. Ladha et al. (1987) recorded a significant increase in the population of phototrophic purple nonsulfur bacteria in soil amended with straw (5t ha^{-1}) in the light, and this increase was greater where the straw was applied at the surface of the flooded soil than where the straw was incorporated into the soil.

In a small-scale laboratory experiment with straw, Kobayashi and Haque (1971) demonstrated a succession of heterotrophic bacteria, phototrophic bacteria and green algae. They found peaks in population sizes of heterotrophic bacteria at about 30 days, followed by phototrophic bacteria at about 60 days, and algae at about 120 days. An explanation for this may be that the growth of the heterotrophic bacteria slowed down due to a decrease in the substrate level and an accumulation of decomposition products which favoured the growth of phototrophic bacteria; in another 3 to 4 weeks, when the C:N ratio of straw declined substantially, the growth of phototrophic bacteria also declined and the growth of green algae took over. Other studies by Matsuguchi and Yoo (1981) found that the surface application of straw to flooded soil promoted growth of the phototrophic bacterium *Rhodopseudomonas* in the first 3 weeks; thereafter, there was a cyanobacterial bloom associated with a very high nitrogenase activity. Despite the prominence of phototrophic microorganisms, Santiago-Ventura et al. (1986) found no significant difference in the N balance of straw-amended soil exposed to light or dark.

Generally, phototrophic purple nonsulfur bacteria occurring in flooded rice soil are not strict anaerobes (Kobayashi, 1982). They can also grow aerobically in the dark, obtaining ATP from a respiratory metabolism of available organic compounds. On the other hand, under anaerobic conditions in the light, the reactions of photosynthesis provide a potentially unlimited supply of ATP and reductants (Stanier et al., 1974).

Phototrophic purple nonsulfur bacteria in pure culture attain maximum nitrogenase activity under light and anaerobic conditions (Kobayashi, 1982). This environment can be provided in the field by intense, highly oxygen-demanding multimicrobial activities induced by surface application of straw (Kobayashi, 1982; Ladha et al., 1987).

Few quantitative data on N_2 fixation by phototrophic bacteria exist, but the semiquantitative data suggest they have significant potential to contribute N to rice-based agriculture. N_2 fixation by phototrophic bacteria could play an important role in rice fields in situations where the conditions are not favourable for good cyanobacterial growth; this is often the case because of limiting factors such as phosphorus deficiency and grazing by invertebrates (Roger and Kulasooriya, 1980). Habte and Alexander (1980) discussed some additional advantages of phototrophic bacteria compared with cyanobacteria: a) faster growth, b) no competition in light utilisation with rice because they utilise light energy in the far red and infrared regions, c) insensitivity to certain herbicides, and d) detoxification of H_2S.

N_2 fixation in dryland soils

Dryland cropping systems include aerobic and anaerobic sites in the same way as flooded systems, although there are differences in scale and arrangement. Intense microbial activity, during the decomposition of crop residues, results in the development of anaerobic and microaerobic microsites in soils, including surface soil (Greenwood and Goodman, 1964; Hill et al., 1990; Parr and Papendick, 1978), and these sites can support N_2 fixation by a wide range of free-living, diazotrophic bacteria, including anaerobic bacteria.

Nitrogen fixation by asymbiotic bacteria has been observed in artificial systems in soils amended with cellulose (Jensen and Swaby, 1941; Kalininskaya, 1972), wheat straw (Delwiche and Wijler, 1956; Halsall and Gibson, 1986, 1989; Lynch and Harper, 1983; Roper and Smith, 1991), oat straw (Brouzes et al., 1969), maize stalks (Abd-el-Malek et al., 1976) and sugar cane trash (Patriquin, 1982) to mention a few. Measurements of N_2 fixed in artificial systems vary greatly. With cellulose as a carbon source, fixation ranged from 2.5–4.5 mg N g^{-1} cellulose (Kalininskaya, 1972) to 14 mg N g^{-1} cellulose (Jensen and Swaby, 1941). Brouzes et al. (1969) found extremely variable rates of N_2 fixation with 2% oat straw (0–70 μg N g^{-1} soil) which, when extrapolated, gave an estimate of annual fixation of 0–200 kg N ha^{-1} $year^{-1}$. Similarly, huge variations in rates of N_2 fixed (2–155 g N ha^{-1} day^{-1}) were found with sugar cane trash by Patriquin (1982), but he estimated that the total N_2 fixed in association with the trash from a cane crop was of the order of 20 kg N ha^{-1}.

Estimates of N_2 fixed in pure culture systems are also extremely variable. For example, Lynch and Harper (1983) measured a N gain of 11.5 mg N g^{-1} wheat straw consumed after incubation of straw inoculated with a N_2-fixing *Clostridium butyricum* and a cellulolytic *Penicillium corylophilum* for a period of 8 weeks. On the other hand, over 38 days, Halsall and Gibson (1986) calculated an efficiency of N_2 fixation by *Azospirillum brasilense* of 72 and 63 mg N g^{-1} wheat straw utilised, in cultures containing respectively, *Cellulomonas gelida* and *Cellulomonas* sp. as the cellulolytic partner. It is likely that culture conditions as well as the cellulolytic and N_2-fixing partners were significant contributing factors to the variable results. In situ measurements of N_2 fixation associated with wheat straw, made at field sites across eastern Australia by Roper (1983) and Roper et al. (1994a), indicated amounts fixed (based on a C_2H_2:N_2 calibration of 3:1) ranging from 1 kg N ha^{-1} over 31 days to 12.3 kg N ha^{-1} over 22 days. The wheat straw content of the soil ranged from 4.3 – 7.2 t ha^{-1}. All measurements were made under conditions where moisture was not limiting (i.e. field capacity), but soil temperatures ranged from as low as 6°C at night to above 36°C during the day. In a longer term field experiment on a vertisol, nitrogenase activity was detected throughout the 10 month period following straw management after harvest, but the level of N_2 fixation was closely linked to temperature and was highest in the summer and early autumn months (Roper et al., 1989). This means that, even if rates of asymbiotic diazotrophic activity are low, particularly in the winter months, there is still potential for a significant input of N over a period of time. In another system, in situ asymbiotic N_2-fixing activity was observed in association with lucerne and its residues, and it was estimated that this amounted to between 3 and 10 kg ha^{-1}, or 3–10% of the total (symbiotic + asymbiotic) N_2 fixed, over a single growing season in south eastern Australia (Roper et al., 1994b). In situ studies with decomposing green sugar cane trash indicated that the highest levels of N_2

fixation were 1.5 kg ha^{-1} $week^{-1}$ (Chapman et al., 1992). All these rates of asymbiotic N_2 fixation are low in comparison to legume systems (Peoples et al., 1995). However, because N contained in the microbial biomass of N_2-fixing bacteria is readily available as a source of N (Lethbridge and Davidson, 1983), it is likely that N_2 fixed asymbiotically can make a significant contribution to a following crop.

Factors affecting N_2 fixation associated with straw

Straw-associated N_2 fixation is modified by mineral N, temperature, moisture, oxygen concentration, soil characteristics and straw management. Both N_2 fixation and straw decomposition are affected directly by these factors, but N_2 fixation is also affected indirectly through straw decomposition. For example, in a field experiment, Roper (1983) observed a positive correlation ($r = + 0.98$; $p < 0.01$) between nitrogenase activity and wheat straw decomposition.

Cereal straw usually has a high C:N ratio (> 70) and therefore, its decomposition may be slow because of the requirement for N by straw-decomposing microorganisms. Straws with higher N contents decompose more quickly than those containing less N (Zielinski, 1980) and when Smith and Peckenpaugh (1986) studied the decomposition of 23 cereal straws they found a direct correlation ($r = + 0.84$, $p < 0.01$) between the N content of the straw and its decomposition. The addition of mineral N to soils amended with straws with high C:N ratios, increases the rate of straw decomposition (Barder and Crawford, 1981; Novak, 1972; Sain and Broadbent, 1977), but it can also decrease the rate of N_2 fixation (Cejudo and Paneque, 1986; Shintani, 1987) through its inhibitory effects on the nitrogenase enzyme (Knowles and Denike, 1974). Both inhibitory and stimulatory effects of mineral N on straw-associated N_2 fixation have been reported (Charyulu and Rao, 1979; Matsuguchi and Yoo, 1981; Rao, 1976) and this indicates that it is necessary to balance the input of mineral N in order to increase straw decomposition without compromising N_2 fixation.

Both straw decomposition and N_2 fixation are significantly affected by temperature. Pal et al. (1975) and Waksman and Gerretsen (1931) showed that straw decomposition increased with increasing temperatures between 7 and 37°C. Roper (1985) observed decomposition at temperatures up to 50°C with the highest activity in the range between 25 and 45°C. Nitrogenase activity, in the field, was found to be positively correlated ($r = + 0.71$, $p < 0.05$) with temperature (Roper, 1983). In glucose-amended soils in the laboratory, Brouzes and Knowles (1973) showed that 37°C was the optimum temperature for nitrogenase activity, whereas Roper (1985) found maximum activities within the range of 20–35°C in soil from one site and 25–45°C in soil from a second site; these temperature ranges reflected the differences in climate between the 2 sites and suggested some adaptation of the microbial populations to the temperatures of their environment.

Moisture and oxygen are interrelated factors and their effects are difficult to separate. Because straw decomposition involves an enormous variety of microorganisms, decomposition can occur over a wide range of moistures, and has been observed in soils as dry as 30% field capacity and up to waterlogged conditions (Roper, 1985). Sorensen (1974) observed that repeated air drying and rewetting of soils resulted in the best rates of decomposition of plant material. With the sensitivity of the nitrogenase enzyme to oxygen and the dependence, at least in part, on moisture to reduce oxygen levels, N_2 fixation is more sensitive to moisture contents. Generally, laboratory incubation studies with soil plus straw have indicated that the highest levels of N_2 fixation occur under waterlogged anaerobic conditions (Brouzes et al., 1969; Rao, 1976; Rice et al., 1967). In laboratory studies with glucose, Roper (1985) found that moistures of between 1.5 and 2 times field capacity resulted in the best rates of N_2 fixation, but the lowest moisture content which supported nitrogenase activity varied according to the characteristics of the soil. Soils containing clays develop microsites of low oxygen tension more readily than sandy soils and N_2 fixation has been observed in clay soils at moistures below 50% field capacity (Roper, 1985). Undisturbed soils in the field support nitrogenase activity at lower moisture contents than in the laboratory, probably because microsites of low oxygen tensions are preserved. In field observations, Roper (1983) found a positive correlation between nitrogenase activity and moisture content, below field capacity, and detected activity at moistures as low as 25% field capacity.

Under uniform conditions of soil moisture and temperature, different soils with similar histories of straw retention produce different rates of nitrogenase activity. pH and clay content are probably the most significant characteristics which modify microbial activity in soils. Roper and Smith (1991) found that in sand culture systems, with microorganisms extracted from soils sampled from wheat-growing areas, straw decomposed efficiently over a wide range of pH, reflecting the broad range of microflora responsible for decomposition. However, in the same experiments, N_2-fixing bacteria had a much more restricted pH range for activity and preferred a pH close to neutral regardless of the pH of the soils from which the N_2-fixing bacteria were derived, indicating that N_2-fixing populations are not always suited to their soil environment. Soil pH controls the uptake of nutrients (Stotzky, 1972) and can restrict the range of N_2-fixing genera such as *Azotobacter* in soil (Alexander, 1961).

Clays affect both straw decomposition and N_2 fixation in soils. Christensen (1987) found that decomposition of straw was higher in a sandy soil than in a clay soil. However, it appears that the type of clay is important and montmorillonite depresses decomposition (Novakova and Sisa, 1984; Roper and Smith, 1991) whereas kaolinite enhances decomposition (Novakova and Sisa, 1984). The restriction of decomposition by some clays can be attributed to the inhibition of respiration by the fine colloids, but this in turn favours the development of microaerobic and anaerobic microsites for nitrogenase activity. The addition of montmorillonite to sand cultures containing microorganisms extracted from soils resulted in enormous increases in N_2-fixing activity and extended the pH range of activity (Roper and Smith, 1991). At the pure culture level, Macura and Pavel (1959) found that N_2 fixation by *Azotobacter* sp. was increased significantly by the presence of montmorillonite. Clays are highly reactive colloidal particles (Baver, 1956) and interact strongly with microorganisms (Marshall, 1975). Clays concentrate nutrients at their surfaces, where microbial activity takes place, and modify local pH (Stotzky, 1972). This may explain the intensified nitrogenase activity and the extension of N_2 fixation to a broader pH range in the presence of clay compared with systems without clay.

Straw management has a significant effect on straw decomposition. Cogle et al. (1987), Douglas et al. (1980), Roper et al. (1989) and Summerell and Burgess (1989) all found that straw decomposed more rapidly when incorporated into the soil than when left on the surface as a mulch. This means that products of decomposition from incorporated straw are more readily available for use by N_2-fixing bacteria in soils. For example, in simulated laboratory experiments, Patriquin (1982) observed a larger and earlier peak of nitrogenase activity in soils with incorporated sugar cane trash than in soil with mulched trash. In the field, Roper et al. (1989) and Roper et al. (1994a) found that nitrogenase activity was best with straw incorporation and that activity decreased in the order straw incorporated > straw mulched > no-tillage. They also showed that the type of incorporation of stubble was important and that straw which was smashed and mixed lightly with the soil near the surface produced significantly better nitrogenase activity than soil in which the straw was incorporated throughout the plough layer (Roper et al., 1989). It is likely that the surface mixing of the straw with soil provided sufficient soil-straw contact, and hence microorganism-straw contact, as well as good aeration to encourage decomposition. In addition, concentration of decomposition products and intensified microbial activity near the surface provided an opportunity for the development of anaerobic and microaerobic microsites (Greenwood and Goodman, 1964).

Ways to enhance straw-associated N_2 fixation

There may be two different situations where the value of plant residue, such as straw, is enhanced as a substrate for N_2 fixation: a) where straw is applied directly to soil, and b) where straw is first allowed to decompose and is used as a compost. One strategy to obtain higher N_2 fixation is to inoculate with efficient microorganisms. Soils maintain stable populations of large numbers of diverse microorganisms which generally resist the establishment of inoculated microorganisms because the inoculants compete poorly with the indigenous population (Ladha, 1986; Nayak et al., 1986). Straw, on the other hand, may not have such established populations of microorganisms and hence, may offer a better chance for the survival of inoculated microorganisms. Therefore, the introduction of selected microorganisms to heaps of straw such as composts or surface straw may be a more workable strategy.

After crop harvest, farmers may pile the straw into heaps in the field to burn. By inoculating with a combination of diazotrophic and cellulolytic microorganisms, the straw could be converted to an efficiently decomposed biofertiliser (compost) that could be incorporated easily into the soil. Composting is essentially a microbiological process. Its efficiency depends on the presence of suitable numbers and kinds of diazotrophic and cellulolytic microorganisms, the composition of straw (C:N ratio), temperature, moisture and aeration. As discussed earlier, cellulolysis and N_2 fixation generally are not combined in the same microorganism. The cellulolytic or primary microorganisms degrade cellulose to cellobiose and glucose. The N_2-fixing or secondary microorganisms, which are unable to hydrolyse cellulose, use cellobiose, glucose, and other free sugars as energy sources. The secondary microorganisms, though dependent on the primary microorganisms, also aid the cellulolytic microorganisms by supplying growth factors (including combined N) and removing free sugars (which normally inhibit cellulose degradation).

In the past, more attention was directed towards the selection and use of cellulolytic microorganisms rather than both cellulolytic and N_2-fixing microorganisms (Subba Rao, 1982) with the result that, during the course of composting, the cellulose and hemicellulose frequently was reduced without any appreciable gain in N (Inoko, 1984). For example, after sampling 105 composts, Inoko (1984) measured on average only 3.9 kg N per t of compost. Yadav and Subba Rao (1980) reported an average increase in N from 0.7% of wheat straw to just 1.6% N in the compost after 12 weeks and following a weight loss of 50%. This represented a gain in total N of 1 kg N t^{-1} straw used. These values are much lower than those obtained in the laboratory experiments reported in Table 2. The major difference in conditions between small-scale laboratory and medium to large-scale compost experiments is possibly the level of aeration. Composting often proceeds under more aerobic conditions with fungi rapidly and completely degrading polysaccharides to CO_2, allowing negligible accumulation of simple carbohydrates.

More recently, co-culture of suitable cellulolytic and N_2-fixing strains has been used in laboratory experiments to improve the efficiency of straw-associated N_2 fixation. Harper and Lynch (1984) found that, in straw decomposition/N_2 fixation systems, the cellulolytic population was dominated by aerobic fungi while the major N_2-fixing organisms were anaerobic bacteria. Based on this finding, Harper and Lynch (1986) inoculated wheat straw with an aerobic, cellulolytic *Trichoderma harzianum* and an anaerobic, N_2-fixing *Clostridium butyricum* and obtained N gains of up to 2 mg N fixed g^{-1} straw used over 4 weeks.

As reported earlier, even higher levels of N_2 fixation were measured by Halsall and Gibson (1985, 1986). Using a sterile sand-straw system with *Cellulomonas gelida* strain 2480 and *Azospirillum brasilense* strain Sp7 (or *Bacillus macerans* strain III), Halsall and Gibson (1985) obtained an efficiency of up to 19 mg N fixed g^{-1} straw consumed over a 30-day period. Subsequently, using the same system with *C. gelida* 2480 (or *Cellulomonas* sp. CSl-17, a mutant strain) and *A. brasilense* Sp7, Halsall and Gibson (1986) reported an efficiency of up to 72 mg N fixed g^{-1} straw consumed after 38 days incubation. *Azospirillum* is a microaerobic diazotroph that can produce xylanase allowing it to use hemicellulose (Halsall et al., 1985). *Cellulomonas* CSl-17 is a mutant strain selected after UV mutagenesis for improved cellulase production and for reduced sensitivity to repression inhibition by cellobiose or glucose (Choudhary et al., 1980). Both *Cellulomonas* strains were able to degrade cellulose at wide ranges of oxygen concentration, but the mutant strain CSl-17 accumulated higher amounts of reducing sugars which were then available for use by the diazotrophic bacteria (Halsall and Gibson, 1986).

The results clearly suggest a potential for improving straw-associated N_2 fixation by the use of selected microbial strains. Lynch (1983) suggested that the efficiency of straw associated N_2 fixation in composts could be increased to as much as 15 mg N g^{-1} total straw by inoculating with suitable microbial strains and, when added to the N already present in the straw, this would represent a total N value of 20 kg N t^{-1} straw. The results of Halsall and Gibson (1986) suggest the target for improvement could be even higher. However, because the ability of inoculants to compete with indigenous microbial populations is limited (Ladha, 1986; Nayak et al., 1986), inoculation is only likely to be successful in composts without established microbial populations.

Other approaches to improve the efficiency of N_2 fixation associated with straw are management of straw and manipulation of soil properties. As indicated earlier, pH near neutral favours N_2 fixation and improvements in nitrogenase activity have been achieved by liming of the soil. For example, in situ measurements in forest systems by Jones and Bangs (1985) showed that raising the pH of the soil to 6 from between 4 and 5 more than doubled the rate of N_2 fixation associated with forest litter. Improvements in straw management offer another means of increasing straw decomposition and associated N_2 fixation. As mentioned earlier, field experiments by Roper et al. (1989) indicated that smashing and lightly mixing the straw with surface soil resulted in the highest rates of N_2 fixation in the systems studied. Further improvements may be achieved by minimising soil disturbance so as to preserve anaerobic and microaerobic microsites for nitrogenase activity by N_2-fixing bacteria whilst at the same time ensuring sufficient soil-straw contact to promote straw decomposition by other soil microorganisms.

Conclusions

Nitrogen fixation by bacteria using crop residues for energy may be a significant source of N in any cropping system where substantial amounts of plant material are left after harvest. Crop residues can be left on the soil, or they can be gathered up, transformed by composting and then spread back onto the soil. In either case naturally occurring, heterotrophic and phototrophic bacteria utilise the straw either directly, by the use of hemicelluloses and simple carbohydrates, or indirectly, following the decomposition of cellulose by decomposer microorganisms. Quantitative data on the contribution of phototrophic and heterotrophic bacteria to the N status of soils are limited and new technologies are needed, particularly in the field, in order to obtain accurate measurements of N_2 fixation (Gibson et al., 1988).

In the laboratory, N gains have been observed as a result of heterotrophic N_2 fixation associated with straw, but the quantities vary. Inoculation of straw with N_2-fixing and cellulolytic organisms, in small-scale systems, has produced substantial improvements in the efficiency of straw-associated N_2 fixation and therefore inoculation has the potential to significantly improve N gains in compost systems. Future research should focus on the selection of suitable cellulolytic and diazotrophic microorganisms and the development of practical methods for the efficient production of straw-biofertiliser. Further gains may be possible if cellulolytic and diazotrophic activity are combined in the same microorganism by introducing into diazotrophic bacteria, genes which encode for polysaccharase functions such as amylase, cellulase, pectinase, and xylanase (Richardson et al., 1991).

Inoculation is only likely to be successful in straw, such as in composts, where microbial populations are not established. In soils, inoculant microorganisms are less likely to survive and be active because of strong competition with the established, indigenous microflora, and therefore alternative strategies are needed to promote the decomposing and N_2-fixing activities of naturally-occurring microbial populations. The use of management protocols, such as liming to neutralise pH and promote microbial activity, straw treatment to ensure good microorganism-straw contact, and altered tillage practices to preserve soil structure and maintain oxygen gradients that support microbial activity, have all resulted in some improvements. Further research into new management systems may result in greater efficiencies.

Increasing N supply through biological N_2 fixation, should reduce the need for inorganic N fertilisers. Therefore, in the long term, provided appropriate strategies can be developed to overcome the negative aspects of straw retention (e.g. the development of appropriate crop rotations to break disease cycles), straw-associated N_2 fixation could provide significant financial savings to producers as well as contribute to the maintenance of soil health.

References

Abd-el Malek Y, Monib M, Gohar M R, Rizk S G and Antoun G G 1976 Decomposition of organic matter under different conditions with special reference to changes in plant nutrients. Soil Organic Matter Studies Symp. IAEA-SM-211/27. pp 183–195. IAEA, Vienna.

Alexander M 1961 Introduction to Soil Microbiology. Wiley, New York. p 472.

Araragi M and Tangcham B 1979 Effect of rice straw on the composition of volatile soil gas and microflora in the tropical paddy field. Soil Sci. Plant Nutr. 25, 283–295.

Ayanaba A and Okigbo B N 1975 Mulching for improved soil fertility and crop production. *In* Organic Material as Fertilizers. Report of the FAO/SIDA Expert Consultation, Rome, December 1974. pp 97–119. Food and Agriculture Organization of the United Nations, Rome.

Barder M J and Crawford D L 1981 Effects of carbon and nitrogen supplementation on lignin and cellulose decomposition by a *Streptomyces*. Can. J. Microbiol. 27, 859–863.

Barrow N J and Jenkinson D S 1962 The effect of water-logging on fixation of nitrogen by soil incubated with straw. Plant and Soil 26, 258–262.

Baver L D 1956 Soil Physics, 3rd edn. John Wiley and Sons, New York.

Brouzes R, Lasik J and Knowles R 1969 The effect of organic amendment, water content, and oxygen on the incorporation of $^{15}N_2$ by some agricultural and forest soils. Can. J. Microbiol. 15, 899–905.

Brouzes R and Knowles R 1973 Kinetics of nitrogen fixation in a glucose-amended, anaerobically incubated soil. Soil Biol Biochem. 5, 223–229.

Cejudo F J and Paneque A 1986 Short-term nitrate (nitrite) inhibition of nitrogen fixation in *Azotobacter chroococcum*. J. Bacteriol. 165, 240–243.

Chapman L S, Halsall D M and Gibson A H 1992 Biological nitrogen fixation and sugarcane. *In* Proceedings of Australian Society of Sugar Cane Technologists. Ed. B T Egan. pp 90–93. Mackay, Queensland.

Charyulu P B B N and Rao R 1979 Nitrogen fixation in some Indian rice soils. Soil Sci. 128, 86–89.

Chatterjee S K and Nandi B 1981 Biodegradation of wheat stubbles by soil micro-organisms and role of the products on soil fertility. Plant and Soil 59, 381–389.

Choudhary N, Gray P P and Dunn N W 1980 Reducing sugar accumulation from alkali pretreated sugar cane bagasse using *Cellulomonas*. Eur. J. Appl. Microbiol. 11, 50–54.

Christensen B T 1987 Decomposability of organic matter in particle size fractions from field soils with straw incorporation. Soil Biol. Biochem. 19, 429–435.

Cogle A L, Strong W M, Saffigna P G, Ladd J N and Amato M 1987 Wheat straw decomposition in subtropical Australia. II. Effect of straw placement on decomposition and recovery of added ^{15}N-urea. Aust. J. Soil Res. 25, 481–490.

Dalton H 1980 The cultivation of diazotrophic microorganisms. *In* Methods for Evaluating Biological Nitrogen Fixation. Ed. F J Bergersen. pp 13–64. John Wiley and Sons, Chichester.

Delwiche C C and Wijler J 1956 Non-symbiotic nitrogen fixation in soil. Plant and Soil 7, 113–129.

Dickinson C H 1974 Decomposition of litter in soil. *In* Biology of Plant Litter Decomposition. Eds. C H Dickinson and G J F Pugh. pp 633–658. Academic Press, London.

Doran J W 1980 Microbial changes associated with residue management with reduced tillage. Soil Sci. Soc. Am. J. 44, 518–524.

Douglas C L, Allmaras R R, Rasmussen P E, Ramig R E and Roager N C 1980 Wheat straw composition and placement effects on decomposition in dryland agriculture of the Pacific Northwest. Soil Sci. Soc. Am. J. 44, 833–837.

Egawa T 1975 Utilization of organic materials as fertilizers in Japan. *In* Organic Material as Fertilizers. Report of the FAO/SIDA Expert Consultation, Rome, December 1974. pp 253–271. Food and Agriculture Organization of the United Nations, Rome.

Flinn J C and Marciano V P 1984 Rice straw and stubble management. *In* Organic Matter and Rice. pp 566–593. International Rice Research Institute, Manila.

Forbes R S 1974 Decomposition of agricultural crop debris. *In* Biology of Plant Litter Decomposition. Eds. C H Dickinson and G J F Pugh. pp 723–742. Academic Press, London.

Gibson A H, Roper M M and Halsall D M 1988 Nitrogen fixation not associated with legumes. *In* Advances in Nitrogen Cycling in Agricultural Ecosystems. Ed. J R Wilson. pp 66–88. CAB International, Wallingford, UK.

Greene R V and Freer S 1986 Growth characteristics of a novel nitrogen-fixing cellulolytic bacterium. Appl. Environ. Microbiol. 52, 982–986.

Greenwood D J and Goodman D 1964 Oxygen diffusion and aerobic respiration in soil spheres. J. Sci. Food Agric. 15, 579–588.

Habte M and Alexander M 1980 Nitrogen fixation by photosynthetic bacteria in lowland rice culture. Appl. Environ. Microbiol. 39, 342–347.

Halsall D M and Gibson A H 1985 Cellulose decomposition and associated nitrogen fixation by mixed cultures of *Cellulomonas gelida* and *Azospirillum* species or *Bacillus macerans*. Appl. Environ. Microbiol. 50, 1021–1026.

Halsall D M and Gibson A H 1986 Comparison of two *Cellulomonas* strains and their interaction with *Azospirillum brasilense* in degradation of wheat straw and associated nitrogen fixation. Appl. Environ. Microbiol. 51, 855–861.

Halsall D M and Gibson A H 1989 Nitrogenase activity by diazotrophs grown on a range of agricultural plant residues. Soil Biol. Biochem. 21, 1037–1043.

Halsall D M, Turner G L and Gibson A H 1985 Straw and xylan utilization by pure cultures of nitrogen-fixing *Azospirillum* spp. Appl. Environ. Microbiol. 49, 423–428.

Harper S H T and Lynch J M 1981 The chemical components and decomposition of wheat straw leaves, internodes and nodes. J. Sci. Food Agric. 43, 1057–1062.

Harper S H T and Lynch J M 1984 Nitrogen fixation by cellulolytic communities at aerobic-anaerobic interfaces in straw. J. Appl. Bacteriol. 57, 131–137.

Harper S H T and Lynch J M 1986 Dinitrogen fixation by obligate and facultative anaerobic bacteria in association with cellulolytic fungi. Curr. Microbiol. 14, 127–131.

Havelka U D, Boyle M G and Hardy R W F 1982 Biological nitrogen fixation. *In* Nitrogen in Agricultural Soils. Agronomy Monograph No. 22. pp 365–422. ASA-CSSA-SSSA, Madison, Wisconsin.

Hill N M, Patriquin D G and Sircom K 1990 Increased oxygen consumption at warmer temperatures favours aerobic nitrogen fixation in plant litters. Soil Biol. Biochem. 22, 321–325.

Imshenetsky A A 1967 Decomposition of cellulose in the soil. *In* The Ecology of Soil Bacteria. An International Symposium. Eds. T R G Gray and D Parkinson. pp 256–269. Liverpool University Press, Liverpool.

Inoko A 1984 Compost as a source of plant nutrients. *In* Organic Matter and Rice. pp 137–145. International Rice Research Institute, Manila.

Jensen H L and Swaby R J 1941 Association between nitrogen-fixing and cellulose-decomposing microorganisms. Nature 147, 147–148.

Jensen V 1981 Heterotrophic micro-organisms. *In* N_2 fixation. Vol. 1: Ecology. Ed. W J Broughton. pp 30–56. Clarendon Press, Oxford.

Jones K and Bangs D 1985 Nitrogen fixation by free-living heterotrophic bacteria in an oak forest: the effect of liming. Soil Biol. Biochem. 17, 705–709.

Kalininskaya T A 1972 Decomposition of cellulose and fixation of nitrogen in the soils of various types. Symp. Biol. Hung. 11, 135–138.

Kimura M, Panichsakpatana S, Wada H and Takai Y 1979 Influences of organic debris and rice root on the nitrogen fixation in the submerged soil. Soil Sci. Plant Nutr. 25, 637–640.

Knowles R 1976 Factors affecting dinitrogen fixation by bacteria in natural and agricultural systems. *In* Proceedings of the 1st International Symposium on N_2 Fixation. Vol 2. Eds. W E Newton and C J Nyman. pp 539–555. Washington State University, Washington DC.

Knowles R and Denike D 1974 Effect of ammonium-, nitrite- and nitrate-nitrogen on anaerobic nitrogenase activity in soil. Soil Biol. Biochem. 6, 353–358.

Kobayashi M 1982 The role of phototrophic bacteria in nature and their utilization. *In* Advances in Agricultural Microbiology. Ed. N S Subba Rao. pp 643–661. Oxford and IBH Publishing Co., New Delhi.

Kobayashi M and Haque M Z 1971 Contribution to nitrogen fixation and soil fertility by photosynthetic bacteria. Plant and Soil, Spec. Vol., 443–456.

Kobayashi M, Takahashi E and Kawaguchi K 1967 Distribution of nitrogen-fixing microorganisms in paddy soils of Southeast Asia. Soil Sci. 104, 113–118.

Ladha J K 1986 Studies on N_2 fixation by free-living and rice plant-associated bacteria in wetland rice field. Bionature 6, 47–58.

Ladha J K, Padre A T, Nayak D N, Garcia M and Watanabe I 1986a Nitrogen fixation by single and mixed heterotrophic bacteria in flooded paddy soils amended with hydrogen peroxide treated straw (hemicellulose). *In* Proceedings of the IV International Symposium on Microbial Ecology. pp 603–608. Ljubljana, Yugoslavia.

Ladha J K, Tirol A C, Daroy M L G, Caldo G, Ventura W and Watanabe I 1986b Plant-associated N_2 fixation (C_2H_2-reduction) by five rice varieties, and relationship with plant growth characters as affected by straw incorporation. Soil Sci. Plant Nutr. 32, 91–106.

Ladha J K, Tirol-Padre A, Daroy M L G, Punzalan G and Watanabe I 1987 The effects on N_2 fixation (C_2H_2 reduction), bacterial population and rice plant growth of two modes of straw application to a wetland rice field. Biol. Fertil. Soils 5, 106–111.

La Rue T A 1977 The bacteria. *In* A Treatise on Dinitrogen Fixation. Section III, Biology. Eds. R W F Hardy and W S Silver. pp 19–62. John Wiley and Sons, New York.

Leschine S B, Holwell K and Canale-Parola E 1988 Nitrogen fixation by anaerobic cellulolytic bacteria. Science 242, 1157–1159.

Lethbridge G and Davidson M S 1983 Microbial biomass as a source of nitrogen for cereals. Soil Biol. Biochem. 15, 375–376.

Lynch J M 1983 Soil Biotechnology – Microbiological Factors in Crop Productivity. Blackwell Scientific Publications, Oxford, London. 191p.

Lynch J M and Harper S H T 1983 Straw as a substrate for cooperative nitrogen fixation. J. Gen. Microbiol. 129, 251–253.

Macura J and Pavel L 1959 The influence of montmorillonite on nitrogen fixation by *Azotobacter*. Folia Microbiol. 4, 82–90.

Magdoff F R and Bouldin D R 1970 Nitrogen fixation in submerged soil-sand-energy material media and the aerobic-anaerobic interface. Plant and Soil 33, 49–61.

Marshall K C 1975 Clay mineralogy in relation to survival of soil bacteria. Ann. Rev. Phytopathol. 13, 357–373.

Matsuguchi T 1979 Factors affecting heterotrophic nitrogen fixation in submerged rice soils. *In* Nitrogen and Rice. pp 207–222. International Rice Research Institute, Manila.

Matsuguchi T and Yoo I D 1981 Stimulation of phototrophic N_2 fixation in paddy fields through rice straw application. *In* Nitrogen Cycling in Southeast Asian Wet Monsoonal Ecosystems. Eds. R Wetselaar, J R Simpson and T Rosswall. pp 18–25. Australian Academy of Science, Canberra.

Nayak D N, Ladha J K and Watanabe I 1986 The fate of marker *Azospirillum lipoferum* inoculated into rice and its effect on growth, yield and N_2 fixation of plants studied by acetylene reduction, $^{15}N_2$ feeding and ^{15}N dilution techniques. Biol. Fertil. Soils 2, 7–14.

Neue H U and Scharpenseel H W 1984 Gaseous products of the decomposing of organic matter in submerged soils. *In* Organic Matter and Rice. pp 311–328. International Rice Research Institute, Manila.

Novak B 1972 Effect of increasing amounts of nitrogen on the microbial transformation of straw in soil. Symp. Biol. Hung. II, 49–53.

Novakova J and Sisa R 1984 Effect of clays on the cellulolytic activity of soil. Zentralbl. Mikrobiol. 139, 505–510.

Pal D, Broadbent F E and Mikkelsen D S 1975 Influence of temperature on the kinetics of rice straw decomposition in soils. Soil Sci. 120, 442–449.

Parr J F and Papendick R I 1978 Factors affecting the decomposition of crop residues by microorganisms. *In* Crop Residue Management Systems. Am. Soc. Agron. Spec. Publ. pp 101–129. Am. Soc. Agron., Madison, Wisconsin.

Patriquin D G 1982 Nitrogen fixation in sugar cane litter. Biol. Agric. Hort. 1, 39–64.

Peoples M B, Herridge D F and Ladha J K 1995 Biological Nitrogen Fixation: an efficient source of nitrogen for sustainable agricultural production? Plant and Soil 174.

Ponnamperuma F N 1984 Straw as source of nutrients for wetland rice. *In* Organic Matter and Rice. pp 117–136. International Rice Research Institute, Manila.

Rao V R 1976 Nitrogen fixation as influenced by moisture content, ammonium sulphate and organic sources in a paddy soil. Soil Biol. Biochem. 8, 445–448.

Rao V R 1978 Effect of carbon sources on asymbiotic nitrogen fixation in a paddy soil. Soil Biol. Biochem. 10, 319–321.

Rao V R 1980 Changes in nitrogen fixation in flooded paddy field soil amended with rice straw and ammonium sulphate. Riso 28, 29–34.

Reddy K R and Patrick W H 1979 Nitrogen fixation in flooded soil. Soil Sci. 128, 80–85.

Rice W A 1979 Influence of the nitrogen content of straw amendments on nitrogenase activity in waterlogged soil. Soil Biol. Biochem. 11, 187–191.

Rice W A and Paul E A 1972 The organisms and biological process involved in asymbiotic nitrogen fixation in waterlogged soil amended with straw. Can J. Microbiol. 18, 715–723.

Rice W A, Paul E A and Wetter L R 1967 The role of anaerobiosis in asymbiotic nitrogen fixation. Can. J. Microbiol. 13, 829–836.

Richardson A E, Connell L R, Halsall D M, Gibson A H and Watson J M 1991 Towards the development of cellulolytic diazotrophs. *In* Proceedings of the 9th Australian Nitrogen Fixation Conference, Australian National University, Canberra, January 1991. Eds. A E Richardson and M B Peoples. pp 11–12. Australian Soc. for Nitrogen Fixation.

Roger P A and Kulasooriya S A 1980 Blue-green algae and rice. International Rice Research Institute, Manila. 112p.

Roger P A and Watanabe I 1986 Technologies for utilizing biological nitrogen fixation in wetland rice: potentialities, current usage, and limiting factors. Fert. Res. 9, 39–77.

Roper M M 1983 Field measurements of nitrogenase activity in soils amended with wheat straw. Aust. J. Agric. Res. 34, 725–739.

Roper M M 1985 Straw decomposition and nitrogenase activity (C_2H_2 reduction): effects of soil moisture and temperature. Soil Biol. Biochem. 17, 65–71.

Roper M M, Gault R R and Smith N A 1994b Contribution to the nitrogen status of soil by free-living nitrogen-fixing bacteria in a lucerne stand. Soil Biol. Biochem. (*In press*).

Roper M M and Halsall D M 1986 Use of products of straw decomposition by N_2-fixing (C_2H_2-reducing) populations of bacteria in three soils from wheat-growing areas. Aust. J. Agric. Res. 37, 1–9.

Roper M M, Marschke G W and Smith N A 1989 Nitrogenase activity (C_2H_2 reduction) in soils following wheat straw retention: effects of straw management. Aust. J. Agric. Res. 40, 241–253.

Roper M M and Smith N A 1991 Straw decomposition and nitrogenase activity (C_2H_2 reduction) by free-living microorganisms from soil: effects of pH and clay content. Soil Biol. Biochem. 23, 275–283.

Roper M M, Turpin J A and Thompson J P 1994a Nitrogenase activity (C_2H_2 reduction) by free-living bacteria in soil in a long-term tillage and stubble management experiment on a vertisol. Soil Biol. Biochem. 26, 1087–1091.

Sain P and Broadbent F E 1977 Decomposition of rice straw in soils as affected by some management factors. J. Environ. Qual. 6, 96–100.

Saito M, Wada H and Takai Y 1977a Microbial ecology of cellulose decomposition in paddy soils. I. Modification of Tribe's cellophane film method and staining methods for the observation of microorganisms growing on cellulose material. Nippon Dojohiryo Gaku Zasshi 48, 313–317 (*In Japanese*).

Saito M, Wada H and Takai Y 1977b Microbial ecology of cellulose decomposition in paddy soils. II. Succession of microorganisms growing on cellulose material. Nippon Dojohiryo Gaku Zasshi 48, 318–322 (*In Japanese*).

Santiago-Ventura T, Bravo M, Daez C, Ventura V, Watanabe I and App A A 1986 Effects of N-fertilizers, straw and dry fallow on the nitrogen balance of a flooded soil planted with rice. Plant and Soil 93, 405–411.

Shintani Y 1987 Effect of nitrite on growth and anaerobic nitrogen fixation in *Bacillus polymyxa*. J. Gen. Appl. Microbiol. 33, 93–97.

Smith J H and Peckenpaugh R E 1986 Straw decomposition in irrigated soil: comparison of twenty-three cereal straws. Soil Sci. Soc. Am J. 50, 928–932.

Sorensen L H 1974 Rate of decomposition of organic matter in soil as influenced by repeated air drying-rewetting and repeated additions of organic material. Soil Biol. Biochem. 6, 287–292.

Staley J T (Ed.) 1989 Bergey's Manual of Systematic Bacteriology. Vol. 3. Williams and Wilkins, Baltimore.

Stanier R, Doudoroff M and Adelberg E A 1974 General Microbiology. MacMillan, India. 873p.

Stewart W D P 1980 Systems involving blue-green algae (cyanobacteria). *In* Methods for Evaluating Biological Nitrogen Fixation. Ed. F J Bergersen. pp 583–635. John Wiley and Sons, Chichester.

Stotzky G 1972 Activity, ecology, and population dynamics of microorganisms in soil. CRC Crit. Rev. Microbiol. 2, 59–137.

Subba Rao N S 1982 Utilization of farm wastes and residues in agriculture. *In* Advances in Agricultural Microbiology. Ed. N S Subba Rao. pp 509–521. Oxford and IBH Publishing Co., New Delhi.

Summerell B A and Burgess L W 1989 Decomposition and chemical composition of cereal straw. Soil Biol. Biochem. 21, 551–559.

Swift M J 1982 Microbial succession during the decomposition of organic matter. *In* Experimental Microbial Ecology. Eds. R G Burns and J H Slater. pp 164–177. Blackwell Scientific Publications, Oxford, UK.

Theander O and Åman P 1978 Chemical composition of some Swedish cereal straws. Swedish J. Agric. Res. 8, 189–194.

Theander O and Åman P 1984 Anatomical and chemical characteristics. *In* Straw and Other Fibrous By-products as Feed. Eds. F Sundstol and O Owen. pp 45–78. Elsevier Publishers, Amsterdam.

Veal D A and Lynch J M 1984 Biochemistry of cellulose breakdown by mixed cultures. Biochem. Soc. Trans. Vol: 1142–1144.

Vostrov I S and Dolgikh Y R 1970 Microflora of submerged soils of the rice field. Izv. Akad. Naukkaz. SSR Ser. Biol. 1, 64–69 (*In Russian*).

Waksman S A and Gerretsen F C 1931 Influence of temperature and moisture upon the nature and extent of decomposition of plant residues by microorganisms. Ecology 12, 33–60.

Wada H, Panichsakpatana S, Kimura M and Takai Y 1979 Organic debris as micro-site for nitrogen fixation. Soil Sci. Plant Nutr. 25, 453–456.

Watanabe I 1984 Anaerobic decomposition of organic matter in flooded rice soils. *In* Organic Matter and Rice. pp 237–258. International Rice Research Institute, Manila.

Watanabe I and Furusaka C 1980 Microbial ecology of flooded rice soils. Adv. Microbiol. Ecol. 4, 125–168.

Watanabe I, Lee K K and Alimagno B V 1978 Seasonal change of N_2-fixing rate in rice field assayed by in situ acetylene reduction technique. I. Experiments in long-term fertility plots. Soil Sci. Plant Nutr. 24, 1–13.

Waterbury J B, Calloway C B and Turner R D 1983 A cellulolytic nitrogen-fixing bacterium cultured from the gland of deshayes in shipworms (Bivalvia: Teredinidae). Science 221, 1401–1403.

Yadav K S and Subba Rao N S 1980 Use of cellulolytic microorganisms in composting. *In* Recycling Residues of Agriculture and Industry. Proceedings of a Symposium at the Punjab Agricultural University. Ed. M S Kalra. pp 267–273. Ludhiana, India.

Yoneyama T, Lee K K and Yoshida T 1977 Decomposition of rice residues in tropical soils. IV. The effect of rice straw on nitrogen fixation by heterotrophic bacteria in some Philippine soils. Soil Sci. Plant Nutr. 23, 287-295.

Yoo I D, Wada H and Takai Y 1982 Paddy Nitrogen. VII. Effects of rice straw application on N_2 fixation in paddy fields with special reference to phototrophic N_2 fixers on surface-placed rice straw. *In* International Seminar on Productivity of Soil Ecosystems. pp 243–248. Nodai Research Institute, Tokyo Univ. Agric.

Yoshida T 1975 Microbial metabolism of flooded soils. *In* Soil Biochemistry, Vol 3. Eds. E A Paul and A D McLaren. pp 83–122. Marcel Dekker Inc., New York.

Zeikus J G 1981 Lignin metabolism and the carbon cycle. Polymer biosynthesis, biodegradation, and environmental recalcitrance. Adv. Microbiol. Ecol. 5, 211–243.

Zielinski J 1980 The effect of nitrogen content on the rate of organic matter decomposition. Pol. Ecol. Stud. 6, 167–182.

Plant and Soil **174:** 225–240, 1995.

Potential and pitfalls of trying to extend symbiotic interactions of nitrogen-fixing organisms to presently non-nodulated plants, such as rice

F.J. de Bruijn[1,2,3], Y. Jing[1,4] and F.B. Dazzo[2,3]
[1]*MSU-DOE Plant Research Laboratory,* [2]*Department of Microbiology,* [3]*NSF Center for Microbial Ecology, Michigan State University, E. Lansing, MI 48824, USA and* [4]*Institute of Botany, Chinese Academy of Sciences, Beijing, China*

Key words: endophytic microbes, endo-symbiosis, *Rhizobium*, rice, root hypertrophies, ultrastructure

Table of contents

Abstract

Introduction

Plant-microbe signalling, nodulation, and symbiotic nitrogen fixation

The oxygen paradox and other potential problems related to microbial nitrogen fixation in novel physiological environments

Nodulation of cereals: What has been tried, what has been published, and is nodulation really necessary to achieve useful levels of nitrogen fixation

Induction of hypertrophies on rice roots and their colonization by microbes

Materials and methods

Induction and microscopic examination of "nodule-like" structures on rice roots

Molecular methods for strain identification

Results and discussion

Abnormal structures on rice roots grown in China

Abnormal structures on rice roots grown at Michigan State University

Acknowledgements

References

Abstract

It has been a long-standing goal in the field of biological nitrogen fixation to extend nitrogen-fixing symbioses to presently non-nodulated cereal plants, such as rice. A number of researchers have recently described the induction of "nodule-like" structures on the roots of cereals primarily by rhizobia, in either the presence or absence of plant cell-wall-degrading enzymes or plant hormones. We briefly review this research and discuss the potential problems associated with the introduction of nitrogen-fixing microbes in novel physiological environments, such as rice roots. The results of experiments carried out in China on the induction of "nodule-like" structures on rice roots by rhizobia are highlighted. In addition, we present preliminary results of a series of experiments designed to repeat and evaluate these results using a variety of microscopic techniques and molecular genetic approaches.

Introduction

In modern agriculture, the replenishment of soil nitrogen most commonly involves extensive application of chemical fertilizers (Peoples et al., 1995), an approach that suffers from several serious drawbacks, including high costs and severe negative environmental impacts. These considerations have stimulated research on alternatives, such as biological nitrogen fixation. Biological nitrogen fixation has been estimated to contribute more than 170 million tons of fixed nitrogen to the biosphere (Earl and Ausubel, 1993). Eighty percent of the stable, biologically fixed nitrogen is a direct result of the symbiotic interaction of members of the Rhizobiaceae and some actinomycetes with leguminous, as well as certain non-leguminous plants.

In the nodules induced on the roots and stems of the host plant, nitrogen fixed by rhizobia or actinomycetes is directly incorporated into nitrogen-rich organic compounds such as amides, amino acids, or ureides and proteins (Schubert, 1986). In contrast to this efficiency, nitrogen fixed by strictly free-living or associative diazotrophs can be rapidly lost by denitrifcation reactions in the biosphere (rhizosphere). Therefore, a considerable amount of research effort has been invested in elucidation of the molecular basis of symbiotic plant-microbe interactions. Two long-term objectives of these studies have been to further improve the efficiency of symbiotic nitrogen fixation and to investigate the potential of extending this beneficial process to presently non-nodulated plants, especially cereals. Recently, a number of claims have been made with regard to the latter, some of which may have been somewhat overstated (Simon Moffatt, 1990). In fact, several of the reports on the supposed induction of (root) "nodules" on nonlegumes, including rice, have been controversial, despite their endorsement in the popular scientific press. This situation has led to concern among scientists in the field of symbiotic nitrogen fixation not to be unrealistic and repeat the mistakes of the early 1970s, when the transfer of the entire package of nitrogen-fixation (*nif*) genes from a free-living diazotroph into plants was widely claimed as "within reach". It has also led to the organization of an International Rice Nodulation Group (Cocking and Davey, 1991) and a small international conference on the potential and prospects of nodulation and nitrogen fixation in cereals such as rice (Khush and Bennett, 1992).

In this paper, we will first review very briefly the present state of knowledge about plant-microbe signalling and the induction of nitrogen-fixing stem and root nodules on legume plants. Second, we will summarize some of the crucial biological problems that would need to be solved to establish the process of symbiotic (endophytic) nitrogen fixation in plants such as cereals. Third, some of the published reports about "nodulation" of cereals will be critically reviewed and the need for the induction of proper nodule structures on presently non-nodulated plants for the purpose of nitrogen fixation will be questioned. Fourth, we will discuss some recent preliminary results from experiments designed to examine a previous report (Li et al., 1991) describing the formation of hypertrophies and nodule-like structures by selected microbes on rice roots, and the presence of microbes in these structures.

Plant-microbe signalling, nodulation, and symbiotic nitrogen fixation

The induction of nitrogen-fixing root and stem nodules on leguminous plants by soil bacteria belonging to the Rhizobiaceae involves fine-tuned interactions between the two symbiotic partners, including multiple regulatory signals that go back and forth between the bacterium and the plant to coordinate expression of gene sets in both partners. These signals include plant factors that induce rhizobial nodulation (*nod*) genes, rhizobial Nod factors that are essential for early stages of nodule ontogenesis, and hormones and other regulatory factors involved in symbiotic nitrogen fixation (for reviews see de Bruijn and Downie, 1991; Dénarié and Cullimore, 1993; Fisher and Long, 1992; Hirsch, 1992). The first step in the plant-microbe signalling pathway leading up to nodulation is the production/secretion of flavonoids, chalcones, conjugated isoflavonoids and betaines by the plant host (see Peters and Verma, 1990; Phillips, 1992; Phillips et al., 1994). These primary signal compounds are responsible for the induction of the nodulation (*nod*) genes of the microbial symbiont, most of which are silent in free-living rhizobia. This induction process involves the specific activation of regulatory factors in the rhizobia (NodD proteins) by particular representatives of the group of plant factors listed above. These activated NodD proteins, in turn, serve as transcriptional activators of other nod genes (see Fischer and Long, 1992).

Once the rhizobial *nod* genes are activated, their gene products are involved in the synthesis of a class of compounds (Nod factors), which serve as the first return signal from the bacteria to their host plant. This class of compounds has been shown to consist of lipo-oligosaccharides of varying lengths, carrying different side groups and substitutes that play major roles in conferring host specificity (see Dénarié and Cullimore, 1993). The Nod factors do not only mediate recognition between the proper rhizobial species and its natural host(s), but also serve as specific morphogens by initiating root hair curling, and cortical cell division in the target root cells (see Dénarié and Cullimore, 1993; Vijn et al., 1993). Similar to the plant derived first signals that activate rhizobial (*nod*) gene expression, the second rhizobial signal molecules (Nod factors) induce specific gene expression in the cognate host plant (see Dénarié and Cullimore, 1993; Vijn et al., 1993). Generally, the plant genes specifically expressed during the formation and functioning of nodules are referred to as nodulin genes (Van Kammen, 1984). They can be divided into several classes, depending on their time point of induction during nodule formation, but the most prominent classes have been named "early and late nodulin genes" (Nap and Bisseling, 1990), and are clearly activated in different ways (see de Bruijn and Schell, 1992). Purified Nod factor has been found to activate early nodulin genes such as Enod5 and Enod12 in the host plant (Horvath et al., 1993), further supporting the proposed role of the Nod factor as an essential signal molecule in nodule ontogeny.

In addition to lipo-oligosaccharide signal molecules, other metabolites have been suggested to transfer microbe-plant signals and to play a role in nodulation. These include hormones (see Cooper and Long, 1994; Bruijn et al., 1992; Hirsch, 1992) and compounds such as N-acetyl glutamic acid (Hollingsworth et al., 1991) and diglycosyl diacylglycerol (Orgambide et al., 1994). With regard to the former, the hormone cytokinin has been proposed to play an important role in signalling required for nodulation, since it activates the early nodulin gene *Enod2* (Dehio and de Bruijn, 1992) and permits otherwise nodulation deficient rhizobia (lacking the common *nod* genes) to induce nodulation on plants (Cooper and Long, 1994). The N-acetyl glutamic acid and diglycosyl diacylglycerol metabolites made by rhizobia were found to elicit root hair deformation and have mitogenic activity on host root cells (Hollingsworth et al., 1991; Orgambide et al., 1994).

Once the nodule is developed and the infecting rhizobia have been released into the plant cell cytoplasm and differentiated into a bacteroid state, the nitrogen fixation process is activated. This activation process is mediated by micro-aerobic induction of the nitrogen fixation (*nif/fix*) genes (David et al., 1990), via the oxygen sensing protein FixL, a hemoprotein, (Gilles-Gonzales et al., 1992) in most cases examined (see Fischer 1994). Concurrent with the onset of nitrogen fixation, a group of late nodulin genes are activated, such as the genes encoding the oxygen carrier protein leghemoglobin (*lb* genes) and genes involved in carbon and nitrogen assimilation (see Nap and Bisseling, 1990). The regulation of the late nodulin genes may involve another type of bacterial signal, since these genes are expressed predominantly in the infected cells of the nodule (de Bruijn and Schell, 1992; Lauridsen et al., 1993; Szczyglowski et al., 1994). The physical presence of (membrane-enclosed) bacteria in the cytoplasm of the infected cells appears to be required and a rhizobial DNA-binding protein interacting with a specific *cis*-acting element in the plant *lb* promoter has been described (de Bruijn et al., 1994; Welters et al.,1993) that may be involved in infected-cell-specific expression of the *lb* genes. If this hypothesis is proven to be true, it would constitute another interesting example of microbe-plant (trans-kingdom) signalling in nodulation.

These studies make it clear that the process of plant-microbe signalling in nodulation is highly complex and that we must develop a clear, basic understanding of the nature and expression of rhizobial and plant genes involved in the nodulation and symbiotic nitrogen fixation process before we attempt to design experiments to extend the symbiotic interaction to presently non-nodulated plants (see de Bruijn and Downie, 1991).

The oxygen paradox and other potential problems related to microbial nitrogen fixation in novel physiological environments

The reduction of dinitrogen to ammonia catalyzed by the enzyme nitrogenase requires large amounts of ATP (up to 40 ATPs/N_2 reduced). Rhizobia generate this ATP via oxidative phosphorylation, requiring a constant O_2 supply. However, nitrogenase is irreversibly denatured at O_2 concentrations exceeding 10 n*M* (Appleby, 1984). This "oxygen paradox" represents an important biological problem when considering the extension of symbiotic interactions to presently non-nodulated plants. In legume nodules, several mechanisms have evolved to deal with the oxygen problem, including the creation of an apparent oxygen barrier in the nodule parenchyma and the high-level synthesis of the oxygen carrier protein leghemoglobin

(Appleby, 1984; Hirsch, 1992). Although some evidence exists for the presence of hemoglobin-type genes in non-legumes (see Bogusz et al., 1988), it remains unclear whether or how the proper physiological conditions for nitrogen fixation can be created in plant tissues not normally infected by nitrogen-fixing organisms.

The same uncertainty applies to other physiological requirements for endosymbiotic or endophytic nitrogen fixation, such as high levels of energy-reducing equivalents, mechanisms to rapidly assimilate the ammonia produced by dinitrogen reduction, and ways to avoid the plant defense response. Again, we must carry out extensive basic physiological studies on these processes before attempting to introduce nitrogen-fixing organisms into novel environments, such as rice roots.

Nodulation of cereals: What has been tried, what has been published, and is nodulation really necessary to achieve useful levels of nitrogen fixation?

Although much basic knowledge about the molecular basis of plant-microbe interactions and the physiology of nitrogen fixation/assimilation is still lacking, researchers worldwide have started to explore the possibility of extending the nitrogen-fixing symbiosis of rhizobia and legumes to presently non-nodulated plants. Several recent reports describe various approaches used to induce nodule-like structures on the roots of rice and other cereals (see Kennedy and Tchan, 1992; Khush and Bennett, 1992; Simon Moffatt, 1990). The infection of "para-nodules", induced by chemicals such as 2,4-D on wheat roots, by rhizobia and free-living diazotrophs has been reported and suggested to be a potential vehicle to achieve nitrogen-fixation in cereals (Bender et al., 1990; Kennedy and Tchan, 1992; Tchan et al., 1991). In fact, nitrogen fixation by *Azospirillum brasilense* and *Azorhizobium caulinodans* in para-nodules induced on wheat roots has been reported (Chen, 1993; Chen et al., 1993; Tchan et al., 1991) and the effect of a variety of microorganisms on the formation of 2,4 D-induced nodule-like structures on wheat roots has been examined (Ridge et al., 1992). Moreover, Christansen-Weniger and Vanderleyden (1994) have shown that ammonium-excreting *Azospirillum* can become established intracellularly in 2-4-D induced paranodules of maize.

The induction of nodule-like structures on and invasion of rice, maize, and wheat seedling roots by *Parasponia* and *Aeschynomene* rhizobia, either spontaneously or after the treatment of roots with cell-wall-degrading enzymes, have also been described (Al-Mallah et al., 1989; Cocking et al., 1990, 1992). It has been proposed that these approaches, involving "nodulation in non-legume crops by rhizobia" may lead to advances in obtaining "nitrogen from the air for non-legume crops" (Cocking and Davey, 1991).

A *Rhizobium* strain has been constructed containing a plasmid carrying a *nodD* allele able to respond to signals produced by rice roots. This strain induces infected nodule-like structures on rice seedlings at a low frequency (Rolfe and Bender, 1990).

Jing et al. (1990) have reported the induction of pseudonodules on barley roots by *Rhizobium astragali* under a permanent magnetic field. Moreover, a high-frequency induction of infected nodule-like structures on rice roots by *Sesbania* rhizobia has been reported (Jing et al., 1992; Li et al., 1991). In fact, acetylene reduction by "rice root nodules" infected by rhizobia and a positive effect of rhizobial infection of rice plants on growth and yield has been suggested (Jing et al., 1990, 1992), and the presence of infection threads and bacteroids in the infected "nodule-like structures" has been proposed (Jing et al., 1992; Li et al., 1992).

Several of these reports are controversial and have not yet been independently confirmed. In some cases, the structures reported have been called "nodules"; without proper experimental support they should probably not even be considered "nodule-like structures". In fact, many of the structures reported appear to be modified lateral roots (Cocking and Davey, 1991; Ridge et al., 1992; see also Kennedy and Tchan, 1992, and below). With regard to nitrogen fixation in these infected structures or positive effects on plant growth, very few of the reported studies have employed ^{15}N-based techniques or have included the proper controls or statistical analyses. Only in the case of wheat "para-nodules", has ^{15}N been used to confirm N_2 fixation (Yu et al., 1993). The ultrastructural analyses of infected cells via various microscopic techniques have also been dubious in several cases; an intact cytoplasm cannot be seen in many of the photographs presented, making it difficult to exclude saprophytic infections of broken cells, to discern whether the infected cells are intact, or to establish whether the infecting bacteria are truly (and stably) maintained intracellularly.

Thus, a lot of unresolved issues remain with regard to these "nodulation" experiments, which need to be carefully (re)examined. The more general question could also be raised whether one actually needs

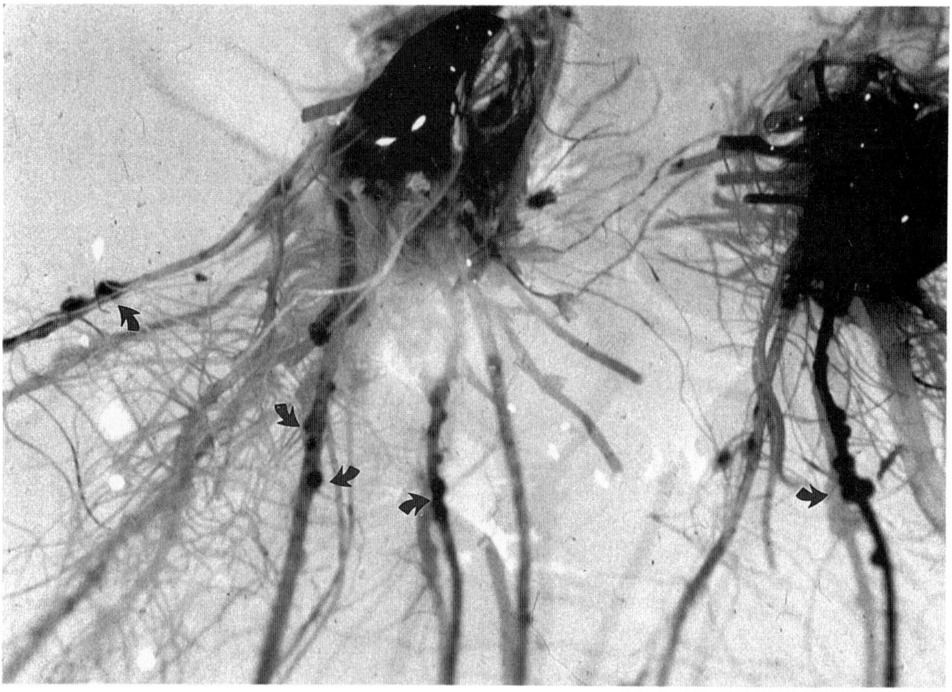

Fig. 1. Rice roots grown in China and containing dark "nodule-like structures" as previously described (Li et al., 1991).

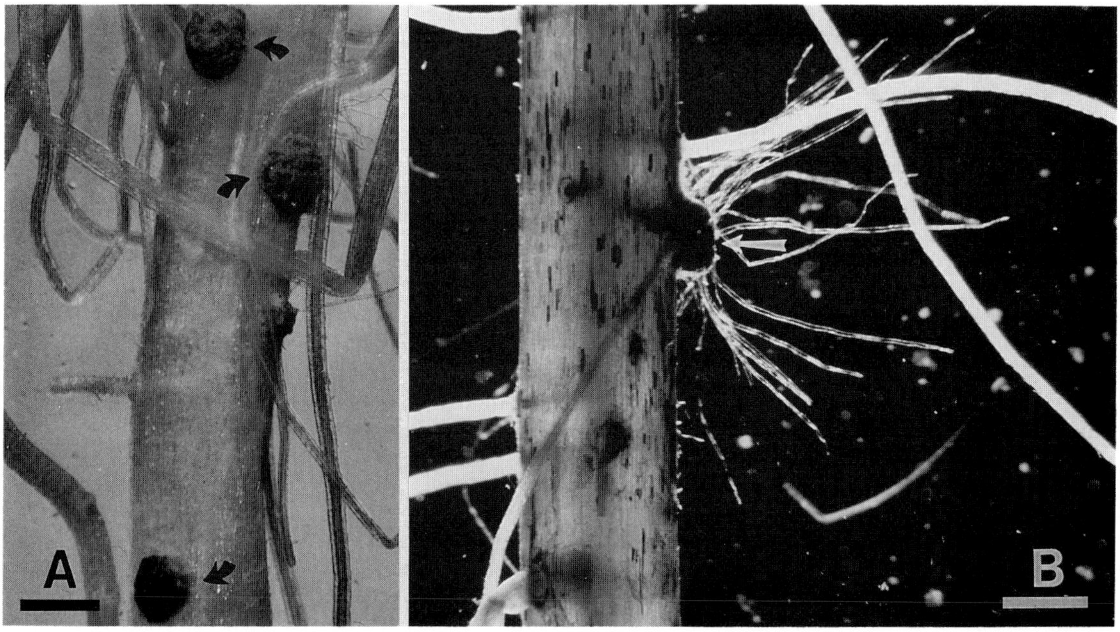

Fig. 2. Stereomicrographs showing the unusual dark hemispheric structures (arrows) attached to the rice root surface. (**A**) Combined transmitted/incident illumination; (**B**) darkfield illumination. *Bar* scale is 1 mm.

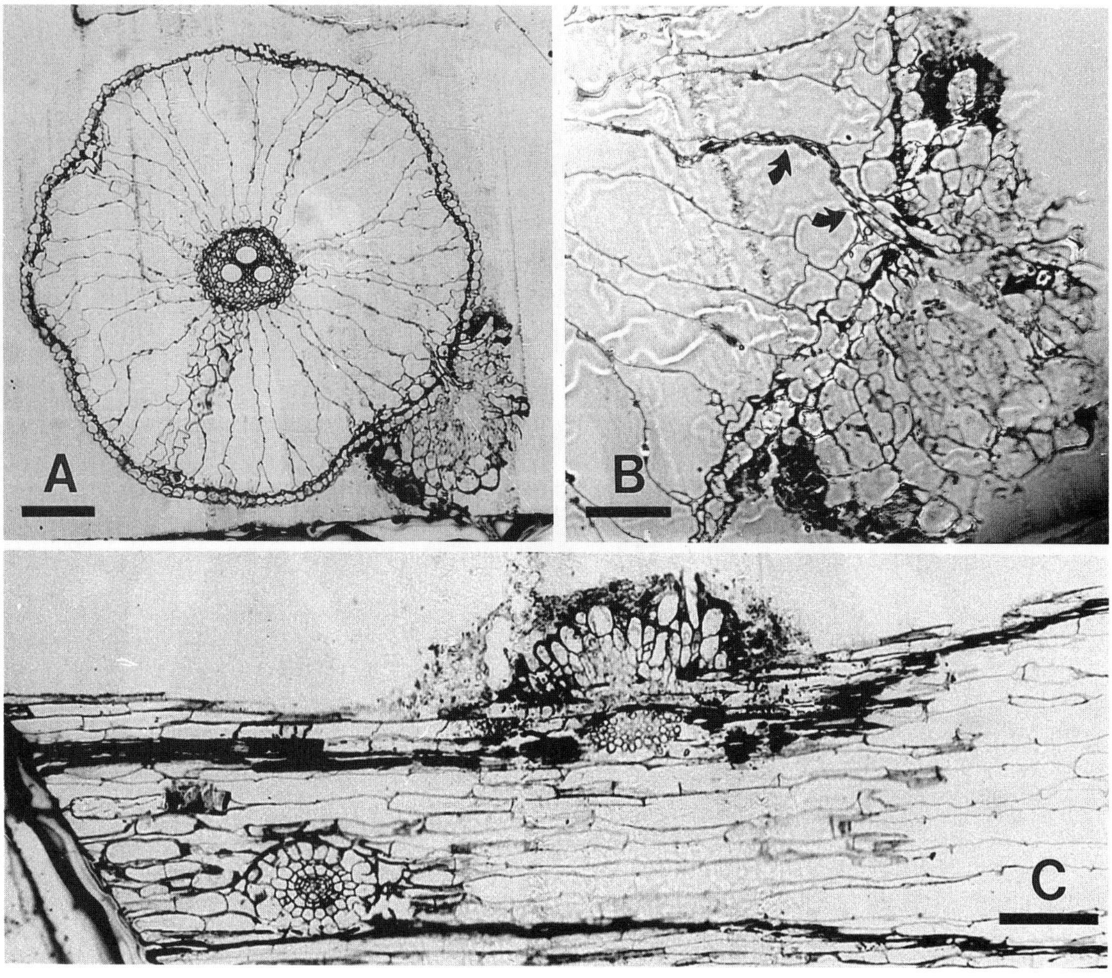

Fig. 3. Brightfield micrographs of the unusual hemispheric structures attached to the rice root surface; 2 μm sections stained with alkaline toluidine blue. (**A**) Rice root in cross-section. (**B**) Higher magnification of the interface between the external hemispheric structures and the root epidermis. Note the septate hyphae (arrows) penetrating the root epidermis and extending into the root cortex. (**C**) Rice root in longitudinal section, showing the external hemispheric dome structure and colonization in the underlying root cortex. *Bar* scales are 100 μm in (A), 50 μm in (B, C).

nodules or nodule-like structures on cereal roots to achieve symbiotic/endophytic nitrogen fixation (see also Kennedy and Tchan, 1992; Ladha et al., 1993). It may, indeed, be sufficient to identify a stable endophyte of rice roots and, if necessary, to engineer this microbe to efficiently fix nitrogen and excrete the fixed nitrogen for use by the plant. This endophyte may not need to be stably maintained intracellularly, like a rhizobial endosymbiont; it may be sufficient if it colonizes the plant root intercellularly, as long as it is able to evade the plant defense responses (Khush and Bennett, 1992; Quispel, 1992).

In this context, it is interesting to note that stable endophytic diazotrophs that contribute substantially to plant growth via nitrogen fixation have been described for other plants, such as sugar cane (*Acetobacter diazotrophicus*; see Boddey et al., 1995; Döbereiner et al., 1993). Moreover, a strain of the nitrogen-fixing bacterium *Azoarcus* has been described that colonizes and spreads systemically in grasses such as rice (Hurek et al., 1994). This bacterium was found to invade roots inter- and intracellularly, to enter and spread through the xylem, and to increase rice yield, although the latter phenotype could not be ascribed to nitrogen fixation (Hurek et al., 1994). Therefore, *Azoarcus* cannot be classified as an endophytic diazotroph, as defined by Döbereiner et al. (1993) and Quispel (1992). It may nevertheless serve as an interesting model organism in

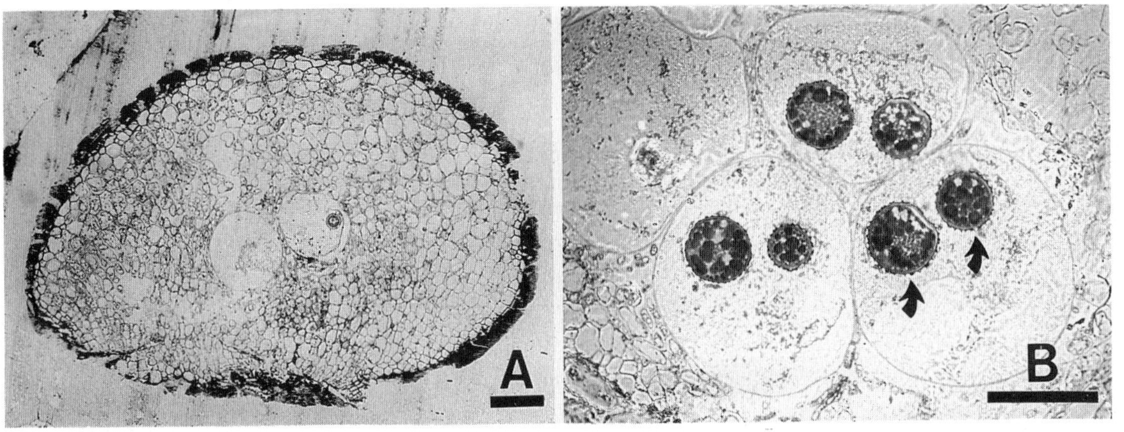

Fig. 4. Brightfield micrographs of the unusual hemispheric structures detached from the rice root; 2 μm sections stained with alkaline toluidine blue. (**A**) Section through the hemispheric structure. (**B**) Higher magnification of the central zone showing the sacks with several ornamented cells (arrow) inside. *Bar* scales are 100 μm in (A, B).

studying the persistence of non-pathogenic bacteria in plant tissues in general and rice root colonization in particular (Hurek et al., 1994).

Induction of hypertrophies on rice roots and their colonization by microbes

The high-frequency induction of nodule-like structures on rice roots by rhizobia isolated from *Sesbania cannabina* nodules (see Fig. 1) and the presence of rhizobial infection (producing thread-like structures) and peribacteroid-like membranes in plant host cells have been reported recently (Li et al., 1991). "Nodulation percentages" of up to 66% on the Chinese rice varieties Lianjiang, Juefu and Jiangxi 80074 were reported, reisolated bacteria were shown to contain *nif* and *nod* genes, and were classified by nutritional requirements in order to show that they were closely related to the input bacteria. They were also found to be able to re-nodulate *S. cannabina* (Li et al., 1991).

This research by Li and colleagues has been considered of particular interest and worthy of further investigation (Khush and Bennett, 1992; Simon Moffatt, 1990). Therefore, experiments were initiated by Dr Jing to re-examine the induction of nodule-like structures described by Li et al. (1991). He used a variety of microscopic techniques to obtain similar structures at Michigan State University (MSU) on a rice variety shown to be highly responsive to bacterial infection (Zhong Xi 8408; Y Jing, unpubl. results) when inoculated with some of the same strains previously used (Li et al., 1991). The preliminary results of these experiments and the molecular analysis of some of the strains employed for these experiments will be presented here.

Materials and methods

Induction and microscopic examination of "nodule-like" structures on rice roots

Rice variety Zhongxi 8408 was grown in China (Jing et al., 1992). Roots bearing the previously described dark "nodule-like -structures" (Li et al., 1991; Fig. 1) were excised, preserved in formalin, and transported to MSU. Root segments were examined directly by stereomicroscopy with transmitted, incident, and darkfield illumination. Short root segments with these hemispherical domes still attached were excised while viewing under the stereomicroscope and processed for plastic embedding using standard methods of specimen preparation, followed by combined light and transmission electron microscopy (Dazzo, 1982). Other formalin-preserved root segments with the dome-shaped structures still attached were processed directly for examination by scanning electron microscopy (Umali-Garcia et al., 1980), and laser scanning confocal microscopy after staining with acridine orange (Dazzo et al., 1993).

In other experiments, rice plants (20–30 seedlings per pot) were grown at MSU in 25–30 cm diameter pots with either sterile sand or soil mixtures in Conviron growth chambers under 15 hr day^{-1} of light (80%

Fig. 5. Scanning electron micrographs of the unusual hemispheric structures attached to the rice root. Note the encrusted surface layer, the internal hyphal network, and the sacks containing several spherical cells with reticulate surface ornamentations. *Bar* scales are 1000 μm in (**A**), 100 μm in (**B**, **C**), and 10 μm in (**D**).

humidity), and inoculated with fresh cultures of rhizobial strains A201, R201 or A3, originally isolated from *Sesbania cannabina* (Y Jing, unpubl.). Periodically a subset of the plants were uprooted and cleaned by running tap water. These fresh roots were examined directly by stereomicroscopy for the presence of abnormal hypertrophies. Root segments containing these structures were excised and processed in three ways for microscopic examination. Some root segments were cleared under vacuum with sodium hypochlorite (40% Chlorox solution), stained with methylene blue solution, and examined by brightfield light microscopy (Hollingsworth et al., 1989; Truchet et al., 1989). Other subsamples of root were fixed, dehydrated, critical-point dried, sputter coated, and examined by scanning electron microscopy (Umali-Garcia et al., 1980), or fixed, dehydrated, embedded in plastic, and thin-sectioned for combined light and transmission electron microscopy (Dazzo, 1982).

Molecular methods for strain identification

Rep-PCR-mediated genomic fingerprinting was used to identify and compare bacterial isolates used for rice root infections and bacteria reisolated from the rice root hypertrophies, as described (de Bruijn, 1992). Nodulation (*nodABC*) and nitrogen fixation (*nifHDK*) gene probes of *Rhizobium meliloti* (pRmSL42; kindly

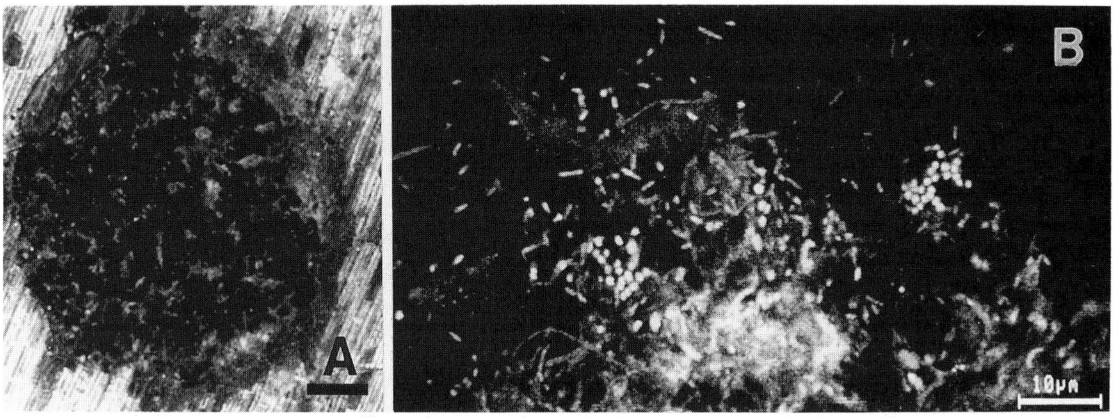

Fig. 6. Laser scanning micrographs of a small hemispherical structure attached to the rice root. (**A**) Top view by transmitted light microscopy. (**B**) Computer reconstructed composite "Z" section of a side view examined in the epifluorescence confocal mode. Shown are surface (top) and internal (below) colonization by bacteria which are stained with acridine orange. *Bar* scales are 50 μm in (A), 10 μm in (B).

provided by S Long) and *Azorhizobium caulinodans* (pRS2; Elmerich et al., 1982), respectively, were used in Southern hybridization experiments in order to verify the presence of the equivalent genes in the original or reisolated bacterial strains, as described by Pawlowski et al. (1987).

Results and discussion

Abnormal structures on rice roots grown in China

Examination of root segments by stereomicroscopy using combined transmitted/incident illumination revealed various brown, hemispherical structures attached to the root surface (Fig. 2A). The same roots viewed by darkfield stereomicroscopy also indicated the presence of several thin, transparent filaments (not root hairs) protruding externally from these dome-shaped structures, and various scattered, dark lesions of the root epidermis that appear to have undergone hypersensitive responses (Fig. 2B). Brightfield light microscopy of thin-sectioned tissue clearly indicated that these dome-shaped structures developed external to the root and did not involve a hypertrophy of the root epidermis, cortex, or endodermis itself (Fig. 3A). Further examination of these cross-sections at higher magnification showed that filamentous, septated cells indicative of fungal hyphae originating from the external dome-shaped structure did penetrate the otherwise intact root epidermis and the underlying root cortex (Fig. 3B).

Root longitudinal sections revealed more extensive colonization of the root cortex just beneath the external dome structures (Fig. 3C). These dome-shaped external structures disintegrated during attempts to clear them with dilute sodium hypochlorite solution. However, they remained intact when pealed off the root and were processed for plastic embedding. Thin-sections of the isolated, larger (presumably older) hemispherical structures revealed an external, acellular encrusted layer, an underlying extensive network of cells, and a central zone containing several large sacks each containing multiple, spherical, thick-walled cells resembling spores with prominent surface ornamentations and internal toluidine-blue staining granules (Fig. 4A,B).

An even clearer indication of the histological organization of these structures was obtained by scanning electron microscopy of their exposed cut surface. This method revealed in greater detail the attachment of these structures to the root, the outer encrusted protective layer, the internal network of filamentous hyphae, and the centrally located sacks containing multiple spherical spore-like cells with reticulate surface ornamentation (Fig. 5A-D). The presence of bacteria within small, hemispherical structures attached to the root was examined by staining with acridine orange solution followed by laser scanning microscopy in the epifluorescence confocal mode. Optical sections revealed a morphologically diverse community of bacteria distributed on the surface and within these hemispherical structures attached to the root (Fig. 6A,B).

It was not possible to match unequivocally the morphology of these hemispherical structures on rice roots

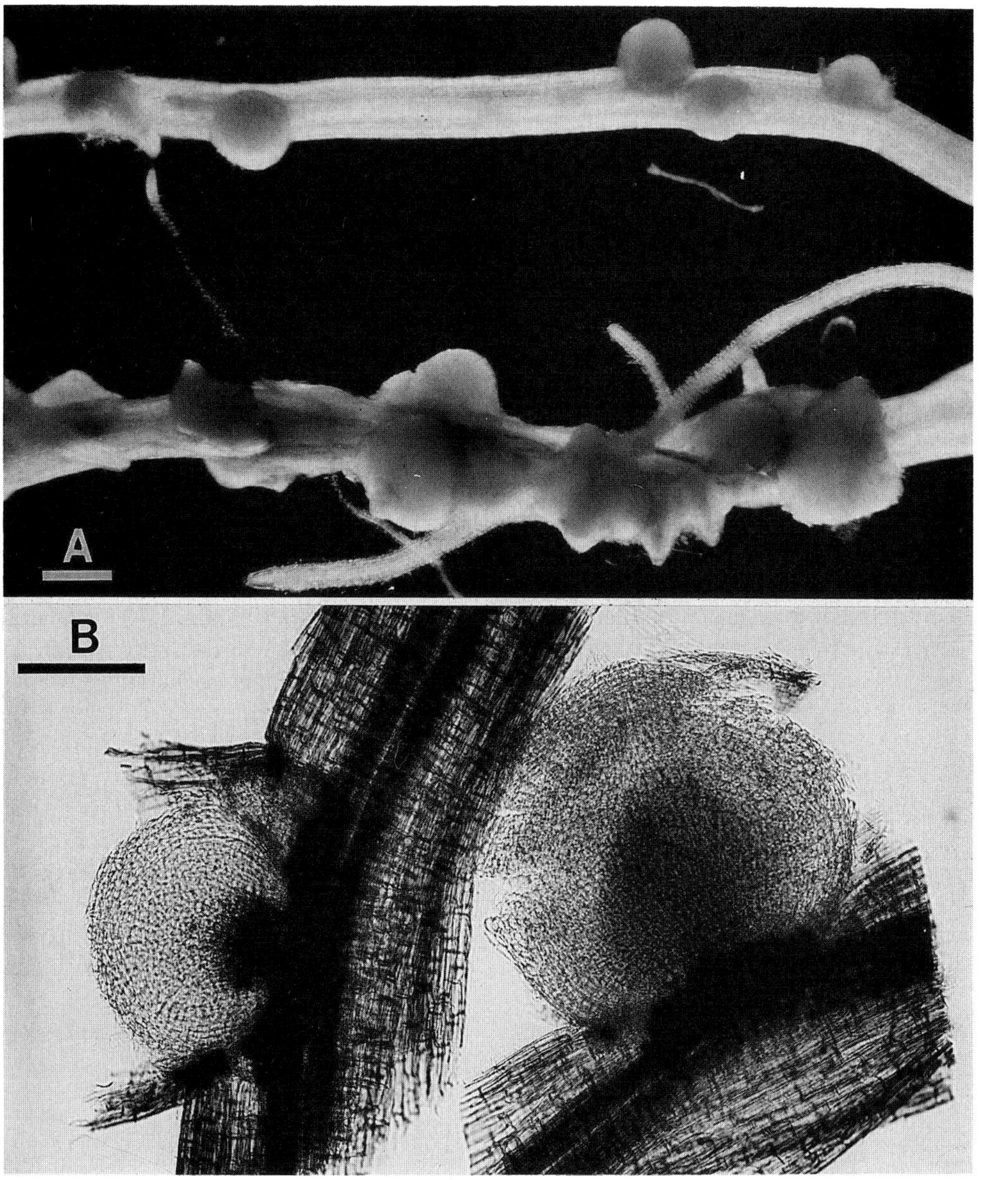

Fig. 7. Unusual modified lateral root meristems on rice viewed by (**A**) darkfield stereomicroscopy of fresh root tissue; (**B**) brightfield light microscopy after clearing in sodium hypochlorite and staining with methylene blue. *Bar* scales are 300 μm in (A), 100 μm in (B).

to any known fungus. Based upon interpretation of photomicrographs by three mycologists who specialize in soil fungi, the following possibilities were suggested: endogomes of chlamydospores of endomycorrhizal fungi; ascocarps of an ascomycete, and a chytrid. It remains a possibility that more than one fungus occupies these hemispheric structures on rice roots, and in addition, they are certainly colonized by a diversity of bacteria.

Abnormal structures on rice roots grown at Michigan State University

Three of the strains used by Li et al. (1991) were also used to infect rice plant roots at MSU. At a low frequency (0.1–0.5% of infected plants) different types of small protruding structures, often present in clusters, were observed on the rice roots. One structure resembling the hypertrophies found on the rice roots in Beijing was identified. A second class of structures

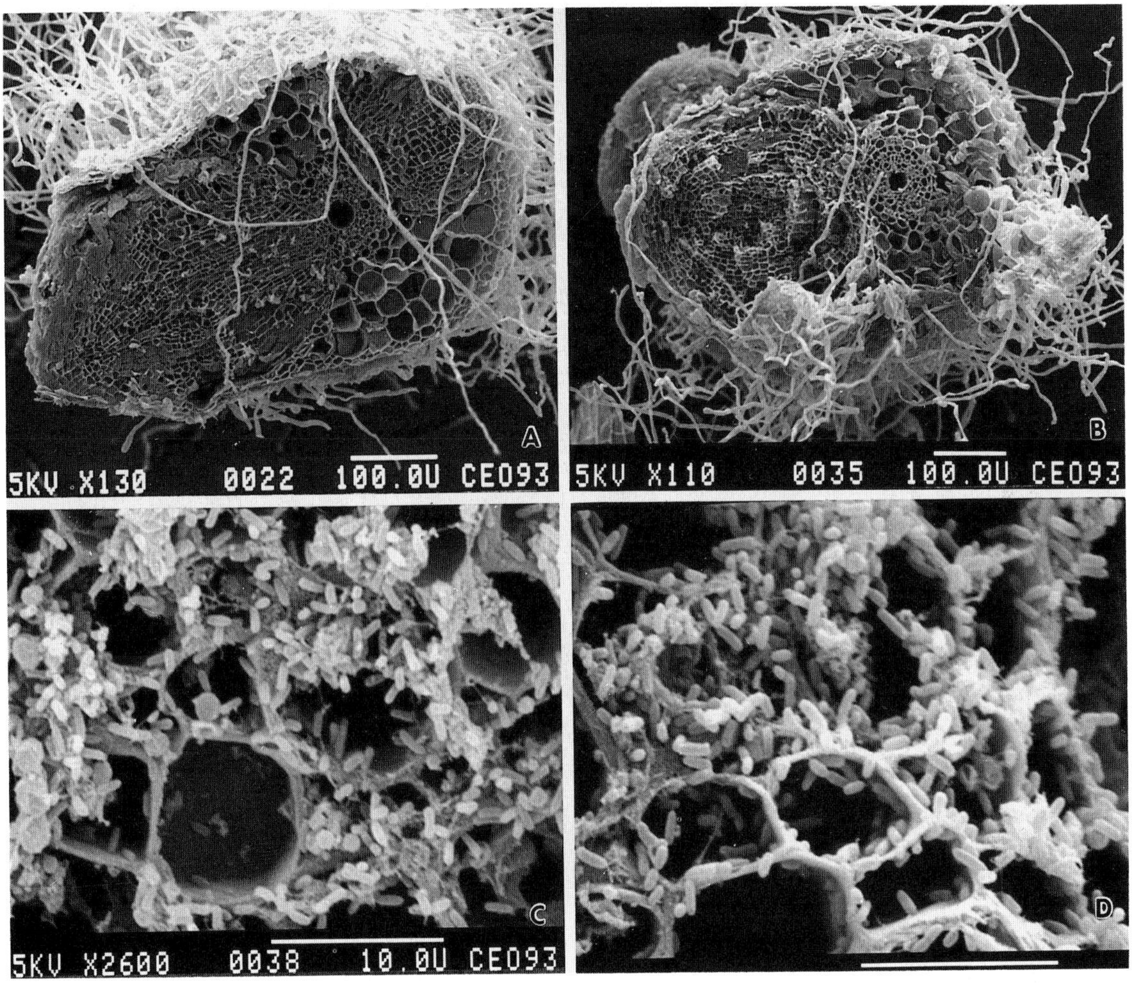

Fig. 8. Scanning electron micrographs of unusual modified lateral root meristems on rice roots. (**A**, **B**) Longitudinal sections of modified lateral meristems cut perpendicular to the parent rice root axis shown in cross-section. (**C**, **D**) Bacterial endorhizosphere colonization of the rice parent root and modified lateral root meristem. *Bar* scales are 100 μm in (A, B), 10 μm in (C, D).

was found to consist of stunted (modified) lateral roots, and these structures were found to be colonized by bacteria.

Stereomicroscopic examination of fresh roots infected by the strains described by Li et al. (1991) revealed infrequent clusters of hemispherical and dome-shaped "pseudonodules" protruding from the surface of thin, young rootlets (Fig. 7A). In contrast to the fungus colonized, hemispheric structures described above, the hemispheric structures illustrated in Figure 7A were derived from the rice root itself. Bright-field microscopy of cleared and stained root segments revealed that these rounded protrusions were short, thick, lateral root meristems with a central vascular system connected to the vascular system of the parent rootlet and an abnormal cortical hypertrophy (Fig. 7B). The histology of these modified lateral roots was confirmed by scanning electron microscopy of the longitudinal cut surface sectioned perpendicular to the parent rootlet axis (Fig. 8A,B).

Scanning electron microscopy at higher magnification revealed an extensive bacterial colonization of the rice endorhizosphere, extending into the vascular system of the parent rice root and these modified lateral roots (Fig. 8C,D). Combined light and transmission electron microscopy confirmed the presence of bacteria in association with these modified lateral root meristems, including their extensive coloniza-

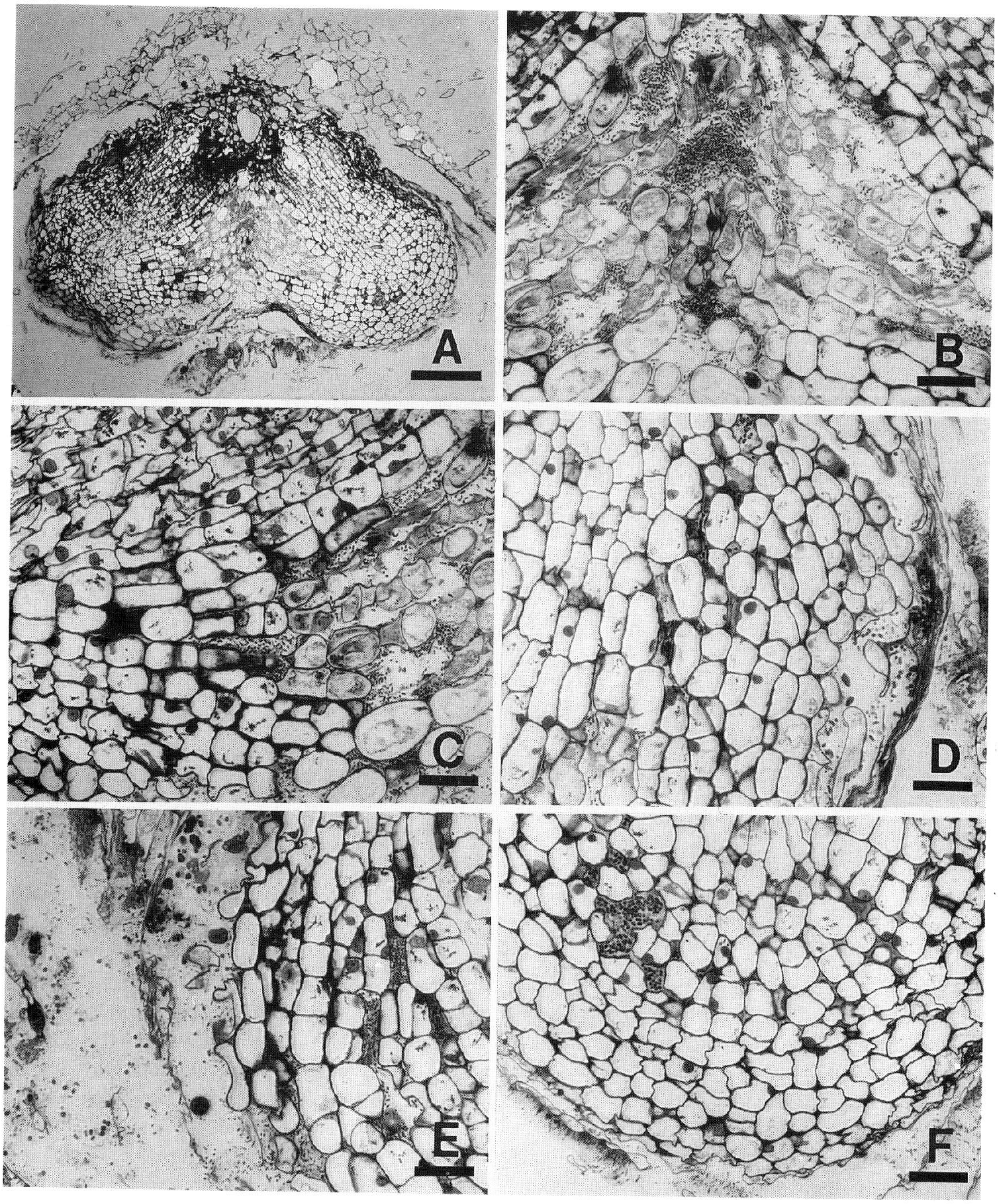

Fig. 9.

tion of the rhizoplane, entry through cracks created at emergence of these modified root structures, intercellular dissemination and colonization, and colonization of dead internal plant cells (Fig. 9A-J). These results relate to the finding of modified lateral roots of rice colonized by *Bradyrhizobium* from *Parasponia* (Cocking et al., 1992) and genetically engineered *R. leguminosarum* bv. trifolii (Rolfe et al., 1992). However,

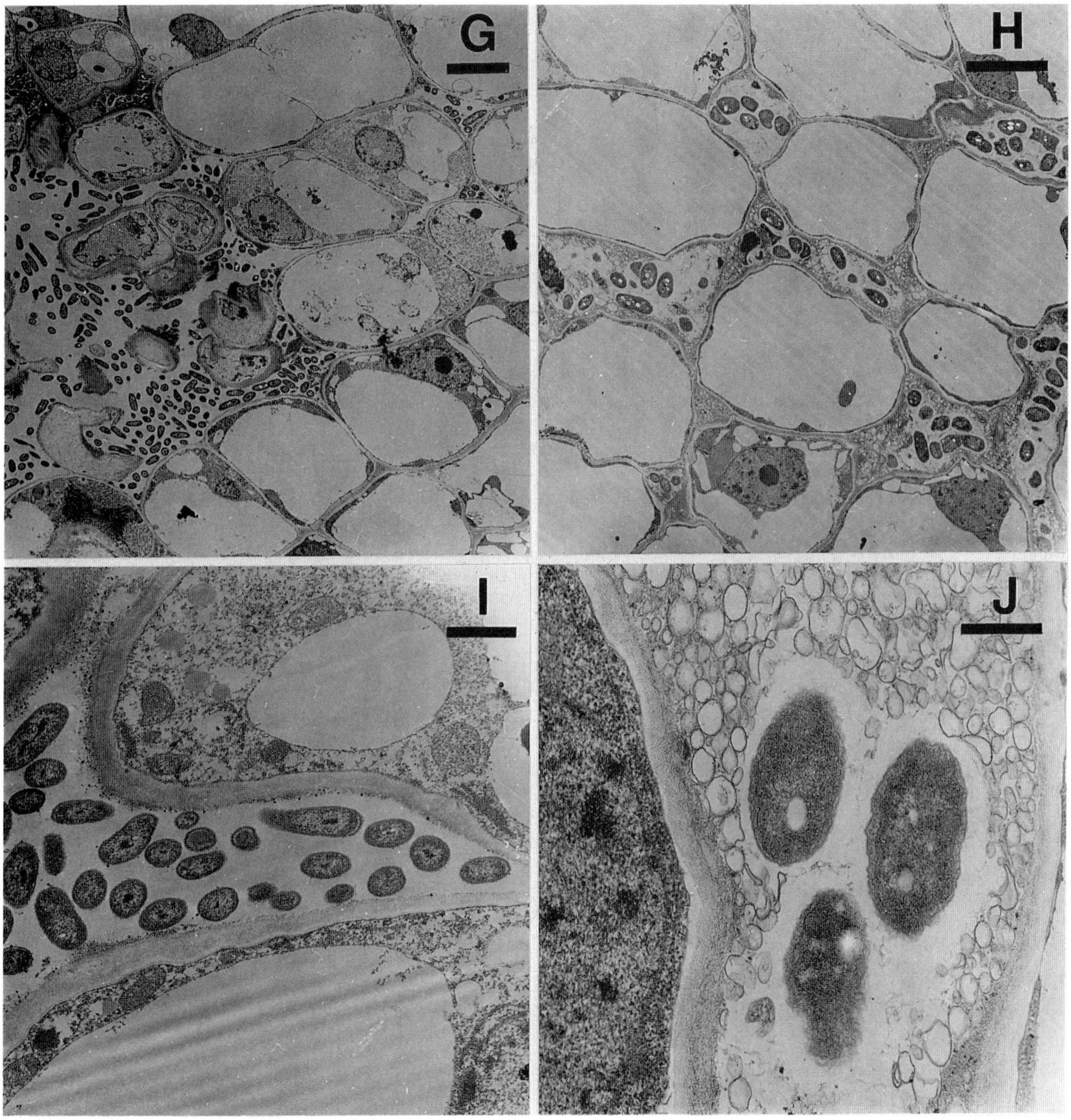

Fig. 9. Combined light microscopy (**A-F**) and transmission electron microscopy (**G-J**) of modified lateral root meristems on rice roots, showing bacterial colonization of the rhizoplane, crack entry, intercellular dissemination, and colonization of dead plant cells. *Bar* scales are 100 μm in (A), 20 μm in (B-F), 5 μm in (G-H), 1 μm in (I), and 0.5 μm in (J).

in contrast to the latter two reports, we have not yet found evidence of a bacterial endosymbiotic state (as would be suggested by membrane-enclosed bacteria within intact host cell cytoplasm). Further studies on plants grown under microbiologically controlled conditions will be necessary to address this issue. In any event, we can already conclude that modified lateral root meristems develop on healthy rice plants without added hormones and that these are invaded by a substantial population of endorhizosphere-colonizing bacteria without the host exhibiting obvious symptoms of disease.

A molecular genetic analysis of the bacterial strains used by Li et al. (1991) and bacteria reisolated from the MSU structures using rep-PCR genomic fingerprinting (de Bruijn, 1992) and hybridization with rhizobial nitrogen fixation (*nif*) and nodulation (*nod*) gene probes, was also initiated and revealed a number of discrepancies with the published results (Li et al., 1991).

For example, experiments with the *nodABC* probe revealed distinct hybridization patterns with the original rhizobial isolate from *Sesbania cannabina* (H18; Li et al., 1991), but the strains A201, A301, R101, and R20 (Y Jing, unpubl.), which were reisolated from nodule-like structures induced on rice roots in China (equivalent to isolates Rr1, Rr2; Li et al., 1991) failed to hybridize with this probe (data not shown). Similar results were obtained with the *nifHDK* probe (data not shown), suggesting that at least some of the reisolated bacterial strains are not (directly) derived from the parental strain, as suggested previously (Li et al., 1991). These results were also confirmed by the rep-PCR genomic fingerprinting studies, which showed that although some of the reisolates reported by Li et al. (1991) or isolated in subsequent experiments in China (Y Jing, unpubl. results) shared common fingerprinting patterns, in most cases they appeared to be unrelated to one another and to the parental *Rhizobium* strain from *Sesbania cannabina* (Maria Schneider, Y Jing and F J de Bruijn, unpubl. results).

Unfortunately, attempts to reisolate bacteria from the hypertrophies (nodule-like structures) generated at MSU were unsuccessful, due to a too severe surface sterilization protocol. Therefore, the relationship of the bacteria seen via the microscope in these structures (see above) to the strain used to inoculate the plants could not be established. These studies suggest that the phenomenon of colonization of rice root hypertrophies reported by Li et al. (1991) and here must be interpreted with extreme caution and may be more complex than initially proposed. More experimentation will be required to prove Koch's postulate and to make the process of rice root infection more reproducible.

In conclusion, although the formation of hypertrophies on rice roots that are infected/colonized by microbes has been documented, little evidence supports their designation as "nodules" (Li et al., 1991) or even nodule-like structures thus far. Moreover, the high frequency of their induction reported by Li et al. (1991) could not be reproduced, and the nature of the strains inducing the colonized hypertrophies and their relationship to the original strains remain unclear. Nevertheless, non-saprophytic colonization of rice root endorhizospheres can be observed in this system, providing clear incentive for further studies in this important area of research.

Acknowledgements

We thank Joann Palma, H Stuart Pankratz, and Joanne Whallon for technical assistance; Jon Stoltzfus for comments on the manuscript; Maria Schneider for technical assistance; Karen Bird for excellent editorial assistance; and J K Ladha, Gary Stacey, Ron Petersen, Barbara Hetrich and Amy Rossman for helpful discussions. The experimental work discussed here was supported by a grant from USAID (RN-5600-G-00-2073-00) to FJdB and Dr J K Ladha (IRRI, The Philippines).

References

Al-Mallah M K, Davey M R and Cocking E C 1989 Formation of nodular structures on rice seedlings by rhizobia. J. Exp. Bot. 40, 473–478.

Appleby C A 1984 Leghemoglobin and *Rhizobium* respiration. Annu. Rev. Plant Physiol. 35, 443–478.

Bender G L, Preston L, Barnard D and Rolfe B G 1990 Formation of nodule-like structures on the roots of the non-legumes rice and wheat. *In* Nitrogen Fixation: Achievements and Objectives. Eds. P M Gresshoff, L E Roth, G Stacey and W E Newton. p 825. Chapman and Hall, New York.

Boddey R M, de Oliveira O C, Urquiaga S, Resi V M, de Olivares F L, Baldani V L D and Döbereiner J 1995 Biological nitrogen fixation associated with sugar cane and rice: contributions and prospects for improvement. Plant and Soil 174.

Bogusz D, Appleby C A, Landsmann J, Dennis E S, Trinick N J and Peacock W J 1988 Functioning hemoglobin genes in non-nodulating plants. Nature 331, 178–180.

Chen T-W 1993 Nitrogen fixation of *Azorhizobium* in artificially induced root para-nodules in wheat. Sci. China 35, 1463–1469.

Chen T-W, Scherer S and Böger 1993 Nitrogen fixation of *Azorhizobium* in artificially induced root para-nodules in wheat. *In* Advances in Molecular Genetics of Plant-Microbe Interactions. Eds. E W Nester and D P S Verma. pp 593–606. Kluwer Academic Publishers, Dordrecht.

Christansen-Weniger C and Vanderleyden J 1994 Ammonium-excreting *Azospirillum* sp. become intracellulariy established in maize (*Zea mays*) para nodules. Biol. Fertil. Soils 17, 1–8.

Cocking E C and Davey M R 1991 Nitrogen from the air for non-legume crops. Chem., London 18 Nov, 831–835.

Cocking E C, Al-Mallah M K, Benson E and Davey M R 1990 Nodulation of non-legumes by rhizobia. *In* Nitrogen Fixation: Achievements and Objectives. Eds. P M Gresshoff, L E Roth, G Stacey and W E Newton. pp 813–823. Chapman and Hall, New York.

Cocking E, Srivastava J, Kothari S and Davey M 1992 Invasion of nonlegume plants by diazotrophic bacteria. *In* Nodulation and Nitrogen Fixation in Rice: Potentials and Prospects. Eds. G Khush and J Bennett. pp 119–121. IRRI, Manila.

David M, Daveran M-L, Batut J, Dedieu A, Domergue O, Ghai J, Hertig C, Boistard P and Kahn D 1988 Cascade regulation of *nif* gene expression in *Rhizobium meliloti*. Cell 54, 671–683.

Dazzo F 1982 Leguminous root nodules. *In* Experimental Microbial Ecology. Eds. R Burns and J Slater. pp 431–466. Blackwell Scientic Publications, Oxford, UK.

Dazzo F, Mateos P, Orgambide G, Philip-Hollingsworth S, Squartini A, Subba-Rao N, Pankratz H S, Baker D, Hollingsworth R and Whallon J 1993 The infection process in the *Rhizobium*-legume symbiosis and visualization of rhizoplane microorganisms by laser scanning confocal microscopy. *In* Trends in Microbial Ecology. Eds. R Guerrero and C Pedros-Alio. pp 259–262. Spanish Society for Microbiology, Barcelona.

De Bruijn F J 1992 Use of repetitive (repetitive extragenic palindromic and enterobacterial repetitive intergenic consensus) sequences and the polymerase chain reaction to fingerprint the genomes of *Rhizobium meliloti* isolates and other soil bacteria. Appl. Environ. Microbiol. 58, 2180–2187.

De Bruijn F J, Chen R, Dehio C, Goel A, Reiser S and Szczyglowski K 1992 Hormones and nodule formation: cytokinin induction of the *Sesbania rostrata* early nodulin gene *Enod2*. *In* Nodulation and Nitrogen Fixation in Rice: Potential and Prospects. Eds. G S Khush and J Bennett. pp 33–40. IRRI, Manila.

De Bruijn F J, Chen R, Fujimoto S Y, Pinaev A, Silver D and Szczyglowski K 1994 Regulation of nodulin gene expression. Plant and Soil 161, 59–68.

De Bruijn F J and Downie J A 1991 Biochemical and molecular studies of symbiotic nitrogen fixation. Curr. Op. Biotech. 2, 184–192.

De Bruijn F J and Schell J 1992 Regulation of plant genes specifically induced in developing and mature nitrogen-fixing nodules: *Cis*-acting elements and *trans*-acting factors. *In* Control of Plant Gene Expression. Ed. D P S Verma. pp 241–258. CRC Press, Boca Raton, Florida.

Dehio C and de Bruijn F J 1992 The early nodulin gene *SrEnod2* from *Sesbania rostrata* is inducible by cytokinin. Plant J. 2, 117–128.

Dénarié J and Cullimore J 1993 Lipo-oligosaccharide nodulation. factors: a new class of signaling molecules mediating recognition and morphogenesis. Cell 74, 951–954.

Döbereiner J, Reis V M, Paula M A and Olivares F 1993 Endophytic diazotrophs in sugar cane, cereals and tuber plants. *In* New Horizons in Nitrogen Fixation. Eds. R Palacios, J Mora and W E Newton. pp 671–676. Kluwer Acad. Publ., Dordrecht, Netherlands.

Earl C D and Ausubel F M 1983 The genetic engineering of nitrogen fixation. Nutr. Rev. 41, 1–6.

Elmerich C, Dreyfus B, Reysset G and Aubert J P 1982 Genetic analysis of nitrogen fixation in a tropical fast-growing Rhizobium. EMBO J. 1, 499–503.

Fischer H-M 1994 Genetic regulation of nitrogen fxation in rhizobia. Microbiol. Rev. 58, 352–386.

Fisher R F and Long S R 1992 *Rhizobium*-plant signal exchange. Nature 357, 655–660.

Gilles-Gonzalez M A, Ditta G S and Helinski D R 1991 A haemoprotein with kinase activity encoded by the oxygen sensor of *Rhizobium meliloti*. Nature 350, 170–172.

Hirsch A M 1992 Tansley review no. 40: Developmental biology of legume nodulation. New Phytol. 122, 211–237.

Hollingsworth R, Squartini A, Philip-Hollingsworth S and Dazzo F 1989 Root hair deforming and nodule initiating factors from *Rhizobium trifolii*. *In* Signal Molecules in Plants and Plant-Microbe Interactions. Ed. B Lugtenberg. pp 387–393. Springer-Verlag, Berlin.

Horvath B, Heidstra R, Lados M, Moerman M, Spaink H P, Promé J-C, van Kammen A and Bisseling T 1993 Lipo-oligosaccharides of *Rhizobium* induce infection-related early nodulin gene expression in pea root hairs. Plant J. 4, 727–733.

Hurek T, Reinhold-Hurek B, van Montagu M and Kellenberger E 1994 Root colonization and systemic spreading of *Azoarcus* sp. strain BH72 in grasses. J. Bacteriol. 176, 1913–1923.

Jing Y, Li G, Jin G, Shan X, Zhang B, Guan C and Li J 1990a Rice root nodules with acetylene reduction activity. *In* Nitrogen Fixation: Achievements and Objectives. Eds. P M Gresshoff, L E Roth, G Stacey and W E Newton. p 829. Chapman and Hall, New York.

Jing Y, Li G and Shan X 1992 Development of nodulelike structure on rice roots. *In* Nodulation and Nitrogen Fixation in Rice: Potentials and prospects Eds. G Khush and J Bennett pp 123–126. IRRI, Manila, Phillippines.

Jing Y, Zhang B T and Shan X Q 1990b Pseudonodules formation on barley roots induced by *Rhizobium astragali*. FEMS Microbiol. Lett. 69, 123–128.

Kennedy I R and Tchan Y-T 1992 Biological nitrogen fixation in non-leguminous field crops: Recent advances. Plant and Soil 141, 93–118.

Khush G S and Bennett J (Eds.) 1992 Nodulation and Nitrogen Fixation in Rice: Potential and Prospects. IRRI Press, Manila,

Ladha J K, Trol-Padre A, Reddy K and Ventura W 1993 Prospects and problems of biological nitrogen fixation in rice production: A critical assessment. *In* New Horizons in Nitrogen Fixation. Eds. R Palacios et al. pp. 677–682. Kluwer Academic Publishers, Dordrecht.

Lauridsen P, Franssen H, Stougaard J, Bisseling T and Marcker K A 1993 Conserved regulation of the soybean early nodulin *Enod2* gene promoter in determinate and indeterminate transgenic root nodules. Plant J. 3, 483–492.

Li G, Jing Y, Shan X, Wang H and Guan C 1991 Identification of rice nodules that contain *Rhizobium* bacteria. Chin. J. Bot. 3, 8–17.

Orgambide G G, Hollingsworth R I and Dazzo F B 1992 Structural characterization of a novel diglycosyl diacylglyceride glycolipid from *Rhizobium trifolii* ANU 843. Carbohydr. Res, 233, 151–159.

Orgambide G G, Philip-Hollingsworth S, Hollingsworth R I and Dazzo F B 1994 Flavone-enhanced accumulation and symbiosis-related biological activity of a diglycosyl diacylglycerol membrane glycolipid from *Rhizobium leguminosarum* biovar *trifolii*. J. Bacteriol. 176, 4338–4347.

Pawlowski K, Ratet P, Schell J and de Bruijn F J 1987 Cloning and characterization of *nifA* and *ntrC* genes of the stem nodulating bacterium ORS571, the nitrogen fixing symbiont of *Sesbania rostrata*: Regulation of nitrogen fixation (*nif*) genes in the free living versus symbiotic state. Mol. Gen. Genet. 206, 207–219.

Peoples M B, Herridge D F and Ladha J K 1995 Biological nitrogen fixation: an efficient source of nitrogen for sustainable agriculture? Plant and Soil 174.

Peters N K and Verma D P S 1990 Phenolic compounds as regulators of gene expression in plant-microbe interactions. Mol. Plant-Micr. Interact. 3, 4–8.

Phillips D A 1992 Flavonoids: Plant signals to soil microbes. *In* Phenolic Metabolism in Plants. Ed. H A Stafford. pp 201–231. Plenum Press, New York, USA.

Phillips D A, Dakora F D, Sande E, Joseph C M and Zon J 1994 Synthesis, release, and transmission of alfalfa signals to rhizobial symbionts. Plant and Soil 161, 69–80.

Quispel A 1992 A search for signals in endophytic microorganisms. *In* Molecular Signals in Plant-Microbe Communications. Ed. D P S Verma. pp 475–491. CRC Press, Boca Raton, FL.

Ridge R W, Bender G L and Rolfe B G 1992 Nodule-like structures induced on the roots of wheat seedlings by the addition of the synthetic auxin 2,4-dichlorophenoxyacetic acid and the effects of microorganisms. Aust. J. Plant Physiol. 19, 481–492.

Rolfe B G and Bender G L 1990 Evolving a *Rhizobium* for non-legume nodulation. *In* Nitrogen Fixation: Achievements and Objectives. Eds. P M Gresshoff, L E Roth, G Stacey and V1r E Newton. pp 829. Chapman and Hall, New York.

Rolfe B, Ride K and Ridge R 1992 *Rhizobium* nodulation of non-legumes. *In* Nodulution and Nitrogen Fixation in Rice: Potentials and Prospects. Eds. G Khush and J Bennett. pp 83–86. IRRI, Manila.

Schubert K R 1986 Products of biological nitrogen fixation in higher plants: synthesis, transport, and metabolism. Ann. Rev. Plant Physiol. 37, 539–574.

Simon Moffatt A 1990 Nitrogen-fixing bacteria find new partners. Science 250, 910–912.

Szczyglowski K, Szabados L, Fujimoto S Y, Silver D and de Bruijn F J 1994 Site-specific mutagenesis of the nodule-infected-cell expression (NICE) element and the AT-rich element (ATRE-BS2*) of the *Sesbania rostrata* leghemoglobin *glb3* promoter. Plant Cell 6, 317–332.

Tchan Y T, Zeman A M M and Kennedy I R 1991 Nitrogen fixation in para-nodules of wheat roots by introduced free-living diazotrophs. Plant and Soil 137, 43–47.

Truchet G Camut S, DeBill F, Ororico R and Vasse J 1989 The *Rhizobium*-legume symbiosis: two methods to discriminate between nodules and other root derived structures. Protoplasma 149, 82–88.

Umali-Garcia M, Hubbell D Gaskins M and Dazzo F 1980 Association of *Azospirillum* with grass roots. Appl. Environ. Microbiol. 39, 219–226.

Van Kammen A 1984 Suggested nomenclature for plant genes involved in nodulation and symbiosis. Plant Mol. Biol. Rep. 2, 43–45.

Vijn L, das Neves L, van Kammen A, Franssen H and Bisseling T 1993 Nod factors and nodulation in plants. Science 260, 1764–1765.

Welters P, Metz B, Felix G, Palme K, Szczyglowski K and de Bruijn F J 1993 Interaction of a rhizobiai DNA-binding protein with the promoter region of a plant leghemoglobin gene. Plant Physiol. 102, 1095–1107.

Yu D, Kennedy I R and Tchan Y T 1993 Verification of nitrogenase activity (C_2H_2 reduction) in *Azospirillum* populated, 2,4-dichlorophenoxyacetic acid induced root structures of wheat. Aust. J. Plant Physiol. 29, 187–195.

Plant and Soil **174:** 241–253, 1995.

New techniques for studying competition by Rhizobia and for assessing nitrogen fixation in the field

Kate J. Wilson, Mark B. Peoples and Richard A. Jefferson
CAMBIA (Center for the Application of Molecular Biology to International Agriculture), GPO Box 3200, Canberra, ACT 2601, Australia and CSIRO Division of Plant Industry, GPO Box 1600, Canberra, ACT 2601, Australia

Key words: bioindicators, legumes, microbial ecology, molecular tools, N_2 fixation, *Rhizobium*

Table of contents

Abstract
Introduction
Reporter genes: The fundamental tool
Choice of reporter genes
Microbial ecology: detection of inoculant strains using simple colour coding
Bioindicators: reporter genes to measure N_2 fixation efficiency in the field
The ureide technique: present use and future potential
Use of bioindicators to measure N_2 fixation efficiency
Engineering response of reporter genes to reflect levels of nitrogenous compounds
  a) Use of bacterial reporters
  b) Use of eukaryotic systems
Conclusions
References

Abstract

One of the key factors limiting the proper assessment and use of rhizobial strains in the field is the lack of suitable methodology to screen the success of individual isolates in competing for nodule occupancy with different cultivars of legumes and in different soil and agronomic conditions. The use of marker genes enables individual rhizobial strains to be identified by a simple colour assay, thus enabling a dramatic increase in throughput of strain screening. One such marker system for rhizobial ecology, the GUS system, is already in use to facilitate rapid screening of rhizobial isolates. Other markers, which will allow the competitive behaviour of several strains to be studied at once, are under development.

Likewise, breeding of the host legume for a high efficiency of nitrogen fixation is hampered by the difficulty in assessing this property. The method which currently gives the highest throughput of analysis, and has been successfully used in soybean breeding programs, is the ureide technique. However, it remains somewhat laborious for use in routine breeding programs. In this paper we discuss the potential use of reporter genes to provide information on the relative levels of ureides and other nitrogenous compounds in plants growing in the field. This would greatly increase the rate at which this trait could be scored, and would thus enable routine assays for increased symbiotic nitrogen fixation for breeding or management purposes in legume crops such as soybean (*Glycine max*) and common bean (*Phaseolus vulgaris*).

Introduction

Everyone who has worked in the field of biological N_2 fixation (BNF) knows what a dramatic effect it can have on plant growth. We have all carried out controlled experiments in the glasshouse and demonstrated the "big plant, little plant" phenomenon — luxuriant healthy green plants that have been inoculated with a compatible *Rhizobium* strain contrasting with slight, pale green plants that are starved of nitrogen (N).

The problem is transferring such spectacular results to the field. There are at least two limitations. Firstly conditions in the field are never entirely devoid of N, and secondly there are usually also some compatible rhizobia present; hence, the big plant-little plant difference is rarely apparent. Nevertheless, even where no major yield increases occur as a result of BNF, some plants may be obtaining more of their N from symbiotic N_2 fixation than their neighbours. In fact, an equally important challenge may be not to increase overall N in the crop dramatically, but to increase the proportion of N that comes from BNF, either through breeding (Herridge and Danso, 1994) or through effective management strategies (Peoples et al., 1995b).

This is difficult to achieve however, because of limitations in methodology. It is easy to measure N_2 fixation in the glasshouse because it is reflected very directly in differences in shoot weight — the N-difference method. Likewise, because we can use single strain inocula and there are no competing rhizobia in vermiculite, sand or perlite, we can determine the fixation efficiency of individual rhizobial strains with ease. The problem comes in transferring these measurements to the field. How do we assess the effect of an inoculant strain in a background of competing rhizobia? How do we measure the proportion of N derived from fixation in a plant that may have no more total N content than its poorly fixing neighbour?

In this paper we discuss novel methodologies based on the use of reporter genes that will greatly expand our ease of measuring these factors in the field and will help us to make better use of this free and renewable source of N in agriculture.

Reporter genes: The fundamental tool

A reporter gene is exactly that — a gene which reports on some biological phenomenon that the researcher wants to gain information about. The original concept was developed in the early 1960s to facilitate the study of gene regulation, and is a very powerful and pervasive one.

Most gene products — proteins — are very difficult to detect or assay. Hence, study of the regulation of a gene — determining when, and in response to which developmental and/or environmental signals it makes its product — is very tedious. The trick underlying the use of reporter genes is to substitute the part of the DNA that codes for the difficult-to-assay protein, with a piece of DNA that specifies a protein which is easy to detect (Fig. 1). This gene is the reporter gene. It almost invariably codes for an enzyme which can be assayed in a simple, standardized, spectrophotometric, fluorometric or colour assay.

This use of reporter genes is possible because the region of DNA coding for the protein is (by and large) physically distinct from that which regulates activity of the gene. Thus, substituting one "coding region" with another (the reporter gene) has little impact on the regulation of the gene. It is then possible to monitor expression of the reporter gene using well defined procedures, and be confident that it will only be produced in response to the same developmental and environmental triggers as the original gene.

Reporter genes were developed by molecular biologists and have primarily been used by them to minutely examine the DNA sequences that determine the when, where, how and why of gene expression. However, because the synthesis of many specific gene products

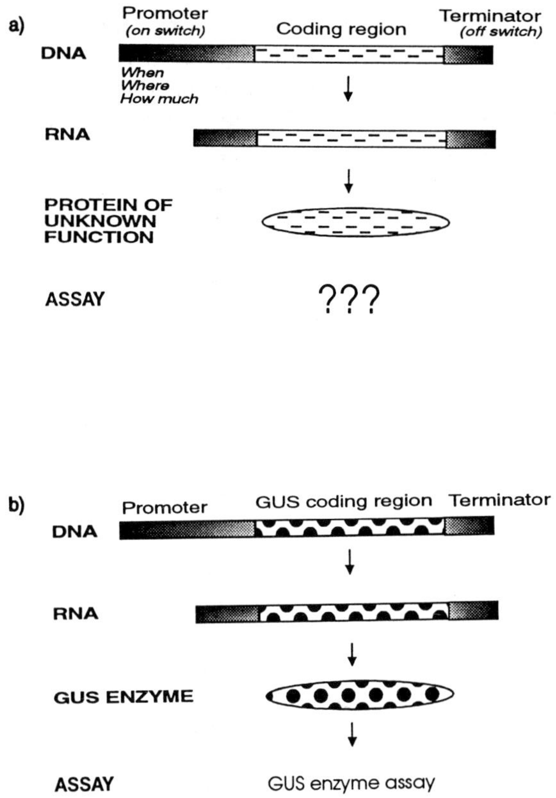

Fig. 1. Structure and functioning of a reporter gene. **a**) represents the synthesis of RNA and protein from a gene of unknown function. It is therefore difficult to assay the resulting protein. In **b**), the coding region of the unknown gene has been substituted with the GUS coding region — the *gusA* reporter gene. This is still regulated in the same way as the original gene, but the assay showing expression of the gene becomes a routine GUS assay.

is known to be responsive to specific environmental signals, we can turn this concept on its head, and use production of the reporter gene as a beacon indicating when a bacterium or a plant has received a particular environmental signal. The reporter genes are then being used as bioindicators.

Bioindicators are already used routinely in diagnostic tests. For example, in mammalian physiology, pregnancy tests rely on detection of the hormone chorionic gonadotropin through an enzyme-linked immunoassay. In effect the hormone is the bioindicator of pregnancy and its presence is being detected through a combination of antigenic detection with an enzyme assay. A reporter gene can be used in a way that is highly analogous — the enzyme is reporting on the production of a protein which is diagnostic of a particular physiological state. The difference is that, rather than having to be coupled to a immunological assay, the reporter enzyme encoded by a reporter gene is produced directly by the organism at the same time and place as other molecules that reflect a specific physiological state.

There are already some examples where a reporter gene has been used to measure an environmental parameter. For example, Heitzer et al. (1992) engineered bacteria which luminesce in response to naphthalene and salicylate bioavailability, and Reches et al. (1994) have developed reporter bacteria to monitor levels of the trace element selenium. Later in this paper we will discuss how reporter genes could be used to report on the efficiency of N_2 fixation.

Choice of reporter genes

The key requirement for an effective reporter gene is that it is easy to assay. Thus most reporter genes encode enzymes with straightforward, quantitative, assays. In the examples being discussed in this paper, the constraints imposed on detection of the reporter gene are quite severe, as we wish to be able to detect their activity at the very least in a simple laboratory which might be available on an agricultural research station, and ideally directly in plants growing in the field. Additionally, the assay must he extremely high throughput and must be inexpensive.

The primary reporter gene used in plant molecular biology at present is *gusA* encoding the enzyme β-glucuronidase or GUS (Jefferson et al., 1987). GUS is a hydrolase which will cleave a very wide range of substrates — almost any aglycone conjugated to D-glucuronic acid in the β configuration. Thus a variety of histochemical and spectrophotometric assays are available. The greatest advantage of GUS is the nearly complete lack of endogenous activity in plants and most agriculturally important organisms, coupled with the simple assays for its activity with spatially restricted chromogenic substrates — leaving insoluble coloured compounds at the site of enzyme activity (Jefferson and Wilson, 1991).

At present, GUS assays must be carried out in a destructive manner, as cells must be permeabilized with destructive detergents to allow entry of hydrophilic GUS substrates. However, modifications are underway to convert GUS to a secreted enzyme such that exogenously applied substrates can gain free access to the enzyme trapped between the cell membrane and the cell wall. This leads in principle to the possibility that enzyme activity could be detected by the simple expedient of painting substrate onto the leaf blade and looking for an easily visible colour change.

A second reporter gene which is under development is another hydrolytic enzyme, an arylsulfatase (ARS), which cleaves a sulfate residue from almost any sulfate-substituted organic ring compound (an arylsulfate). ARS is produced by a wide variety of animals and microbes, but has not been detected in plants, and thus will serve as a second reporter gene with no background activity in plants. Most importantly, the enzyme appears to be naturally secreted and localized to the cell wall in yeast strains which we have isolated. Addition of substrate to cells that express ARS does not kill or inhibit growth of the cells, presumably because the substrates do not have to penetrate the living cell and its membrane system to come in contact with the enzyme. Thus, once the gene is isolated and characterized, this will provide a second reporter system that will allow in vivo detection of the enzyme.

As well as using secreted reporter enzymes to detect reporter gene activity in living cells, another approach is to use reporter enzymes which act on endogenous substrates within the cell to give rise to a coloured — and hence easily visible — product. A potential source of such genes is from bacteria which are naturally pigmented. For example, a single gene has been isolated from *Rhodococcus* which confers production of the pigment indigo on the laboratory bacterium *Escherichia coli* using the endogenous substrate indole (created from tryptophan by *E. coli* tryptophanase; Hart et al., 1992). Another bacterial pigment system with potential as a reporter system is the violacein operon from *Chromobacterium violaceum* (Pemberton et al., 1991). A set of six genes has been identified which confers production of violacein on *E. coli* from the substrates tryptophan and oxygen. The role of each gene is not yet known, but subsets of the genes will give rise to different coloured compounds, and a single gene from this pathway can lead to production of an unknown blue pigment in *E. coli* (D Quiggin, P Keese and L Graf, pers. commun.).

The relative advantages of both approaches have yet to be tested. The advantage of using a secreted reporter enzyme which requires exogenous substrate is that the plants are largely unaffected, assuming that there are no natural substrates for the enzymes (which is the case for GUS and arylsulfatase). On the other hand, the assay still requires some kind of action to be performed on each plant. The use of reporter genes which act on endogenous compounds means that a plant — or plant parts — will naturally change colour in response to the chosen physiological or environmental parameter. This allows for extremely inexpensive and rapid screening, but has the disadvantage that the assay could be cumulative — depending on how stable the coloured product is. It is also possible that metabolism of endogenous substrates by the reporter gene will have an adverse effect on the plant. Nevertheless, this type of reporter system could be invaluable in designing sacrificial "sentinel" plants which are scattered throughout the field and report visually on particular physiological stresses.

Having introduced the concept of reporter genes and their use to report on physiological and environmental assays, we will now discuss one example — rhizobial ecology — where they are already in use — and subsequently the possibility of using reporter genes to enhance the ureide assay for efficiency of N_2 fixation.

Microbial ecology: detection of inoculant strains using simple colour coding

One of the key limiting factors in ecological analysis of field performance of inoculated rhizobial strains is the difficulty in distinguishing the introduced strain from indigenous rhizobia. Current methodologies, including various antigenic techniques and the use of antibiotic resistance markers, all require individual analysis of single nodules and hence are labour intensive and limit the amount of data that can be gathered. This is true also of other molecular biological approaches that determine nodule occupancy by analysis of the DNA present in a nodule. This labour intensity means that sample sizes analysed are generally very small — typically 20-30 nodules per plot (e.g. Somasegaran and Hoben, 1985) in a plot that might contain 10^3–10^4 nodules in total, and, coupled with the fact that there is known horizontal (e.g. Wollum and Cassel, 1984) and vertical (e.g. McDermott and Graham, 1989) variation in the profile of strains occupying nodules, this means that obtaining highly accurate values for nodule occupancy is extremely difficult (see Wilson, 1995) Perhaps as importantly, the laboriousness of current assays means that there is no routine screening procedure to assess how successful a given highly effective strain would be in diverse soil types or with novel cultivars of a host plant.

We have developed a system which uses the GUS reporter system to report on the presence or absence of a specific strain in a nodule (Wilson et al., 1991;

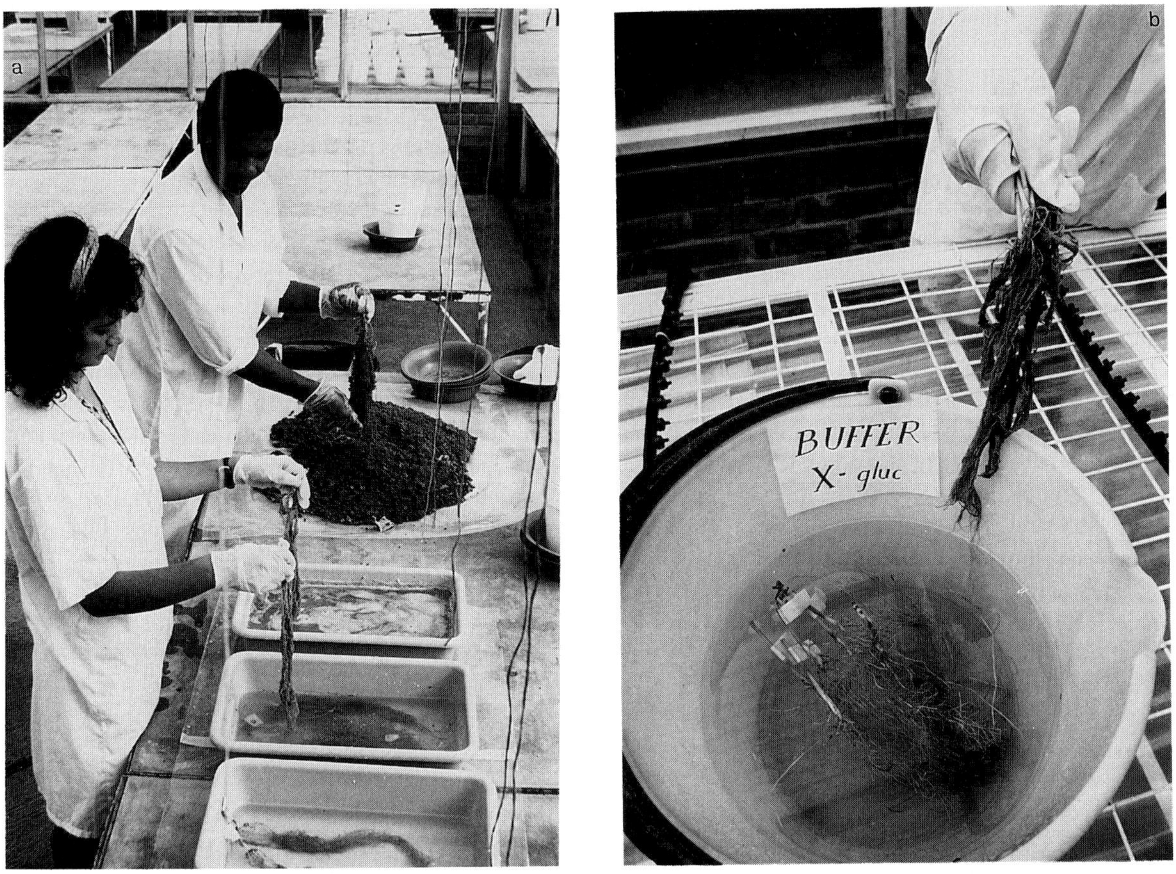

Fig. 2. Reporter gene assays in a bucket. Roots from soil-grown *Phaseolus vulgaris* plants inoculated with GUS-marked strains of *Rhizobium etli* are first washed (**a**) to remove excess soil, then assayed directly in X-Gluc buffer in a bucket (**b**) to determine nodule occupany by the inoculant strain at CIAT in Colombia (photograph courtesy of Doug Beck).

Wilson, 1995) Here again, the reporter gene is being asked to report visually on something that is otherwise difficult to detect and assay.

First a rhizobial strain of interest is marked with GUS in a simple bacterial mating procedure which requires only routine microbiological equipment. The assay for nodule occupancy then becomes trivial: the whole root system of an inoculated plant is incubated in buffer containing an indigogenic GUS substrate and nodules occupied by the marked rhizobia are detected by virtue of a simple colour change There are now a number of indigogenic substrates that give rise to differently coloured products These include: 5-bromo-4-chloro-3-indolyl glucuronide (X-GlcA or X-Gluc) which gives rise to a blue coloured precipitate on cleavage by GUS; 5-bromo-6-chloro-3-indolyl glucuronide (Magenta-GlcA), giving rise to a magenta colour; and 6-chloro-3-indolyl glucuronide (Salmon-GlcA) giving rise to a salmon colour. Thus nodules occupied by a GUS-marked strain can be detected as blue, magenta or pink-coloured (see Figure 1 in Wilson, 1995; and front cover of this volume).

As there is no endogenous GUS activity in either the plant or in any rhizobial strain tested so far, development of the characteristic colour is completely diagnostic of nodule occupancy by the inoculant strain. Moreover the assay is so straightforward that at CIAT (Centro Internacional de Agricultura Tropical) in Colombia, this assay is carried out by immersing whole root systems in buckets containing the GUS substrate X-GlcA (Fig. 2). The use of buckets as a reaction vessel for a "molecular" reaction is clearly novel, and illustrates the extent to which tools based on molecular biology can really be taken to field-level, large-scale analysis. This assay greatly increases the number of nodules that can be analysed for nodule occupancy — it now becomes logistically feasible to analyse all the nodules on all the plants sampled from a plot — and also retains information regarding the positional location of nodules induced by the inoculant strain along the root.

The assay has so far been used with *Bradyrhizobium* strains that nodulate cowpea (*Vigna unguiculata* (Wilson et al., 1991)) and soybean (*Glycine max*) (G Hardarson, pers. commun.), and with *Rhizobium* sp. inoculated onto siratro (*Macroptilium atropurpureum*) and pigeonpea (*Cajanus cajan*) (Wilson, 1995; Wilson et al., 1995) and various *R. tropici* (Streit et al., 1992; Sessitsch et al., 1995) and *R. etli* strains (Streit et al., 1995) on *Phaseolus vulgaris*. Examination of the competitive ability of marked strains indicates that derivatives can rapidly be identified with competitive ability that differs little from the wild-type (Sessitsch et al., 1995; Streit et al., 1995), and that these marked strains then allow rapid, and highly reproducible competition assays. Enhancements underway for the near future include the development of additional marker genes that will give rise to differently coloured products; these will allow multi-strain competition trials by such straightforward assays, and will also allow ready assessment of dual strain occupancy of nodules.

As well as being used for rhizobial nodule occupancy studies, marker genes can be used for studies of rhizosphere colonization. For example, the GUS system allows detection of some of the earliest stages of root colonization and infection (de Boer and Djordjevic, 1995). It has recently been used in this manner to monitor infection of roots by the fungus *Fusarium oxysporum* (Couteaudier et al., 1993) and of tomatoes (*Lycopersicon esculentum*) by the fungus *Cladosporium fulvum* (Oliver et al., 1993) and to study colonization of maize (*Zea mays*) para-nodules by an *Azospirillum* species (Christiansen-Weniger and Vanderleyden, 1993). In an analogous manner, the *lacZ* gene encoding β-galactosidase was recently used to monitor colonization of wheat (*Triticum aestivum*) roots by *A. lipoferum* — although here it is necessary to first treat the roots so that the endogenous β-galactosidase activity in the plant is eliminated (Katupitiya et al., 1995).

Perhaps the most powerful demonstration of the practical utility of the marker gene approach to date is the use of GUS-marked *Cladosporium fulvum* as a bioassay to rapidly screen a mutagenized population derived from a fungus-resistant tomato line, and to readily identify susceptible derivatives (Hammond-Kosack et al., 1994). These examples all demonstrate that there is also great potential for using GUS and other marker genes in studies of associative N_2 fixation between diazotrophs and non-legumes such as cereals or C4 grasses (Boddey et al., 1995).

Bioindicators: reporter genes to measure N_2 fixation efficiency in the field

The use of legume plants in cropping systems can serve to greatly reduce the need for application of chemical nitrogenous fertilizers. The potential agronomic, economic and environmental benefits are great, particularly in less developed countries where economic constraints are often the overwhelming factor — if a farmer simply cannot afford to buy chemical fertilizers, inclusion of legumes in the cropping system may be the only possibility for maintaining, let alone increasing, soil N fertility. However, no legume plant derives 100% of its N from atmospheric N_2 unless it is grown in a totally N-free environment in the greenhouse; rather a proportion varying from 0 to 90% is usually derived from fixation and the remainder is taken up from the pool of available N in the soil, so exploiting soil N reserves just like any other plant. Clearly the extent to which N benefits can be derived from use of legumes depends on the proportion of N which is actually derived from BNF. However, in the field it may not be possible to easily identify poorly-fixing plants from actively fixing neighbours on the basis of crop growth or N content alone unless the soil is depleted of plant-available N (Table 1).

There are many factors governing the extent of N_2 fixation in a legume crop. One critical factor is the availability of compatible, effective and competitive rhizobia in the soil, whether indigenous or inoculated. The second key component is the genotype of the plant. The proportion of nitrogen (P_{fix}) derived from fixation for a field-grown crop can vary immensely even within a single species (see Giller and Wilson, 1991; Peoples and Herridge, 1990; Peoples et al., 1995a) indicating that there is strong genetic potential for improving the contribution of BNF. However, emphasis on N_2 fixation as one of the important traits in breeding programs for legumes has been relatively rare despite its link with yield potential in infertile soils. This could be for a number of reasons, but one of the overriding limitations is the difficulty in actually measuring the amount of N_2 fixation (Herridge and Danso, 1995).

The one technique which can be readily and relatively inexpensively used to evaluate N_2 fixation in most tropical legumes involves the measurement of

Table 1. Effect of prior management and inoculation with *Bradyrhizobium japonicum* on the accumulation of crop nitrogen, the proportion of plant nitrogen derived from N_2 fixation (P_{fix}) and the amount of N_2 fixed by soybean (*Glycine max*)[a]

Pre-treatment	Inoculation rate (bacteria seed^{-1})	Total crop N (kg N ha^{-1})	N_2 fixation: P_{fix} (%)	N_2 fixation: Amount[c] (kg N ha^{-1})
Cropped to cereal (18)[b]	0	108	0	0
	5×10^7	267	56	149
Fallow (38)[b]	0	246	0	0
	5×10^7	347	9	32

[a] Data of Bergersen et al. (1989).
[b] Levels of soil mineral N (mg N g dry soil^{-1}) in the top 10 cm of soil at the time of sowing soybean.
[c] Amount of N_2 fixed = (Crop N)) $\times \frac{(P_{fix})}{100}$.

ureides. This technique relies on the observation that in certain legumes pathways of assimilation of N derived from fixation and from soil are different: ammonia derived from symbiotic fixation is converted into the ureides, allantoin and allantoic acid, in the nodule and then transported to the shoot in this chemical form in the transpiration stream; in contrast, N taken up from the soil, which is primarily nitrate, is transported either directly as nitrate or is assimilated into the amino acids asparagine or glutamine in the root prior to transport (Herridge and Peoples, 1990). Therefore, xylem sap composition changes from one dominated by ureides in fully symbiotic plants to one dominated by nitrate and amino acids in poorly nodulated plants utilizing soil N for growth. The proportion of total sap N formed by ureides has been shown to be a reliable indicator of the P_{fix} for a number of legume species (Table 2), and comparable field estimates of N_2-fixation can be obtained using either ureide or ^{15}N-based methodologies (Herridge et al., 1990).

The ureide technique: present use and future potential

The application of the ureide technique to measurement of N_2 fixation is discussed extensively in Peoples et al. (1989a). In essence, a sample of xylem sap containing N being transported from root to shoot is collected, and the amounts of the different N-containing compounds within that sample are measured. The sample can be obtained in one of three ways: root-bleeding sap, shoot tissue extract, or vacuum extracted sap (Herridge and Peoples, 1990). Approximately 150 samples can be collected per day with a team of six people working in the field. The samples are then taken to the laboratory and the determinations of the different N-solutes are done using spectrophotometric methods. A different set of colorimetric reactions have to be carried out on each sample to determine separately ureides and nitrate, and if necessary amino acids.

The ureide technique has been successfully applied to breeding programs involving soybean (Herridge and Danso, 1995). The procedure was used to identify individual lines and crosses with enhanced capacity to fix atmospheric N_2 in the presence of soil nitrate. Using the currently existing technique, a complete set of analyses can be undertaken for about 800 plants in four weeks (Herridge and Danso, 1995). To be truly useful to a legume breeder wishing to select for improvements in symbiotic performance, it would be necessary to be able to screen 1,000–2,000 plants per day (10–20,000 per fortnight) (D Herridge, pers. commun.). One way to achieve this would be to build on the great power of the ureide technique for measuring N_2-fixation by simplifying the assay procedure using reporter genes so that such rapid screening is possible. The ureide assay would then become an even more powerful tool for breeding of high N_2-fixing legumes.

Another powerful application of the use of bioindicators would be to use transgenic "sentinel plants" scattered among the main crop in the field to report

Table 2. Legume species where the relative abundance of ureides in xylem sap have been correlated to the proportion of plant nitrogen derived from N_2 fixation (P_{fix}) estimated using ^{15}N techinques

Species		Reference
Crop legumes		
Soybean	*Glycine max*	McClure et al. (1980)
		Herridge and Peoples (1990)
Common bean/dry bean	*Phaseolus vulgaris*	Peoples and Herridge (1990)
		Hansen et al. (1993)
Pigeonpea	*Cajanus cajan*	Peoples et al. (1989b)
Cowpea	*Vigna unguiculata*	Pate et al. (1980)
		Peoples and Herridge (1990)
Rice bean	*V. umbellata*	Rerkasem et al. (1988)
Green gram	*V. radiata*	Peoples and Herridge (1990)
Black gram	*V. mungo*	Peoples and Herridge (1990)
Forages and legume cover crops		
Calopogonium caeruleum		Faizah and Peoples, unpubl.
Desmodium ovalifolium		
Macroptilium atropurpureum		
Shrub legumes		
Desmodium rensonii		Herridge et al. (1995)
Codariocalyx gyroides		

on levels of BNF. In this case, the main crop would not be engineered with the reporter genes, and the sentinel plants would be used for rapid evaluation of the impact of management on N_2 fixation. It is clear that it may be possible to impose a range of management strategies to manipulate and enhance the proportion of N derived from BNF (Peoples et al., 1995b), and evaluation of the impacts of different strategies could be streamlined using such sentinel plants.

Use of bioindicators to measure N_2 fixation efficiency

In this example the strategy would be to develop reporter genes to report on the levels of different nitrogenous compounds in the sap of an N_2-fixing plant. As discussed earlier, bioindicators can be either endogenous molecules or enzymes that result from the expression of an introduced gene. Thus, in the conventional ureide assay, allantoin and allantoic acid are bioindicators indicating the reliance of a legume on N_2-fixation for growth. An alternative route, which could both speed the assays and make them more reliable and also would avoid the dangerous or toxic chemicals normally required for colorimetric assays in the laboratory, would be the use of reporter genes whose expression would reflect the relative levels of the different nitrogenous compounds.

Engineering response of reporter genes to reflect levels of nitrogenous compounds

Two different reporter gene strategies would be possible. One would be to engineer reporter bacteria either bacteria which report on the levels of nitrogenous compounds in a tissue extract, or endosymbiotic bacteria which could report on the levels of these compounds directly in the plant. The second approach, and the long-term goal, would be to engineer the plants themselves to report directly on their N status while growing in the field.

The concentration of amino-N in sap does not vary much with varying dependence on N_2 fixation (Herridge and Peoples, 1990), so the relative ureide index, the indicator of P_{fix} (see Peoples et al., 1989a), can be based on the two N-solutes which are most sensitive to changes in the source of N for growth:

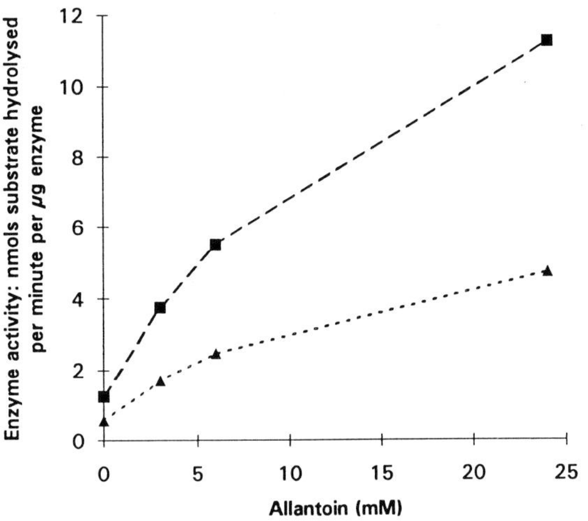

Fig. 3. Induction of allantoinase and hypothetical induction of a GUS reporter gene in response to different levels of allantoin (figure based on Venkateswara Rao et al., 1990). –▲– allantoinase activity; –■– GUS activity.

$$\text{Relative ureide index}(\%) = \frac{4(\text{ureide}) \times 100}{4(\text{ureide}) + \text{nitrate}} \quad (1)$$

Hence the most important compounds to measure are the ureides and nitrate. The approach therefore would be to engineer reporter systems to respond quantitatively to the levels of these two compounds.

a) Use of bacterial reporters

The bacterial approach is most likely to succeed in the short term, because N metabolism and its regulation are better understood in bacteria than in eukaryotes, and because it would not involve production and use of transgenic plants.

Little work has been carried out on ureide metabolism in bacteria. There is one report on allantoin metabolism in *Bradyrhizobium* strain NC92, a symbiont of peanut and pigeonpea as well as other legume crops (Venkateswara Rao et al., 1990). Allantoinase activity was identified in this species and activity of the enzyme was reported to be inducible by allantoin. Importantly for the reporter gene strategy, the extent of induction was proportional to the concentration of allantoin in the medium over a range from 0.3 to 24 m*M* allantoin (Fig. 3). Since levels of ureides reported in xylem sap of nodulated soybean, cv Bragg, ranged from 0.3 to 2.5 m*M* (Herridge and Peoples, 1990), the implication is that the bacterial gene regulatory sequences would be sensitive enough to respond to physiological levels of ureides.

By contrast, a great deal of work has been carried out on general N metabolism in bacteria, particularly in enteric bacteria and in rhizobial species (reviewed in Merrick, 1992). The system is highly complex and is affected (in different species) by factors such as whether bacteria are growing aerobically or anaerobically, and what N sources are available. In *E. coli*, induction of nitrate reductase is primarily regulated by the ambient oxygen tension, being stimulated in microaerobic conditions (2% oxygen). However, it is also responsive to levels of nitrate (Dong et al., 1992), and the genetics of *E. coli* is sufficiently well understood that it would be a relatively simple matter to break these two control circuits apart and engineer an *E. coli* strain in which nitrate reductase was induced directly in response to levels of nitrate. If a reporter gene was engineered to respond to nitrate in the same way, such a strain could then act as the sensor of the soil N component in the xylem sap of a plant.

As well as using existing knowledge to identify genes which are induced in response to the presence of specific nitrogenous compounds, it is possible to use a "promoter-probe" approach to screen for genes which respond in the desired way. A reporter gene without DNA regulatory elements is introduced into a bacterium — *E. coli*, or perhaps a *Rhizobium* strain, and the bacteria are screened for ones in which the desired response is obtained. For example, a *Rhizobium* strain with an introduced promoterless GUS gene could be screened in the presence and absence of allantoin and allantoic acid for strains which made blue colonies, indicating induction of the reporter gene, in the presence of these ureides, but not in their absence.

Whichever approach is followed, construction of bacterial strains carrying easily detectable and distinguishable reporter genes which respond to ureides, and to nitrate, is certainly feasible. It would then be necessary to fine tune the system so that activation of these genes parallels levels of the inducer compounds over a certain range, and to ensure that the induction is not affected by other molecules which might be present in the plant sap. It would also be important to combine the two reporter genes in a single indicator bacterial strain so that the two reporter gene activities directly reflected relative levels of ureides and nitrate and complications were not introduced by varying bacterial population sizes. The end result should be a bacterial indicator strain which allows rapid, reliable and safe analysis of relative ureide index.

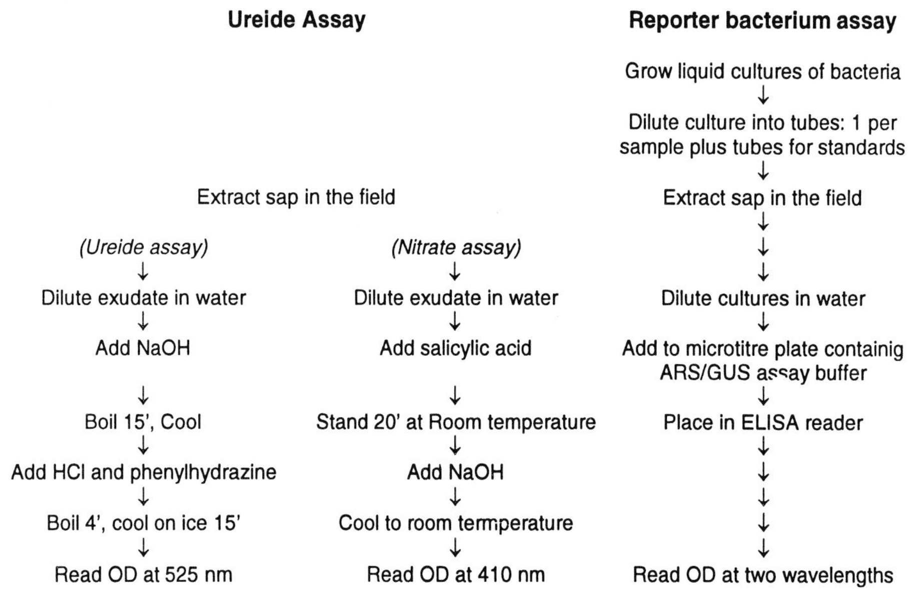

Fig. 4. A comparison of the conventional ureide assay and the steps involved in a reporter bacterium assay.

The procedure for using a reporter bacterial strain to analyse relative ureide index could be envisaged as follows. Prior to the assay, a culture of the reporter bacteria would be grown up, diluted in fresh medium to a concentration low enough to enable fresh growth, and distributed among enough tubes to allow one tube per sample of sap. The sap would be collected in the field as normal for the ureide assay, and a given volume would be placed into a tube containing the reporter bacterial strain. The assay would effectively begin at this point, as the bacteria would then be being incubated in a given concentration of allantoin, allantoic acid and nitrate, and would grow and induce the reporter enzymes during transport from the field to the laboratory. In contrast to the conventional ureide assay, where the speed of transport of the sap sample to the laboratory or storage on ice is important so that the key compounds do not break down prior to analysis, the warm field conditions that are often prevalent will actually be an advantage as they will facilitate growth of the bacteria and expression of the reporter enzymes.

Once in the laboratory, aliquots of the bacterial/xylem exudate mix would be added to a non-toxic buffer containing a colorigenic substrate for each reporter enzyme. Ideally, if substrates with different absorption maxima are available, both enzymes could be assayed in a single reaction vessel. Alternatively, two parallel reactions could be set up, one for each reporter enzyme. The reaction(s) would then be allowed to proceed until colour that is clearly visible to the eye develops. At that point, the reaction would be stopped and the absorbance read on a spectrophotometer. Calculation of the rate of hydrolysis of substrate for each of the two reporter enzymes will reflect the level of induction of each reporter gene, and hence the approximate concentration of the inducer compounds. Since both reporter genes would be in a single bacterial strain which was responding to a single sap sample, the ratio of these two reporter gene activities would directly reflect the relative levels of ureides and nitrate. For example, if GUS induction reflects ureides, and ARS induction reflects nitrate, the relative ureide index can be calculated as: $(4 \times \text{GUS}/(4 \times \text{GUS} + \text{ARS})) \times 100$.

The steps involved in the current procedure and a reporter gene one are compared in Figure 4. Advantages are: the reduced number of steps; the fact that the assay now depends on measurement of the reporter enzymes which are robust rather than that of the signature compounds, which are fragile; the complete avoidance of toxic or dangerous compounds. The laboratory infrastructure required is very similar to the conventional ureide and nitrate assays. However, because this procedure has fewer steps, it lends itself more readily to automation using microtitre plates and automatic reading and calculation of the data. Given that it is possibly to carry out 96 assays in a conventional microtitre dish, if these are processed at the rate of 50 per day using an ELISA reader, this gives a rate of assay of 4,800 samples per day, approximately 60 times higher than the maximal rate of conventional ureide analysis. In this case, it is the rate of sampling of xylem sap which becomes the limiting factor, rather than the rate of analysis of the sap contents.

b) Use of eukaryotic systems

A longer-term goal would be to engineer plants to report directly on their N status in the field. In this case, prokaryotic regulatory circuits might be adapted for use in eukaryotes — there are precedents. However, a far more powerful approach would be to identify genetic regulatory circuits in the plants themselves that respond to the different N sources, and to link these to reporter genes that would function to give distinct colour changes in living plants.

There is a lot of background data indicating that this approach is highly feasible. As far as measurement of nitrate goes, both fungal and higher plant genes encoding the enzyme nitrate reductase are known to be regulated in response to the available levels of nitrate, and hence these regulatory circuits could be adapted to construct gene fusions that would detect different levels of nitrate (e.g. Banks et al., 1993; Cheng et al., 1991).

Genes which respond to ureide levels have not been identified in plants (Winkler et al., 1988), but precedents exist in yeast. *Saccharomyces cerevisiae* has the ability to use allantoin (or allantoic acid) as a sole N source, degrading it in five enzymatic steps to ammonia, glyoxylate and carbon dioxide (the first intermediate is allantoic acid), and the regulation of this pathway has been extensively studied. The enzymes of allantoin degradation and utilization were shown to be induced in response to the presence of allantoin (Cooper and Lawther, 1973), and, while strictly-speaking the inducer is allophanate, the last intermediate in the degradative pathway, the net effect is induction of these enzyme activities in response to levels of allantoin and allantoic acid. There is also an allantoin permease, induction of which is directly caused by the presence of allantoin (Sumrada et al., 1978). This regulation has subsequently been shown to be transcriptionally mediated i.e. the mRNA and hence protein are not made in the absence of inducer, and the DNA sequences required for this response have been identified (Cooper et al., 1987; Yoo and Cooper, 1989). These data indicate that the strategy of looking for analogous regulatory sequences that respond to ureides in plants is highly plausible.

There are numerous methods for isolation of genes whose expression is modulated relative to environmental or physiological parameters. All of them rely on obtaining plant material from each of the two opposite conditions which are under consideration. For example, to identify genes that respond to ureide levels, ureide-producing plants such as soybean would be grown that were either 100% dependent on BNF (maximum ureide production) or which had no N_2 fixation activity (minimum ureide condition). RNAs can then be prepared from both sets of plants and used in a variety of molecular protocols e.g. subtractive cDNA libraries (Duguid and Dinauer, 1990) or arbitrarily primed PCR fingerprinting of RNA (Welsh et al., 1992), to identify mRNAs which are produced specifically under the target conditions (high ureides, in this example). This in turn will lead to identification of the genes and the regulatory sequences, and would enable construction of the appropriate reporter gene fusions.

Thus it would be possible to develop transgenic plants containing two different reporter constructs. One, fused to the GUS gene, produces GUS under high nitrate conditions. The second, fused to the ARS gene, produces ARS under high ureide conditions. Since substrates are available for each which give rise to different colours, it is possible to envisage painting a mixture of the two substrates onto a stem and petiole and looking to see whether it turns predominantly blue (high GUS, hence high nitrate and low fixation) or red (high ARS, hence high ureide and high fixation). Such plants might be used directly in a breeding program, or as sentinel plants to indicate the effect of different farm management practices on the extent of N_2 fixation.

Clearly a vast amount of work would have to be undertaken to fine-tune such bioindicator reporter genes to accurately reflect relative ureide levels, but the examples given here clearly illustrate that at least the basis for this approach exists in the amount of information already available about genetic circuits that respond to N compounds in eukaryotes.

Conclusions

Molecular biology is now reaching a stage of maturity where it can provide tools to facilitate analysis of population dynamics and physiological responses of microbes and plants in the natural environment, and hence can facilitate decision-making regarding inoculation, management or breeding strategies. This is practically demonstrated by the use of marker gene technology by rhizobial ecologists to rapidly screen the competitive ability of a number of potential inoculum strains for *Phaseolus vulgaris*. In addition, the concept is extended to the future scenario of using reporter genes to report on the extent of N_2 fixation in a crop, either by way of bioindicator bacteria, or directly in transgenic sentinel plants. This concept moves away from the idea of using genetic manipulation only to engineer solutions, but rather uses it as a tool to aid our understanding of the natural environment.

References

Banks G R, Shelton P A, Kanuga N, Holden D W and Spanos A 1993 The *Ustilago maydis nar l* gene encoding nitrate reductase activity, sequence and transcriptional regulation. Gene 131, 68–78.

Bergersen F J, Brockwell J, Gault R R, Morthorpe L, Peoples M B and Turner G L 1989 Effects of available soil nitrogen and rates of inoculation on nitrogen fixation by irrigated soybeans and evaluation of ^{15}N methods for measurement. Aust. J. Agric. Res. 40, 763–80.

Boddey R M, de Oliveira O C, Urquiaga S, Reis V M, de Olivares F L, Baldani V L D and Döbereiner J 1995 Biological nitrogen fixation associated with sugar cane and rice: contributions and prospects for improvement. Plant and Soil 174.

Cheng C L, Acedo G N, Dewdney J, Goodman H M and Conkling M A 1991 Differential expression of the two *Arabidopsis* nitrate reductase genes. Plant Physiol. 96, 275–279.

Christiansen-Weniger C and Vanderleyden J 1993 Ammonium-excreting *Azospirillum* sp. become intracellularly established in maize *Zea mays* para-nodules. Biol. Fertil. Soils 17, 1–8.

Cooper T G and Lawther R P 1973 Induction of the allantoin degradative enzymes in *Saccharomyces cerevisiae* by the last intermediate of the pathway. Proc. Natl. Acad. Sci. USA 70, 2340–2344.

Cooper T G, Chisholm V T, Cho H J and Yoo H S 1987 Allantoin transport in *Saccharomyces cerevisiae* is regulated by two induction systems. J. Bacteriol. 169, 4660–4667.

Couteaudier Y, Daboussi M J, Eparvier A, Langin T and Orcival J 1993 The GUS gene fusion system (*Escherichia coli* β-D-glucuronidase gene), a useful tool in studies of root colonization by *Fusarium oxysporum*. Appl. Environ. Microbiol. 59, 1767–1773.

De Boer M H and Djordjevic MA 1995 The inhibition of infection thread development in the cultivar-specific interaction of *Rhizobium* and subterranean clover is not caused by a hypersensitive response. Protoplasma (*In press*).

Dong X R, Li S F and DeMoss J A 1992 Upstream sequence elements required for NarL-mediated activation of transcription from the *narGHJI* promoter of *Escherichia coli*. J. Biol. Chem. 267, 14122–14128.

Duguid J R and Dinauer M 1990 Library subtraction of in vitro cDNA libraries to identify differentially expressed genes in scrapie infection. Nucl. Acids Res. 18, 2789–2792.

Giller K E and Wilson K J 1991 Nitrogen Fixation in Tropical Cropping Systems. CAB International, Wallingford, Oxon, UK. 313p.

Hammond-Kosack K E, Jones D A and Jones J D G 1994 Identification of two genes required in tomato for full Cf9-dependent resistance to *Cladosporium fulvum*. Plant Cell 6, 361–374.

Hansen A P, Rerkasem B, Lordkaew S and Martin P 1993 Xylem-solute technique to measure N_2 fixation by *Phaseolus vulgaris* L; calibration and sources of error. Plant and Soil 150, 223–231.

Hart S, Kock K R and Woods D R 1992 Identification of indigo-related pigments produced by *Escherichia coli* containing a cloned *Rhodococcus* gene. J. Gen. Microbiol. 138, 211–216.

Heitzer A, Webb O F, Thonnard J E and Sayler G S 1992 Specific and quantitative assessment of naphthalene and salicylate bioavailability by using a bioluminescent catabolic reporter bacterium. Appl. Environ. Microbiol. 58, 1839–1846.

Herridge D F and Danso S K A 1995 Enhancing crop legume N_2 fixation through selection and breeding. Plant and Soil 174.

Herridge D F and Peoples M B 1990 Ureide assay for measuring nitrogen fixation by nodulated soybean calibrated by ^{15}N methods. Plant Physiol. 93, 495–503.

Herridge D F, Bergersen F J and Peoples M B 1990 Measurement of nitrogen fixation by soybean in the field using ureide and natural ^{15}N abundance methods. Plant Physiol, 93, 708–716.

Herridge D F, Palmer B, Nurhayati D P and Peoples M B 1995 Application of xylem solute ureide method for measuring N_2 fixation in tree legumes. Soil Biol. Biochem. (*In press*).

Jefferson R, Kavanagh T A and Bevan M W 1987 GUS fusions: β-glucuronidase as a sensitive and versatile gene fusion marker in higher plants. EMBO J. 6, 3901–3907.

Jefferson R A and Wilson K J 1991 The GUS gene fusion system. *In* Plant Molecular Biology Manual. Eds. S Gelvin, R Schilperoort and D P Verma. pp B14/1–B14/33. Kluwer Academic Publ., Dordrecht.

Katupitiya S, New P B, Elmerich C and Kennedy I R 1994 Improved nitrogen fixation in 2 4-D treated wheat roots associated with *Azospirillum lipoferum*; studies of colonization using reporter genes. Soil Biol. Biochem. 27, 447–452.

McClure P R, Israel D W and Volk R J 1980 Evaluation of the relative ureide content of xylem sap as an indicator of N_2 fixation in soybeans. Plant Physiol. 66, 720–725.

McDermott T R and Graham P H 1989 *Bradyrhizobium japonicum* nodule occupancy and acetylene reduction in the soybean root system. Appl. Environ. Microbiol. 55, 2493–2498.

Merrick M 1992 Regulation of nitrogen fixation genes in free-living and symbiotic bacteria. *In* Biological Nitrogen Fixation. Eds. G Stacey, H J Evans and R H Burris. pp 835–876. Routledge Chapman and Hall, New York.

Oliver R P, Farman M L, Jones J D G and Hammond-Kosack K E 1993 Use of fungal transformants expressing β-glucuronidase activity to detect infection and measure hyphal biomass in infected plant tissues. Mol. Plant-Micr. Interact. 6, 521–525.

Pate J S, Atkins C A, White S T, Rainbird R M and Woo K C 1980 Nitrogen nutrition and xylem transport of nitrogen in ureide-producing grain legumes. Plant Physiol. 65, 961–965.

Pemberton J M, Vincent K M and Penfold R J 1991 Cloning and heterologous expression of the violacein biosynthesis gene cluster from *Chromobacterium violaceum*. Curr. Microbiol. 22, 355–358.

Peoples M B, Faizah A W, Rerkasem B and Herridge D F 1989a Methods for evaluating nitrogen fixation by nodulated legumes in the field. Aust. Council for Int. Agric. Res., Canberra. 76p.

Peoples M B, Hebb D M, Gibson A H and Herridge D F 1989b Development of the xylem ureide assay for the measurement of nitrogen fixation by pigeonpea *Cajanus cajan* L. Millsp. J. Exp. Bot. 40, 535–542.

Peoples M B and Herridge D F 1990 Nitrogen fixation by legumes in tropical and sub-tropical agriculture. Adv. Agron. 44, 155–223.

Peoples M B, Herridge D F and Ladha J K 1995a Biological nitrogen fixation, an efficient source of nitrogen for sustainable agricultural production? Plant and Soil 174.

Peoples M B, Ladha J K and Herridge D F 1995b Enhancing legume N_2 fixation through plant and soil management. Plant and Soil 174.

Reches M, Zhao C and Engelberg-Kulka H 1994 A bioassay based on recombinant DNA technology for determining selenium concentrations. Appl. Environ. Microbiol. 60, 45–50.

Rerkasem B, Rerkasem K, Peoples M B, Herridge D F and Bergersen F J 1988 Measurement of N_2 fixation in maize *Zea mays* L. — ricebean *Vigna umbellata* [Thunb] Ohwi and Ohashi intercrops. Plant and Soil 108, 125–135.

Sessitsch A, Jjemba P K, Hardarson G, Akkermans A D L and Wilson K J 1995 The *gusA* marker gene as a new tool for competition studies of *Rhizobium*. Soil Biol. Biochem. (*Submitted*).

Somasegaran P and Hoben H J 1985 Methods in legume-Rhizobium technology, NifTAL Project University of Hawaii. 367p.

Streit W, Kosch K and Werner D 1992 Nodulation competitiveness of *Rhizobium leguminosarum* bv *phaseoli* and *Rhizobium tropici* strains measured by glucuronidase (*gus*) gene fusions. Biol. Fertil. Soils 14, 140–144.

Streit W, Botero L, Werner D and Beck D 1995 Competition for nodule occupancy on *Phaseolus vulgaris* by *Rhizobium etli* and *Rhizobium tropici* can be efficiently monitored in an ultisol during the early stages of growth using a constitutive GUS gene fusion. Soil Biol. Biochem. (*In press*).

Sumrada R, Zacharski C A, Turoscy V and Cooper T G 1978 Induction and inhibition of the allantoin permease in *Saccharomyces cerevisiae*. J. Bacteriol. 135, 498–510.

Venkateswara Rao N, Subhash Reddy R and Sivarama Sastry K 1990 An inducible allantoinase and the specific toxicity of parabanic acid on allantoin utilization in a *Rhizobium* species. Curr. Microbiol. 20, 115–119.

Welsh J, Chada K, Dalal S S, Cheng R, Ralph D and McClelland M 1992 Arbitrarily primed PCR fingerprinting of RNA. Nucl. Acids Res. 20, 4965–4970.

Wilson K J 1995 Molecular techniques for the study of rhizobial ecology in the field. Soil Biol. Biochem. 27, 501–514.

Wilson K J, Giller K E and Jefferson R A 1991 β-glucuronidase GUS operon fusions as a tool for studying plant-microbe interactions. *In* Advances in Molecular Genetics of Plant-Microbe Interactions. Vol 1. Eds. H Hennecke and D P S Verma. pp 226–229. Kluwer Academic Publ., Dordrecht.

Wilson K J, Sessitsch A, Corbo J C, Giller K E, Akkermans A D L and Jefferson R A 1995 β-glucuronidase (GUS) transposons for ecological and genetic studies of rhizobia and other Gram-negative bacteria. Microbiology (*In press*).

Winkler R G, Belvins D G, Polacco J C and Randall D D 1988 Ureide catabolism in nitrogen-fixing legumes. Trends Biochem. Sci. 13, 97–100.

Wollum A G and Cassel K 1984 Spatial variability of *Rhizobium japonicum* in two North Carolina soils. Soil Sci. Soc. Am. J. 48, 1082–1086.

Yoo H S and Cooper T G 1989 The DAL7 promoter consists of multiple elements that cooperatively mediate regulation of the gene's expression. Mol. Cell. Biol. 9,3231–3243.

Plant and Soil **174:** 255–277, 1995.

Future benefits from biological nitrogen fixation: An ecological approach to agriculture

K. E. Giller and G. Cadisch
Department of Biological Sciences, Wye College, University of London, Wye, Ashford, Kent TN25 5AH, UK

Key words: agricultural development, agroecology, agroforestry, farming systems, fertilizers, green manures, legumes, soil degradation

Table of contents

Abstract
Introduction
Future targets for research on BNF
  The problem of soil degradation
  The role of BNF in providing the future needs of an expanding population
Approaches and timescales for increasing inputs from BNF
Ecological considerations on current approaches for improving BNF
  Solving environmental constraints
  Effects of N on N_2-fixation
  Inoculation with rhizobia
  Breeding for enhanced N_2-fixation
  Engineering of cereals for N_2-fixation
  Introduction of new legumes into farming systems
Farmers' success in experimentation?
Nitrogen-fixing trees in arid lands
Legumes for soil-fertility for smallholder farmers
Managing fixed nitrogen: Optimizing use efficiency
  The role of fixed N in long-term soil fertility
Experimentation or extension
Conclusions
  Appropriate solutions tailored to farmers' needs
  Future benefits: Ecological approaches to integrated systems
  Momentum to favour BNF
  Research agenda
Acknowledgements
References

Abstract

Strategies for the enhancement and exploitation of biological nitrogen fixation are assessed with attention to the likely timescales for realization of benefits in agriculture. Benefits arising from breeding of legumes for N_2-fixation and rhizobial strain selection have less potential to increase inputs of fixed N than alleviation of environmental stresses or changes in farming systems to include more legumes. Genetic engineering may result in substantial enhancement of N_2-fixation, particularly if the ability to fix N_2 is transferred to other crops but these are long-term goals. Immediate dramatic enhancements in inputs from N_2-fixation are possible simply by implementation of existing technical knowledge. Apart from the unfortunate political and economic barriers to the use of agricultural inputs, better communication between researchers and farmers is required to ensure proper focus of research and development of appropriate technologies. Legumes must be considered within the context of the farming systems within which they are grown and not in isolation. Proper integration of legumes requires a good understanding of the role of the legume within the system and a better understanding of the relative contributions of N sources and of the fates of fixed N.

Introduction

The biological fixation of atmospheric dinitrogen is a 'free' source of N for agriculture. In the more developed countries the interest in low-external-input agriculture is driven largely by overproduction and concerns of the environmental effects of intensive agrochemical use. In the less-developed countries interest in low-external-input agriculture is often a necessity, fuelled by the lack of access to high input approaches, due either to local or national economics. The other papers in this volume and in previous reviews (Giller and Wilson, 1991; Ladha et al., 1992) have dealt in detail with the many ways in which N_2-fixation can contribute to agriculture, and with various exciting aspects of current research. Our discussion therefore attempts to evaluate the timescales and likely benefits of research on BNF, to assess the desirability of the various research goals and finally to deal with the thorny problem of how this research can actually be translated into practice.

In discussing the exploitation of BNF in agriculture we concentrate on both the problems encountered under farmers' conditions and the most promising approaches by which our exploitation of BNF may be enhanced in the future. An overriding conclusion of this review is that BNF cannot be considered in isolation, but must be examined within the context of the farming systems in which legumes are utilized.

Future targets for research on BNF

The problem of soil degradation

Soil degradation is the result of a number of processes of which the most obvious, and the most severe (as it results in the loss of the soil itself) is soil erosion. During erosion of soil, the topsoil is preferentially removed leading to an enrichment of organic matter (and many nutrients) in the eroded soil which is lost and a concomitant reduction in the concentration of nutrients in the soil remaining. Even where soil erosion is not a great problem there is a continual loss of nutrients from agricultural soils due to removal in crops, leaching, and in the case of N in particular, gaseous losses. Thus degradation of soil fertility must be considered an inevitable consequence of agriculture unless careful attention is paid to nutrient conservation and replenishment, through inorganic and organic fertilizers, recycling of agricultural wastes and of course biological N_2-fixation.

Estimates of soil nutrient losses in sub-Saharan Africa (Stoorvogel et al., 1993; Table 1), Asia and Latin America (Tandon, 1993) suggest that there is a net removal of between 20 and 70 kg ha^{-1} of N from agricultural land each year, and that these losses are likely to increase. Against this background it is not surprising that FAO in its study 'World Agriculture: Toward 2000' suggested that fertilizer use will have to be increased in tropical agriculture to prevent soil fertility exhaustion (Alexandratos, 1988). In the case of nutrients such as phosphorus and potassium there is little option but to use fertilizers (whether in their raw rock forms or in purified, more reactive forms) to

Table 1. Nutrient balances (1982–1984 and projected for 2000) of N, P and K for the arable land of some sub-saharan countries (from Stoorvogel et al., 1993)

Country	N		P		K	
	1982–1984	2000	1982–1984	2000	1982–1984	2000
Benin	−14	−16	− 1	− 2	− 9	−11
Botswana	0	− 2	1	0	0	− 2
Cameroon	−20	−21	− 2	− 2	−12	−13
Ethiopia	−41	−47	− 6	− 7	−26	−32
Ghana	−30	−35	− 3	− 4	−17	−20
Kenya	−42	−46	− 3	− 1	−29	−36
Malawi	−68	−67	−10	−10	−44	−48
Mali	− 8	−11	− 1	− 2	− 7	−10
Nigeria	−34	−37	− 4	− 4	−24	−31
Rwanda	−54	−60	− 9	−11	−47	−61
Senegal	−12	−16	− 2	− 2	−10	−14
Tanzania	−27	−32	− 4	− 5	−18	−21
Zimbabwe	−31	−27	− 2	2	−22	−26

balance these losses if soil fertility is to be maintained. In the case of N, losses can be balanced by both N fertilizers and BNF. But simply to balance the current losses of N, with current food production and fertilizer use (against a backdrop of increasing population), inputs from BNF would have to increase *on average* by more than 30 kg N ha^{-1} (Table 1).

Introducing more legumes into farming systems can in itself help to reduce erosion losses, for instance where a legume understorey is maintained in plantation crops. Where losses from erosion are not high, such as under pastures with a good maintenance of soil cover, productivity declines in degraded pastures have been attributed to immobilization of N (Robbins et al., 1989). This is an example where nutrient availability, rather than nutrient loss can lead to degradation of soil fertility. Here N_2-fixation by legumes can play an important role in providing legume litter with improved quality for decomposition. Maintaining productive systems can also reduce exploitation of new marginal lands such as those dominated in many regions by rainforests.

Soil degradation can also occur as a result of pollution. Of particular note with regard to N_2-fixation is pollution of soils with heavy metals. At concentrations too small to cause phytotoxic effects, long-term heavy metal contamination of soils results in extinction of both free-living N_2-fixing organisms (Brookes et al., 1986; Lorenz et al., 1992) and *Rhizobium* (Giller et al., 1989; McGrath et al., 1988) which can lead to complete suppression of N_2-fixation. As heavy metals are known to persist in soils for thousands of years this can be considered an irreversible form of soil degradation. Research here is needed to advise on regulations to ensure that soil contamination is prevented.

The role of BNF in providing the future needs of an expanding population

Sustainability of agricultural production has become the focus of agricultural research and policy in the past decade, but with a variety of interpretations as to what is considered to be sustainable. If sustainability is interpreted strictly in terms of balanced nutrient budgets then unproductive, subsistence agriculture on inherently infertile soils may be interpreted as 'sustainable', but may in fact go little further than maintaining production at a level close to the bare minimum for subsistence; in essence 'recycling poverty' (Greenland, 1994). A widely adopted definition of sustainability is that of the Brundtland commission: " ... agricultural production will be sustainable only if the resource base is ...sustained, enhanced and, where it has been diminished or destroyed, restored" (WCED, 1987). Our working definition of sustainability therefore includes the need for equitability emphasised by Conway (1987) which necessitates a certain degree of agricultural productivity in order to meet more than just the very basic needs of subsistence farmers. What is an acceptable or optimal degree of productivity is impossible to generalise across environments and farming systems, but agricultural development must at least aim towards "increasing production and alleviating drudgery and poverty" (Dudal and Deckers, 1993). Where land area is not limiting for agriculture, total farm production may be much more important than productivity per unit area and optima for production may fall at a lower point on the response curve. Where land areas are at more of a premium the use of nutrient input rates to ensure 80% of maximum production often maximises the return in increased production per unit of input (and in doing so optimises the efficiency of input use) and can perhaps be a useful guide as to what is long term, optimal production (Sanchez and Salinas, 1981).

In essence then production has to be at more than a base level to be acceptable and must be increased in many areas to meet the needs of expanding populations. FAO anticipated that increases in production will be due partly to an increase in the area of cropped land, but mainly due to increases in crop yields (Alexandratos, 1988). This therefore represents an enormous challenge for biological N_2-fixation to provide a major part of the additional N required to sustain and enhance agricultural productivity.

Approaches and timescales for increasing inputs from BNF

Biological N_2-fixation contributes to productivity both directly, where the fixed N_2 is harvested in grain or other food for human or animal consumption, or indirectly, by contributing to the maintenance or enhancement of soil fertility in the agricultural system by adding N to the soil. The different approaches which can be used to enhance inputs from BNF can be classified into those which address environmental constraints to N_2-fixation, those concerned with *Rhizobium* (or other N_2-fixing bacteria), with the host legume (or non-legume) or those which address changes in the farming system (Table 2). The purpose of this table is to indicate the probable benefits and when they are likely to be realised.

The technology required to make substantial improvements in the inputs from BNF is already available for implementation in many cases – for example in many parts of the tropics BNF by legumes is severely limited by phosphorus deficiency and could be dramatically improved by use of P fertilizers (George et al., 1995; Thomas, 1995; Wani et al., 1995). Such an analysis is patently simplistic in that it takes no account of

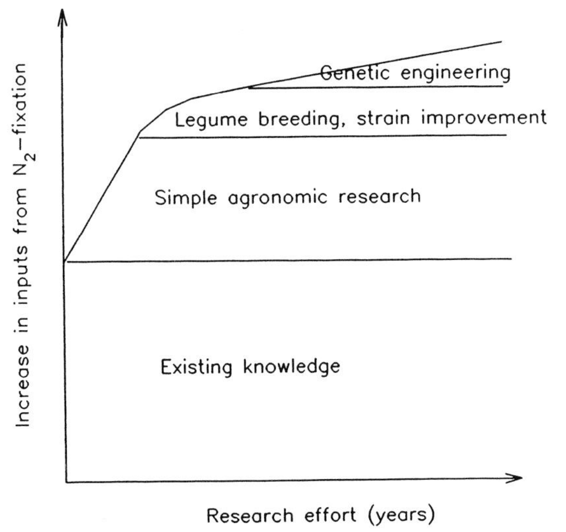

Fig. 1. The likely relationship between investment in research effort and the possible increases in inputs from biological N_2-fixation as indicated in Table 2.

whether the inputs are feasible or available to farmers (e.g. irrigation water in the case of drought), but it does serve to highlight the potential gains and the timescales over which they could be realised. We believe that the amounts of N_2 fixed in agriculture under current cropping systems could be dramatically increased by implementation of current agronomic knowledge with immediate benefits if phosphorus fertilizers were made widely available. This would require all of the economic, transport and knowledge barriers to implementation of simple fertilizer technology to be removed.

We can see that the greatest immediate improvements to inputs from BNF are likely to be achieved with relatively simple technologies for which the technical understanding is already sufficient. Gains from bacterial strain improvement or from plant breeding for enhanced N_2-fixation are likely to be fairly modest and longer term than the gains in the amounts of N_2 fixed due to the introduction of legumes into new areas (Fig. 1). The creation of N_2-fixing cereal crops, whether by induction of nodulation or by genetic engineering of cereal crops, is a very distant goal (Bennet and Ladha, 1992). It is, however, a goal which we can imagine would have enormous benefits in agriculture.

The fact that certain research strategies are unlikely to realise their goals within a few years does not mean that research which will not give immediate pay-offs should not be funded: any sensible research strategy should have short, medium and long-term goals.

Table 2. Approaches to enhancement of the role of biological nitrogen fixation in agriculture, the potential benefits which might accrue and the timescale for improvements given our current technical understanding

Problem/Target	Research approach	Potential for enhancement of inputs from N_2-fixation	Time scale for improvement
Environmental constraints			
– Soil acidity	Liming/fertilizer use, green manure use	High	Immediate
– Water stress	Irrigation	High	Immediate
	Drought tolerance/avoidance	Medium-high	Medium-long
– Nutrient deficiencies	Fertilizer use	High	Immediate
– High soil-N status	Crop rotation	Medium	Immediate
Rhizobium			
– Absence of compatible strains	Inoculate	High	Immediate
– Small/poorly effective indigenous population	Inoculate	Medium-high	Immediate
– Effective population present	Strain selection	Low-medium	Medium
	Genetic engineering	Low-medium	Medium
Host Legume			
– No nodulation due to lack of compatible rhizobium	Selection for promiscuity	High	Medium
– Effectively nodulated legume	Breeding for increased N_2-fixation	Low-medium	Medium
	Genetic engineering	Low-medium	Long
–Environmental constraints	Selection/breeding for adaptation	Medium-high	Medium
Non-legumes			
– 'Associative' N_2-fixation	Strain selection	Low	Medium
	Genetic engineering	Low	Long
– Lack of N_2-fixation	Nodulation/N_2-fixation	High	Long
	Crop engineering for N_2-fixation	High	Long
Farming system			
– New crop/product/forage	Legume introduction	High	Immediate
– Poor soil fertility	Legume introduction	High	Immediate
	Crop residue management	High	Immediate

Indeed what is obvious is that we have sufficient technical knowledge to make dramatic increases in the amounts of N_2 fixed in agriculture, but that what is lacking is a strategy for implementation of this knowledge. This point will be addressed later in this paper but first we will discuss some of the various research approaches which have been used in more detail.

Ecological considerations on current approaches for improving BNF

The literature abounds with examples where legumes have been found to fix a large proportion, often 60–80%, of their N under experimental field conditions (Giller and Wilson, 1991). Such rates may be achieved in farmers' fields, but only where environmental factors do not operate to limit N_2-fixation (Peoples et al., 1995). The main environmental factors which limit BNF are soil nutrient deficiencies, or factors associated with soil acidity, large concentrations of plant-available N in the soil and moisture availability. Whereas breeding or genotype selection for enhanced BNF may be expected to result in increases in inputs in the order of 10% above that of existing cultivars, alleviation of environmental constraints can result in the difference between virtually no input from N_2-fixation and a very substantial contribution from N_2-fixation. In this section we take a more detailed look at the research strategies outlined in Table 2.

Solving environmental constraints

Liming or selection for acid tolerance

Where environmental conditions cannot readily be altered, emphasis must be placed on maximising our exploitation of the ecological adaptation of the different components of the agricultural system. An excellent example where this has been achieved is with the selection of both pasture legumes and effective rhizobia for the acid, aluminium-rich soils of the South American savannas which must be one of the most hostile soil environments (see Giller and Wilson, 1991; Thomas, 1995). The available germplasm has not been fully explored and therefore the most promising approach for tropical pastures is still selection rather than breeding. The enormous variability of rhizobia which exists in tropical soils has yet to be fully exploited but promising results with selection of rhizobia for pasture legumes in acid soils (see below) indicates that further gains could be made in this area for other grain or tree legumes.

Grain legumes also vary widely in their adaptation to acid soils (Sanchez and Salinas, 1981) and thus crop selection by farmers has already been governed to a large extent by the adaptation of crops. Yet even where well-adapted germplasm is available, chronic nutrient deficiencies of a whole range of major and minor nutrients must be alleviated to ensure productivity. Liming in acid soils can be governed by the need to supply calcium and to reduce aluminium saturation to an acceptable level rather than to raise the soil pH (Kamprath, 1970; Oates and Kamprath, 1983). This will minimise the requirements for lime which is a bulky commodity for which transport costs quickly become prohibitive given the enormous distances over which it must be moved in many countries.

High-input versus low-input approaches to solving nutrient deficiencies

There is an ongoing discussion between those who urge the importance of greater fertilizer use to stimulate agricultural production in developing countries and of those who argue that sustainable agriculture might best be achieved if greater emphasis is given to organic methods. For a sustainable system, in the long-term we will have to consider not only the use of inorganic fertilizers but also of efficient recycling of nutrients in agricultural and human wastes and of course inputs from N_2-fixation. But it is essential to realise that the problems of the less-industrialised nations differ from those of highly-developed countries. Given the inherent poor fertility of many tropical soils, the high intensity of rains leading to further nutrient losses and the urgent need for productive cropping, then in the vast majority of cases purely organic methods are insufficient to meet the immediate nutrient demands of agricultural production. N_2-fixation is the only process by which essential nutrients are created in situ; the efficiency of use of other nutrients can be optimised but losses of nutrients in agricultural produce must be replaced. In this light the debate between the rights and wrongs of high and low input approaches becomes rather redundant and what is needed is the optimal, efficient use of available resources combining the sensible approaches to nutrient recycling and soil conservation to retain resources within cropping systems whilst recognising the need for judicious and efficient use of fertilizers. Before fertilizers can be used efficiently in many developing countries greater attention is required to making different nutrients available separately and supplying different blends of fertilizers to better match the requirements of different crops and soils. Often only compound fertilizers containing a large proportion of N are available and these are clearly inappropriate for use with legumes.

Economic considerations may make N_2-fixation a primary N source for resource poor farmers but there is more to gain than only 'free N', such as benefits from addition of organic matter in the case of green manures or disease breaking by including legumes into crop rotations. However, for biological N_2-fixing systems to provide a substantial amount of N to the system it is essential to ensure good legume growth and this often requires the use of fertilizers. Putting biological N_2-fixation to work effectively is certainly not an issue of no inputs if a system is to extend beyond basic self-sufficiency. Henzell (1988) stated that improved

Table 3. Nitrogen content and N_2-fixation in a *Sesbania rostrata*/multipurpose cowpea intercrop grown in a farmers' field with or without recommended rates of lime and PK fertilizers and the actual N benefits to a subsequent crop of rice (JF McDonagh, B Toomsan, V Limpinuntana and KE Giller; unpubl. results)

Fertilizer treatment	Total legume N (kg ha^{-1})	N from N_2-fixation in legumes		N recovered in rice	
		%	(kg N ha^{-1})	(%)	(kg N ha^{-1})
Control	36	51	20	23	5
Lime	36	47	17	38	10
PK	54	53	29	24	9
Lime + PK	107	76	84	33	24

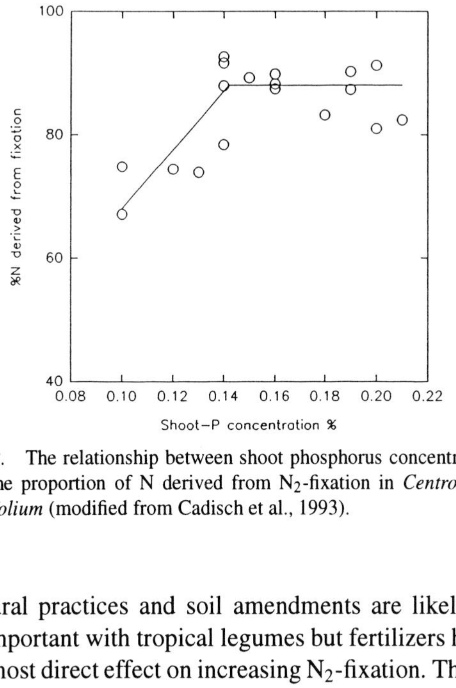

Fig. 2. The relationship between shoot phosphorus concentration and the proportion of N derived from N_2-fixation in *Centrosema acutifolium* (modified from Cadisch et al., 1993).

cultural practices and soil amendments are likely to be important with tropical legumes but fertilizers have the most direct effect on increasing N_2-fixation. This is especially true in farmers' fields where deficiencies of major and micro-nutrients frequently limit nodulation and N_2-fixation (Table 3).

In particular phosphorus deficiency is a factor commonly restricting the realisation of the potential of N_2-fixation by legumes mainly through reduced yield but also lowered tissue-%N and reductions in the proportion of N derived from fertilizer which occur with severe P deficiency (Fig. 2; Thomas, 1995). In developed countries the exploration of more efficient mechanisms to gather soil-P reserves, for example the use of vesicular-arbuscular mycorrhizas VAM or improved cultivars is viable due to often excessive use of P fertilizers in the past. In tropical countries P sources have often to be added to balance the long-term reductions in soil P reserves due to nutrient export in crops, and many tropical soils have a poor capacity to supply P for crop growth due to the strong sorption of P. There are however alternatives to the use of soluble P fertilizers for certain conditions such as highly reactive or partially acidulated rock-phosphates. In fact N_2-fixing systems are well suited to utilize rock-phosphates due to their net root proton excretion (soil acidification effect) and being a greater sink for Ca than non-legumes which thus helps to solubilize the rock-P. Potentially this may make the use of rock-P feasible at a higher pH (or in soils with a low buffering capacity) than the recommended pH < 5.5. The soil acidifying effect of legume roots can be utilized as a tool to favour legumes in associations. Rajan et al. (1991) reported that rock-P stimulated white clover growth more than ryegrass growth and hence increased the legume proportion in the pasture. This may also be important in intercrop systems, where better P availability and the presence of mycorrhiza can favour growth of the grain legume component in a maize-bean intercrop (Fig. 3).

Legumes are generally highly dependent on mycorrhizas for P uptake due to their coarse root systems. Inoculation of legumes with VAM can thus improve the utilization of low grade rock-P in acid soils with strong capacity to bind P (e.g. Cabala-Rosand and Wild, 1982). However, the problems of commercial production of VAM inoculum make use of VAM inoculation primarily a target for intensive vegetable systems or tree nurseries at present. Franco et al. (1992) reported great success in the recuperation of heavily degraded areas in Brazil using a package consisting of leguminous trees seedlings inoculated with rhizobia, VAM and supplied with rock-P and compost.

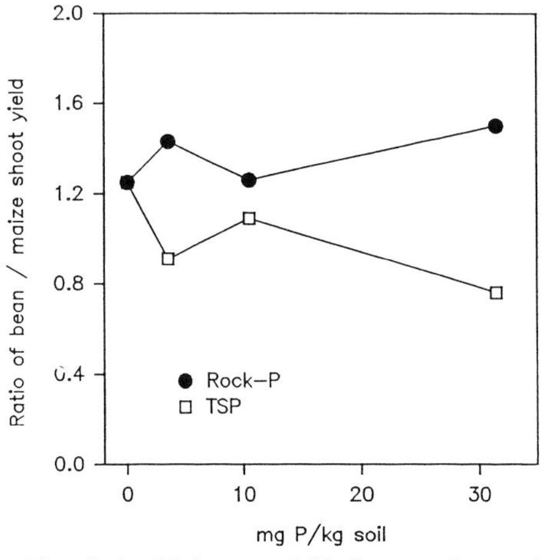

Fig. 3. The relationship between yield of common bean and maize in an intercrop as affected by phosphorus source (M Fürst and G Cadisch, unpubl. results).

Effects of N on N_2-fixation

Development of nitrate-insensitive legumes

In developed countries where fertilizers are freely used the major constraint to nodulation may often be the large pool of available N in the soil and this is reflected in the emphasis of research in developed countries on overcoming the suppression of nodulation and N_2-fixation by combined N (Herridge and Danso, 1995). Research has concentrated on the development of plant genotypes able to nodulate in the presence of large concentrations of nitrate in the soil. Different approaches have been used such as the screening of the sensitivity of different legumes, or different genotypes of the same legume, or the use of mutagenesis to create new variability in this respect. Genotypes insensitive to nitrate have been identified in pea (*Pisum sativum;* Jacobsen, 1984), common bean (*Phaseolus vulgaris*; Park and Buttery, 1988) and soyabean (*Glycine max;* Carroll et al., 1985). The 'supernodulating' genotypes of soyabean (Carroll et al., 1985) have attracted enormous media attention. As their name suggests, these supernodulating genotypes nodulate abundantly irrespective of the prevailing soil nitrate concentration. In fact they have lost the ability to control the number of developing infections (the process of autoregulation), which is one of the most important characters which serves to distinguish the symbiosis between rhizobia and legumes from a parasitic infection by bacteria.

Given the changing emphasis of agricultural policy away from maximizing production and towards minimizing possible negative environmental effects, one can question whether trying to extend N_2-fixation in soils rich in available nitrate is a wise approach. Losses of nitrate from soils by leaching or by denitrification are greatest when there is a large pool of nitrate left free in the soil (Addiscott et al., 1991; George et al., 1993). Any approach which will result in larger pools of unutilized nitrate in soils is likely to exacerbate problems of nitrate loss. It is thus hard to reconcile the enhancement of N_2-fixation in the presence on soil nitrate with public concern surrounding gaseous emissions of nitrous oxides from soils and pollution of groundwaters with nitrate.

Management of soil-N use in crop mixtures

Perhaps a more sensible and 'ecologically-sound' way of dealing with the 'problem' of excessive plant-available N for legume N_2-fixation is to exploit it by using crop combinations. Legumes in the tropics are often intercropped with cereals. Cereal crops usually germinate and establish effective root systems more rapidly than legumes with the result that the soil N concentration will be reduced such that nodulation and N_2-fixation are not inhibited. Small amounts of available soil or fertilizer N have often been shown to have a stimulatory effect on legume nodulation and N_2-fixation, the so-called 'starter effect'. This is mainly due to the stimulatory effect of N on growth and plant establishment during the period between root emergence and the onset of active N_2-fixation. The advantage in N yield in intercrops compared with sole crops can thus be attributed to the sparing effect of legumes (an example of weak competition for resources between the two crops) by which the N yield and efficiency of N use is better in the intercrop on an area basis than with the comparable sole crops. The role of direct transfer of N from legumes to companion crops appears to be of minor importance compared to the importance of sparing effects (Giller et al., 1991). Legume growth and N_2-fixation may in fact be reduced in intercrops by competition for resources such as light by tall cereals (e.g. Nambiar et al., 1983), but even so intercrops often give substantial yield advantages compared with the sole crops. Improved harvest technologies may in future make intercropping more attractive in intensive agriculture.

Management of soil-N use in mixed grass-legume pastures

The use of grass-legume pastures is another example where we can maximise N utilization within a system with the grass competing strongly for available N and thus controlling excessive soil mineral-N concentrations. Indeed, legumes in pasture production systems often fix a greater proportion of their N due to competition from the associated grasses for available soil N. The potential amounts fixed are easily enough to provide sufficient protein for animal production, but the use of legumes in pastures in both tropical and temperate regions is limited at present.

Inoculation with rhizobia

The greatest success in terms of modified agricultural practices arising from scientific research on BNF has undoubtedly been the development of rhizobial inoculants. The use of inoculants has allowed the successful introduction of legumes to new farming systems where compatible rhizobia were absent from the soils. Examples include the introduction of soyabean in the USA and Brazil, and the introduction of several pasture legumes in Australia. But despite years of research on inoculation the only example where there is widespread adoption of rhizobial inoculants in developing countries is that of soyabean. This is also one of the easiest research problems to tackle, due to the relative (but by no means absolute) specificity of soyabean for rhizobia. Even in some parts of South-east Asia, for instance North-east Thailand where soyabeans have become an important crop recently, inoculation is necessary to ensure good yields as compatible rhizobia are absent (Toomsan, 1990). The rapid expansion in soyabean cultivation by small-scale farmers in this region of Thailand has been possible due to the supply of inoculants produced by the Department of Agriculture in Bangkok, which has at times been unable to keep pace with demand.

Breeding for promiscuity

Part of the 'problem' of the need for inoculation of soyabean has undoubtedly been due to the intensive breeding of what was first introduced to agriculture in North America as a forage crop into a high-yielding oilseed crop *from a narrow genetic base.* Subsequent introduction of these high-yielding varieties into South America and other parts of the world thus entailed attention to ensuring their nodulation and N_2-fixation. In Brazil, the approach used was to grow without fertilizer N and to select rhizobial strains able to establish and fix N_2 under the acid soil conditions in which soyabean was grown. In Africa an alternative approach was adopted which is referred to as breeding for promiscuity.

In 1977 genotypes able to nodulate with indigenous rhizobia were reported from Tanzania (Chowdhury, 1977). As a result of this observation a strategy was developed to breed high-yielding, larger-seeded but promiscuous soyabeans which would nodulate with indigenous rhizobia. Despite some success in achieving nodulation of soyabean in Nigerian soils without inoculation, the 'promiscuous' genotypes developed in fact nodulated with a small subset of the indigenous rhizobia such that they often need to be inoculated to achieve good nodulation (Eaglesham, 1989). Another breeding programme for promiscuous soyabeans was conducted in southern Africa. This programme was designed to select well-adapted soyabeans which could nodulate abundantly with indigenous soil rhizobia in Zambia and led to the release of two varieties Magoye and Hernon 147 (Jahaveri and Joshi, 1987). These varieties have been widely adopted by smallholders in the north of Zambia and in southern Malawi (S Carr, pers. commun.). The promiscuous soyabeans yield around 1–1.5 t ha^{-1} under farmers' conditions which is comparable with average yields of their staple crop maize (1–2 t ha^{-1}) without fertilizer use.

Despite intense local demand for these soyabean varieties among smallholder farmers in Malawi there has been resistance to their distribution from researchers and extension workers. This is due to the scientific view that other varieties can give better yields, but this is true only when they are supplied with the correct inoculum. At present inoculants are only produced in Malawi on a small-scale for research and demonstration, and there is no mechanism for supply of inoculants to farmers. When limited amounts of inoculant are provided the recommendation in Malawi is that the inoculant is applied to the seed using sugar as a 'sticker', but sugar is such a scarce and highly prized commodity in Malawi that no farmer will readily bury it in the soil, even if only a small quantity is required. Under such circumstances the promiscuous soyabeans must be the most appropriate choice even though 'potential' yields could be higher using different, higher yielding varieties with inoculation.

By contrast in Zambia an inoculum production facility was established close to Lusaka in the early 1980's which has been successful in producing inoculants for soyabean. It was recently reported that inoculants which have been readily adopted by commercial farmers such that 23,000 ha of soyabeans were grown with inoculation in the 1991/2 growing season (BNF Bulletin, 1993). The same article questioned why inoculants are not widely adopted among smallholders in Zambia. At the time of the review a packet of inoculum cost US $3.60, and yet the average annual income of a Zambian smallholder was in the region of US $400 (Celis et al., 1991). This, coupled with the problems of distribution and dissemination of inoculants over the large distances in Zambia, and additional problems of soil infertility in many of the regions of Zambia occupied largely by smallholder farmers provides a fairly clear indication why inoculants are not widely used on small farms. Once farmers have seed of the promiscuous soyabeans they will save seed from one year to the next and have no extra cash input involved with soyabean cultivation. Surely this is an example where one technology, the use of rhizobial inoculants, is appropriate in one case (that of large-scale, commercial farmers) and another technology, the use of promiscuous soyabeans is appropriate for another group of farmers, the smallholders. A hopeful scenario may be envisaged that as farmers livelihoods and incomes improve, then their buying power and access to inputs such as rhizobial inoculants will increase such that they will then be able to achieve the higher potential yields offered with improved technology.

Selection of rhizobial strains

If large populations of compatible, effective rhizobia are present then responses to inoculation will often not be found using standard available inocula. This has been used as the basis for a model for predicting inoculation responses based on the availability of N and number of rhizobia in the soil (Thies et al., 1991). But the lack of inoculation responses when large populations of rhizobia are present is largely due to the very limited strain selection which has been done. Even in the case of some major grain legume crops a 'recommended' strain may simply represent a strain selected as the best nodulating and N_2-fixing strain under axenic, glasshouse conditions (in 'Leonard jars') compared with a limited number of isolates collected from different regions or countries. No plant breeder would claim to have selected an 'elite' genotype after taking a small number of randomly collected accessions (in the case of rhizobia often a number less than 100) and comparing their performance under laboratory or glasshouse conditions. Only very few strains ever reach the field testing stage in rhizobial assessment. If we consider that in one gram of soil there are often between 1000–10000 rhizobia compatible with a single crop, in one kg 10^6-10^7, in one hectare of topsoil 10^{13} or more! It is clear that there is enormous genetic diversity of rhizobia which has yet to be explored or exploited, and this also highlights the potential difficulties in establishing an inoculant strain.

The potential benefits of a more rigorous approach to strain screening have been demonstrated by the work of Sylvester-Bradley et al. (1983, 1991). By screening strains in undisturbed cores of soil it was possible to select strains which gave substantial increases in growth and N_2-fixation with a number of different tropical pasture legumes in the field in acid Oxisol soils. The soils in which these strains were evaluated had large indigenous populations of rhizobia, so that in order to give a response in plant N yield an inoculant strain had to be good competitor, (to displace the indigenous population), be well-adapted to the hostile acid soil environment and of course be a good N_2 fixer in order to cause an inoculation response. If a similar approach is used for rigorous strain selection for other legumes then substantial benefits in N_2-fixation may be found.

Breeding for enhanced N_2-fixation

Until recently legume breeding programmes have generally been conducted at high fertility sites, or with luxurious amounts of nutrients, including N fertilizer which will have led to selection against N_2-fixation. Several breeding programmes using a variety of selection methods have resulted in enhancement of BNF in grain legumes (see Bliss, 1993; Graham and Temple, 1984; Herridge and Danso, 1995; Miller and Fernandez, 1985). The simplest method of selection for N_2-fixation is to select for total N yield under conditions of poor N availability in the field. It is obviously important to maintain attention in any breeding programme for a specific trait to other important agronomic factors such as environmental adaptation and disease resistance. The total amount of N_2 fixed by a legume is not under the control of a few genes, but is a function of dry matter yield and as such is a result of the complex interaction of many physiological processes determined by the plant genotype and the environment. Thus breeding for enhanced N_2-fixation under specific conditions is not likely to yield a genotype of widespread usefulness. Rio Tibagi, a genotype of *Phaseolus vulgaris* regarded as a poor N_2-fixing genotype in Brazil

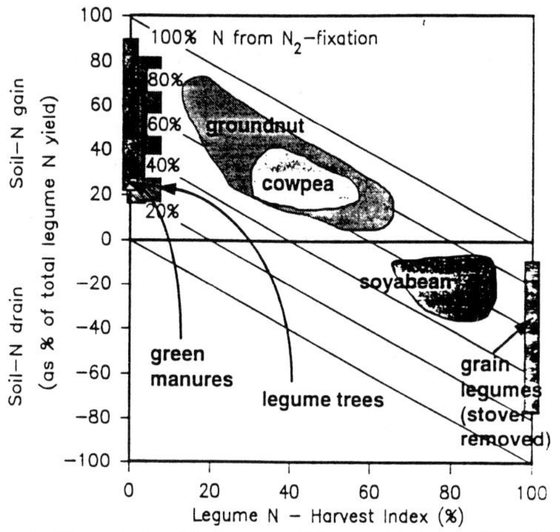

Fig. 4. The relationship between N harvest index and the %N from N_2-fixation in legumes required to give a net benefit to the cropping system (after Giller et al., 1994).

(Duque et al., 1985) was found to be among the best nodulators and N_2-fixers in Colombia (Kipe-Nolt and Giller; unpubl. results). This emphasises the need for selection of genotypes for specific agroecological environments, particularly in legumes with such a diverse environmental adaptation as *Phaseolus vulgaris.*

Engineering of cereals for N_2-fixation

This topic is discussed in detail by de Bruijn (1995) and it is clear that enormous physiological and biochemical obstacles need to be surmounted before the goal of N_2-fixing cereal crops will be realised. The potential benefits of N_2-fixing rice varieties, for example, are enormous. It should be recognised, however, that if cereals fix their own N this may lead to increased dependence on monocultures rather than rotations with legumes and other crops which may bring with it other problems for agriculture.

Introduction of new legumes into farming systems

Contributions of legumes to soil fertility in rotations with crops

The capabilities of grain legumes to contribute to soil fertility differ enormously. The greater the efficiency of the legume in translocating N to the harvested grain the smaller the contribution of N to the system. In this respect groundnut, which has green residues which contain large amounts of N at harvest, can supply much more N for subsequent crops than other grain legumes such as soyabean (Giller et al., 1994; Fig. 4). If legume stover is not returned to the soil at harvest then there will be a significant removal of soil N from the system by the legume crop (Wani et al., 1995), although this is often less than growth of a comparable cereal. Much of the apparent benefit of legumes in crop rotations may simply be due to the legume 'sparing' soil N by fixing the majority of the N removed at harvest, rather than the legume directly contributing N to the soil (Peoples et al., 1995). Such considerations serve to emphasise the benefits of using legumes as green manures, where all of the substantial amount of N_2 fixed can be used to enhance soil fertility directly (Becker and Ladha, 1995). It is important to remember that if the N_2-fixation by the legume is constrained by an adverse environment then the contribution of the legume to soil fertility will also be minimal.

Despite repeated demonstrations of the usefulness of green manures, whether herbaceous or woody legumes, in fixing large quantities of N and enhancing soil fertility their adoption remains very limited. Legumes and the aquatic fern *Azolla* have been used traditionally for maintaining soil fertility as green manures but their use has declined since N fertilizers became widely available. This is partly due to additional management required for using green manures, intensification of cropping which means that the land is continually required for production and the availability of off-farm employment. Continued applications of urea or ammonia-based N fertilizers lead to soil acidification and are unsustainable in the absence of liming and soil organic matter management. Government subsidies support the use of inorganic N fertilizers (Becker and Ladha, 1995; Tandon, 1993) and more imaginative policies for agricultural support would undoubtedly favour the use of green manures more, though whether this would be sufficient stimulus for widespread return to green manures is unclear.

Promising avenues? Agroforestry with N_2-fixing trees

Substantial emphasis has been placed on the potential of exploiting fast-growing, N_2-fixing trees for the enhancement of soil fertility in agroforestry systems over the past decade. Research has been focused on hedgerow intercropping or alley cropping in which trees are grown within arable fields and are pruned periodically to provide green manure for soil amendment (Sanginga et al., 1995). Many development organisations have been promoting alley cropping for adoption by smallholders in the humid and sub-humid tropical regions of Africa over the past few years. But recently researchers have been expressing doubts as to the general usefulness and acceptability of this technology for farmers in many regions. This is an example where an innovation which is intrinsically appealing and successful under controlled research station conditions has been rapidly recommended without thorough evaluation by farmers.

It has always been a source of some surprise to us that non-nodulating tree legumes have been popular in many field experiments exploring the use of leguminous trees in alley cropping. In particular, *Senna (Cassia) siamea* has widely been used in alley cropping experiments throughout the tropics and other species have more local interest, for instance *S. spectabilis* in countries of southern Africa. Both of these species, which used to be placed in the genus *Cassia*, are unable to nodulate (Allen and Allen, 1981) and do not fix N_2 (Ladha et al., 1993). These are trees commonly grown as ornamentals in the tropics but it is of note here that in 1936 it was recommended that these species should not be used in agricultural fields: "... all leguminous plants of *Cassia* genus, without nodules, are harmful to the main crops with which they are associated. Their use as a cover crop or as a shade is not recommended." (IIA, 1936). If the N_2-fixing and non-nodulating trees have comparable rates of production, either the N_2-fixing tree is not fixing substantial amounts of N or the non-nodulating tree is extracting more N from the soil. What matters here is where is that extra N coming from; is it being extracted from the surface soil horizons at a time that will be in competition with the alley crop? Or is it being recovered from depths where otherwise it would be lost from the system with the roots acting as a safety-net (van Noordwijk, 1989). In the latter case, there is no problem of competition with the crop although the implications for long-term soil fertility will again be different than if a nodulated tree legume was actually contributing N from N_2-fixation to the system. Results of recent investigations in Malawi indicate that *S. spectabilis* has a very large density of fine roots in the surface horizons of the soil throughout the alleys (O Itimu and K E Giller, unpubl. results) which is likely to cause substantial competition for nutrients and water with the alley crop.

Nodulated, N_2-fixing trees can also compete strongly with crops under conditions of moisture stress such that there can be negative effects on crop growth. Even in sub-humid climates there can be problems during dry spells. For instance, *Gliricidia sepium* is a tree widely used on field boundaries throughout the tropics and it is used in Indonesia as a support for black pepper, but farmers think it 'dries out the soil' when rainfall is limited (T Fairhurst, pers. commun.). From the information available to date it seems that most leguminous trees explore the nutrient-rich topsoil for N and other nutrients. The ideal agroforestry tree which can fix N_2 abundantly and derive nutrients only at depth has yet to be found.

Leguminous trees have been planted in some regions in grass strips on steep slopes as part of soil conservation measures with some acceptance, but no universal recommendations can be made about the wider applicability of technologies involving introduction of trees into farming systems.

Productive pastures: Introduction of forage legumes

In temperate regions of developed countries the use of mixed pastures will increase again with the current trend to extensification. This should be easily possible since there is sufficient knowledge of appropriate animal and meadow management for a successful use of such mixed swards. Under tropical conditions, legumes have so far played only a minor role in pastures, but there is a need for increased use of N_2-fixing systems as many rangelands are becoming degraded and ongoing deforestation of rainforests is partly due to cattle farmers who abandon degraded pure grass pastures. Oversowing of tropical savannas and rangelands with palatable legumes appears not to be a viable option because cattle preferentially graze the legumes which leads to poor persistence (Fisher et al., 1992). Additionally, it appears that native savanna grasses could not respond to the improved N conditions unless other nutrients such as P, K and Ca are also supplied. There is also the danger that the savanna grasses cannot withstand the heavy grazing pressure often associated with the improved nutritional conditions which may thus lead to further degradation unless all components are carefully controlled.

The problem of legume persistence is also evident in most improved tropical pastures since the grasses possess a C_4 photosynthetic pathway and are thus highly aggressive in competition with introduced legumes. The proportion of legumes in grazed pastures needed to sustain the N balance has been estimated to be just over 10% (as % of herbage dry matter production) for temperate pastures (Sheehy, 1989) whilst for tropical pastures estimates range from 13–45% depending on pasture utilization (Cadisch et al., 1994; Thomas, 1992). However, legume presence in mixed pastures decreases often to < 10% after 2–3 years. To avoid competition problems it has been suggested that protein banks with legume monocultures should be used. This approach avoids persistence problems, but does not maximise N_2-fixation since through the recycling of the high quality legume litter material and animal excreta soil mineral N concentrations will progressively rise and thereby suppress N_2-fixation. There is thus an urgent need to tackle the problem of legume persistence in pastures. This will need a multidisciplinary effort to select appropriate legumes, alleviate environmental constraints, and identify adequate animal management options (see Thomas, 1995). For the more humid tropics there now exist some more persistent stoloniferous legumes like *Desmodium ovalifolium* and *Arachis pintoi* which are tolerant to trampling and have their growing points below-ground, protected from grazing. For drier areas alternatively more woody species or legume 'cocktails' appear to be promising as are legumes which contain large amounts of tannins which reduce palatability and thereby avoid grazing pressure on the legume. The presence of 'anti-quality' factors such as polyphenols however, also may reduce the rate of litter decomposition (Fox et al., 1990) and thereby reduce the ability of high quality legume litter material to alleviate N immobilization induced by grass litter with a high C:N ratio. However, not all polyphenols are equally active in affecting decomposition (Handayanto et al., 1994) and this may help to select legume materials with a reduced acceptance by cattle, but which still decompose readily to contribute directly to increased N turnover in mixed pastures. The amounts of such legumes required would be less in the case of unpalatable legumes since N losses are reduced if less legume N is cycled through the animal excreta (Cadisch et al., 1994).

The use of agropastoral systems to recuperate the estimated 50 million ha degraded pure grass pastures has been rapidly adopted by farmers in the Cerrados of Brazil. This was stimulated by the development of acid-tolerant rice lines which are sown together with the new pasture (Kluthcouski et al., 1991). Unfortunately, currently most farmers adopt this technique only to recuperate their pure grass pastures but rarely include legumes. However, this system presents a unique economically-viable opportunity to introduce N_2-fixing legumes into pastures. Tropical farmers readily adopt new grasses such as *Brachiaria* spp., but adoption of adapted legume material by farmers is frustratingly slow to date. While legume benefits are apparent to researchers, farmers perceive these benefits slowly and see the use of legumes as an additional complication for pasture management. Consequently, such new materials require focus and promotion as new products or novel crops and not just as 'new cultivars' (Ferguson, 1992). The speed of adoption of new material will ultimately also depend on the commercial availability of seeds which is currently a bottleneck.

Farmers' success in experimentation?

In many cases where new legumes have been introduced into farming systems, the specific importance of BNF is hard to evaluate per se. Often legumes have been introduced due to the benefit arising from a specific agricultural product, whether as a source of grain or fodder. There are however, some examples where the introduction of legumes has been for a large part due to their ability to fix atmospheric N_2.

Nitrogen fixing trees in arid lands

An interesting example which serves to emphasise the need for a full understanding of the system is that of the islands of fertile soil found under N_2-fixing trees (mainly of the Mimosoideae) in semi-arid and arid environments. In northern India *Prosopis cineraria* trees are planted and maintained along field boundaries and within fields used for arable crops (Tejwani, 1987) and *Faidherbia albida* is widely distributed in cropping fields throughout the Sahel and parts of southern Africa (Vandenbeldt, 1991). Crop yields are often double beneath the tree than outside the canopy and this is largely due to the improved organic matter content of the soil, and the associated improvements in moisture retention and the supply of N and other nutrients. Of these trees *Faidherbida albida* is of particular interest in that it is deciduous, but in contrast to most deciduous trees in arid areas it sheds its leaves at the beginning of the rainy season. The adaptive significance of this peculiar phenology is a mystery as it effectively maximises evapotranspirational losses, but it has the advantage of reducing interception of light thus reducing competition with crops growing underneath the trees. The best documented role of *F. albida* is in fields in West Africa where it has been traditionally maintained in farmers' fields again with islands of fertile, productive soil around the trees (Dancette and Poulain, 1969). Similar effects are seen beneath *Acacia* trees which comprise a large proportion of the trees in savannas of the Australia, East Africa and the Sahel

(Kessler and Breman, 1994). These beneficial effects of the building of soil fertility have been attributed to the N_2-fixing ability of the trees, but does this simple analysis stand up to closer scrutiny?

The enrichment of soils under the canopies of trees is not confined to N_2-fixing trees; many other trees such as the Baobab (*Adansonia digitata*) in East Africa (Belsky et al., 1993) have similar effects. The beneficial effects of the trees can be measured in terms of both the nutrient content of the soil and the productivity of the understorey vegetation. Successful establishment of *F. albida* trees was highly correlated with the more fertile sites in the soil and with termite mounds, suggesting that the trees only established where the soil was inherently more fertile. Once a tree is established the inputs of litter from falling leaves, the inputs of droppings of birds and animals which tend to congregate under the trees and inputs from fine roots all contribute nutrients to the soil beneath the trees. The trees therefore become very efficient in recycling the nutrients which continue to accumulate beneath the canopy and organic matter contents of the soil increase.

As yet there is no direct evidence for a substantial role of N_2-fixation in most of these cases. What studies have been done to examine nodulation have found little evidence for abundant nodulation in the field, even in young trees (Vandenbeldt, 1991). In some areas *F. albida* trees roots have been reported as deep as 40 m where they are probably reaching a permanent water table. Where a deep water table is reached by the deep roots of trees, then nodulation may occur at depth. Nodulation at depths of 4–6 m has been shown with *Prosopis* growing in the Sonoran Desert and large populations of rhizobia have been found at the water table (Virginia et al., 1986). Large populations of compatible *Bradyrhizobium* have been found at depths of 30–35 m below the soil surface under *Faidherbia* trees (Dupuy and Dreyfus, 1992) and as roots have been reported to reach such depths it is likely that nodules are formed at the deep water table in such environments.

Where root systems of trees in arid lands have been examined in detail non-legume trees tend to have much more substantial root systems which may extend up to 50 m from the trees (Soumaré et al., 1993). This may indicate that different strategies of nutrient acquisition are at work with the different species, the non-legumes drawing nutrients from a larger area whilst the nodulated legumes fixing N. The result of both strategies is nutrient enrichment around the tree base, but the long-term implications for soil fertility are obviously different.

The importance of a proper understanding of the sources of N in such systems is that we cannot predict the long-term implications for soil fertility without this knowledge, and this is emphasised by the example given above. Attempts to intensify production based on the use of non-legume trees in such systems may founder as the area available for nutrient capture for each tree is reduced, whereas if legume trees are fixing their N then a degree of intensification of tree planting may have long-term feasibility.

Legumes for soil-fertility for smallholder farmers

To what extent has research led to increases in inputs from N_2-fixation in fields of small-scale farmers in the tropics? An interesting example from the tropics is that of the spontaneous adoption of the velvet bean (*Mucuna pruriens* var. *utilis*) in Atlantic Honduras in a relay-rotation with maize (Buckles et al., 1992). The *Mucuna* is sown in the fields some 40–50 days after sowing of the maize and grows vigorously to produce a heavy mat of organic matter after maize is harvested. At the time for next maize sowing the *Mucuna* has matured and set seed, from which it can regenerate after it is slashed back and the maize planted in holes made through the thick mulch. In this example the role of the farmer and researcher is difficult to elucidate. The advent of this practice can be traced to earlier widespread use of *Mucuna* for interplanting with maize in the southern USA (Tracy and Coe, 1918), and a probable introduction of the velvet bean to Honduras by immigrant farmers some 20 years ago. The legume was incorporated into the cropping system by farmers so successfully that a spontaneous diffusion of the technology took place, such that now some 70% of farmers practice this method in some areas.

Legume cover crops in plantations

Research conducted on the management of rubber and oil palm plantations in Malaysia in the early part of this century led to the widespread adoption of cover legumes to protect the soil. Often a mixture of *Pueraria phaseoloides*, *Centrosema pubescens* and *Calopogonium caeruleum* was found to be successful and even though the legumes generally died out within five years of plantation establishment, benefits from using cover crops in terms of reduced requirements for N fertilizers could be seen in enhanced production many years later (Broughton, 1977). The annual amounts of N_2-fixation have been estimated to be in the range of

Table 4. Residual benefit and utilization of mineralized ^{15}N labelled legume residue N in temperate agriculture under field conditions

Crop residues	Plant part/ timing	N added as green manure (kg N ha^{-1})	N recovery by first catch crop (%)	Accumulated N recovery after (x) years (%)	Residue N released but not utilized[a] by first crop (kg N ha^{-1}(%))	N loss	Reference
Medicago	Whole	48	20-28	<33 (2)	<4 (<30%)	-	Ladd et al. (1983)
littoralis	Whole	33-97	17	21 (2)	5-13 (45%)	16	Ladd and Amato (1986)
Lentil	Whole	51	19 (9)[c]	-	9 (48%)	13	Bremer and van Kessel (1992)
	Straw	45	6 (-18)[c]	-	1 (1%)	4	Bremer and van Kessel (1992)
	Whole	85-104	12-27	-	-	14-40	Janzen et al. (1990)
Clover	Whole	78-81	20-24	-	38-46 (71%)	6-11	Müller and Sundman (1988)
Subteranean	Whole	70	18	-	32 (71)	5	Müller and Sundman (1988)
clover	Whole	60	11-20		30 (71-81%)	8-28	Müller (1988)
Alfalfa	Whole	112	14-29	15-30 (2)	33-47 (25-46%)	25-46	Harris and Hesterman (1990)
	Whole	65	11	18 (3)	-	-	Ta and Farris (1990)
Field bean	Whole	73	17		26 (68%)	1	Müller and Sundman (1988)
	Root+ stover	88	7	19 (2)	-	-	Senaratne and Hardarson (1988)
Lathyrus	Whole	76-110	9-26	11-28 (2)	-	17-54	Janzen et al. (1990)
Pea	Autumn	134	14	20 (3)	35 (64%)	-	Jensen (1994)
	Spring[b]	134	6	9 (3)	46 (70%)	-	Jensen (1994)
	Root+ stover	34	26	44 (2)	-	-	Senaratne and Hardarson (1988)
Soybean	Leaves	150	9 (4)[c]	-	-	68	Bergersen et al. (1992)
	Stems	55	2 (2)[c]	-	-	89	Bergersen et al. (1992)
	Roots	11	3 (145)[c]	-	-	70	Bergersen et al. (1992)
Urea	-	25-75	46	49 (2)	14-41 (56%)	20	Ladd and Amato (1986)
$(NH_4)_2SO_4$	-	50	16-44	18-46 (2)	28-42 (56-84%)	29-63	Janzen et al. (1990)
	-	20-40	39	-	-	-	Ta and Faris (1990)

[a] Residue N not utilized = N added × % N released/100 - N recovered in first catch crop.
[b] Sowing time of catch crop, residue incorporated in previous autumn.
[c] N difference method.

150 kg ha^{-1} (Giller and Wilson, 1991) although these estimates were based on standing crops of the cover crop and do not allow for the heavy mulch of decomposing leaves which rapidly develops under the cover crop. The input from decomposing leaves under *Calopogonium* over a three month period was equivalent to the standing biomass which suggests that inputs from cover crops in plantations may be substantially greater (Van Noordwijk and Purnomisid, 1992). Driving for kilometres through a plantation during the establish-

Table 5. Residual benefit and utilization of mineralized residue N of tropical green manures under field conditions

N source	Amount of N added (kg N ha^{-1})	Crop yield increase (%)	Fertilizer equivalent (kg N ha^{-1})	N recovery by first catch crop (%)	Accumulated N recovery after (x) years (%)	Residue N released but not utilized[a] by first crop (kg N ha^{-1}(%))	Estimated N loss (kg N ha^{-1})	Reference
Groundnut	100-130	65	75	6-10[b] (12-21)[c]	-	77-97 (91%)	50	McDonagh et al. (1993)
	130	25-32	-	<37[c]	-	>63 (>56%)	-	Suwanarit et al. (1986)
Cowpea	31-33	-	40-50	11-28[b]	44-73 (2)	18-21 (66-85%)	-	Sisworo et al. (1990)
	70	52	-	41[c]	-	29 (50%)	-	Morris et al. (1986)
Soyabean	29-38	-	-	9-14[b]	23 (2)	23-27 (92%)	-	Sisworo et al. (1990)
	165	37	-	<31[c]	-			Suwanarit et al. (1986)
Mungbean	110	35	80	33[c]	-	53 (60%)	-	Morris et al. (1986)
Urea	-	-	60/90	9-36[b]	23/50 (2)	18-59 (58-70%)	-	Sisworo et al. (1990)
	-	35	80	33[c]	-	54 (68%)	-	Morris et al. (1986)

[a] Residue N not utilized = N added × % N released from litterbags/100 - N recovered in first catch crop.
[b] ^{15}N estimates.
[c] N difference method.

Table 6. Residual benefit and utilization of mineralized pruning N in tropical agroforestry systems under field conditions

Pruning materials	N added as green manure (kg N ha^{-1})	DM Yield increase (%)	Fertilizer equivalent (kg N ha^{-1})	N recovery by first catch crop (%)	Accumulated N recovery after (x) years (%)	Residue N released but not utilized[a] by first crop (kg N ha^{-1}(%))	Reference
Gliricidia	94	120	-	12[b](29)[c]	27 (1)	75 (88%)	Haggar et al. (1993)
	180	35-52	≥45	21[c]	-	133 (78%)	Tian et al. (1993)
Leucaena	178	40-71	>45	45[c]	-	115 (72%)	Tian et al. (1993)
	450	84-200	78	8-13[c]	-	344 (88%)	Sanginga et al. (1986)
	370	38-104	-	2-6[c]	-	235 (94%)	Mulongoy and van der Meersch (1988)
Sesbania	194	40	>60	31[c]	-	93 (60%)	Becker et al. (1991)
Erythrina	153	110	-	9[b](17)[c]	12 (1)	100 (87%)[d]	Haggar et al. (1993)
Urea	-	-	60	58[c]	-	25 (42%)	Becker et al. (1991)

[a] Residue N not utilized = N added × % N released from litterbags/100 - N recovered in first catch crop.
[b] ^{15}N estimates.
[c] N difference method.
[d] ^{15}N released from mulch.

ment phase and stopping to examine the knee-deep luxuriance of the cover crop is enough to convince anyone of the unparalleled benefits of N_2-fixation in this system. The advantages of using the legumes are no doubt partly due to their ability to form ground cover rapidly, reduce weeding problems and to reduce soil erosion, thereby reducing losses of nutrients, but the capacity for rapid growth in nutrient poor soils with proper phosphorus fertilization is due to their ability to fix N_2. Indeed, legume cover crops have proved so successful in this role that sowing of cover crops is the standard practice for plantation establishment in Southeast Asia. Substantial areas of tree crops such as rubber or oil palm in Indonesia are produced by smallholder farmers on their own land and such tree-crop based systems can result in wealth-generating and productive agriculture for the rural poor.

Managing fixed nitrogen: Optimizing use efficiency

Given the high quality (low C:N ratio) generally found in legume green manure residues it is surprising how little of this N is utilized by subsequent crops. In temperate regions only some 9–29% of the added green manure N was recovered by the first catch crop (Table 4). In tropical agricultural systems similar values of 6–28% for green manures and only 9–12% of the N in legume tree prunings (^{15}N estimates) was recovered (Tables 5 and 6). Estimates based on N balance data usually give better recoveries (up to 45%) than ^{15}N methods, but the available data shows that crop recovery of N from legume residues was poorer than recovery of fertilizer N. The suggestion that N not utilized by the first crop will be available to subsequent crops has yet to be proven. Current, limited data clearly show that recoveries of legume N in the second and later crops are generally small (2–15%) under both tropical and temperate conditions.

High quality legume residues decompose quickly; litterbag studies suggest that approximately 50–80% of the added residue N is released under temperate conditions and 70–95% under tropical conditions during the first subsequent crop cycle. Based on litterbag data or direct ^{15}N mineralization studies we have calculated the amount and the proportion of residue N released but not utilized by the first catch crop under field conditions (Tables 4–6). These calculations suggest that in temperate regions up to 50 kg N ha^{-1}, in tropical green manures systems up to 100 kg N ha^{-1} and in agroforestry systems even up to 300 kg N ha^{-1} of the N which is available for crop uptake is not utilized by the first catch crop. Thus there is a large potential for improving the efficiency of use of legume N in cropping systems. Better utilization of the resources is also important to prevent pollution as N not utilized represents also the amount of N potentially subject to losses.

There are different management options which may improve utilization of N in crop residues. Attention is currently being focused on improving the synchronization between crop N demand and the timing of net N mineralization. Marked differences in recovery of pea residues by different catch crops have been observed, with winter barley recovering 12% more than winter wheat (Jensen, 1994b). Simple management strategies such as burying residues can help to reduce the significant losses which may occur due to ammonia volatilization losses. Up to 14% of the N added in lentil stover was lost within 14 days, and losses from fresh green manure placed above the soil may be even greater (Costa et al., 1990; Janzen and McGinn, 1991). Incorporating green manure into soil can virtually eliminate such ammonia losses.

The role of fixed N in long-term soil fertility

Recovery of N from legume residues in the second and further subsequent crops is very poor and is usually < 1% by the third crop. Jensen (1994a) found that only 1–2% of the residual organic ^{15}N from pea residues was mineralizable after 2 years of decomposition in the field. This suggests that the remaining residue N is present in rather recalcitrant soil organic matter, but long-term data on recovery of N from legume residues is lacking (see Tables 4–6).

Whether legume residues can be used to 'build' or augment soil organic matter contents of degraded soils is unclear. The proportional utilization of mineralized N is often better from poor quality legume materials such as lentil straw (Bremer and van Kessel, 1992) or non-legume materials (Bremer and van Kessel, 1992; Wagger et al., 1985) and thus a higher N conservation effect may be achieved with such materials in the long-term. With plant material releasing N rapidly, as in the case of high quality legume leaf materials, a better yield response may be obtained in the first year but also a higher proportion seems to be lost. Additionally, Tian et al. (1993) suggested that plant residues with low C:N ratio and lignin content enhance crop performance through direct nutritional contributions, whereas low quality (high C:N, and lignin) plant residues do so more through effects of mulching. The mixing of residues of different quality may be one approach to managing rates of decomposition and N release which can help to optimise the efficiency which with fixed N_2 is used.

Experimentation or extension

Simple innovations in agriculture, such as the introduction of new varieties, fertilizers or rhizobial inoculum can rapidly lead to widespread adoption of the changes. Innovations which involve making more substantial changes within the farming system such as adding a legume green manure, or changing the way in which grain legume residues are used may often be highly successful on a research station or on a demonstration plot, but fail to achieve widespread adoption. A comment often made is that farmers are conservative and not ready to adopt new ideas, but in practice many examples can be found where farmers have adopted new innovations rapidly when the benefit can clearly be seen. The case cited above concerning the adoption of *Mucuna* as a green manure in Honduras is an example of this. Here lessons can be learned from the farmer, with the role of the researcher being to facilitate transfer of a successful technology to other comparable farming systems and there is now intense interest in extending this technique to areas with similar farming systems in Central America.

Nor are farmers unwilling to invest huge amounts of cooperative labour in ameliorating degraded lands. In one of the most hostile, arid environments for crop production in northern Burkina Faso farmers work to overcome the ecological constraints of barren, capped soils by constructing planting pits, collecting and transporting organic manures to the fields and building semi-permeable stone bunds to aid retention and infiltration of the sparse rainfall (Batterbury, 1993). Obviously such harsh environments offer few alternatives for young farmers ready to set up their own household, and land security is necessary to repay the initial investment needed to recuperate the land. But this is an important example which illustrates the possibilities for rehabilitation of degraded soils by farmers.

Research scientists often play scant attention to the farming systems context of their work, and it must be remembered that it is the farmer who takes the risks when trying new methods. There is widespread realisation now that successful innovation for change in agriculture must begin with the farmer (Chambers et al., 1989) and this is true in both less-industrialised and highly industrialised countries. There is a fashion in development projects to assess farmers' needs by all too rapid 'rapid rural appraisal' or participatory rural appraisal. Such techniques are undoubtedly useful tools in gaining understanding of farming systems but cannot substitute for longer cooperation in experimentation between farmers and researchers.

There is a danger if enthusiastic researchers or development workers recommend technologies before they have been fully evaluated by farmers. For instance, some legume trees such as *Sesbania sesban* or *Erythrina* spp. are initially very productive and appear to be useful species for alley cropping but in fact are short-lived, particularly under intensive pruning. If the wrong tree is widely recommended before it has been well-tested this may cause resistance to further testing of a management strategy which might be succesful when a different tree species is used. It is the farmers who are taking the risks and risks of trying new technologies may be great for resource poor farmers. It is interesting in this context to remember that researchers in Latin America still bemoan the 'loss of credibility' of tropical pasture legumes with farmers due to the early introduction of pasture legumes which had been highly successful in Australia, but were poorly adapted to the aluminium rich soils of South America. Although well-adapted legumes are now available, their adoption is inhibited due to the earlier premature enthusiasm which led to a technology being offered without thorough testing under farmers' conditions.

Many of these arguments concerning adoption and extension of technologies are true for all aspects of agricultural research, and not specific solely to N_2-fixation. But because symbiotic BNF involves a complex interaction between *Rhizobium* strains, legume genotypes and the environment it is particularly sensitive to stresses and necessitates a careful approach.

Conclusions

Appropriate solutions tailored to farmers needs

In the discussion of the advantages of legume inoculation compared with breeding for promiscuity there are powerful scientific arguments to suggest that inoculation is the approach most likely to result in maximising N_2-fixation. However, the circumstances for small holder farmers prevailing in southern Africa suggest that the use of the promiscuous soyabean varieties is more appropriate at present. Poor farmers in Rwanda even have problems maintaining favoured varieties of *Phaseolus* beans as they are forced to eat their seed stocks at times of the year when food is limited (Sperling and Loevinsohn, 1993). Technologies to help such farmers necessitate costs to be kept to a minimum.

Awareness is also required that driving forces for farmers' acceptance of technologies are likely to be governed by different priorities than the researcher. The adoption of *Mucuna* by farmers in Honduras owes much to the ablity of the thick mat of green manure to control weeds, and this is a strategy which has been recommended for promoting green manure use in Asia (van der Heide and Hairiah, 1989). Intercropping of pre-rice green manures with a highly-valued vegetable legume is a more attractive option than using the green manure alone (Giller et al., 1994). In Rwanda, smallholder farmers are keen to grow legume trees such as *Calliandra calothyrsus* and *Sesbania sesban* to provide stakes to support climbing beans (Giller and Wilson, 1991). The best 'selling point' for a technology for which the main purpose in the mind of a researcher is to improve soil fertility may therefore well be different for the farmer.

Future benefits: Ecological approaches to integrated systems

Emphasis must be placed on maximising our exploitation of the ecological adaptation of the different components of the agricultural system to growth under hostile environments. The enormous variability which exists in rhizobia in tropical soils, or in tree legumes for agroforestry, is largely unexploited. It is thus essential that applied experimentation recognises and attempts to solve constraints of the environment on N_2-fixation, by focusing on research under farmers' conditions and by working in concert with scientists from other disciplines.

In order to maximise utilization of the environment emphasis must be placed on diversity within agricultural systems. Diverse assemblages, whether as intercrops, mixed pasture swards or agroforestry systems have the overall advantage of greater interception and more efficient utilization of resources. Combining biological approaches such as the use of rhizobia with mycorrhizas (together with slowly-solubilised rock phosphates) have additional advantages. The benefits of diverse systems do not come from N_2-fixation alone, but allow efficient use of the fixed N_2 within the system. The challenge here is how to manage diverse systems to optimise their productivity and to tailor the farming system to the environment. This is essentially what farmers have always done, but is also a lesson the agricultural researcher must always bear in mind.

Momentum to favour BNF?

In the more industrialised nations the driving force behind changes in agricultural practices are overproduction and environmental concerns, which have led to an upsurge of interest in low-input and organic farming. By definition such practices rely intensively on efficient recycling and on renewable resources. It is likely that increased dependence will be seen on biological means of maintenance of soil fertility in the future, which must be centred on BNF. In Europe this will almost certainly mean an increase in the amount of legume/grass pasture sown in arable/ley rotations as the most effective means of restoring and enhancing soil fertility.

The subsidy of N fertilizers is a significant disincentive to the use of biological methods to enhance N availability such as the use of green manures. In Asia the increased availability of fertilizers has been matched by dramatic increases in production which as part of the green revolution technologies have been instrumental in feeding the growing population. It is probably not possible to produce the yields on the scale required by reliance solely on biological N_2-fixation, but there is evidence for declining rice yields under highly intensive rice production which are thought to be related to organic matter quality. Legume green manures may therefore have a role to play in integrated systems of N management which involve the use of both fixed, organic N and fertilizer N for rice production.

In many parts of Africa yields are still so appallingly small such that they could readily be doubled with reliance solely on biological N_2-fixation as an N source. The urgency of yield improvement and the lack of adequate distribution perhaps dictates that removal of fertilizer subsidies would be criminal, although at present it is generally the better-off farmers who actually use fertilizers and thus benefit from what fertilizer subsidies there are. Unfortunately in many African countries where large potential benefits might be accrued from BNF the basic inputs such as P fertilizers necessary to allow N_2-fixation to work are beyond the reach of the farmers. Until this problem is solved there is little possibility for increasing the inputs from N_2-fixation substantially.

Research agenda

The deadline for development of new technologies based on research programmes is often set for 'within the next 5–10 years'; often because of the shared need of both funding agency and scientist to suggest that they are conducting research of immediate importance in terms of the economic impact of their research. Such a time-scale can cynically be interpreted as being just long enough to ensure the appointment of a successor so that should there be a payoff the cudos can be claimed by the present incumbent scientist, but should things not materialise there is always the option for claiming that the follow-up to the initial scientist was not appropriate. In many cases a deadline of 5–10 years is perhaps not an unrealistic prediction of what is possible, but there is a disjunction in our research strategies which leads to scientists having research goals which are far-removed from those of farmers. Often only lip-service has been paid to farmers as a justification of research but the majority of research was (and often still is) done to pursue topics of scientific and personal interest rather than to develop on-farm technology.

One of the greatest problems for technology development is that scientists working in international development research have (and are often forced to have) two driving agenda: one to do novel research and produce scientific papers which will be applauded by their peers and the second to produce applicable technologies for farmers. The second agenda can often be little more than a justification for sponsors. If the farmer is the target then exploratory on-farm research involving the farmer must be the starting point for any programme for agricultural improvement, and yet this is often not the case.

Technical knowledge is available which if implemented could result in immediate dramatic increases in the inputs from N_2-fixation based on simple technologies such as solving mineral nutrient deficiencies through fertilizer use. Policies for development of sustainable agriculture must recognise that strategies for productive agriculture necessitate a fusion of methods based on recycling and biological N_2-fixation, together with carefully measured fertilizer inputs designed to optimise nutrient conservation and the efficiency with which applied nutrients are used.

When there are promising technologies which deserve introduction to new regions for testing by farmers then more attention must be paid to adapting innovations based on introduction of legumes to the needs of the farmer. This can best be done by introducing new ideas or species and allowing their diffusion into existing systems by a more natural process of evolution rather than the conventional 'top-down' approach. The full benefits of N_2-fixation cannot be realised until more emphasis is placed on fitting legume-based production to farming systems, with due attention to the farmers' needs.

This is not only an academic problem but a problem of great importance in terms of providing sustainable livelihoods and means of production for subsistence farmers who are likely to face increasing problems in the future. We do not have the luxury of time to wait for new genetic approaches to produce results but need to use all the knowledge currently available in appropriate and imaginative ways.

Acknowledgements

Many of the ideas presented here have been developed during discussions with colleagues of whom Simon Batterbury, Malcolm Blackie, Thomas Fairhurst, Barry Smithson and Kate Wilson deserve particular mention. Errors or omissions remain our responsibility and we would be delighted to receive comments regarding this paper as part of what is an ongoing and extremely important debate.

References

Addiscott T M, Whitmore A P and Powlson D S 1991 Farming, Fertilizers and the Nitrate Problem. CAB International,Wallingford. 170p.

Alexandratos N 1988 World Agriculture: Toward 2000. An FAO Study. Belhaven Press, London.

Allen O N and Allen E K 1981 The Leguminosae. A Source Book of Characteristics, Uses and Nodulation. Macmillan, Madison. 812p.

Batterbury S P J 1993 Planners or Performers? Indigenous Dryland Farmers in Northern Burkina Faso. ODI AgriculturalAdministration (Research and Extension) Network Paper 42., London. pp 18–31.

Becker M and Ladha J K 1995 Green manure technology: Potential, usage and limitations. Plant and Soil 174.

Becker M, Diekmann K H, Ladha J K, de Datta S K and Ottow J C G 1991 Effect of NPK on growth and nitrogen fixation of *Sesbania rostrata* as a green manure for lowlandrice (*Oryza sativa* L.). Plant and Soil 132, 149–158.

Belsky A J, Mwonga S M and Duxbury J M 1993 Effects of widely spaced trees and livestock grazing on understoryenvironments in tropical savannas. Agrofor. Syst. 24, 1–20.

Bennet J and Ladha J K 1992 Introduction: feasibility of nodulation and nitrogen fixation in rice. *In* Nodulation and Nitrogen Fixation in Rice. Potentials andProspects. Eds. G S Khush and J Bennet. pp 1–14, IRRI, Los Baños.

Bergersen F J, Turner G L, Gault R R, Peoples M B, Morthorpe L J and Brockwell J 1992 Contributions of nitrogen in soybean crop residues to subsequent crops and to soils. Aust. J. Agric. Res. 43, 155–169.

Bliss F A 1993 Breeding common bean for improved biological nitrogen fixation. Plant and Soil 152, 71–80.

BNF Bulletin 1993 BNF means million $ for Zambia: Who benefits? pp 1–2.

Bremer E and van Kessel C 1992 Plant-available nitrogen from lentil and wheat residues during a subsequent growing season. Soil Sci. Soc. Am. J. 56, 1155–1160.

Brookes P C, McGrath S P and Heijnen C E 1986 Metal residues in soils previously treated with sewage sludge and their effects on growth and nitrogen-fixation by blue-green algae. Soil Biol. Biochem. 18, 345–353.

Broughton W J 1977 Effect of various covers on soil fertility under *Hevea brasiliensis* Muell. Arg. and on growth of the tree. Agro-Ecosyst. 3, 147–170.

Buckles D, Ponce I, Sain G and Medina G 1992 Tierra cobarde se vuelve valiente: Uso y difusión del frijol de abono (*Mucuna deeringiana*) en las laderas del Litoral Atlántico de Honduras. CIMMYT, Mexico.

Cabala-Rosand P and Wild A 1982 Direct use of low grade phosphate rock from Brazil as fertilizer. II. Effects of mycorrhizal inoculation and nitrogen source. Plant and Soil, 65, 363–373.

Cadisch G, Sylvester-Bradley R and Nösberger J 1993 Effects of phosphorus and potassium on N_2 fixation (^{15}N dilution) of field grown *Centrosema acutifolium* and *C. macrocarpum*. Field Crops Res. 31, 329–340.

Cadisch G, Schunke R M and Giller K E 1994 Nitrogen cycling in a pure grass pasture and a grass-legume mixture on a red Latosol in Brazil. Trop. Grassl. 28, 43–52.

Carroll B J, McNeill D L and Gresshoff P M 1985 Isolation and properties of soybean [*Glycine max* (L.) Merr.] mutants that nodulate in the presence of high nitrate concentrations. Proc. Natl. Acad. Sci. USA 82,4162–4166.

Celis R, Milimo J and Wanmali S 1991 Adopting Improved Farm Technology: A Study of Smallholder Farmers' in Eastern-Province, Zambia. IFPRI, Wageningen, 105p.

Chambers R, Pacey A and Thrupp L A (Eds) 1989 Farmer First. Farmer Innovation and Agricultural Research. IT Publications, London.

Chowdhury M S 1977 Response of soybean to *Rhizobium* inoculation at Morogoro, Tanzania. *In* Biological Nitrogen Fixation in Farming Systems of the Humid Tropics. Eds. A Ayanaba and P J Dart. pp 245–253. John Wiley and Sons, London.

Conway G D 1987 The properties of agroecosystems. Agric. Syst. 24, 95–117.

Costa F J S A, Bouldin D R and Suhet A R 1990 Evaluation of N recovery from mucuna placed on the surface or incorporated in a Brazilian oxisol. Plant and Soil 124, 91–96.

Dancette C and Poulain J F 1969 Influence de l'*Acacia albida* sur les facteurs pedoclimatiques et les rendements des cultures. Sols Afr. 13, 197–239.

De Bruijn F J 1995 Potential and pitfalls of trying to extend symbiotic interactions of nitrogen-fixing organisms to presently non-nodulated plants, such as rice. Plant and Soil 174.

Dudal R and Deckers J 1993 Soil organic matter in relation to soil productivity. *In* Soil Organic Matter Dynamics and Sustainability of Tropical Agriculture. Eds. K Mulongoy and R Merckx. pp 377–380. John Wiley and Sons, Chichester, New York.

Dupuy N and Dreyfus B 1992 Presence of bradyrhizobia under *Acacia albida. In Faidherbia albida* in the West African Semi-Arid Tropics. Ed. R J Vandenbeldt. pp 145–148. ICRISAT/ICRAF, Andhra Pradesh, Nairobi.

Duque F F, Neves M C P, Franco A A, Victoria R L and Boddey R M 1985 The response of field grown *Phaseolus vulgaris* to *Rhizobium* inoculation and the quantification of N_2-fixation using ^{15}N. Plant and Soil 88, 333–343.

Eaglesham A R J 1989 *Rhizobium* inoculants. *In* Microbial Inoculation of Crop Plants. Eds. R Campbell and R M MacDonald. pp. 29–48. IRL Press, Oxford. UK.

Ferguson J E 1992 Experiences at the interface of research and development with tropical pastures. *In* Pastures for the Tropical Lowlands: CIAT's Contribution. pp101–120. CIAT, Cali, Colombia.

Fisher M J, Lascano C E, Vera R R and Rippstein G 1992 Integrating the native savanna resource with improved pastures. *In* Pastures for the Tropical Lowlands:CIAT's Contribution. pp 75–100. CIAT, Cali, Colombia.

Fox R H, Myers R J K and Vallis I 1990 The nitrogen mineralization rate of legume residues in soils as influenced by their polyphenol, lignin and nitrogen contents. Plant and Soil 129, 251–259.

Franco A A, Campello E F, da Silva E M R and Faria S M 1992 Revegetaçâo de solos degradados. ComunicadoTecnico No. 9, 11p. EMBRAPA-CNPBS, Rio de Janeiro.

George T, Ladha J K, Buresh R J and Garrity D P 1993 Nitrogen-dynamics during the aerobic soil phase in lowland rice-based cropping systems. Soil Sci. Soc. Am. J. 57, 1526–1532.

George T, Cassman L G and Singleton P W 1995 Manipulating legume nitrogen fixation in low phosphorus tropical upland cropping systems. Plant and Soil 174.

Giller K E and Wilson K J 1991 Nitrogen Fixation in Tropical Cropping Systems. CAB International, Wallingford. 313p.

Giller K E, McGrath S P and Hirsch P R 1989 Absence of nitrogen fixation in clover grown on soil subject to longterm contamination with heavy metals is due to survival of only ineffective *Rhizobium*. Soil Biol. Biochem. 21, 841–848.

Giller K E, Ormesher J and Awah F M 1991 Nitrogen transfer from *Phaseolus* bean to intercropped maize measured using ^{15}N-enrichment and ^{15}N-isotope dilution methods. Soil Biol. Biochem. 23, 339–346.

Giller K E, McDonagh J F and Cadisch G 1994 Can Nitrogen Fixation Sustain Agriculture in the Tropics? *In* Soil Science and Sustainable Land Management in theTropics. Eds J K Syers and D L Rimmer. pp 173–191. CAB International, Wallingford.

Giller K E, McDonagh J F, Toomsan B and Limpinuntana V 1994 Legumes in the cropping systems of North-east Thailand. *In* Soil Management in Sustainable Agriculture. Eds. H Cook and H Lee. pp 95–103. Wye College Press, Wye, Kent, UK.

Graham P H and Temple S R 1984 Selection of improved nitrogen fixation in *Glycine max* (L.) Merr. and *Phaseolus vulgaris* L. Plant and Soil 82, 315–327.

Greenland D J 1994 Soil science and sustainable land management. *In* Soil Science and Sustainable Land Management in the Tropics. Eds. J K Syers and D L Rimmer. pp 1–15. CAB International, Wallingford.

Haggar J P, Tanner E V J, Beer J W and Kass D C L 1993 Nitrogen dynamics of tropical agroforestry and annual cropping systems. Soil Biol. Biochem. 25, 1363–1378.

Handayanto E, Cadisch G and Giller K E 1994 Nitrogen release from prunings of legume hedgerow trees in relation to quality of the prunings and incubation method. Plant and Soil, 160, 237–248.

Harris G H and Hesterman O B 1990 Quantifying the nitrogen-contribution from alfalfa to soil and two succeeding crops using nitrogen-15. Agron. J. 82, 129–134.

Henzell E F 1988 The role of biological nitrogen fixation research in solving problems in tropical agriculture. Plant and Soil, 108, 15–21.

Herridge D F and Danso S K A 1995 Enhancing legume N_2fixation through selection and breeding. Plant and Soil 174.

International Institute of Agriculture 1936 Use of Leguminous Plants in Tropical Countries as Green Manure, Cover and Shade. The Institute, Rome. 262p.

Jacobsen E 1984 Modification of symbiotic interaction of pea (*Pisum sativum* L.) and *Rhizobium leguminosarum* by induced mutations. Plant and Soil 82, 427–438.

Jahaveri F and Joshi J 1987 Soyabean production by small scale farmers in Zambia. *In* Soyabeans in Southern Africa. Ed. D L Cole. pp 9–12. University of Zimbabwe, Harare.

Janzen H H, Bole J B, Biederbeck V O and Slinkard A E 1990 Fate of N applied as green manure or ammonium fertilizer to soil subsequently cropped with spring wheat at three sites in Western Canada. Can. J. Soil Sci. 70, 313–323.

Janzen H H and McGinn, S M 1991 Volatile loss of nitrogen during decomposition of legume green manure. Soil Biol. Biochem. 23, 291–297.

Jensen E S 1994a Dynamics of mature pea residue nitrogen turnover in unplanted soil under field conditions. Soil Biol. Biochem. 26, 455–464.

Jensen E S 1994b Availability of nitrogen in ^{15}N-labelled mature pea residues to subsequent crops in the field. Soil Biol. Biochem. 26, 465–472.

Kamprath E J 1970 Exchangeable aluminium as a criterion for liming leached mineral soils. Soil Sci. Soc. Am. Pro. 34, 252–254.

Kessler J J and Breman H 1994 Woody Plants in Semi-arid West Africa. Springer Verlag, Heidelberg. 300p.

Kluthcouski J, Pacheco A R, Teixeira S M and Oliveira E T 1991 Renovaçâo de pastagens de Cerrado com arroz. I – Sistema Barreirâo. Bullettin Technico No. 33. EMBRAPA-CNPAF, Goiânia, Goias, Brazil.

Ladd J N and Amato M 1986 The fate of nitrogen from legume and fertilizer sources in soils successively cropped with wheat under field conditions. Soil Biol. Biochem. 18, 417–425.

Ladd J N, Amato M, Jackson R B and Butler J H A 1983 Utilization by wheat crops of nitrogen from legume residues decomposing in soils in the field. Soil Biol. Biochem. 15, 231–238.

Ladha J K, George T and Bohlool B B (eds) 1992 Biological Nitrogen Fixation for Sustainable Agriculture. Kluwer Academic Publ., Dordrecht. 209p.

Ladha J K, Peoples M B, Garrity D P, Capuno V T and Dart P J 1993 Estimating dinitrogen fixation of vegetation using the nitrogen-15 natural abundance method. Soil Sci. Soc. Am. J. 57, 732–737.

Lorenz S E, McGrath S P and Giller K E 1992 Assessment of free-living nitrogen fixation activity as a biological indicator of heavy metal toxicity in soil. Soil Biol. Biochem. 24, 601–606.

McDonagh J F, Toomsan B, Limpinuntana V and Giller K E 1993 Estimates of the residual nitrogen benefit of groundnut to maize in Northeast Thailand. Plant and Soil 154,267–277.

McGrath S P, Brookes P C and Giller K E 1988 Effects of potentially toxic metals in soil derived from past applications of sewage sludge on nitrogen fixation by *Trifolium repens* L. Soil Biol. Biochem. 20, 415–424.

Miller J C and Fernandez G C J 1985 Selection for enhanced nitrogen fixation in cowpea. *In* Cowpea Research, Production and Utilization. Eds. S R Singh and K O Rachie. pp 318–325. John Wiley and Sons, New York.

Morris R A, Furoc R E and Dizon M A 1986 Rice responses to a short-duration green manure. II. N recovery and utilization. Agron. J. 78, 413–416.

Müller M M 1988 The fate of clover-derived nitrogen ^{15}N during decomposition under field conditions: Effect of soil type. Plant and Soil 105, 141–147.

Müller M M and Sundman V 1988 The fate of nitrogen (^{15}N) released from different plant materials during decomposition under field conditions. Plant and Soil 105, 133–139.

Mulongoy K and van der Meersch M K 1988 Nitrogen contribution by leucaena (*Leucaena leucocephala*) prunings to maize in an alley cropping system. Biol. Fertil. Soils 6, 282–285.

Nambiar P T C, Rao M R, Reddy M S, Floyd C N, Dart P J andWilley R W 1983 Effect of intercropping on nodulation and N_2-fixation by groundnut. Exp. Agric. 19,79–86.

Oates K M and Kamprath E J 1983 Soil acidity and liming: II. Evaluation of using aluminium extracted by various chloride salts for determining lime requirements. Soil Sci. Soc. Am. J. 47, 690–692.

Park S J and Buttery B R 1988 Nodulation mutants of white bean (*Phaseolus vulgaris* L.) induced by ethyl-methane sulphonate. Can. J. Plant. Sci. 68, 199–202.

Peoples M B, Herridge D F and Ladha J K 1995 Biological nitrogen fixation: an efficient source of nitrogen for sustainable agriculture? Plant and Soil 174.

Rajan S S S, Fox R L, Saunders W M H and Upsdell M 1991 Influence of pH, time and rate of application on phosphate rock dissolution and availability to pastures. I Agronomic benefits. Fert. Res. 28, 85–93.

Robbins G B, Bushell J J and Kannna P K 1989 Nitrogen immobilization in decomposing litter contributes to productivity decline in aging pastures of green panic (*Panicum maximum* var. trichoglume). J. Agric. Sci., Camb. 113, 401–406.

Sanchez P A and Salinas J G 1981 Low input technology for managing Oxisols and Ultisols in tropical America. Adv. Agron. 34, 280–400.

Sanginga N, Mulongoy K and Ayanaba A 1986 Inoculation of *Leucaea leucocephala* (Lam.) de Wit with *Rhizobium* and its nitrogen contribution to a subsequent maize crop. Biol. Agric. Hort. 3, 347–352.

Sanginga N, Vanlauwe B and Danso S K A 1995 Management of biological N_2 fixation in alley cropping systems: Estimation and contribution to nitrogen balance. Plant and Soil 174.

Senaratne R and Hardarson G 1988 Estimation of residual N effect of faba bean and pea on two succeeding cereals using ^{15}N methodology. Plant and Soil 110, 81–89.

Sheehy J E 1989 How much dinitrogen fixation is required in grazed grassland? Ann. Bot. 64, 159–161.

Sisworo W H, Mitrosuhardjo M M, Rasjid H and Myers R J K 1990 The relative roles of N fixation, fertilizer, crop residues and soil in supplying N in multiple cropping systems in a humid, tropical upland cropping system. Plant and Soil 121, 73–82.

Soumaré A, Groot J J R, Kone D and Radersma S 1993 Structure spatiale du systéme racinaire de deux arbes du Sahel: *Acacia seyal* et *Sclerocarya birrea*. Instituted'Economie Rurale, Bamoko, Mali and AB-DLO, Wageningen, The Netherlands. 37p.

Sperling L and Loevinsohn M E 1993 The dynamics of adoption: Distribution and mortality of bean varieties among small farmers in Rwanda. Agric. Syst. 41, 441–453.

Stoorvogel J J, Smaling E M A and Janssen B H 1993 Calculating soil nutrient balances in Africa at different scales I.Supra-national scale. Fert. Res. 35, 227–235.

Suwanarit A, Suwannarat C and Chotechanungmanirat S 1986 Quantities of fixed N and effects of grain legumes on following maize, and N and P status of soil as indicated by isotopes. Plant and Soil 93, 249–258.

Sylvester-Bradley R, Ayarza M A, Mendez J E and Moriones R 1983 Use of undisturbed soil cores for evaluation of *Rhizobium* strains for inoculation of tropical forage legumes in a Colombian Oxisol. Plant and Soil 74, 237–247.

Sylvester-Bradley R, Mosquera D, Asakawa M and Sanchez G 1991 Promiscuity and responses to rhizobial inoculation of tropical kudzu (*Pueraria phaseoloides*). Field Crops Res. 27, 267–279.

Ta T C and Faris M A 1990 Availability of N from ^{15}N-labelled alfalfa residues to three succeeding barley crops under field conditions. Soil Biol. Biochem. 22, 835–838.

Tandon H L S 1993 Soil fertility and fertiliser use – an overview of research for increasing and sustaining soil productivity. *In* The Integration of Natural and Man-made Chemicals in Sustainable Agriculture. Ed. H L S Tandon. pp 1–32. CASAFA-ISSS-TWAS, New Delhi.

Tejwani K G 1987 Agroforestry practices and research in India. *In* Agroforestry: Realities, Possibilities and Potentials. Ed. H L Gholz. p 225. Martinus Nijhoff Publ., Dordrecht.

Thies J E, Singleton P W and Bohlool B B 1991 Modelling symbiotic performance of introduced rhizobia in the field by use of indices of indigenous population size and nitrogen status of the soil. Appl. Environ. Microbiol. 57, 29–37.

Thomas R J 1992 The role of the legume in the nitrogen cycle of productive and sustainable pastures. Grass Forage Sci. 47, 133–142.

Thomas R J 1995 Role of legumes in providing N for sustainable tropical pasture systems. Plant and Soil 174.

Tian G, Kang B T and Brussaard L 1993 Mulching effect of plant residues with chemically contrasting compositions on maize growth and nutrients accumulation. Plant and Soil 153, 179–187.

Toomsan B 1990 Groundnut microbiology research at Khon Kaen University. *In* Groundnut Improvement Project at Khon Kaen University. Report of Work for 1986–1988. Ed. A Patanothai. pp 89–111. Khon Kaen University, Thailand.

Tracy S M and Coe H S 1918 Velvet Beans. Farmers' Bulletin 962. USDA, Washington.

Vandenbeldt R J 1991 Rooting systems of western and southern African *Faidherbia albida* (Del.) A. Chev. (syn. *Acacia albida* Del.) – a comparative analysis with biogeographic implications. Agrofor. Syst. 14, 233–244.

Van der Heide J and Hairiah K 1989 The role of green manures in rainfed farming systems in the humid tropics. ILEIA 5, 11–13.

Van Noordwijk M 1989 Rooting depth in cropping systems in the humid tropics in relation to nutrient use efficiency. *In* Nutrient Management for Food Crop Production inTropical Farming Systems. Ed. J van der Heide. pp 129–169. Institute of Soil Fertility, Haren, The Netherlands.

Van Noordwijk M and Purnomisid P 1992 Litter fall in cassava-based cropping systems in Lampung. Agrivita 15, 29–33.

Virginia R A, Jenkins M B and Jarrell W M 1986 Depth of root symbiont occurrence in soil. Biol. Fertil. Soils 2,127–130.

WCED 1987 Our Common Future. Oxford University Press, Oxford.

Wagger M G, Kissel D E and Smith S J 1985 Mineralization of nitrogen from nitrogen-15 labeled crop residues under field conditions. Soil Sci. Soc. Am. J. 49, 1220–1226.

Wani S P, Rupela O P and Lee K K 1995 Sustainable agriculture in the semiarid tropics through biological nitrogen fixation in green legumes. Plant and Soil 174.

Plant and Soil **174:** 279–286, 1995.

Research evaluation and impact analysis of biological nitrogen fixation

M.C.S. Bantilan and C. Johansen
International Crops Research Institute for the Semi-Arid Tropics (ICRISAT), Patancheru, Andhra Pradesh 502 324, India

Key words: chickpea, genetics of BNF, impact analysis, nitrogen fixation, nodulation variants, *Rhizobium* inoculation, technology adoption

Table of contents

Abstract
Introduction
Limited adoption of inoculation technology
  Assessment of "need to-inoculate"
  Inadequate demonstration of inoculation technology
  Quality control of inoculant production
  Difficulties of *Rhizobium* inoculation in the tropics
  Economics of *Rhizobium* inoculation technology
The research-adoption-impact continuum
Considerations for attracting support for BNF research
  Impact analysis
  Benefit/cost analysis
  Management and genetic options
  Inoculation technology
  Outlook for N fertilizer
Conclusion
Acknowledgements
References

Abstract

Although viable *Rhizobium* inoculation technology for cultivated legumes has long been available, there has been little sustained adoption of this technology in tropical regions. Reasons contributing to this include inadequate demonstration of the technology, presence of adequate native rhizobia, high soil mineral nitrogen levels which suppress nitrogen fixation, inadequate quality control of *Rhizobium* inoculum and difficulties of inoculating under tropical conditions. In order to ensure a better adoption rate of existing or emerging biological nitrogen fixation (BNF) technologies, it is proposed that future research and development efforts better focus on the research-adoption-impact continuum. The salient features of this approach are described in this paper, using the example of recently developed nodulation variants in chickpea as a potential means of increasing BNF in this crop. It is suggested that previous experience with *Rhizobium* inoculation technology is amenable to ex-post impact analysis to analyze bottlenecks, and that ex-ante impact analysis should be built into on-going or planned BNF research, to better ensure that technology adoption occurs.

Introduction

The oil crisis of the early 1970s, and the consequent price escalation of nitrogenous fertilizers sparked off a "BNF-boom" in research that lasted through to the early 1980s. Optimistic claims were made concerning substitution of fertilizer nitrogen (N) by biologically fixed N, and funds flowed to support research in biological nitrogen fixation (BNF). In the 1990s, however, there seems to be little residual effect of this BNF-boom in the fields of resource-poor farmers in developing countries, particularly in South Asia. One reason is, of course, that the oil crisis prompted the discovery of vast new oil and gas reserves, and that N fertilizer prices have generally stabilized at levels affordable (with or without government subsidy) to all but the poorest of farmers. Another reason is that the promising prospects concerning BNF in the 1970s have not been realized in the form of adoption of BNF technology by resource-poor farmers to any significant extent.

In India in particular, there have been several large-scale schemes to introduce *Rhizobium* inoculation for the major legume crops (Verma and Bhattacharyya, 1992), but there is little evidence of any widespread adoption of this technology by farmers. This situation exists despite evidence from many experiments showing significant responses to inoculation (e.g. Kumar Rao, 1990) and calculations of economic viability (Verma and Bhattacharyya, 1992). This contrasts with the situation in countries such as Australia (Roughley and Pulsford, 1982) or Canada (Rennie and Hynes, 1993), where BNF research has led to widespread adoption of *Rhizobium* inoculation technology. But there the circumstances were different, with mainly introduced temperate legume species requiring specific strains of rhizobia, and with large-scale, mechanized (thus simplifying *Rhizobium* inoculation procedures) and commercialized farming systems. The only example of large-scale, sustained adoption of *Rhizobium* inoculation technology that we are aware of in Asia is the case of soybean in Thailand (Chanaseni and Kongngoen, 1992). Here also there appears to be a need for introduction of specific rhizobia to match the introduced soybean cultivars; native rhizobia do not always adequately nodulate these cultivars.

With this history of promise offered, but little evidence of delivery in Asia (apart from Thailand), it is not unnatural that research administrators are somewhat wary of new proposals for BNF research targetted at improving the lot of small, resource-poor farmers. With respect to agriculturally-important legumes, applied research in BNF has previously been overwhelmingly directed towards *Rhizobium* inoculation technology, with the aim of enhancing infection and N_2 fixation by addition of superior *Rhizobium* strains. Other options, such as manipulation of agronomic practices to favor BNF or genetically altering the plant to increase the symbiotic activity, have received less attention in the applied sense. In this paper we attempt to summarize the reasons for limited adoption of inoculation technology in Asia, suggest means of evaluating BNF research and measuring its impact at the level of farmers' fields, and give some specific suggestions for future research and development approaches.

Limited adoption of inoculation technology

Various reasons have become apparent for limited adoption by farmers of experimentally proven *Rhizobium* inoculation technology, in the tropics generally and in South Asia in particular.

Assessment of "need-to-inoculate"

Recommendations to inoculate are often of a universal nature (e.g. Jeswani and Baldev, 1990), to be applied across diverse environments and legume species, without apparent recognition of well established and marked site-to-site and legume species, and even cultivaral, differences in inoculation response. It is sometimes argued that, as all such differences in response cannot possibly be known or understood, inoculation may be regarded as an "insurance policy" with a low premium. However, before farmers, or anyone else, would be prepared to invest in "insurance" an understanding of the risks or forgone opportunities of not using the technology is needed. For effective extension of BNF technology, it appears necessary to define more carefully than hitherto the probability of an inoculation response for a specific situation. Some major factors determining response to inoculation are as follows:

- Absence or inadequate numbers of rhizobia in the soil, native or introduced, that can effectively nodulate the target legume. Tropical legumes are largely promiscuously nodulated by *Bradyrhizobium* which are ubiquitous in soils where these legumes normally grow; hence the limited response of these legumes to *Rhizobium* inoculation (Date, 1977).
- Whether indeed the natural variation in the rhizobial germplasm has been adequately examined to identify truly superior strains for particular situations.
- Even moderate levels of soil mineral N inhibit nodulation (Harper and Gibson, 1984), which is not overcome by addition of more rhizobia through inoculation.
- There are large differences between and within legume species in the degree to which they can meet their own N needs through fixation.
- Other plant growth limiting factors strongly interact with nitrogen fixation.
- The quality of the *Rhizobium* inoculum and the effectiveness of the inoculation technique.

The INLIT (International Network of Legume Inoculation Trials) approach (Davis et al., 1985) of NifTAL (Nitrogen Fixation by Tropical Agricultural Legumes), University of Hawaii, remains a valid approach to determine the need-to-inoculate. Treatments consist of an uninoculated control, an inoculated treatment, a treatment with "optimum" N fertilizer, and presence or absence of another major limiting factor for the legume (usually phosphorus). As multilocational field trials are expensive, various preliminary tests can give an indication as to likely response. An example is the use of simple models relating inoculation responsiveness to most probable number (MPN) of effective rhizobia and level of soil mineral N (Singleton et al., 1992).

Inadequate demonstration of inoculation technology

Activities in BNF technology have often remained within the discipline of "soil microbiology" with inadequate interaction with other disciplines, let alone extension personnel. There is little evidence that the demonstration and extension process for BNF technology, to accompany other improved practices, has been thoroughly thought through and effectively applied on a farm-scale basis.

Quality control of inoculant production

In the tropics, there are few cases where *Rhizobium* inoculant production systems of consistently adequate quality have been established and maintained over a reasonable period. Shortcomings and their suggested remedies have been described by Thompson (1984, 1991).

Difficulties of Rhizobium *inoculation in the tropics*

High temperatures typical of tropical and sub-tropical environments mitigate against continued viability of *Rhizobium* in carrier packets, even if their numbers had been adequate initially (Somasegaran et al., 1984). Normal sowing times of legumes in these regions, at the beginning and end (for crops to be grown on residual soil moisture) of a summer rainy season, are normally hot periods (ambient maximum temperatures >30°C) where exposure of cultures to lethal temperatures during the storage and inoculation processes is almost unavoidable, even with refrigeration available. Further, if the inoculum is prepared as non-sterile, higher temperatures may favor competitors of *Rhizobium*. More work is needed to develop robust procedures that would minimize adverse high temperature effects.

Economics of Rhizobium *inoculation technology*

Although calculations of the economic viability of introducing inoculation technology have been done and high rates of return asserted (e.g. Verma and Bhattacharyya, 1992), these calculations often have deficiencies. For example, production costs are often subsidized by government agencies and personnel costs are sometimes ignored; actual costs are therefore underestimated. There can be mis-calculation of expected returns, based on inoculation responses extrapolated over regions, and costs in terms of time or skill required for effective inoculation at the normally busy time of sowing are often overlooked. A more thorough, and more conservative, accounting is desirable to convincingly present likely returns on investment in *Rhizobium* inoculation technology.

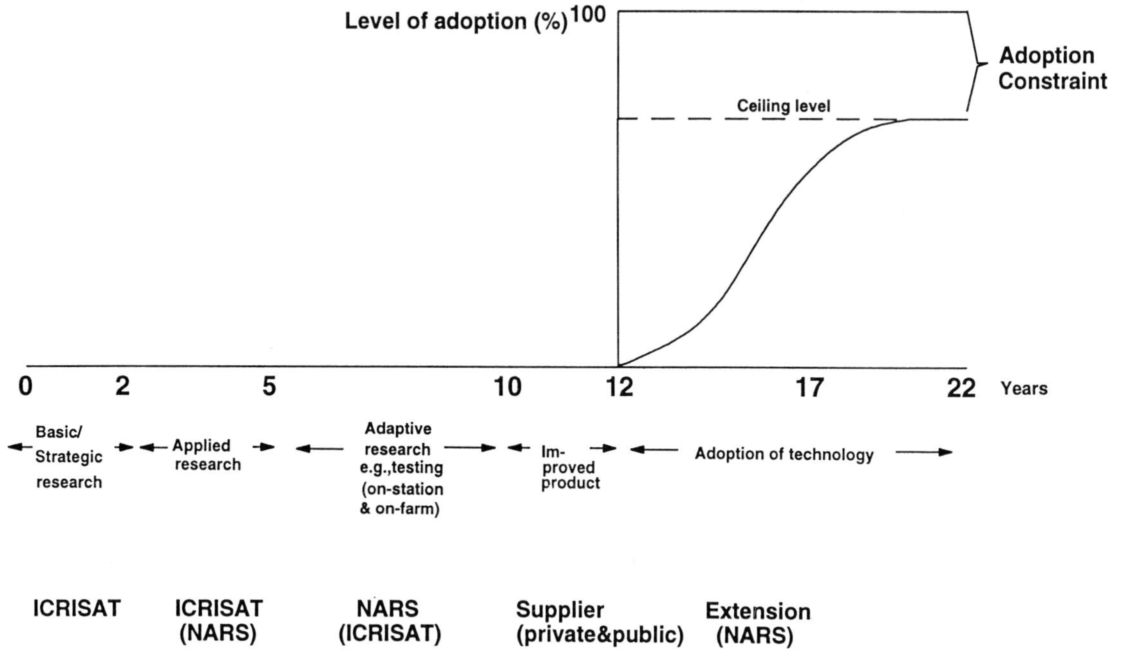

Fig. 1. Schematic representation of the research, development and adoption process over time, indicating relative involvement of ICRISAT and national agricultural research systems (NARS).

The research-adoption-impact continuum

It is suggested that future proposals for BNF research, that claim to be ultimately directed towards farmers' fields, be considered in the light of the entire continuum from basic research to impact assessment. Particularly with an increasing scarcity of resources, more and more the bottom line of any research undertaking is its impact or likely impact. To facilitate the conduct of impact assessment, both ex-post (after the event) and ex-ante (before the event), an understanding of the whole research process is essential.

The research-evaluation continuum may be systematically viewed by using a general framework (Davis et al., 1994) which traces the development of the different components of the research process, its output and logical consequences. The conceptualization of the framework starts with the consideration of research investments which are used to fund a specific research project, designed to develop a new technology for use by farmers. If the research project does successfully achieve its objectives, it usually generates research output in the form of, first, some new knowledge and then a change in the technology for use by farmers. To be more specific, the application of science-based technologies resulting from BNF research is expected to bring about increases in yield and product quality from crops presently planted or from those which are subsequently planted. BNF research is also expected to improve the efficiency of input use via agronomic practices and crop management. Ultimately, the above changes in the production and consumption environment are translated into upgrading of the welfare of farmers who use the technology as well as of consumers who use the final products. Further, the environmental benefits of greater reliance on BNF in cropping systems are elaborated below and elsewhere in this volume.

Before the final benefits of research accrue to society (i.e. producers and consumers), two important conditions must be met. First, the research undertaken must be successful in achieving its targeted objectives. This introduces the notion of probability of success or relative research capability, relating to the risk involved in most research which could mean that an intended technological improvement may not indeed eventuate, even after a significant period of experimentation or investigation. Second, the potential increase in production promised by a new technology is ultimately achieved only when the technology is adopted and utilized by farmers. If the technology does not result in an improvement in some way over existing technologies then farmers are unlikely to use it. In this case the technology, although developed, is redundant. Even if it does unambigously result in improved conditions some farmers may still not adopt it. Several reasons may explain this, one of which is that there may be reluctance among farmers to deviate from well-tried, and in their opinion proven, practices. This condition necessitates the consideration of rates of technology adoption and the factors constraining it.

The measurement of the welfare gain to society is incomplete if it does not take into account the externalities which the technology involves. The externality consideration in this framework may either be negative or positive. Classic examples of negative externalities are human-induced soil erosion in agriculture and detrimental effects of chemical-based technology. The long list of effects of the latter example includes the deleterious effect of pesticides on the health of farmers and their families, the transmittal of chemical residues through the food chain to consumers, the toxic effect of chemicals on animals like fish, shrimps, frogs and helpful insects in the farmers' fields, the contamination of ground and surface waters, and the reduction of microorganism populations in the soil that help sustain soil fertility.

The positive externalities are incorporated within the above framework through consideration of the concept of spillover effects. Three types of spillover effects are possible (Bantilan and Davis, 1991). The first type involves across-location spillovers wherein a technology developed through research for one product in a specific location can be adapted to improve the production efficiency of the same product in other locations (geo-political or agro-ecological). The consideration of this type of spillover effect is relevant because the applicability of the new technology may not be the same for all locations as these locations refer to production environments differentiated by agronomic, climatological and ecological factors.

The second type of spillover effect refers to across-commodity applicability of the technology developed. For example, a cultural management technique developed specifically for groundnut production may also potentially improve the efficiency of production of other legumes.

The nature of the first two types of spillover effects reflects the direct applicability of a technology across different locations/production environments and across different commodities. Thus, they are referred to as direct spillover effects.

A third type of spillover effect is referred to as the indirect or price spillover effects. Because technological change for a particular commodity in a specific location brings forth increased supply which may cause price changes, then the price effect on other locations (if the commodities are traded) or its price effect on related commodities may have significance. This is particularly relevant when the elasticities of the product demand are relatively small and/or the rate of product transformation among commodities is significant.

Another factor which can influence welfare gains due to research is existing government policies. These policies influence the production and/or consumption of a commodity, or inputs used to produce it. They can influence both the benefits flowing from research and the distribution of these benefits.

The welfare effects which can flow from a research effort can vary significantly among research efforts, regions, and commodities. Choices among research options are likely to be influenced by the magnitude and distribution of these effects. Which ones are important requires clarification. For example, if two regions are part of one country and if the total national welfare gain is the objective of the research institutions, then a measure of the research impact of this objective is provided by adding all the gains (or losses) of all sectors. If, however, the objective is to maximize gains to poor farmers only, the subset of welfare changes to this particular sector is added to give a measure of how well the research option may satisfy this objective. Estimates of these welfare changes, if quantified, can be summarized in a form suitable to assist decision-makers in setting research priorities or other allocation decisions. This information is combined with other information before decision-makers make final judgments about allocation decisions.

Other aspects for consideration are: a) effect on income distribution and poverty; b) food security; c) human capital development; d) institution building and strengthening of national programs; e) employment generation; f) sustainability and environmental impact; and g) implications on policy change.

It is clear that a spectrum of considerations has to be taken into account with regard to the assessment of a research project, both ex-post with respect to impact achieved by a completed project and ex-ante with respect to impact likely to be achieved by a proposed project. It is equally clear that a detailed understanding of the components of the research-evaluation continuum is necessary in arriving at a quantitative assessment of impact. What follows is a sketch of the type of information needed, both ex-post and ex-ante, in the assessment of an example of BNF research that is directed towards proving N_2 fixation ability of chickpea (Rupela and Johansen, 1992). Improving N_2 fixation potential of chickpea cultivars involves the following activities:

< –Stage 1– >	< –Stage 2– >	< –Stage 3– >	< –Stage 4– >
Concept–ualization	Development of high–nodulating lines	On–farm testing	Extension and adoption
1988	1990	1995	1998 Year

Stage 1 involved the development of the concept of genetic alteration of the plant for better nodulation, through selections within existing cultivars (Rupela and Johansen, 1992). This stage led to the basic concepts and methodology for the development of the improved technology. Stage 2 involves actual conduct of the prescribed selection procedure to identify lines with superior N_2 fixation capability and their validation in on-station experiments. Stage 3 involves on-farm validation of the value of the selections. Stages 1, 2 and 3 represent the basic, applied and adaptive research components in the development of this technology.

Stage 4 is the demonstration, extension and adoption of the technology among farmers. The process underlying the adoption of technologies (Bantilan, 1993) is represented by the curve in Figure 1, in which adoption-related variables are highlighted: adoption lags, rate of adoption and ceiling level of adoption. Introduction of a new technology is not usually met with immediate adoption. The gestation period between the generation of a technology and its adoption varies by sector, commodity and type of technology. There are farmers who adopt only after the effects have been convincingly demonstrated. Reluctance among farmers to adopt a technology may be due to difficulty in its use, unavailability of the inputs required, market uncertainty, price fluctuations or preference for very low management crop technology. Thus, a sigmoid adoption curve is usually used to illustrate the adoption process; where the level of adoption is initially low, rises at an increasing rate after sufficient diffusion is attained, and finally reaches a ceiling level of adoption.

The quantitative assessment of impact requires some basic data. Data on the key factors involved in the various stages of the research process (Fig. 1) are needed to estimate the expected impact of BNF research. An important feature of the BNF research example described above is that the expected research/adoption lag is about 10 years. This represents the time it takes until the envisioned technology is achieved, validated and made available to farmers. The probability of achieving the expected research results (probability of success) has to be estimated, particularly for stages 1, 2 and 3. Estimates on the rate of technology adoption and ceiling level of adoption, which may vary considerably among types of farmers, have to be obtained. The cost of the actual implementation of the research in the first three stages should be taken into account in the overall assessment of benefit/cost ratios for the research endeavor.

Considerations for attracting support for BNF research

Impact analysis

It is suggested that proposals for BNF research and development would be much more attractive to research administrators and donors if it could be clearly shown how proposed activities fit into the entire research-adoption-impact continuum. They need to be based on sound calculations of expected gains from research and other parameters of the adoption curve. Considering previous limited adoption of BNF technology there is scope for adoption constraint studies, to pinpoint bottlenecks. Impact analysis should be built into any proposed project. Improvement of BNF would seem a readily quantifiable candidate for this suggested systematic and holistic approach, as ammenable data sets are likely to be available.

Benefit/cost analysis

An important first step in ex-ante impact analysis, as a basis for a project proposal, is a rigorous benefit/cost projection. A prime requirement is to establish, for particular target legumes and cropping systems, the actual gains to be expected from improving BNF above an existing level, in comparison to achieving these gains by using mineral N fertilizer. This firstly requires assessment of the extent to which the legume can meet its N needs through fixation. Essentially, need-to-inoculate studies (see above) supplemented by more detailed studies on rates and time of application of N fertilizer, can accomplish this (although there would inevitably be some difficulties of interpretation related to fertilizer N-use efficiency and N metabolism within the plant). Further, the residual value of legumes, in terms of equivalents of N fertilizer applied to a subsequent crop, needs to calculated. Also, relative value of N derived from either fertilizer or organic matter sources needs to be estimated, from the viewpoint of environment protection and sustainability of cropping systems. These data provide a baseline against which to estimate gains that can be expected from further improving BNF as a result of research or by direct application of known technologies. With discounting for factors such as probability of success, time lags and ceiling rate of adoption, reasonable estimates can be made for costs and benefits of a suggested research and/or development effort (Davis et al., 1987, 1994; Edwards and Freebairn, 1984; Norton and Davis, 1981; McKenney et al., 1991).

Management and genetic options

A careful evaluation is needed of management (primarily inoculation technology) and genetic options for enhancing BNF, in view of the new genetic options being proposed (e.g. Rupela and Johansen, 1992, 1994). If we can genetically alter the plant to better accept native rhizobia in an effective symbiosis, especially within existing cultivars, that would both meet the legumes' N needs as well as leave substantial residual N, then the aforementioned problems of inoculation technology can to some extent be bypassed. But, this assessment does depend on knowledge of to what extent the target legumes are currently limited by N, as explained in the previous section.

Inoculation technology

If it is decided that further pursuit of *Rhizobium* inoculation technology is viable then the shortcomings discussed earlier need to be comprehensively addressed.

Outlook for N fertilizer

The popularity of BNF research, and hence the degree of funding for it, is directly and closely related to the relative (compared with other agricultural inputs) price of N fertilizer. More emphasis should be given to comprehensive comparisons of BNF enhancement versus use of N fertilizer. This not only involves relative input costs, in relation to benefits expected, but also adverse consequences of use of either N source. For example, reliance on N fertilizer can result in soil acidification, N leaching losses and eutrophication of water bodies. But reliance on BNF can also lead to soil acidification (e.g. by proton excretion from legume roots (Marschner, 1986)) and inflexibility of cropping systems (particularly if legumes are a low value cropping option).

Conclusion

The need for shifting the balance from fertilizer derived N to N derived from BNF to meet the N nutritional needs of crop plants is as imperative as it ever was. In addition to well established management options for doing this, such as *Rhizobium* inoculation technology, there are increasingly feasible options becoming available for genetic enhancement of the host legumes' ability to fix N. However, the relatively poor adoption record of long-established BNF technologies, and inoculation technology in particular, suggest that caution is needed in preparing project proposals for research aimed at enhancing BNF. We thus advocate use of ex-ante impact analyses for development of such proposals, with careful estimation of benefit/cost ratios. Further, impact analysis should be written into future research proposals such that movement along the research-adoption-impact continuum can be monitored, any necessary mid-course adjustments made, and ex-post impact assessments done.

Acknowledgements

We thank Drs D McDonald, O P Rupela and J V D K Kumar Rao, ICRISAT, Dr P W Singleton, NifTAL and Dr J K Ladha, IRRI, for helpful comments on the manuscript.

References

Bantilan M C S 1993 Technology adoption and economic assessment of on-farm adaptive research. *In* Approaches to On-farm Research in Asia: Summary Proceedings of the Regional Workshop on On-Farm Adaptive Research, 18–20 February 1993, Ho Chi Minh City, Vietnam. pp 21–27. ICRISAT, Patancheru, India.

Bantilan M C S and Davis J S 1991 Across-commodity spillover effects of research and opportunity costs in a multi-product production environment. ACIAR/ISNAR Project Paper No. 30. Australian Center for International Agricultural Research, Canberra, Australia/International Service for National Agricultural Research, The Hague, Netherlands. 24p.

Chanaseni C and Kongngoen S 1992 Extension programs to promote rhizobial inoculants for soybean and groundnut in Thailand. Can. J. Microbiol. 38, 594–597.

Date R A 1977 Inoculation of tropical pasture legumes. *In* Exploiting the Legume-*Rhizobium* Symbiosis in Tropical Agriculture. Proceedings of a Workshop, August, 1976. University of Hawaii, College of Tropical Agriculture, Miscellaneous Publication No. 145. pp 293–311. University of Hawaii, Honolulu, USA.

Davis J S, Bantilan M C S and Ryan J G 1994 Development of information systems to support research decision making: An overall perspective. *In* Designing Information Systems to Support Priority Assessments in Agricultural Research: Concepts and Practices for International and National Institutions. Eds. J S Davis and J G Ryan. Monograph No. 17. Australian Center for International Agricultural Research (ACIAR), Canberra, Australia. (*In press*).

Davis J S, Oram P A and Ryan J G 1987 Assessment of Agricultural Research Priorities: An International Perspective. Australian Center for International Agricultural Research, Monograph No. 7, Canberra, Australia.

Davis R J, Cady F B, Wood C L and Chan C P Y 1985 Design and analysis of an international experimental network: legume inoculation trials in the NifTAL Project, the INLIT experience. Research Series 042, College of Tropical Agriculture and Human Resources, University of Hawaii. University of Hawaii, Honolulu, USA. 43p.

Edwards G W and Freebairn J W 1984 The gains from research into tradeable commodities. Am. J. Agric. Econ. 66 (324), 41–49.

Harper J E and Gibson A H 1984 Differential nodulation tolerance to nitrate among legume species. Crop Sci. 24, 797–801.

Jeswani L M and Baldev B 1990 Advances in Pulses Production Technology. Indian Council of Agricultural Research, New Delhi, India. 190p.

Kumar Rao J V D K 1990 Pigeonpea: nitrogen fixation. *In* The Pigeonpea. Eds.Y L Nene, S D Hall and V K Sheila. pp 233–256. CAB Int., Wallingford, UK.

Marschner H 1986 Mineral Nutrition in Higher Plants. Academic Press, Harcourt Brace Jovanovich, Publishers, London. 674p.

McKenney D W, Davis J S, Turnbull J W and Searle S D 1991 The impact of Australian Tree Species Research in China. ACIAR Economic Assessment Series No. 12. Australian Centre for International Agricultural Research, Canberra. 18p.

Norton G W and Davis J S 1981 Evaluating returns to agricultural research: a review. Am. J. Agric. Econ. 63, 685–99.

Rennie R J and Hybes R K 1993 Scientific and legislative control of legume inoculants for lentil and field pea. J. Prod. Agric. 6(4), 569–74.

Roughley R J and Pulsford D J 1982 Production and control of legume inoculants. *In* Nitrogen Fixation in Legumes. Ed. J M Vincent. pp 193–209. Academic Press, Sydney, Australia.

Rupela O P and Johansen C 1992 Genetic variation for nodulation within chickpea cultivars. Paper presented in the Research Coordination Meeting of IAEA "The Use of Isotopes in Increasing the Yield and Nitrogen Fixation Capabilities of Common Grain Legumes in Asia", 13–17 July 1992, Harbin, China.

Rupela O P and Johansen C 1994 Prospects for developing high mineral N tolerant N_2-fixation in legumes — A proposal. *In* Proceedings of the International Conference of Commission IV, International Society of Soil Science, 1–3 Dec 1992, Dhaka, Bangladesh (*In press*).

Singleton P, Thies J and Bohlool B B 1992 Useful models to predict response to legume inoculation. *In* Biological Nitrogen Fixation and Sustainability of Tropical Agriculture. Eds. K Mulongoy, M Gueye and D S C Spencer. pp 245–256. John Wiley and Sons, Chichester, UK.

Somasegaran P, Reyes V G and Hoben H J 1984 The influence of high temperatures on the growth and survival of *Rhizobium* spp. in peat inoculants during preparation, storage and distribution. Can. J. Microbiol. 30, 23–30.

Thompson J A 1984 Production and quality control of carrier-based legume inoculants. Information Bulletin No. 17. ICRISAT, Patancheru, India. 37p.

Thompson J A 1991 Legume inoculant production and quality control. *In* Expert Consultation on Legume Inoculant Production and Quality Control, Rome, 19–21 March, 1991. pp 15–32. FAO, Rome.

Verma L N and Bhattacharyya P 1992 Production, distribution and promotion of biofertilisers. *In* Fertilisers, Organic Manures, Recyclable Wastes and Biofertilisers. Ed. H L S Tandon. pp 132–147. Fertilizer Development and Consultation Organization, New Delhi.

Author index

Ali, M., 181
Baldani, V.L.D. 195
Bantilan, M.C.S., 279
Becker, M., 181
Boddey, R.M. 195
Bottomley, P.J., 143
Brockwell, J., 143
Cadisch, G., 255
Danso, S.K.A., 51, 119
Dazzo, F.B., 225
de Bruijn, F.J., 225
de Olivares, F.L. 195
de Oliveira, O.C., 195
Döbereiner, J. 195
Giller, K.E., 255
Herridge, D.F., 3, 51, 83
Jefferson, R.A. 241
Jing, Y., 225
Johansen, C., 279
Ladha, J.K., 3, 83, 181, 211
Lee, K.K., 29
Peoples, M.B., 3, 83, 241
Reis, V.M. 195
Roper, M.M., 211
Rupela, O.P., 29
Sanginga, J., 119
Thies, J.E., 143
Thomas, R.J., 103
Urquiaga, S. 195
Vanlauwe, B., 119
Wani, S.P., 29
Wilson, K.J. 241

Developments in Plant and Soil Sciences

1. J. Monteith and C. Webb (eds.): *Soil Water and Nitrogen in Mediterranean-type Environments*. 1981 ISBN 90-247-2406-6
2. J. C. Brogan (ed.): *Nitrogen Losses and Surface Run-off from Landspreading of Manures*. 1981 ISBN 90-247-2471-6
3. J. D. Bewley (ed.): *Nitrogen and Carbon Metabolism*. 1981 ISBN 90-247-2472-4
4. R. Brouwer, I. Gašparíková, J. Kolek and B. C. Loughman (eds.): *Structure and Function of Plant Roots*. 1981 ISBN 90-247-2510-0
5. Y. R. Dommergues and H. G. Diem (eds.): *Microbiology of Tropical Soils and Plant Productivity*. 1982 ISBN 90-247-2624-7
6. G. P. Robertson, R. Herrara and T. Rosswall (eds.): *Nitrogen Cycling in Ecosystems of Latin America and the Caribbean*. 1982 ISBN 90-247-2719-7
7. D. Atkinson, K. K. S. Bhat, M. P. Coutts, P. A. Mason and D. J. Read (eds.): *Tree Root Systems and Their Mycorrhizas*. 1983 ISBN 90-247-2821-5
8. M. R. Sarić and B. C. Loughman (eds.): *Genetic Aspects of Plant Nutrition*. 1983 ISBN 90-247-2822-3
9. J. R. Freney and J. R. Simpson (eds.): *Gaseous Loss of Nitrogen from Plant-Soil Systems*. 1983 ISBN 90-247-2820-7
10. United Nations Economic Commission for Europe (ed.): *Efficient Use of Fertilizers in Agriculture*. 1983 ISBN 90-247-2866-5
11. J. Tinsley and J. F. Darbyshire (eds.): *Biological Processes and Soil Fertility*. 1984 ISBN 90-247-2902-5
12. A. D. L. Akkermans, D. Baker, K. Huss-Danell and J. D. Tjepkema (eds.): Frankia *Symbioses*. 1984 ISBN 90-247-2967-X
13. W. S. Silver and E. C. Schröder (eds.): *Practical Application of* Azolla *for Rice Production*. 1984 ISBN 90-247-3068-6
14. P. G. L. Vlek (ed.): *Micronutrients in Tropical Food Crop Production*. 1985 ISBN 90-247-3085-6
15. T. P. Hignett (ed.): *Fertilizer Manual*. 1985 ISBN 90-247-3122-4
16. D. Vaughan and R. E. Malcolm (eds.): *Soil Organic Matter and Biological Activity*. 1985 ISBN 90-247-3154-2
17. D. Pasternak and A. San Pietro (eds.): *Biosalinity in Action*. Bioproduction with Saline Water. 1985 ISBN 90-247-3159-3
18. M. Lalonde, C. Camiré and J. O. Dawson (eds.): Frankia *and Actinorhizal Plants*. 1985 ISBN 90-247-3214-X
19. H. Lambers, J. J. Neeteson and I. Stulen (eds.): *Fundamental, Ecological and Agricultural Aspects of Nitrogen Metabolism in Higher Plants*. 1986 ISBN 90-247-3258-1
20. M. B. Jackson (ed.): *New Root Formation in Plants and Cuttings*. 1986 ISBN 90-247-3260-3
21. F. A. Skinner and P. Uomala (eds.): *Nitrogen Fixation with Non-Legumes* (Proceedings of the 3rd Symposium, Helsinki, 1984). 1986 ISBN 90-247-3283-2
22. A. Alexander (ed.): *Foliar Fertilization*. 1986 ISBN 90-247-3288-3
23. H. G. v.d. Meer, J. C. Ryden and G. C. Ennik (eds.): *Nitrogen Fluxes in Intensive Grassland Systems*. 1986 ISBN 90-247-3309-X
24. A. U. Mokwunye and P. L. G. Vlek (eds.): *Management of Nitrogen and Phosphorus Fertilizers in Sub-Saharan Africa*. 1986 ISBN 90-247-3312-X
25. Y. Chen and Y. Avnimelech (eds.): *The Role of Organic Matter in Modern Agriculture*. 1986 ISBN 90-247-3360-X
26. S. K. De Datta and W. H. Patrick Jr. (eds.): *Nitrogen Economy of Flooded Rice Soils*. 1986 ISBN 90-247-3361-8
27. W. H. Gabelman and B. C. Loughman (eds.): *Genetic Aspects of Plant Mineral Nutrition*. 1987 ISBN 90-247-3494-0
28. A. van Diest (ed.): *Plant and Soil: Interfaces and Interactions*. 1987 ISBN 90-247-3535-1

Developments in Plant and Soil Sciences

29. United Nations Economic Commission for Europe and FAO (eds.): *The Utilization of Secondary and Trace Elements in Agriculture.* 1987 ISBN 90-247-3546-7
30. H. G. v.d. Meer, R. J. Unwin, T. A. van Dijk and G. C. Ennik (eds.): *Animal Manure on Grassland and Fodder Crops.* Fertilizer or Waste? 1987 ISBN 90-247-3568-8
31. N. J. Barrow: *Reactions with Variable-Charge Soils.* 1987 ISBN 90-247-3589-0
32. D. P. Beck and L. A. Materon (eds.): *Nitrogen Fixation by Legumes in Mediterranean Agriculture.* 1988 ISBN 90-247-3624-2
33. R. D. Graham, R. J. Hannam and N. C. Uren (eds.): *Manganese in Soils and Plants.* 1988 ISBN 90-247-3758-3
34. J. G. Torrey and J. L. Winship (eds.): *Applications of Continuous and Steady-State Methods to Root Biology.* 1989 ISBN 0-7923-0024-6
35. F. A. Skinner, R. M. Boddey and I. Fendrik (eds.): *Nitrogen Fixation with Non-Legumes* (Proceedings of the 4th Symposium, Rio de Janeiro, 1987). 1989 ISBN 0-7923-0059-9
36. B. C. Loughman, O. Gašparíková and J. Kolek (eds.): *Structural and Functional Aspects of Transport in Roots.* 1989 ISBN 0-7923-0060-2; Pb 0-7923-0061-0
37. P. Plancquaert and R. Haggar (eds.): *Legumes in Farming Systems.* 1990 ISBN 0-7923-0134-X
38. A. E. Osman, M. M. Ibrahim and M. A. Jones (eds.): *The Role of Legumes in the Farming Systems of the Mediterranean Areas.* 1990 ISBN 0-7923-0419-5
39. M. Clarholm and L. Bergström (eds.): *Ecology of Arable Land – Perspectives and Challenges.* 1989 ISBN 0-7923-0424-1
40. J. Vos, C. D. van Loon and G. J. Bollen (eds.): *Effects of Crop Rotation on Potato Production in the Temperate Zones.* 1989 ISBN 0-7923-0495-0
41. M. L. van Beusichem (ed.): *Plant Nutrition – Physiology and Applications.* 1990 ISBN 0-7923-0740-2
42. N. El Bassam, M. Dambroth and B.C. Loughman (eds.): *Genetic Aspects of Plant Mineral Nutrition.* 1990 ISBN 0-7923-0785-2
43. Y. Chen and Y. Hadar (eds.): *Iron Nutrition and Interactions in Plants.* 1991 ISBN 0-7923-1095-0
44. J. J. R. Groot, P. de Willigen and E. L. J. Verberne (eds.): *Nitrogen Turnover in the Soil-Crop System.* 1991 ISBN 0-7923-1107-8
45. R. J. Wright, V.C. Baligar and R. P. Murrmann (eds.): *Plant-Soil Interactions at Low pH.* 1991 ISBN 0-7923-1105-1
46. J. Kolek and V. Kozinka (eds.): *Physiology of the Plant Root System.* 1992 ISBN 0-7923-1205-8
47. A. U. Mokwunye (ed.): *Alleviating Soil Fertility Constraints to Increased Crop Production in West Africa.* 1991 ISBN 0-7923-1221-X; Pb 0-7923-1222-8
48. M. Polsinelli, R. Materassi and M. Vincenzini (eds.): *Nitrogen Fixation* (Proceedings of the 5th Symposium, Florence, 1990). 1991 ISBN 0-7923-1410-7
49. J.K. Ladha, T. George and B.B. Bohlool (eds.): *Biological Nitrogen Fixation for Sustainable Agriculture.* 1992 ISBN 0-7923-1774-2
50. P.J. Randall, E. Delhaze, R.A. Richards and R. Munns (eds.): *Genetic Aspects of Plant Mineral Nutrition.* 1993 ISBN 0-7923-2118-9
51. K.S. Kumarasinghe and D.L. Eskew (eds.): *Isotopic Studies of Azolla and Nitrogen Fertilization of Rice.* 1993 ISBN 0-7923-2274-6
52. F.A. Bliss and G. Hardarson (eds.): *Enhancement of Biological Nitrogen Fixation of Common Baen in Latin America.* 1993 ISBN 0-7923-2451-X
53. M.A.C. Fragoso and M.L. van Beusichem (eds.): *Optimization of Plant Nutrition.* 1993 ISBN 0-7923-2519-2
54. N.J. Barrow (ed.) *Plant Nutrition - From Genetic Engineering to Field Practice.* 1993 ISBN 0-7923-2540-0
55. A.D. Robson (ed.): *Zinc in Soils and Plants.* 1993 ISBN 0-7923-2631-8
56. A.D. Robson, L.K. Abbott and N. Malajczuk (eds.): *Management of Mycorrhizas in Agriculture, Horticulture and Forestry.* 1994 ISBN 0-7923-2700-4

Developments in Plant and Soil Sciences

57. P.H. Graham, M.J. Sadowsky and C.P. Vance (eds.): *Symbiotic Nitrogen Fixation.* 1994 ISBN 0-7923-2781-0
58. F. Baluška, M. Čiamporová, O. Gašparíkova and P.W. Barlow (eds.): *Structure and Function of Roots.* 1995 ISBN 0-7923-2832-9
59. J. Abadía (ed.): *Iron Nutrition in Soils and Plants.* 1995 ISBN 0-7923-2900-7
60. P.S. Curtis, E.G. O'Neill, J.A. Teeri, D.R. Zak and K.S. Pregitzer (eds.): *Belowground Responses to Rising Atmospheric CO_2.* Implications for Plants, Soil Biota, and Ecosystem Processes. 1994 ISBN 0-7923-2901-5
61. P.C. Struik, W.J. Vredenberg, J.A. Renkema and J.E. Parlevliet (eds.): *Plant Production on the Threshold of a New Century.* 1994 ISBN 0-7923-2903-1
62. L.O. Nilsson, R.F. Hüttl and U.T. Johansson (eds.): *Nutrient Uptake and Cycling in Forest Ecosystems* (CEC/IUFRO Symposium, Halmstad, Sweden, 1993). 1995 ISBN 0-7923-3030-7
63. H.P. Collins, G.P. Robertson and M.J. Klug (eds.): *The Significance and Regulation of Soil Biodiversity* (International Symposium, East Lansing, 1993). 1995 ISBN 0-7923-3138-9
64. R.A. Date, N.J. Grundon, G.E. Rayment and M.E. Probert (eds.): *Plant-Soil Interactions at Low pH: Principles and Management* (3rd International Symposium, Brisbane, 1993). 1995 ISBN 0-7923-3198-2
65. J.K. Ladha and M.B. Peoples (eds.): *Management of Biological Nitrogen Fixation for the Development of More Productive and Sustainable Agricultural Systems* (15th Congress of Soil Science, Acapulco, Mexico, 1994). 1995 ISBN 0-7923-3413-2; Pb 0-7923-3414-0

Kluwer Academic Publishers – Dordrecht / Boston / London